Seismic Waves in Laterally Inhomogeneous Media

Edited by
Ivan Pšenčík
Vlastislav Červený

Springer Basel AG

Reprint from Pure and Applied Geophysics
(PAGEOPH), Volume 159 (2002), No. 7-8

Editors:

Ivan Pšenčík
Academy of Sciences of the Czech Republic
Geophysical Institute
Boční II, 1401
14131 Praha 4
Czech Republic
e-mail: ip@ig.cas.cz

Vlastislav Červený
Charles University
Faculty of Mathematics & Physics
Dep. of Geophysics
Ke Karlovu 3
12116 Praha 2
Czech Republic
e-mail: vcerveny@seis.karlov.mff.cuni.cz

A CIP catalogue record for this book is available from the Library of Congress, Washington D.C., USA

Deutsche Bibliothek Cataloging-in-Publication Data

Seismic waves in laterally inhomogeneous media / ed. by Ivan Pšenčík ; Vlastislav Červený. - Springer Basel AG 2002
   (Pageoph topical volumes)
   ISBN 978-3-7643-6677-3   ISBN 978-3-0348-8146-3 (eBook)
   DOI 10.1007/978-3-0348-8146-3

© 2002 Springer Basel AG
Originally published by Birkhäuser Verlag, Basel - Boston - Berlin 2002
Printed on acid-free paper produced from chlorine-free pulp

9 8 7 6 5 4 3 2 1

Contents

Pure appl. geophys. 159 (2002) 1401–1402
0033–4553/02/081401–01 $ 1.50 + 0.20/0

❙ Pure and Applied Geophysics

Preface

This special issue of *Pure and Applied Geophysics* contains some of the contributions presented at the workshop ***Seismic Waves in Laterally Inhomogeneous Media V,*** which was held at the Castle of Zahrádky, Czech Republic, June 5–9, 2000. The workshop was organized by the Geophysical Institute of the Academy of Sciences of Czech Republic, Prague and the Faculty of Mathematics and Physics of the Charles University, Prague. As the previous workshops, organized under the same name in 1978, 1983, 1988 and 1995, this one also was devoted mainly to the theoretical and computational aspects of seismic wave propagation in complex laterally varying, isotropic or anisotropic, layered and block structures.

The special issue starts with papers which are mostly focused on the ray computation of seismic wavefields in complex 3-D and 2-D structures, and on related problems. Among others, the following topics are discussed: coupling of quasi-shear waves in inhomogeneous weakly anisotropic media, ray perturbation techniques, testing of ray results by comparing them with results of the reflectivity method, smoothing of the model for effective ray computations and application of ray methods in seismic exploration. The issue then continues with papers discussing applicability of directional wavefield decomposition, phase space, path integral and parabolic equation methods to seismic waves. Several papers are devoted to the problems of attenuation and scattering of seismic waves in realistic structures. Papers discussing various types of inversion problems are presented in the end of the issue.

We take this opportunity to thank all the authors for contributing to this issue and the reviewers, listed below, for their patient cooperation and assistance.

Referees of the special issue: T. Alkhalifah, P. Bakker, M. Bouchon, J. Brac, K. Bube, S. Buske, J. Carcione, V. Červený, Ch. Chapman, R. Coates, J. Costa, P. Daley, J. Dellinger, A. Druzhinin, N. Ettrich, V. Farra, W. Figueiro, D. Gajewski, Ch. Hanitzsch, A. Hanyga, H. Helle, M. de Hoop, S. Horne, J. Hudson, B. Ioss, E. Iversen, B.L.N. Kennett, A. Kiratzi, L. Klimeš, D. Kosloff, E. Krebes, G. Lambaré, E. Landa, S. Le Bégat, H. Magistrale, F. Mainardi, V. Maupin, J. Mechie, W. Menke, A. Michelini, P. Moczo, J.T. Moser, T. Müller, A. Norris, R.L. Nowack, G. Pratt, B. Romanowicz, G. Ruempker, J. Schleicher, P. Spudich, E. Tessmer, C. Thomson, M. Tygel, V. Vavryčuk, R. Wu, K. Yomogida.

May 2001

Ivan Pšenčík
Vlastislav Červený

Ivan Pšenčík
Academy of Sciences of the Czech
Republic
Geophysical Institute
Boční II, 1401
14131 Praha 4
Czech Republic
e-mail: ip@ig.cas.cz

Vlastislav Červený
Charles University
Faculty of Mathematics & Physics
Dep. of Geophysics
Ke Karlovu 3
12116 Praha 2
Czech Republic
e-mail: vcerveny@seis.karlov.mff.cuni.cz

Pure appl. geophys. 159 (2002) 1403–1417
0033–4553/02/081403–15 $ 1.50 + 0.20/0

┃ Pure and Applied Geophysics

Coupled Anisotropic Shear-wave Ray Tracing in Situations where Associated Slowness Sheets Are Almost Tangent

P. M. BAKKER[1]

Summary—For shear waves in anisotropic media a frequency-dependent transport equation is derived, which involves coupling of both shear modes if the slowness vector points in a direction for which the associated slowness surfaces are (almost) tangent. This is achieved by averaging the Hamiltonians for both uncoupled shear modes. No assumption on weak anisotropy is made, and the shear polarisation is perpendicular to the *P*-wave polarisation for the actual slowness direction instead of an isotropically approximated *P*-wave polarisation. These are the main differences compared with the so-called zero-order quasi-isotropic approach.

Key words: Anisotropy, ray tracing, shear wave.

1. Introduction

Reliable modelling of polarisations for shear waves in anisotropic media is relevant in the context of imaging mode-converted (OBC/OBS) data in cases where not only *P-SV*, but also *P-SH* conversions take place. In such situations it may be necessary to use the polarisation direction for consistent continuation of the different wave modes and for appropriate imaging. Also in the context of shear-wave splitting analysis, the polarisation of shear waves is of primary importance.

The conventional method of computing anisotropic polarisation vectors and dynamic ray-tracing propagator matrices for shear waves breaks down because of singularities in situations where the slowness vector points in a direction for which the slowness sheets of the two shear modes are (almost) tangent. Such situations typically occur at kiss-singularities (e.g., in a TI-medium for a slowness direction parallel to the axis of symmetry), in cases where the anisotropy partly vanishes (e.g., for HTI- or VTI-media with equal horizontal and vertical *SH*-velocity the *SH*- and *SV*-slowness sheets are tangent along a circle), or when the anisotropy vanishes entirely. It has been shown by various authors (CHAPMAN and SHEARER, 1989;

[1] Shell International Exploration and Production B.V., Research and Technical Services, Volmerlaan 8, P.O. Box 60, 2280 Ab Rijswijk, The Netherlands. E-mail: p.m.bakker@siep.shell.com

COATES and CHAPMAN, 1990; PŠENČÍK 1998; ZILLMER *et al.*, 1998) that in these circumstances the individual anisotropic shear modes, which result from the ray method, are not meaningful to describe the complete wavefield for a finite bandwidth. Frequency-dependent coupling of the two shear modes gives a more realistic approximation of the wavefield in which the typical rapid rotation of the polarisation vector (for kiss-singularities) is avoided. The approach to deal with these problems, known as quasi-isotropic (QI) approximation, is usually based on the assumption of weak anisotropy, and an isotropic background model is chosen for reference. Thus, the dimensionless anisotropy parameters of the medium should be at most of order $O(c/\omega L)$, where c is a reference velocity, L is a length scale, and ω the frequency.

The motivation of this paper is to avoid the assumption of weak anisotropy. Here the analysis is restricted to the propagation in a $O(c/\omega L)$-neighbourhood of the singularity (with regards to space and slowness). Instead of tracing a reference ray through the isotropic background medium, it is proposed to trace a reference ray based on a Hamiltonian which is the average of two Hamiltonians for both uncoupled shear waves. Analysis shows that the particle displacement of the shear wave can be approximated by solving a coupled set of frequency-dependent transport equations along such a reference ray. In the QI-approach the geometrical spreading is controlled by the isotropic background medium. Here we find that the geometrical spreading is averaged between the (different) geometrical spreading factors associated with the uncoupled shear modes. For instance, when a shear wave is propagated virtually in the direction of the symmetry axis of a TI-medium, we find here that the geometrical spreading is averaged between the divergence of the group velocity of the pure SH-mode and that of the SV-mode, depending on the direction of the polarisation. If the anisotropy is not weak, then the difference between these two divergence terms can be of order $O(1)$, which makes this theory different from the QI-theory.

Another difference with the QI-approach is that here the shear-wave polarisation is in the plane perpendicular to the P-wave polarisation for the actual slowness direction, whereas the zero-order QI-approach models the shear-wave polarisation in the plane perpendicular to the P-wave polarisation in the isotropic background medium (on the assumption of weak anisotropy). If the anisotropic medium deviates from isotropy by an order of $O(1)$, then the polarisations in this paper are (at least formally) different from what is obtained by a QI-approximation. A (perhaps exotic) example is that of an elliptically TI-medium for which the slowness sheets of both shear modes are completely overlapping (i.e., the situation with Thomsen's parameters given by $\varepsilon = \delta \neq 0$ and $\gamma = 0$). In this case, the SV-polarisation for a normalised slowness direction $(n_1, 0, n_3)^T$ has the direction of $(-n_3(c_{13} + c_{44}), 0, n_1(c_{33} - c_{44}))^T$ (where the c_{IJ} are the Voigt notations of the stiffnesses as usual) and the SH-polarisation is given by $(0, 1, 0)^T$. If $\delta \neq 0$, this SV-polarisation cannot be in the direction of $(-n_3, 0, n_1)^T$ which is normal to the slowness vector.

As the treatment in this paper is restricted to an area in which the slowness sheets of both pure modes are almost tangent, i.e., slowness vectors and group velocity vectors are nearly equal for both modes, it leaves open what has to be done when a ray propagates out of (or into) such a area. One may envisage a method of matching this coupling theory to regular anisotropic ray tracing with splitted shear waves by decomposition or orthogonal projection of the polarisation vectors. However, such an approach will be rather complicated because of the fact that the shear-wave coupling is an interplay of energy exchange and phase or travel-time differences between the pure shear modes. In practical situations (and rather weak anisotropy) one may take the formal restrictions of this coupling theory for granted, and try if global application is successful. These aspects are beyond the scope of this paper. Thus far we have no numerical experience with this coupling theory.

The paper is structured as follows. Section 2 describes some of the basic concepts of anisotropic ray tracing. In Section 3 the averaged common ray tracing system is defined, and several properties of the eigensystem of the Christoffel matrix are analysed in the proximity of a tangent point of the slowness sheets. Section 4, which is highly technical, leads to the coupled transport equation, which is symmetrised in Section 5. In order to achieve a symmetric form, a suitable basis must be integrated, analogous – but more intricate – than the more conventional ray-centred basis for isotropic media (cf. POPOV and PšENčíK 1978; PšENčíK 1998).

2. The Anisotropic Ray-tracing System

Using the summation convention, the frequency-space domain formulation of wave propagation in anisotropic media is given by

$$-\rho\omega^2 u_i = (c_{ijkl}u_{k,l})_{,j} \tag{2.1}$$

where $\mathbf{u} = (u_1, u_2, u_3)^T$ is the displacement vector, $\{c_{ijkl}\}$ is the stiffness tensor, ρ is the density, and ω is the frequency. The zero-order ray approximation is of the form

$$\mathbf{u}(\mathbf{x}, \omega) = \mathbf{U}(\mathbf{x}) \exp(i\omega\tau(\mathbf{x})) , \tag{2.2}$$

where $\mathbf{U}$ is a frequency independent amplitude vector, and τ is a travel-time function. When the representation (2.2) is substituted into equation (2.1), requiring the term of order $(-i\omega)^2$ to vanish in the power series in $1/(-i\omega)$, then $\mathbf{U}$ has to be an eigenvector of the Christoffel matrix $\boldsymbol{\Gamma}$,

$$(\boldsymbol{\Gamma} - \mathbf{I})\mathbf{U} = \mathbf{0} , \tag{2.3}$$

with $\Gamma_{ik} = \frac{1}{\rho}c_{ijkl}p_j p_l$, where $\mathbf{p} = \nabla\tau$ and $\mathbf{p} = (p_1, p_2, p_3)^T$ (where the superscript ^{T} denotes transposition). Moreover, the associated eigenvalue of $\boldsymbol{\Gamma}$ is equal to 1, which implicitly defines the length of $\mathbf{p}$ for a selected eigenmode. Note that $\boldsymbol{\Gamma}$ is symmetric positive definite as a consequence of the stability properties of a valid stiffness tensor.

Proceeding along these lines one obtains three different wave modes, associated with the three eigenvalues $G^{(n)}(\mathbf{x}, \mathbf{p})$, which can be written as

$$G^{(n)}(\mathbf{x}, \mathbf{p}) = \frac{1}{\rho} c_{ijkl} p_j p_l g_i^{(n)} g_k^{(n)} \text{ (no summation with respect to } n) , \qquad (2.4)$$

where $\mathbf{g}^{(n)}$ are the orthonormal eigenvectors of $\mathbf{\Gamma}$, $n = 1, 2, 3$ (see ČERVENÝ, 1972; ČERVENÝ and FIRBAS, 1984). This holds if the three eigenvalues of $\mathbf{\Gamma}$ are mutually different for a given length of $\mathbf{p}$. However, if two of these eigenvalues are (approximately) equal then the eigenvector directions are ambiguous.

For each of the wave modes we may consider a phase velocity sheet, spanned by the vectors $v_{ph}^{(n)} \mathbf{v}$, or a slowness sheet spanned by the vectors $\mathbf{p}^{(n)} = (1/v_{ph}^{(n)}) \mathbf{v}$, where $\mathbf{v}$ is a unit vector, and $v_{ph}^{(n)}$ is such that $G^{(n)}(\mathbf{x}, (1/v_{ph}^{(n)}) \mathbf{v}) = 1$.

Now let $(\mathbf{x}_0, \mathbf{p}_0)$ be a point where two of these slowness sheets (for $n = 1, 2$) are tangent. Then we have

$$G^{(1)}(\mathbf{x}_0, \mathbf{p}_0) = G^{(2)}(\mathbf{x}_0, \mathbf{p}_0) \qquad (2.5)$$

and

$$\frac{\partial G^{(1)}}{\partial p_i}(\mathbf{x}_0, \mathbf{p}_0) = \frac{\partial G^{(2)}}{\partial p_i}(\mathbf{x}_0, \mathbf{p}_0), \quad i = 1, 2, 3 . \qquad (2.6)$$

The kinematic ray-tracing system reads

$$\frac{dx_i^{(n)}}{dt} = \frac{1}{2} \frac{\partial G^{(n)}}{\partial p_i}, \quad \frac{dp_i^{(n)}}{dt} = -\frac{1}{2} \frac{\partial G^{(n)}}{\partial x_i} , \qquad (2.7)$$

with $\tau(\mathbf{x}(t)) = t$. Hence, at a point $(\mathbf{x}_0, \mathbf{p}_0)$ and in its region where the slowness sheets of the two (quasi-) shear modes are (almost) tangent, the associated group velocities are (almost) equal.

Obviously, this is the case for isotropic media in which case $G^{(1)} - G^{(2)}$ vanishes with all its partial derivatives. For TI-media such a tangent point (kiss singularity) occurs for a slowness vector along the rotational axis of symmetry (which may vary with $\mathbf{x}$). In the vicinity of such tangent points the eigenvectors $\mathbf{g}^{(1)}$ and $\mathbf{g}^{(2)}$ may change rapidly with the variation of the Christoffel matrix. This may be due to analytically singular behaviour at the tangent point, or it may be a consequence of numerical errors. In the vicinity of the kiss-singularity, with slowness vector parallel to the rotational axis of symmetry, such analytically singular behaviour occurs: both eigenvectors are virtually in the plane perpendicular to the axis of symmetry; one of these (associated with the SV wave) along the radially oriented orthogonal projection of the slowness vector onto this plane, the other (for the SH wave) orthogonal to this projection. In the symmetry plane perpendicular to the rotational axis of symmetry, the two shear wave sheets are tangent along a circle if Thomsen's parameter γ vanishes. In that case the eigenvectors behave analytically smoothly with $\mathbf{p}$, although numerical computation of the eigenvectors may be unstable if no explicit expressions are used. Generally speaking, if the two shear-related eigenvalues are close, then spatial variation may critically influence the polarisation directions of the individual shear modes. Such behaviour, with wildly varying changes of polarisations, is a consequence of the ray method, rather than a physical phenomenon for the total wavefield.

The asymptotical analysis of this paper is valid if the two shear modes are almost tangent, implying that both phase- and group-velocities of the two modes are practically equal. Conical diffraction points, where cones of group-velocity vectors originate from a singular contact point at the slowness sheets, are not covered by this theory.

3. An Approximate Common Ray-tracing System for Both Shear Modes in the Proximity of a Tangent Point

On the assumption that the P-wave related eigenvalue $G^{(3)}$ of the Christoffel matrix is well separated from the other two shear-wave related eigenvalues $G^{(1)}$ and $G^{(2)}$, we consider the behaviour of $G^{(1)} - G^{(2)}$ in the neighbourhood of a point $(\mathbf{x}_0, \mathbf{p}_0)$ where the sheets of these two eigenmodes are assumed to be tangent.

The eigenvector $\mathbf{g}^{(3)}$ varies smoothly with $(\mathbf{x}, \mathbf{p})$, provided that its orientation is chosen consistently. Hence, a smoothly varying orthonormal basis $\{\mathbf{f}^{(1)}, \mathbf{f}^{(2)}\}$ can be chosen for the orthogonal complement of $\mathbf{g}^{(3)}$, and $G^{(1)}$, $G^{(2)}$ are the eigenvalues of the matrix

$$\hat{\boldsymbol{\Gamma}} = \begin{pmatrix} f_i^{(1)} \Gamma_{ik} f_k^{(1)} & f_i^{(1)} \Gamma_{ik} f_k^{(2)} \\ f_i^{(2)} \Gamma_{ik} f_k^{(1)} & f_i^{(2)} \Gamma_{ik} f_k^{(2)} \end{pmatrix} . \tag{3.1}$$

The matrix $\hat{\boldsymbol{\Gamma}}$ is a restriction of the Christoffel matrix to the two-dimensional space of eigenvectors associated with $G^{(1)}$ and $G^{(2)}$. Let γ_1 and γ_2 be defined by

$$\gamma_1 = f_i^{(1)} \Gamma_{ik} f_k^{(1)} - f_i^{(2)} \Gamma_{ik} f_k^{(2)} \tag{3.2}$$

$$\gamma_2 = 2 f_i^{(1)} \Gamma_{ik} f_k^{(2)} , \tag{3.3}$$

which are smoothly varying functions of $(\mathbf{x}, \mathbf{p})$. Then

$$G^{(1)} - G^{(2)} = \pm \sqrt{\gamma_1^2 + \gamma_2^2} , \tag{3.4}$$

and the associated (non-normalised) eigenvectors of $\hat{\Gamma}$ are expressed in $\{\mathbf{f}^{(1)}, \mathbf{f}^{(2)}\}$-coordinates by $(\gamma_1 + \sqrt{\gamma_1^2 + \gamma_2^2}, \gamma_2)^T$ and $(-\gamma_2, \gamma_1 + \sqrt{\gamma_1^2 + \gamma_2^2})^T$. The branch selection in eq. (3.4) is briefly discussed at the end of this section. These expressions show that the variations of the eigendirections are controlled by changes in γ_1/γ_2 while $\gamma_1 \to 0$ and $\gamma_2 \to 0$ for $(\mathbf{x}, \mathbf{p}) \to (\mathbf{x}_0, \mathbf{p}_0)$. At a tangent point $(\mathbf{x}_0, \mathbf{p}_0)$, where eqs. (2.5) and (2.6) are valid, we have $\gamma_I(\mathbf{x}_0, \mathbf{p}_0) = 0$ and $\partial \gamma_I / \partial p_j (\mathbf{x}_0, \mathbf{p}_0) = 0$, for $j = 1, 2, 3$ and $I = 1, 2$.

Furthermore, for all $(\mathbf{x}, \mathbf{p})$ we have

$$\frac{\partial \mathbf{f}^{(I)}}{\partial p_j} = \alpha_j^{IJ} \mathbf{f}^{(J)} + \alpha_j^{I3} \mathbf{g}^{(3)}, \quad \text{with } \alpha_j^{IJ} = -\alpha_j^{JI} \text{ for } I, J = 1, 2 . \tag{3.5}$$

Hence, at the tangent point $(\mathbf{x}_0, \mathbf{p}_0)$ we get

$$f_i^{(1)} \frac{\partial \Gamma_{ik}}{\partial p_j} f_k^{(2)} = \frac{\partial}{\partial p_j} \left(f_i^{(1)} \Gamma_{ik} f_k^{(2)} \right) = \frac{1}{2} \frac{\partial \gamma_2}{\partial p_j} = 0 \tag{3.6}$$

and

$$f_i^{(1)} \frac{\partial \Gamma_{ik}}{\partial p_j} f_k^{(1)} - f_i^{(2)} \frac{\partial \Gamma_{ik}}{\partial p_j} f_k^{(2)} = \frac{\partial}{\partial p_j} \left(f_i^{(1)} \Gamma_{ik} f_k^{(1)} - f_i^{(2)} \Gamma_{ik} f_k^{(2)} \right) = \frac{\partial \gamma_1}{\partial p_j} = 0 . \tag{3.7}$$

Equations (2.5) and (3.6) imply that the group velocities for both shear modes, $\frac{1}{2} \partial G^{(I)} / \partial p_i$ for $I = 1, 2$, are equal at the tangent point. We shall exploit this by defining an approximate common ray-tracing system for both shear modes. Let a Hamiltonian be defined by

$$H(\mathbf{x}, \mathbf{p}) = \tfrac{1}{2} \{ G^{(1)}(\mathbf{x}, \mathbf{p}) + G^{(2)}(\mathbf{x}, \mathbf{p}) \} , \tag{3.8}$$

with the associated ray-tracing system, obtained from the Hamiltonian equation $H(\mathbf{x}, \mathbf{p}) = 1$,

$$\frac{dx_i}{dt} = \frac{1}{2} \frac{\partial H}{\partial p_i}, \quad \frac{dp_i}{dt} = -\frac{1}{2} \frac{\partial H}{\partial x_i} , \tag{3.9}$$

with $\tau(\mathbf{x}(t)) = t$.

If the wave propagation on the basis of this system starts with a smooth, nonsingular, initial wavefront, then propagation of this wavefront over a short distance in the vicinity of a tangent point by integrating the system (3.9) is kinematically an acceptable approximation for propagation of the initial wavefront by integrating the system (2.7) for $n = 1, 2$ simultaneously. As the curvatures of the slowness sheets for both individual shear modes, and hence the geometrical spreading, can be different at the tangent point $(\mathbf{x}_0, \mathbf{p}_0)$, we cannot expect that the geometrical spreading of the common rays describes the geometrical spreading of the total wavefield adequately. Furthermore, if the slowness vector $\mathbf{p}$ deviates from a tangent point then both shear modes exhibit splitting of the initially common wavefront. Consequently, the approximate common ray-tracing system might be useful only in a neighbourhood of a tangent point. Indeed, for $(\mathbf{x}, \mathbf{p})$ in a $O(1/\omega)$-proximity of $(\mathbf{x}_0, \mathbf{p}_0)$ we have $\|\hat{\mathbf{\Gamma}} - \mathbf{I}\| = O(1/\omega)$ (in the following we take for granted that the symbol $O(1/\omega)$ implicitly contains factors which make the dimensions correspond).

Note that the quantities γ_1 and γ_2 depend smoothly on $(\mathbf{x}, \mathbf{p})$ if the medium parameters vary smoothly. However, this does not generally imply that

$(G^{(1)} - G^{(2)})(\mathbf{x}_0, \mathbf{p})$ is twice continuously differentiable with respect to $\mathbf{p}$ at the tangent point. If $\gamma_I = 0$ and $\partial\gamma_I/\partial p_j = 0$ for $j = 1, 2, 3$ at $(\mathbf{x}_0, \mathbf{p}_0)$, then $\gamma_I(\mathbf{x}_0, \mathbf{p})$ behaves locally as a quadratic form of $\mathbf{p} - \mathbf{p}_0$ (up to second order). Hence $G^{(1)} - G^{(2)}$ behaves as the square-root of the sum of two squares of such quadratic forms. For any fixed direction of $\mathbf{p} - \mathbf{p}_0$ this implies that $G^{(1)} - G^{(2)}$ behaves as a quadratic form of the distance between $\mathbf{p}$ and $\mathbf{p}_0$. Analytic continuation of $G^{(1)} - G^{(2)}$ across the tangent point may require a change of the sign in eq. (3.4). Numerically this is hard to achieve reliably, which makes consistent shear-wave ray tracing of separate shear modes problematic in these circumstances. Also the computation of $d\mathbf{p}^{(n)}/dt$ suffers from these difficulties. It requires specific relationships between the second-order $\mathbf{p}$ derivatives of γ_1 and γ_2 (which will not be made explicit here) to have a quadratic form of $\mathbf{p} - \mathbf{p}_0$ for $G^{(1)} - G^{(2)}$, uniformly for all directions of $\mathbf{p} - \mathbf{p}_0$, without a sign reversal in eq. (3.4). A sufficient condition for this to be the case is that γ_1 and γ_2 locally behave as positive definite quadratic forms.

The branch selection problems mentioned here are circumvented in the proposed averaged ray-tracing scheme (3.9), because $G^{(1)} + G^{(2)} = \mathrm{tr}(\hat{\mathbf{\Gamma}})$ behaves smoothly with $(\mathbf{x}, \mathbf{p})$.

4. The Transport Equation Related to Averaged Travel Time

A wavefield $\mathbf{u}(\mathbf{x}, \omega)$ can always be written as $\mathbf{u}(\mathbf{x}, \omega) = \mathbf{U}(\mathbf{x}, \omega)\exp(i\omega\tau(\mathbf{x}))$ where $\tau(\mathbf{x})$ is the averaged 'travel-time' function associated with the common ray-tracing system (3.9). Differentiation of this expression gives

$$u_{k,l} = (i\omega\tau_{,l}U_k + U_{k,l})\exp(i\omega\tau) , \tag{4.1}$$

and

$$u_{k,lj} = \{(i\omega)^2\tau_{,l}\tau_{,j}U_k + i\omega\tau_{,l}U_{k,j} + i\omega\tau_{,lj}U_k + i\omega\tau_{,j}U_{k,l} + U_{k,lj}\}\exp(i\omega\tau) . \tag{4.2}$$

Substitution of the results (4.1) and (4.2) into the wave equation (2.1) gives

$$\begin{aligned}(i\omega)^2\{\rho U_i - \tau_{,l}\tau_{,j}c_{ijkl}U_k\}\\-i\omega\{c_{ijkl}(\tau_{,l}U_{k,j} + \tau_{,j}U_{k,l} + \tau_{,lj}U_k) + c_{ijkl,j}\tau_{,l}U_k\}\\-\{c_{ijkl}U_{k,lj} + c_{ijkl,j}U_{k,l}\} = 0 .\end{aligned} \tag{4.3}$$

In a $O(1/\omega)$-neighbourhood of a tangent point of the slowness sheets we seek a solution of type

$$\mathbf{U}(\mathbf{x}, \omega) = a(\mathbf{x})\mathbf{f}^{(1)}(\mathbf{x}, \mathbf{p}) + b(\mathbf{x})\mathbf{f}^{(2)}(\mathbf{x}, \mathbf{p}) + \frac{c(\mathbf{x})}{i\omega}\mathbf{g}^{(3)}(\mathbf{x}, \mathbf{p}) , \tag{4.4}$$

where $\mathbf{p} = \nabla\tau(\mathbf{x})$, such that eq. (4.3) is solved approximately, i.e. its right-hand side is of order $O(1)$ for $\omega \to \infty$. We are primarily interested in the zero order terms a

and b, and we shall derive a system of differential equations for these coefficients that can be integrated along the common ray. The first-order term with the coefficient c is introduced only to enforce that the orthogonal projection onto $\mathbf{g}^{(3)}$ of the left-hand side of eq. (4.3) is of order $O(1)$ if c is chosen appropriately. This will allow us to restrict the analysis to the orthogonal projection onto the span of $\mathbf{f}^{(1)}$ and $\mathbf{f}^{(2)}$.

We introduce the vectorial differential operator $\mathbf{N} = (N_1, N_2, N_3)^T$ by

$$N_i \mathbf{U} = \frac{1}{\rho} c_{ijkl}(\tau_{,l} U_{k,j} + \tau_{,j} U_{k,l} + \tau_{,lj} U_k) + \frac{1}{\rho} c_{ijkl,j} \tau_{,l} U_k \ . \tag{4.5}$$

The third term on the left-hand side of eq. (4.3) will be ignored as it is of order $O(1)$, and in the second term we may disregard the component $(i\omega)^{-1} c(\mathbf{x}) \mathbf{g}^{(3)}(\mathbf{x}, \mathbf{p})$ of $\mathbf{U}(\mathbf{x}, \omega)$. Then the leading terms of eq. (4.3) result into

$$(i\omega)^2 (\mathbf{I} - \mathbf{\Gamma})\left(a\mathbf{f}^{(1)} + b\mathbf{f}^{(2)} + \frac{c}{i\omega}\mathbf{g}^{(3)} \right) - (i\omega)\{\mathbf{N}(a\mathbf{f}^{(1)} + b\mathbf{f}^{(2)})\} = O(1) \ . \tag{4.6}$$

If we take the inner-product with $\mathbf{f}^{(1)}$ or $\mathbf{f}^{(2)}$, then the component $c/i\omega\mathbf{g}^{(3)}$ cancels out, because

$$\mathbf{f}^{(K)T}(\mathbf{I} - \mathbf{\Gamma})\mathbf{g}^{(3)} = 0 \quad \text{for } K = 1, 2 \ .$$

In the $O(1/\omega)$ of the tangent point we have $\|(\hat{\mathbf{\Gamma}} - \mathbf{I})\mathbf{f}^{(I)}\| = O(1/\omega)$. Thus, if a and b have been determined, then we may choose $c = \mathbf{g}^{(3)T}{}^S\mathbf{N}(a\,\mathbf{f}^{(1)} + b\,\mathbf{f}^{(2)})/(1 - G^{(3)})$, such that the left-hand side of eq. (4.6) is orthogonal to $\mathbf{g}^{(3)}$. Moreover, c is bounded for $\omega \to \infty$ provided that $G^{(3)} \neq 1$, i.e., if the P-wave phase velocity is well separated from the S-wave phase velocities.

Thus, it remains to be shown that a and b can be chosen appropriately, such that the inner-products of $\mathbf{f}^{(1)}$ and $\mathbf{f}^{(2)}$ with the left-hand side of eq. (4.6) are of order $O(1)$. In the sequel we shall ignore the so-called additional term $i\omega^{-1}c\mathbf{g}^{(3)}$.

Let A be a smoothly varying scalar field depending on $(\mathbf{x}, \mathbf{p})$. By substituting $A\mathbf{f}^{(L)}$ for $\mathbf{U}$ in eq. (4.5), taking the inner-product with $\mathbf{f}^{(K)}$, and by rearranging the various terms we find

$$\mathbf{f}^{(K)T}\mathbf{N}(A\mathbf{f}^{(L)}) = A\frac{1}{\rho}\left\{ \left(c_{ijkl}\tau_{,l}f_i^{(K)}f_k^{(L)} \right)_{,j} + c_{ijkl}\left(\tau_{,j}f_i^{(K)}f_{k,l}^{(L)} - \tau_{,l}f_{i,j}^{(K)}f_k^{(L)} \right) \right\}$$
$$+ \frac{1}{\rho}c_{ijkl}(\tau_{,l}A_{,j} + \tau_{,j}A_{,l})f_i^{(K)}f_k^{(L)} \ . \tag{4.7}$$

Using the relation

$$\frac{\partial \Gamma_{ik}}{\partial p_j} = \frac{1}{\rho}(c_{ijkl}p_l + c_{ilkj}p_l), \quad \text{with } p_l = \tau_{,l} \ , \tag{4.8}$$

we rewrite the last term on the right-hand side of eq. (4.7) as

$$\frac{1}{\rho}c_{ijkl}(\tau_{,l}A_j + \tau_{,j}A_{,l})f_i^{(K)}f_k^{(L)} = f_i^{(K)}\frac{\partial\Gamma_{ik}}{\partial p_j}f_k^{(L)}A_j \ . \tag{4.9}$$

Provided that the medium properties are sufficiently smooth, the eqs. (3.6) and (3.7) show that

$$f_i^{(K)}\frac{\partial\Gamma_{ik}}{\partial p_j}f_k^{(L)} = O(1/\omega) \quad \text{for } K \neq L \tag{4.10}$$

with $(\mathbf{x}, \mathbf{p})$ in a $O(1/\omega)$-environment of a tangent point $(\mathbf{x}_0, \mathbf{p}_0)$. For $K = L$ we have

$$f_i^{(K)}\frac{\partial\Gamma_{ik}}{\partial p_j}f_k^{(K)} = 2\frac{dx_j}{dt} + O(1/\omega), \quad K = 1, 2 \text{ (no summation)} \ . \tag{4.11}$$

Hence, from eq. (4.9) we find that within an $O(1/\omega)$-environment of $(\mathbf{x}_0, \mathbf{p}_0)$ the last term on the right-hand side of eq. (4.7) is equal to $2\,dA/dt + O(1/\omega)$.

For $K = L$ the second term within {} on the right-hand side of eq. (4.7) drops out, so

$$\mathbf{f}^{(K)T}\mathbf{N}(A\mathbf{f}^{(K)}) = A\frac{1}{\rho}\left(c_{ijkl}\tau_{,l}f_i^{(K)}f_k^{(K)}\right)_{,j} + 2\frac{dA}{dt} + O(1/\omega) \quad \text{for } K = 1 \text{ or } 2 \ . \tag{4.12}$$

For $K \neq L$ eq. (4.7) leads to

$$\mathbf{f}^{(K)T}\mathbf{N}(A\mathbf{f}^{(L)}) = A\frac{1}{\rho}\left\{\left(c_{ijkl}\tau_{,l}f_i^{(K)}f_k^{(L)}\right)_{,j} + c_{ijkl}\left(\tau_{,j}f_i^{(K)}f_{k,l}^{(L)} - \tau_{,l}f_{i,j}^{(K)}f_k^{(L)}\right)\right\} + O(1/\omega) \ . \tag{4.13}$$

Now the transport equation can be formulated as a coupled system of differential equations for a and b along the common ray

$$2\begin{pmatrix} da/dt \\ db/dt \end{pmatrix} = -\begin{pmatrix} \mathbf{f}^{(1)T}\mathbf{N}\mathbf{f}^{(1)} & \mathbf{f}^{(1)T}\mathbf{N}\mathbf{f}^{(2)} \\ \mathbf{f}^{(2)T}\mathbf{N}\mathbf{f}^{(1)} & \mathbf{f}^{(2)T}\mathbf{N}\mathbf{f}^{(2)} \end{pmatrix}\begin{pmatrix} a \\ b \end{pmatrix}$$
$$-(i\omega)\begin{pmatrix} \mathbf{f}^{(1)T}(\mathbf{\Gamma} - \mathbf{I})\mathbf{f}^{(1)} & \mathbf{f}^{(1)T}(\mathbf{\Gamma} - \mathbf{I})\mathbf{f}^{(2)} \\ \mathbf{f}^{(2)T}(\mathbf{\Gamma} - \mathbf{I})\mathbf{f}^{(1)} & \mathbf{f}^{(2)T}(\mathbf{\Gamma} - \mathbf{I})\mathbf{f}^{(2)} \end{pmatrix}\begin{pmatrix} a \\ b \end{pmatrix} + O(1/\omega) \ . \tag{4.14}$$

This coupled transport equation (4.14) is analogous to equation (36) in PŠENČÍK (1998), which is based on the assumption of weak anisotropy with a travel-time function defined by a nearby isotropic medium which should not deviate more than $O(1/\omega)$ from the actual anisotropic medium. In order to show the agreement, we interpret eq. (4.14) for an isotropic situation. The factors $\mathbf{f}^{(K)T}(\mathbf{\Gamma} - \mathbf{I})\mathbf{f}^{(L)}$ in eq. (4.14) take the place of the factors $\beta^{-2}B_{KL}$ in Pšenčík's equation (36), where β is the isotropic shear velocity of the background medium. The effect of coupling these frequency-dependent terms is nicely explained in Pšenčík's paper. For high frequencies these coupling terms predominantly cause time splitting of the fast and slow shear wave. The splitting occurs consistently with the eigenvalue decomposition of the matrix $(\mathbf{\Gamma} - \mathbf{I})$, restricted to the span of $\mathbf{f}^{(1)}$ and $\mathbf{f}^{(2)}$. For extremely low frequencies this term is subdominant, and for medium range frequencies it generates a combination of time splitting and energy exchange between the components along $\mathbf{f}^{(1)}$ and $\mathbf{f}^{(2)}$.

The first term on the right-hand side of eq. (4.14) replaces the isotropic geometrical spreading found in equation (36) of Pšenčík's quasi-isotropic treatment. Indeed, for an isotropic medium the diagonal terms $\mathbf{f}^{(K)T}\mathbf{N}\mathbf{f}^{(K)}$ are equal to the divergence of a density-normalised group velocity, $1/\rho \operatorname{div}(\rho\mathbf{v}_{gr}) = 1/\rho \operatorname{div}(\rho\beta^2\nabla\tau)$, which is equal to $1/\rho\{(\rho\beta^2)_{,j}\tau_{,j} + (\rho\beta^2)\tau_{,jj}\}$. The latter factor is found as the isotropic geometrical spreading factor in Pšenčík's equation (36). After some manipulation one finds for the off-diagonal terms of eq. (4.14) in the isotropic situation

$$\mathbf{f}^{(K)T}\mathbf{N}\mathbf{f}^{(L)} = \beta^2\tau_{,j}\left(f_k^{(K)}f_{k,j}^{(L)} - f_k^{(L)}f_{k,j}^{(K)}\right) + \beta^2\tau_{,j}\left(f_k^{(K)}f_{j,k}^{(L)} - f_k^{(L)}f_{j,k}^{(K)}\right) , \qquad (4.15)$$

for $(K, L) = (1, 2)$ or $(2, 1)$. The first term on the right-hand side of eq. (4.15) is equal to

$$f_k^{(K)}\frac{df_k^{(L)}}{dt} - f_k^{(L)}\frac{df_k^{(K)}}{dt} = 2f_k^{(K)}\frac{df_k^{(L)}}{dt}$$

and the second term is equal to $-\beta^2\left\{\tau_{,jk}\left(f_k^{(K)}f_j^{(L)} - f_k^{(L)}f_j^{(K)}\right)\right\} = 0$. Therefore, in the isotropic situation we have

$$\mathbf{f}^{(K)T}\mathbf{N}\mathbf{f}^{(L)} = 2f_k^{(K)}\frac{df_k^{(L)}}{dt}, \quad (K, L) = (1, 2) \text{ or } (2, 1) , \qquad (4.16)$$

which corresponds to Pšenčík's equation (36).

In non-isotropic circumstances the first term on the right-hand side of equation (4.14) accounts for differences in geometrical spreading, as defined by the curvatures of the slowness sheets of the individual shear modes, see BAKKER (1996) and ČERVENÝ (2001). In fact, the four radii of curvatures of the two slowness sheets become mixed.

Another difference between the zero-order QI-approach and the treatment in this paper is that the shear-wave polarisations vary in different planes. The zero-order QI-approach leads to shear polarisation in a plane perpendicular to the P-wave polarisation in the isotropic background field (with equal direction of slowness), which is justified on the assumption of weak anisotropy. The treatment in this paper leads to a zero-order shear-wave polarisation (i.e., with $c = 0$ in equation (4.4)) which is exactly perpendicular to the anisotropic P-wave polarisation with the same direction of slowness. In the introduction we have seen an example of elliptic anisotropy with completely overlapping slowness sheets, for which the SV-polarisation is not nearly orthogonal to the direction of slowness.

Although the coupled transport equation (4.14) looks like a system of normal differential equations along the common reference ray, it still has features of a partial differential equation, because the terms $\mathbf{f}^{(K)T}\mathbf{N}\mathbf{f}^{(L)}$ contain spatial partial derivatives of the basis vectors $\mathbf{f}^{(L)}$. In Section 5 we shall derive a system of normal differential equations which can be integrated to yield suitable basis vectors $\mathbf{f}^{(1)}$ and $\mathbf{f}^{(2)}$. For this basis the coupled transport equation (4.14) can be expressed within $O(1/\omega)$-accuracy without the necessity of using spatial derivatives of the basis vectors.

5. A Basis for which the Coupled Transport Equation is Symmetric

In the isotropic state we have derived the relation (4.16) for the off-diagonal terms of the first term on the right-hand side of eq. (4.14). A convenient orthonormal basis $\{\mathbf{f}^{(1)}, \mathbf{f}^{(2)}\}$ of the isotropic shear-wave polarisation plane can be obtained if one integrates these vectors such that $d\mathbf{f}^{(K)}/dt$ is in the direction of the P-wave polarisation vector, see PšENČÍK (1998). Then the off-diagonal terms $\mathbf{f}^{(K)T}\mathbf{N}\mathbf{f}^{(L)}$ in eq. (4.14) vanish, and the first matrix on the right-hand side of the coupled transport equation becomes a scalar times the identity matrix. This is consistent with the fact that the isotropic geometrical spreading factor for shear waves is uniform for all directions of polarisation.

For the anisotropic case we shall show that a basis $\{\mathbf{f}^{(1)}, \mathbf{f}^{(2)}\}$ can be chosen, such that the matrix with elements $\mathbf{f}^{(K)T}\mathbf{N}\mathbf{f}^{(L)}$ becomes symmetric. As the second, frequency-dependent part of eq. (4.14) is also symmetric, this implies that the dominant term of the entire coupled transport equation is symmetric for this basis. Moreover, this basis allows for computation of the elements $\mathbf{f}^{(K)T}\mathbf{N}\mathbf{f}^{(L)}$ without using spatial derivatives of the vectors $\mathbf{f}^{(L)}$.

The orthonormality of the vectors $\{\mathbf{f}^{(1)}, \mathbf{f}^{(2)}, \mathbf{g}^{(3)}\}$ ensures the existence of coefficients c_j^{KJ} for which

$$f_{i,j}^{(K)} = c_j^{KJ} f_i^{(J)} + c_j^{K3} g_i^{(3)}, \quad \text{with } c_j^{KJ} = -c_j^{JK} \quad \text{for } K, J = 1, 2 \ . \tag{5.1}$$

Similar to the notations in GAJEWSKI and PšENČÍK (1990) we introduce the coefficients $A_j^{K3(x)}, A_j^{K3(p)}$ and $A_j^{KL(p)}$ by the definitions

$$A_j^{K3(x)} = f_i^{(K)} \frac{\partial \Gamma_{ik}}{\partial x_j} g_k^{(3)}, \quad A_j^{K3(p)} = f_i^{(K)} \frac{\partial \Gamma_{ik}}{\partial p_j} g_k^{(3)}, \quad A_j^{KL(p)} = f_i^{(K)} \frac{\partial \Gamma_{ik}}{\partial p_j} f_k^{(L)} \ , \tag{5.2}$$

where the partial derivatives are in the six-dimensional phase-space domain.

Somewhat analogous to results in JECH and PšENČÍK (1989), we derive expressions for the coefficients c_j^{K3}. Since $\mathbf{g}^{(3)}$ is an eigenvector of the Christoffel matrix, perpendicular to $\mathbf{f}^{(K)}$, we have

$$\left(f_i^{(K)} \Gamma_{ik} g_k^{(3)} \right)_{,j} = 0 \tag{5.3}$$

which is expanded to

$$f_{i,j}^{(K)}\Gamma_{ik}g_k^{(3)} + f_i^{(K)}\Gamma_{ik}g_{k,j}^{(3)} + f_i^{(K)}\Gamma_{ik,j}g_k^{(3)} = 0 \ . \tag{5.4}$$

Substituting (5.1) and a similar expression for $g_{k,j}^{(3)}$, and by rewriting the spatial derivatives $\Gamma_{ik,j}$ in terms of derivatives in the phase-space domain, one obtains

$$(\hat{\Gamma}_{KL} - G^{(3)}\delta_{KL})c_j^{L3} = A_j^{K3(x)} + A_m^{K3(p)}\tau_{,mj} \ , \tag{5.5}$$

where $\hat{\Gamma}_{KL}$ are the entries of the restricted Christoffel matrix as defined in equation (3.1), and $\tau_{,mj}$ are the second-order spatial derivatives of travel time as defined by the averaged ray-tracing system (3.9). For each j, equation (5.5) is a 2×2-system for the coefficients c_j^{13} and c_j^{23} (non-singular if the P velocity is well separated from the S velocities). Within a $O(1/\omega)$-vicinity of a tangent point of the S-wave slowness sheets $(\hat{\Gamma}_{KL} - G^{(3)}\delta_{KL})$ can be approximated by $(G^{(S)} - G^{(3)})\delta_{KL} + O(1/\omega)$, where $G^{(S)}$ stands for $G^{(1)}$, $G^{(2)}$, or an averaged value.

Applying the relation (4.7) with $A = 1$, we find that equation (4.13) is satisfied exactly, without the $O(1/\omega)$-term. The result can be rewritten as

$$\mathbf{f}^{(K)T}\mathbf{N}\mathbf{f}^{(L)} = \frac{1}{\rho}\left\{\left(c_{ijkl}\tau_{,l}f_i^{(K)}f_k^{(L)}\right)_{,j} + c_{ijkl}\left(f_k^{(K)}f_{i,j}^{(L)} - f_{i,j}^{(K)}f_k^{(L)}\right)\tau_{,l}\right\}$$

$$= \frac{1}{\rho}\left\{\left(c_{ijkl}\tau_{,l}\right)_{,j}f_i^{(K)}f_k^{(L)} + c_{ijkl}\left(f_k^{(K)}f_{i,j}^{(L)} + f_i^{(K)}f_{k,j}^{(L)}\right)\tau_{,l}\right\} \ . \tag{5.6}$$

Inserting (5.1) we derive

$$\mathbf{f}^{(K)T}\mathbf{N}\mathbf{f}^{(L)} = \frac{1}{\rho}\left\{\left(c_{ijkl}\tau_{,l}\right)_{,j}f_i^{(K)}f_k^{(L)}\right.$$

$$\left. + c_{ijkl}\left(c_j^{LJ}f_i^{(J)}f_k^{(K)} + c_j^{L3}g_i^{(3)}f_k^{(K)} + c_j^{LJ}f_k^{(J)}f_i^{(K)} + c_j^{L3}g_k^{(3)}f_i^{(K)}\right)\tau_{,l}\right\} \ , \tag{5.7}$$

which can be written as

$$\mathbf{f}^{(K)T}\mathbf{N}\mathbf{f}^{(L)} = \frac{1}{\rho}\left(c_{ijkl}\tau_{,l}\right)_{,j}f_i^{(K)}f_k^{(L)} + c_j^{LJ}A_j^{KJ(p)} + c_j^{L3}A_j^{K3(p)} \ . \tag{5.8}$$

Now we shall show that $\mathbf{f}^{(K)T}\mathbf{N}\mathbf{f}^{(L)}$ is symmetric for an appropriate choice of the basis vectors. Thus, we require $\mathbf{f}^{(1)T}\mathbf{N}\mathbf{f}^{(2)} - \mathbf{f}^{(2)T}\mathbf{N}\mathbf{f}^{(1)} = 0$, which yields

$$\frac{1}{\rho}\left(c_{ijkl}\tau_{,l}\right)_{,j}\left(f_i^{(1)}f_k^{(2)} - f_i^{(2)}f_k^{(1)}\right) + c_j^{21}A_j^{11(p)} - c_j^{12}A_j^{22(p)} + c_j^{23}A_j^{13(p)} - c_j^{13}A_j^{23(p)} = 0 \ . \tag{5.9}$$

From the averaged Hamiltonian (3.8) and the associated ray-tracing system (3.9), and using the equality $c_j^{12} = -c_j^{21}$, we obtain

$$c_j^{21} A_j^{11(p)} - c_j^{12} A_j^{22(p)} = -c_j^{12}\left(A_j^{11(p)} + A_j^{22(p)}\right) = -4c_j^{12}\frac{dx_j}{dt} \ . \tag{5.10}$$

We also have the relation

$$4c_j^{12}\frac{dx_j}{dt} = 4\frac{\partial f_i^{(1)}}{\partial x_j}f_i^{(2)}\frac{dx_j}{dt} = 4\frac{df_i^{(1)}}{dt}f_i^{(2)} \ . \tag{5.11}$$

Since $c_{ijkl} = c_{klij}$ and $\tau_{,lj} = \tau_{,jl}$, the first term on the left-hand side of eq. (5.9) is equal to $\frac{1}{\rho}c_{ijkl,j}\tau_{,l}\big(f_i^{(2)}f_k^{(1)} - f_i^{(1)}f_k^{(2)}\big)$. Thus, we arrive at

$$4\frac{df_i^{(1)}}{dt}f_i^{(2)} = \frac{1}{\rho}c_{ijkl,j}\tau_{,l}\left(f_i^{(1)}f_k^{(2)} - f_i^{(2)}f_k^{(1)}\right) + c_j^{23}A_j^{13(p)} - c_j^{13}A_j^{23(p)} \ . \tag{5.12}$$

Equation (5.12) shows how the basis $\{\mathbf{f}^{(1)}, \mathbf{f}^{(2)}\}$ should be integrated in order to make the coupled transport equation (4.14) symmetric. The vector $d\mathbf{f}^{(1)}/dt$, which should be perpendicular to $\mathbf{f}^{(1)}$ due to the orthonormality of the basis, is completed by its projection onto $\mathbf{g}^{(3)}$. We have

$$\frac{df_i^{(1)}}{dt}g_i^{(3)} = c_j^{13}\frac{dx_j}{dt} \ . \tag{5.13}$$

The coefficients c_j^{13} and c_j^{23} are determined by eq. (5.5). Note that all quantities on the right-hand sides of eqs. (5.12) and (5.13) are attributes of the ray which is traced kinematically by integrating the system (3.9), except from the terms involving $A_m^{K3(p)}\tau_{,mj}$, which contribute to c_j^{K3}.

We shall show that, again within $O(1/\omega)$-accuracy in the proximity of a tangent point of the S-wave slowness sheets, the occurrence of the second order derivatives $\tau_{,mj}$ can be avoided in the integration of the basis vectors. In the differential equation (5.12), the contribution of the terms $c_j^{23}A_j^{13(p)} - c_j^{13}A_j^{23(p)}$ can be approximated by

$$c_j^{23}A_j^{13(p)} - c_j^{13}A_j^{23(p)} = \frac{\left(A_j^{23(x)} + A_m^{23(p)}\tau_{,mj}\right)A_j^{13(p)}}{\left(G^{(2)} - G^{(3)}\right)}$$
$$- \frac{\left(A_j^{13(x)} + A_m^{13(p)}\tau_{,mj}\right)A_j^{23(p)}}{\left(G^{(1)} - G^{(3)}\right)} + O(1/\omega) \ . \tag{5.14}$$

Since $G^{(1)} - G^{(2)} = O(1/\omega)$ in a $O(1/\omega)$-vicinity of a tangent point of the slowness sheets, we obtain for this approximation

$$c_j^{23}A_j^{13(p)} - c_j^{13}A_j^{23(p)} = \frac{A_j^{23(x)}A_j^{13(p)}}{\left(G^{(2)} - G^{(3)}\right)} - \frac{A_j^{13(x)}A_j^{23(p)}}{\left(G^{(1)} - G^{(3)}\right)} + O(1/\omega) \ . \tag{5.15}$$

Applying a similar approximation to the coefficient c_j^{13} in eq. (5.14), we find

$$\frac{df_i^{(1)}}{dt} g_i^{(3)} = \frac{(A_j^{13(x)} + A_m^{13(p)} \tau_{,mj})\, dx_j}{G^{(1)} - G^{(3)}} \frac{dx_j}{dt} + O(1/\omega)$$

$$= \left\{ A_j^{13(x)} \frac{dx_j}{dt} + A_m^{13(p)} \frac{dp_m}{dt} \right\} / (G^{(1)} - G^{(3)}) + O(1/\omega)$$

$$= (f_i^{(1)} \frac{d\Gamma_{ik}}{dt} g_k^{(3)}) / (G^{(1)} - G^{(3)}) + O(1/\omega) \ . \tag{5.16}$$

We now return to the terms $\mathbf{f}^{(K)T} \mathbf{N} \mathbf{f}^{(L)}$ in the coupled transport equation (4.14), for which we use the symmetrized form of eq. (5.8),

$$\mathbf{f}^{(K)T} \mathbf{N} \mathbf{f}^{(L)} = \frac{1}{2\rho} (c_{ijkl} \tau_{,l})_{,j} \left(f_i^{(K)} f_k^{(L)} + f_i^{(L)} f_k^{(K)} \right)$$

$$+ \frac{1}{2} \left(c_j^{LJ} A_j^{KJ(p)} + c_j^{KJ} A_j^{LJ(p)} + c_j^{L3} A_j^{K3(p)} + c_j^{K3} A_j^{L3(p)} \right) \ . \tag{5.17}$$

The term $c_j^{LJ} A_j^{KJ(p)} + c_j^{KJ} A_j^{LJ(p)}$ can be shown to be of order $O(1/\omega)$ in a $O(1/\omega)$– vicinity of a tangent point of the slowness sheets. For $(K, L) = (1, 2)$ this term is equal to $c_j^{21}(A_j^{11(p)} - A_j^{22(p)})$, and for $(K, L) = (1, 1)$ or $(2, 2)$ this term is equal to $(-1)^K 2 c_j^{21} A_j^{12(p)}$. The equations (4.11) and (4.10) establish that these terms are of order $O(1/\omega)$, indeed.

When these terms are left out, the leading term of the coupled transport equation is formulated in a form which is suitable for numerical integration along the reference ray, provided that the second order derivatives $\tau_{,mj}$ (which also occur implicitly in the coefficients c_j^{L3} and c_j^{K3}) are precalculated by dynamic ray tracing for the averaged Hamiltonian (3.8).

6. Conclusion

A symmetrised, frequency-dependent, coupled transport equation has been derived which enables smooth propagation of anisotropic shear-wave polarisation nearby a point $(\mathbf{x}_0, \mathbf{p}_0)$ at which the shear-wave slowness sheets are tangent (and hence the group-velocities of both modes equal). Differences with the weakly anisotropic zero-order QI-approach of PšENČÍK (1998) are twofold: the geometrical spreading is averaged between the generally different geometrical spreading factors as obtained for the uncoupled shear modes, and the shear-wave polarisation is in the plane perpendicular to that of the P-wave polarisation for the actual direction of slowness, rather than perpendicular to the P-wave polarisation in an isotropic reference model.

Whether these differences are significant in practical circumstances remains to be seen. We have not addressed the issue of smoothly matching the coupled frequency-dependent shear-wave ray tracing with regular uncoupled anisotropic dynamic ray tracing for areas where the shear-wave modes are sufficiently separated.

Acknowledgement

The author thanks Shell EP Technology Applications and Research for permission to publish this paper. Ivan Pšenčík, Leen Roozemond and an anonymous reviewer are gratefully acknowledged for their valuable comments. In particular I am grateful to Ludek Klimeš, who contributed to the results and presentation of Section 5.

REFERENCES

BAKKER, P.M. (1996), *Theory of Anisotropic Dynamic Ray Tracing in Ray-centred Coordinates*, Pure appl. geophys. *148*, 583–589.

ČERVENÝ, V. (1972), *Seismic Rays and Ray Intensities in Inhomogeneous Anisotropic Media*, Geophys. J. R. Astr. Soc. *29*, 1–13.

ČERVENÝ, V. *Dynamic ray tracing in inhomogeneous anisotropic media*, Chapter in *Seismic Waves in Complex 3-D Structures* (Cambridge University Press 2001).

ČERVENÝ, V. and FIRBAS, P. (1984), *Numerical Modelling and Inversion of Travel Times of Seismic Body Waves in Inhomogeneous Anisotropic Media*, Geophys. J. R. Astr. Soc. *76*, 41–51.

CHAPMAN, C.H. and SHEARER, P.M. (1989), *Ray Tracing in Azimuthally Anisotropic Media-II. Quasi-shear Wave Coupling*, Geophys. J. *96*, 65–83.

COATES, R.T. and CHAPMAN, C.H. (1990), *Quasi-shear Wave Coupling in Weakly Anisotropic 3-D Media*, Geophys. J. Int. *103*, 301–320.

GAJEWSKI, D. and PŠENČÍK, I. (1990), *Vertical Seismic Profile Synthetics by Dynamic Ray Tracing in Laterally Varying Layered Anisotropic Structures*, J. Geoph. Res. *95*, 11,301–11,315.

JECH, J. and PŠENČÍK, I. (1989), *First-order Perturbation Method for Anisotropic Media*, Geophys. J. Int. *99*, 369–376.

POPOV, M.M. and PŠENČÍK, I. (1978), *Computation of Ray Amplitudes in Laterally Inhomogeneous Media with Curved Interfaces*, Studia Geoph. Geod. *22*, 248–258.

PŠENČÍK, I. (1998), *Green's Functions for Inhomogeneous Weakly Anisotropic Media*, Geophys. J. Int. *135*, 279–288.

ZILLMER, M., KASHTAN, B.M., and GAJEWSKI, D. (1998), *Quasi-isotropic Approximation of Ray Theory for Anisotropic Media*, Geophys. J. Int. *132*, 643–653.

(Received September 5, 2000, revised December 13, 2000, accepted April 9, 2001)

Pure appl. geophys. 159 (2002) 1419–1435
0033–4553/02/081419–17 $ 1.50 + 0.20/0

▌Pure and Applied Geophysics

Numerical Algorithm of the Coupling Ray Theory in Weakly Anisotropic Media

Petr Bulant[1] and Luděk Klimeš[1]

Summary — The paper is devoted to the numerical calculation of the frequency-dependent complex-valued vectorial amplitudes of S waves in weakly anisotropic media by the coupling ray theory. An efficient and accurate method of numerical integration of the coupling equation is proposed, and the accuracy of the method is estimated in order to control the integration step so that the relative error in the wavefield amplitudes due to the integration is kept below a given limit. Several quasi-isotropic approximations of the coupling ray theory are briefly discussed and a numerical example is presented.

Key words: Coupling ray theory, anisotropy, travel times, amplitudes, Green function.

1. Introduction

There are two different high-frequency asymptotic ray theories, the *isotropic ray theory* assuming equal velocities of both S-wave polarizations and the *anisotropic ray theory* assuming both S-wave polarizations strictly decoupled.

In the isotropic ray theory, the S-wave polarization vectors do not rotate about the ray, whereas in the anisotropic ray theory, they coincide with the eigenvectors of the Christoffel matrix which may rotate rapidly about the ray.

In "weakly anisotropic" models, at moderate frequencies, the S-wave polarization tends to stay unrotated round the ray but is partly attracted by the rotation of the eigenvectors of the Christoffel matrix. The intensity of the attraction increases with frequency.

The isotropic and anisotropic ray theories are thus limiting cases and the gap between them has to be filled. A ray theory providing continuous transition between the isotropic and anisotropic ray theories was proposed by COATES and CHAPMAN (1990) and is called the *coupling ray theory*. There are many possible modifications and approximations of the coupling ray theory. For example, the reference ray may be calculated in different ways (BAKKER, 2002), the Christoffel matrix may be approximated by its quasi-isotropic projections onto the plane perpendicular to the

[1] Department of Geophysics, Charles University, Ke Karlovu 3, 121 16 Praha 2, Czech Republic. E-mails: bulant@seis.karlov.mff.cuni.cz, klimes@seis.karlov.mff.cuni.cz

reference ray and onto the line tangent to the reference ray (PŠENČÍK, 1998; ČERVENÝ, 2001), travel times corresponding to the anisotropic ray theory may be approximated in several ways, e.g., by linear quasi-isotropic perturbation with respect to the density-normalized elastic parameters (PŠENČÍK, 1998; ČERVENÝ, 2001), etc. Several quasi-isotropic approximations of the coupling ray theory are briefly mentioned in Appendix A.

For comparison of isotropic, anisotropic and coupling ray theories with the exact solution in a simple model, refer to BULANT et al. (2000). The comparison also includes the quasi-isotropic approximation of the coupling ray theory by PŠENČÍK (1998), especially the effect of the quasi-isotropic approximation of travel times by linear perturbation with respect to the elastic parameters. The effect of the quasi-isotropic approximation of the Christoffel matrix is demonstrated on the numerical example in this paper.

This paper contains notes regarding the numerical calculation of the frequency-dependent complex-valued S-wave polarization vectors of the coupling ray theory. The method of numerical integration of the coupling equation, proposed by ČERVENÝ (2001), is applied to the coupling equation derived by COATES and CHAPMAN (1990), with emphasis on numerical implementation. The method of integration does not require the calculation of the angular velocity $d\varphi/d\tau$ of the rotation of the eigenvectors of the Christoffel matrix along the reference ray and does not require $d\varphi/d\tau$ to be smooth or finite along the reference ray. This is an important property because the angular velocity of the rotation is undefined in the singular regions of the two equal eigenvalues of the Christoffel matrix.

The accuracy of the method of numerical integration of the coupling equation is estimated in Section 6 in order to control the integration step so that the relative error in the wavefield amplitudes due to the integration is kept below a given limit, which is of principal importance for numerical applications.

The calculation of P-wave travel-time perturbations from a reference isotropic model to an anisotropic model and the calculation of the S-wave coupling ray theory travel time and amplitude corrections have been coded and added to the Fortran 77 package CRT (KLIMEŠ, 1998; BUCHA and KLIMEŠ, 1999; BUCHA et al., 2000). A simple numerical example of synthetic seismograms calculated by the coupling ray theory is presented.

2. Coupling Ray Theory for S Waves

Assume a curve in phase space, hereinafter called the "reference ray", parameterized by reference travel time τ, with reference slowness vectors $p_i(\tau)$ known at all its points $x_j(\tau)$. The reference ray should be close to the ray of the wave under study. In particular, if shear-wave coupling in weakly anisotropic media is investigated and the same reference ray is used for both S-wave polarizations, the reference ray should be close to the high-frequency approximations of the rays of both S waves (BAKKER, 2002).

Using the reference slowness vectors, we can calculate reference Christoffel matrices

$$\Gamma_{jk}(\tau) = p_i(\tau)\, a_{ijkl}(x_m(\tau))\, p_l(\tau) \tag{1}$$

and their eigenvectors $g_{i1}(\tau)$, $g_{i2}(\tau)$, $g_{i3}(\tau)$ along the reference ray. Assume that eigenvectors $g_{i1}(\tau)$ and $g_{i2}(\tau)$ correspond to S waves and that they vary continuously along the reference ray. The continuity is not required in the regions where the corresponding two eigenvalues are equal. Let us denote $\tau_1(\tau)$ and $\tau_2(\tau)$ the travel times corresponding to polarizations $g_{i1}(\tau)$ and $g_{i2}(\tau)$, respectively. They may be approximated by quadratures along the unperturbed reference ray,

$$\frac{d\tau_1}{d\tau} = [\Gamma_{jk}g_{j1}g_{k1}]^{-\frac{1}{2}}\,, \qquad \frac{d\tau_2}{d\tau} = [\Gamma_{jk}g_{j2}g_{k2}]^{-\frac{1}{2}}\,. \tag{2}$$

For the derivation of (2) refer to Appendix B. The coupling ray theory solution u_i of the elastodynamic equations may then be expressed, for S waves, as a linear combination of the anisotropic ray theory solutions (COATES and CHAPMAN, 1990, eq. 15),

$$u_i = \sum_{M=1}^{2} g_{iM} A_M r_M \exp(i\omega\tau_M)\,, \tag{3}$$

where $A_M = A_M(\tau)$ are the complex-valued scalar amplitudes in the high-frequency approximation corresponding to polarizations g_{iM}. Because the amplitudes are calculated for the selected system of reference rays, using the relevant dynamic ray tracing along the reference ray (ČERVENÝ, 1972), they are identical, $A_M \equiv A = A(\tau)$. The coupling ray theory equation for complex-valued amplitude factors $r_M = r_M(\tau)$ reads (COATES and CHAPMAN, 1990, eq. 30)

$$\frac{d}{d\tau}\begin{pmatrix} r_1 \\ r_2 \end{pmatrix} = \frac{d\varphi}{d\tau}\begin{pmatrix} 0 & \exp(i\omega[\tau_2(\tau) - \tau_1(\tau)]) \\ -\exp(i\omega[\tau_1(\tau) - \tau_2(\tau)]) & 0 \end{pmatrix}\begin{pmatrix} r_1 \\ r_2 \end{pmatrix}\,, \tag{4}$$

where

$$\frac{d\varphi}{d\tau} = g_{k2}\frac{dg_{k1}}{d\tau} = -g_{k1}\frac{dg_{k2}}{d\tau} \tag{5}$$

is the angular velocity of the eigenvector rotation. For several possible modifications and approximations of the above formulation of the coupling ray theory refer to Appendix A.

For high frequencies, $\exp(i\omega[\tau_2(\tau) - \tau_1(\tau)])$ in equation (4) may oscillate rapidly. These oscillations have minimal impact on the solution at high frequencies because they cancel out after each period and the derivative of the solution is zero on average

at high frequencies. However, these oscillations may either considerably reduce the
accuracy, or increase the cost of the numerical integration.

In addition, travel times τ_M in equation (4) depend on the initial conditions.
Equation (4) is thus much less suitable for defining the propagator matrix than
equations with purely local coefficients.

The oscillation and global character of the coefficient matrix of equation (4) is
caused by different travel times τ_M with respect to which the amplitude factors r_M are
defined. Let us introduce new amplitude factors a_M, both related to the same
"average" travel time $\bar{\tau}$ (ČERVENÝ, 2001),

$$r_1 \exp(i\omega\tau_1) = a_1 \exp(i\omega\bar{\tau}) \ , \qquad r_2 \exp(i\omega\tau_2) = a_2 \exp(i\omega\bar{\tau}) \ . \tag{6}$$

Inserting $r_M = a_M \exp(i\omega[\bar{\tau} - \tau_M])$ into equation (4), we obtain equation

$$\frac{d}{d\tau}\begin{pmatrix} a_1 \\ a_2 \end{pmatrix} = \left[\begin{pmatrix} 0 & 1 \\ -1 & 0 \end{pmatrix} \frac{d\varphi}{d\tau} - i\omega \begin{pmatrix} \frac{d(\bar{\tau} - \tau_1)}{d\tau} & 0 \\ 0 & \frac{d(\bar{\tau} - \tau_2)}{d\tau} \end{pmatrix} \right] \begin{pmatrix} a_1 \\ a_2 \end{pmatrix} \ . \tag{7}$$

It seems reasonable to simplify equation (7) by choosing

$$\bar{\tau}(\tau) = \tfrac{1}{2}[\tau_1(\tau) + \tau_2(\tau)] \ . \tag{8}$$

Equation (7) then takes the form

$$\frac{d}{d\tau}\begin{pmatrix} a_1 \\ a_2 \end{pmatrix} = \left[\begin{pmatrix} 0 & 1 \\ -1 & 0 \end{pmatrix} \frac{d\varphi}{d\tau} - \begin{pmatrix} i & 0 \\ 0 & -i \end{pmatrix} \frac{\omega}{2}\frac{d(\tau_2 - \tau_1)}{d\tau} \right] \begin{pmatrix} a_1 \\ a_2 \end{pmatrix} \ . \tag{9}$$

3. Propagator Matrix

Propagator matrix Π^g of equation (9), defined as

$$\Pi^g_{MN}(\tau, \tau_0) = \frac{\partial a_M(\tau)}{\partial a_N(\tau_0)} \ , \tag{10}$$

is a complex-valued 2×2 matrix satisfying equation

$$\frac{d}{d\tau}\Pi^g = \left[\begin{pmatrix} 0 & 1 \\ -1 & 0 \end{pmatrix} \frac{d\varphi}{d\tau} - \begin{pmatrix} i & 0 \\ 0 & -i \end{pmatrix} \frac{d\epsilon}{d\tau} \right] \Pi^g \ , \tag{11}$$

directly following from equation (9). Here

$$\epsilon(\tau) = \tfrac{1}{2} \ \omega[\tau_2(\tau) - \tau_1(\tau)] \ . \tag{12}$$

Propagator matrix Π^g is symplectic and unitary, but we do not need to make use of
these properties in this paper.

It is difficult to integrate equation (11) by the Runge-Kutta or another numerical
method which requires derivative $d\varphi/d\tau$ along the reference ray to be calculated,

because this derivative is undefined in the singular regions of the two equal eigenvalues of Christoffel matrix (1). For a more detailed discussion of the application of the Runge-Kutta methods to coupling equation (11) refer to Appendix D. The method of integration proposed in this paper does not require the calculation of derivative $d\varphi/d\tau$ and does not require the derivative to be smooth or finite along the reference ray.

Since Π^g is a propagator matrix satisfying the chain rule, it may be numerically calculated as the product of propagator matrices Π^g corresponding to reasonably small segments of the reference ray (ČERVENÝ, 2001). Frequency-dependent propagator matrices along individual small ray segments may be approximated with various degrees of accuracy and efficiency.

One of the possible means of approximating the local solutions, derived in more detail by the method of mean coefficients by ČERVENÝ (2001), is suggested in this paper.

4. Analytic Solution in a Special Case

Assume that $d\varphi/d\tau = constant$ $(d\epsilon/d\tau)$, i.e., that $(\varphi - \varphi_0)/(\epsilon - \epsilon_0) = constant$. This assumption is used to derive a simple analytic solution of equation (11) within a small finite segment along the reference ray. The analytic solution will then be used to approximate the exact solution for variable ratio $(\varphi - \varphi_0)/(\epsilon - \epsilon_0)$, and the accuracy of the approximation will be studied in Section 6.

For $d\varphi/d\tau = constant$ $(d\epsilon/d\tau)$, equation (11) takes the form

$$\frac{d}{d\tau}\Pi^g = \mathbf{A}\frac{d\alpha}{d\tau}\Pi^g ,\tag{13}$$

with constant matrix

$$\mathbf{A} = \left[\begin{pmatrix} 0 & 1 \\ -1 & 0 \end{pmatrix}(\varphi - \varphi_0) - \begin{pmatrix} i & 0 \\ 0 & -i \end{pmatrix}(\epsilon - \epsilon_0)\right]\alpha^{-1}\tag{14}$$

and parameter $\alpha = \alpha(\tau, \tau_0)$ defined as

$$\alpha = \sqrt{(\varphi - \varphi_0)^2 + (\epsilon - \epsilon_0)^2} .\tag{15}$$

Here we have put $\varphi = \varphi(\tau)$, $\varphi_0 = \varphi(\tau_0)$, $\epsilon = \epsilon(\tau)$ and $\epsilon_0 = \epsilon(\tau_0)$. Since

$$\mathbf{A}^2 = -\mathbf{1} ,\tag{16}$$

the solution of equation (13) with unit initial conditions may be expressed in the form of

$$\Pi^g(\tau, \tau_0) = \exp(\mathbf{A}\alpha) = \mathbf{1}\cos(\alpha) + \mathbf{A}\sin(\alpha) \ . \tag{17}$$

Note that equation (16), unlike (17), is valid exactly also for variable ratio $(\varphi - \varphi_0)/(\epsilon - \epsilon_0)$.

5. Approximation for Short Ray Segments

We now use matrix (17) with (14) and (15) as the approximation for short ray segments, suggest how to calculate $\varphi - \varphi_0$ and $\epsilon - \epsilon_0$, and finally estimate the accuracy of the approximation.

Assume that vectors $g_{jM}^0 = g_{jM}(\tau_0)$ are mapped onto vectors $g_{jM} = g_{jM}(\tau)$ in this way: (a) the vectors are rotated in plane span$\{g_{j1}^0, g_{k2}^0\}$, (b) the vectors are rotated round the line of intersection of planes span$\{g_{j1}^0, g_{k2}^0\}$ and span$\{g_{j1}, g_{k2}\}$, (c) the vectors are rotated in plane span$\{g_{j1}, g_{k2}\}$. The sum of the rotation angles in planes span$\{g_{j1}^0, g_{k2}^0\}$ and span$\{g_{j1}, g_{k2}\}$ is then given by

$$\Delta\varphi = \varphi - \varphi_0 = \arctan\left(\frac{g_{k1}g_{k2}^0 - g_{k2}g_{k1}^0}{g_{k1}g_{k1}^0 + g_{k2}g_{k2}^0}\right) \ , \tag{18}$$

see (5). For the derivation of (18) refer to Appendix C.

Assume that travel times τ_M, $M = 1, 2$, can be obtained by trapezoidal quadratures of equations (2),

$$\Delta\tau_M = \tau_M(\tau) - \tau_M(\tau_0) \approx \frac{1}{2}\left[\frac{d\tau_M}{d\tau}(\tau) + \frac{d\tau_M}{d\tau}(\tau_0)\right]\Delta\tau \ . \tag{19}$$

Equations (8) and (12) then yield

$$\Delta\bar{\tau} = \frac{1}{2}\left(\Delta\tau_2 + \Delta\tau_1\right) \tag{20}$$

and

$$\Delta\epsilon = \epsilon - \epsilon_0 = \frac{1}{2}\,\omega\left(\Delta\tau_2 - \Delta\tau_1\right) \ . \tag{21}$$

6. Accuracy of the Approximation

Equation (17) has been derived for constant ratio $(\varphi - \varphi_0)/(\epsilon - \epsilon_0)$ within the individual short ray segments, but is used as the numerical approximation of the exact solution for variable ratio $(\varphi - \varphi_0)/(\epsilon - \epsilon_0)$. We shall now estimate the accuracy of this numerical integration of coupling equation (11) in order to control the integration step so that the relative error in the wavefield amplitudes due to the integration is kept below a given limit.

The derivative of equation (17) along the reference ray reads

$$\frac{\mathrm{d}}{\mathrm{d}\tau}\Pi^g = [-1\sin(\alpha) + \mathbf{A}\cos(\alpha)]\frac{\mathrm{d}\alpha}{\mathrm{d}\tau} + \frac{\mathrm{d}\mathbf{A}}{\mathrm{d}\tau}\sin(\alpha) \ , \tag{22}$$

the matrix inverse to (17) is

$$[\Pi^g]^{-1} = 1\cos(\alpha) - \mathbf{A}\sin(\alpha) \ , \tag{23}$$

and their product is

$$\frac{\mathrm{d}}{\mathrm{d}\tau}\Pi^g\,[\Pi^g]^{-1} = \mathbf{A}\frac{\mathrm{d}\alpha}{\mathrm{d}\tau} + \frac{\mathrm{d}\mathbf{A}}{\mathrm{d}\tau}\cos(\alpha)\sin(\alpha) - \frac{\mathrm{d}\mathbf{A}}{\mathrm{d}\tau}\mathbf{A}[\sin(\alpha)]^2 \ . \tag{24}$$

The derivatives of α and $\mathbf{A}$ along the reference ray are

$$\frac{\mathrm{d}\alpha}{\mathrm{d}\tau} = \left[(\varphi - \varphi_0)\frac{\mathrm{d}\varphi}{\mathrm{d}\tau} + (\epsilon - \epsilon_0)\frac{\mathrm{d}\epsilon}{\mathrm{d}\tau}\right]\alpha^{-1} \tag{25}$$

and

$$\frac{\mathrm{d}\mathbf{A}}{\mathrm{d}\tau} = \left\{\begin{pmatrix} 0 & 1 \\ -1 & 0 \end{pmatrix}\left[\frac{\mathrm{d}\varphi}{\mathrm{d}\tau} - (\varphi - \varphi_0)\frac{\mathrm{d}\alpha}{\mathrm{d}\tau}\alpha^{-1}\right] - \begin{pmatrix} i & 0 \\ 0 & -i \end{pmatrix}\left[\frac{\mathrm{d}\epsilon}{\mathrm{d}\tau} - (\epsilon - \epsilon_0)\frac{\mathrm{d}\alpha}{\mathrm{d}\tau}\alpha^{-1}\right]\right\}\alpha^{-1} \ . \tag{26}$$

Insertion of (25) and (15) into (26) yields

$$\frac{\mathrm{d}\mathbf{A}}{\mathrm{d}\tau} = \left[\begin{pmatrix} 0 & 1 \\ -1 & 0 \end{pmatrix}(\epsilon - \epsilon_0) + \begin{pmatrix} i & 0 \\ 0 & -i \end{pmatrix}(\varphi - \varphi_0)\right]\left[(\epsilon - \epsilon_0)\frac{\mathrm{d}\varphi}{\mathrm{d}\tau} - (\varphi - \varphi_0)\frac{\mathrm{d}\epsilon}{\mathrm{d}\tau}\right]\alpha^{-3} \ . \tag{27}$$

The product of (27) and (14) is

$$\frac{\mathrm{d}\mathbf{A}}{\mathrm{d}\tau}\,\mathbf{A} = \begin{pmatrix} 0 & i \\ i & 0 \end{pmatrix}\left[(\epsilon - \epsilon_0)\frac{\mathrm{d}\varphi}{\mathrm{d}\tau} - (\varphi - \varphi_0)\frac{\mathrm{d}\epsilon}{\mathrm{d}\tau}\right]\alpha^{-2} \ . \tag{28}$$

Now we subtract the matrix multiplying Π^g on the right-hand side of (11) from matrix $\mathbf{A}\,(\mathrm{d}\alpha/\mathrm{d}\tau)$ appearing on the right-hand side of (24). The coefficient with matrix $\begin{pmatrix} 0 & 1 \\ -1 & 0 \end{pmatrix}$ is

$$(\varphi - \varphi_0)\alpha^{-1}\frac{\mathrm{d}\alpha}{\mathrm{d}\tau} - \frac{\mathrm{d}\varphi}{\mathrm{d}\tau} = -(\epsilon - \epsilon_0)\left[(\epsilon - \epsilon_0)\frac{\mathrm{d}\varphi}{\mathrm{d}\tau} - (\varphi - \varphi_0)\frac{\mathrm{d}\epsilon}{\mathrm{d}\tau}\right]\alpha^{-2} \tag{29}$$

and the coefficient with matrix $\begin{pmatrix} i & 0 \\ 0 & -i \end{pmatrix}$ is

$$-(\epsilon - \epsilon_0)\alpha^{-1}\frac{d\alpha}{d\tau} + \frac{d\epsilon}{d\tau} = -(\varphi - \varphi_0)\left[(\epsilon - \epsilon_0)\frac{d\varphi}{d\tau} - (\varphi - \varphi_0)\frac{d\epsilon}{d\tau}\right]\alpha^{-2} . \tag{30}$$

Now we may insert equations (27) – (30) into (24) to calculate the deviation of the approximation from equation (11),

$$\frac{d}{d\tau}\Pi^g[\Pi^g]^{-1} - \left[\begin{pmatrix} 0 & 1 \\ -1 & 0 \end{pmatrix}\frac{d\varphi}{d\tau} - \begin{pmatrix} i & 0 \\ 0 & -i \end{pmatrix}\frac{d\epsilon}{d\tau}\right]$$

$$= \left\{\left[\begin{pmatrix} 0 & 1 \\ -1 & 0 \end{pmatrix}(\epsilon - \epsilon_0) + \begin{pmatrix} i & 0 \\ 0 & -i \end{pmatrix}(\varphi - \varphi_0)\right]\left[\cos(\alpha)\frac{\sin(\alpha)}{\alpha} - 1\right]\right.$$

$$\left. - \begin{pmatrix} 0 & i \\ i & 0 \end{pmatrix}[\sin(\alpha)]^2\right\}\left[(\epsilon - \epsilon_0)\frac{d\varphi}{d\tau} - (\varphi - \varphi_0)\frac{d\epsilon}{d\tau}\right]\alpha^{-2} . \tag{31}$$

The relative error in Π^g per one step along a ray is then approximately matrix $\begin{pmatrix} 0 & i \\ i & 0 \end{pmatrix}$ times

$$\Delta\delta \approx -\int_{\tau_0}^{\tau}\left[(\epsilon - \epsilon_0)\frac{d\varphi}{d\tau} - (\varphi - \varphi_0)\frac{d\epsilon}{d\tau}\right]d\tau = -\int_{\varphi_0}^{\varphi_0 + \Delta\varphi}\left[(\epsilon - \epsilon_0) - (\varphi - \varphi_0)\frac{d\epsilon}{d\varphi}\right]d\varphi . \tag{32}$$

Approximating $\epsilon - \epsilon_0$ by the second-order Taylor expansion with respect to $\varphi - \varphi_0$, we arrive at

$$(\epsilon - \epsilon_0) - (\varphi - \varphi_0)\frac{d\epsilon}{d\varphi} \approx -\frac{1}{2}\frac{d^2\epsilon}{d\varphi^2}(\varphi - \varphi_0)^2 . \tag{33}$$

Within the second-order approximation (33), we can analytically integrate (32),

$$\Delta\delta \approx \frac{d^2\epsilon}{d\varphi^2}\frac{(\Delta\varphi)^3}{6} \approx \frac{(\Delta\varphi)^2}{6}\Delta\frac{d\epsilon}{d\varphi} \approx \frac{\Delta\varphi\Delta\tau}{6}\Delta\frac{d\epsilon}{d\tau} = \frac{\Delta\varphi\Delta\tau}{12}\omega\left(\Delta\frac{d\tau_2}{d\tau} - \Delta\frac{d\tau_1}{d\tau}\right) . \tag{34}$$

To keep the error in Π^g along the whole ray of length τ (measured in travel time) below the maximum specified limit, δ, we require

$$|\Delta\delta| \leq \frac{\Delta\tau}{\tau}\delta , \tag{35}$$

i.e.,

$$\boxed{|\Delta\varphi|\left|\Delta\frac{d\tau_2}{d\tau} - \Delta\frac{d\tau_1}{d\tau}\right| \leq \frac{12}{\omega\tau}\delta ,} \tag{36}$$

see (2). Note that, in addition to inequality (36), step $\Delta\varphi$ in the eigenvector rotation must also be sufficiently smaller than $45°$,

$$|\Delta\varphi| \ll \frac{\pi}{4} ,$$ (37)

in order to reliably follow the selected eigenvector g_{kM} along the reference ray. Condition (37) need not be satisfied in the regions where the corresponding two eigenvalues are equal, i.e., where $\Delta\epsilon = 0$ within the required numerical accuracy. For example, there is no restriction on the selection of the eigenvectors of the Christoffel matrix in the isotropic parts of the model, while the proposed integration of the coupling equation is accurate. Inequality (36) can be satisfied in smooth models without problems, independently of the degree of anisotropy. The numerical integration with the integration step controlled by equations (36) and (37) has been coded and numerically tested.

7. Numerical Example

The calculation of P-wave travel-time perturbations from a reference isotropic model to an anisotropic model and the calculation of the above described S-wave coupling ray theory travel-time and amplitude corrections along the isotropic reference rays have been coded in subroutine file wan.for and added to the Fortran 77 package CRT (KLIMEŠ, 1998; BUCHA and KLIMEŠ, 1999; BUCHA et al., 2000). The subroutines are called by program green.for if the model is anisotropic. In this case, program green.for generates frequency-dependent elementary Green functions (for S waves), further converted by program greenss.for into the response functions corresponding to a particular source radiation pattern. In the above cited versions of the code, the quasi-isotropic projection described in Subappendix A.2 is applied at the source and receiver points.

A 1-D anisotropic model QI (model WA rotated by $45°$) was provided by PŠENČÍK and DELLINGER (2001) who performed the coupling ray theory calculations using the programs of package ANRAY and compared the results with the anisotropic reflectivity calculations. The normalized elastic parameters a_{ijkl} at the surface (zero depth) are

$$\begin{pmatrix} 14.48500 & 4.52500 & 4.75000 & 0.00000 & 0.00000 & -0.58000 \\ & 14.48500 & 4.75000 & 0.00000 & 0.00000 & -0.58000 \\ & & 15.71000 & 0.00000 & 0.00000 & -0.29000 \\ & & & 5.15500 & -0.17500 & 0.00000 \\ & & & & 5.15500 & 0.00000 \\ & & & & & 5.04500 \end{pmatrix}$$ (38)

and at the depth of 1 length unit they are

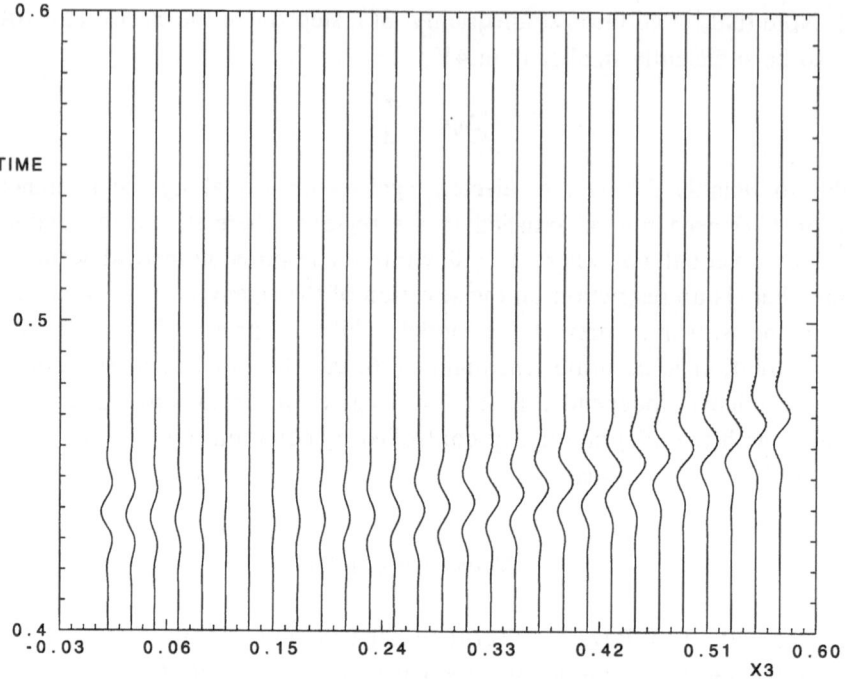

Figure 1

The first (radial) component of the coupling ray theory synthetic seismograms is shown as the solid line.
The dotted seismograms (somewhat obscured in this figure but better visible in Fig. 2) correspond to the
quasi-isotropic approximation of the Christoffel matrix, see also Figures 2 and 3.

$$\begin{pmatrix} 22.08963 & 6.90063 & 7.24375 & 0.00000 & 0.00000 & -0.88450 \\ & 22.08953 & 7.24375 & 0.00000 & 0.00000 & -0.88450 \\ & & 23.95775 & 0.00000 & 0.00000 & -0.44225 \\ & & & 7.86138 & -0.26688 & 0.00000 \\ & & & & 7.86138 & 0.00000 \\ & & & & & 7.69363 \end{pmatrix} . \tag{39}$$

The reference isotropic model is given by

$$v_{\mathrm{P}}^2 = 15.00, \qquad v_{\mathrm{S}}^2 = 5.10 \tag{40}$$

at the surface, and

$$v_{\mathrm{P}}^2 = 23.00, \qquad v_{\mathrm{S}}^2 = 7.79 \tag{41}$$

at the depth of 1 length unit. All the above values are interpolated linearly with
depth. The homogeneous density is $\rho = 1$. The synthetic seismograms, corresponding
to vertical force $\mathbf{F} = (0, 0, 100)^{\mathrm{T}}$ at position $(50, 50, 0)^{\mathrm{T}}$, are calculated at 29 receivers
$(51, 50, 0.010)^{\mathrm{T}}$, $(51, 50, 0.030)^{\mathrm{T}}$, $(51, 50, 0.050)^{\mathrm{T}}, \ldots, (51, 50, 0.570)^{\mathrm{T}}$ located in a

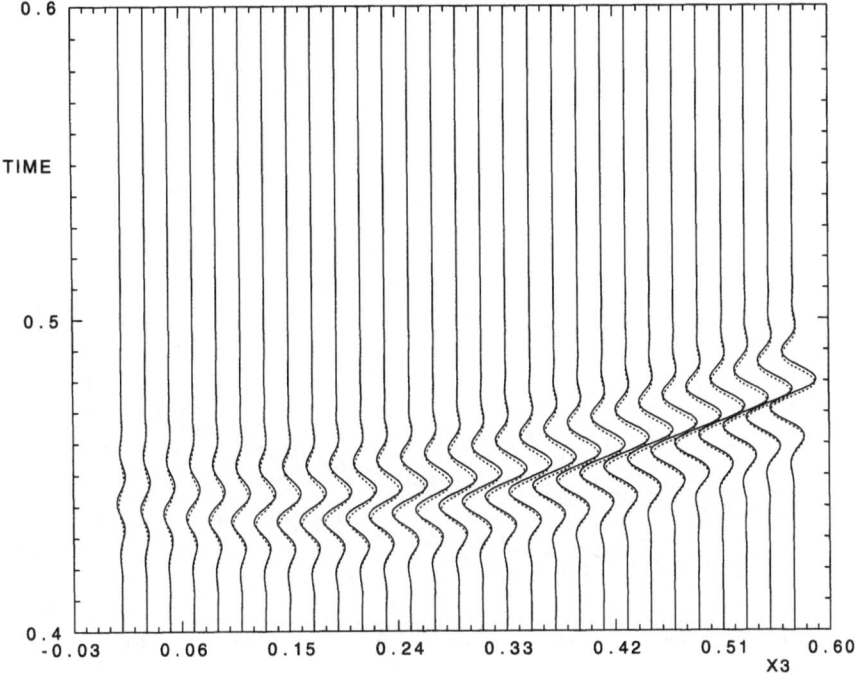

Figure 2

The second (transverse) component of the synthetic seismograms. Note that this component is zero in the one-dimensional reference isotropic model.

vertical well. The source time function is the Gabor signal $\cos(2\pi ft)\exp[-(2\pi ft/4)^2]$ with reference frequency $f = 50\,\text{Hz}$, band-pass filtered by the cosine filter given by frequencies $0\,\text{Hz}$, $5\,\text{Hz}$, $60\,\text{Hz}$ and $100\,\text{Hz}$.

The resulting coupling ray theory seismograms, modified by the quasi-isotropic projection of Subappendix A.2, are plotted in Figures 1, 2 and 3 as a solid line. The maximum polarization error of the quasi-isotropic projection is 0.061 radians in this example. For comparison, the seismograms calculated according to the quasi-isotropic approximation of Subappendix A.3 (corresponding to the seismograms calculated by package ANRAY) are plotted as a dotted line. The effect of the quasi-isotropic approximation of Subappendix A.4 is negligible in this example. On the other hand, the effect of the quasi-isotropic approximation of Subappendix A.4 has been demonstrated by BULANT et al. (2000) in a simple model in which the quasi-isotropic approximation of Subappendix A.3 does not affect the results.

The model is named QI and the data for packages CRT and ANRAY may be found on compact disks KLIMEŠ (1998), BUCHA and KLIMEŠ (1999) and BUCHA et al. (2000), together with the Fortran 77 source code of the packages. For comparison with the isotropic and anisotropic ray theory seismograms and for a more detailed discussion and description of this model refer to PŠENČÍK and DELLINGER (2001).

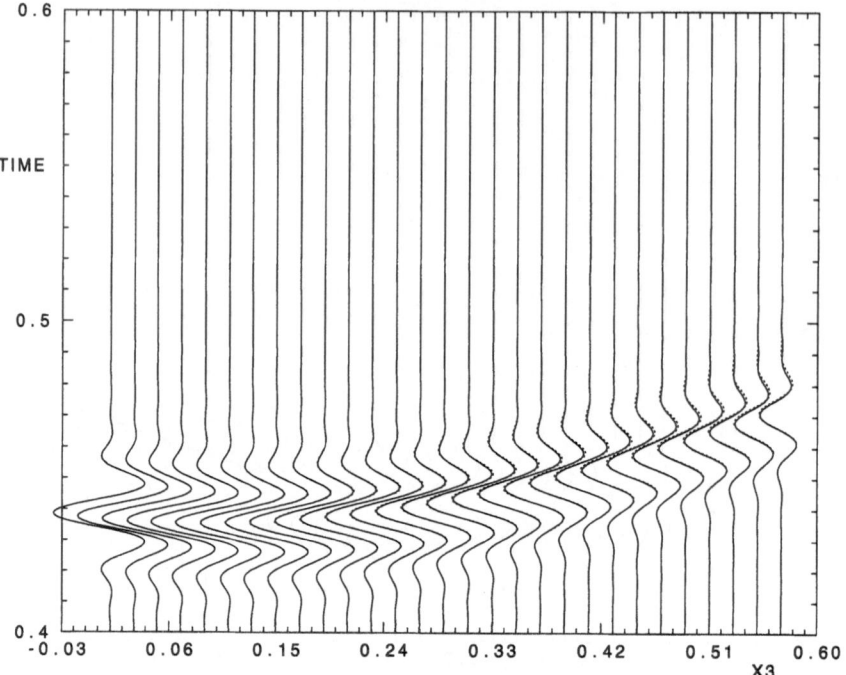

Figure 3
The third (vertical) component of the synthetic seismograms.

8. Conclusions

The proposed numerical algorithm for the coupling ray theory is very efficient and accurate, with controlled accuracy. It requires negligible computational time because the coupling equation is solved only along the previously calculated two-point rays. In weakly anisotropic media, the coupling ray theory should be considerably more accurate than the isotropic and anisotropic ray theories, although its actual accuracy and limits of applicability are still unknown.

Appendix A: Quasi-isotropic Approximations of the Coupling Ray Theory

A.1. Selection of the Reference Ray

The isotropic ray theory is always the limiting case of the coupling ray theory for decreasing anisotropy at fixed frequency. On the other hand, the high-frequency limit of the coupling ray theory at fixed anisotropy depends on the choice of the reference ray, and even on the choice of the *system* of reference rays, because the amplitudes are determined by the paraxial reference rays.

If we choose the *anisotropic ray theory reference ray* and select the initial polarization corresponding to the reference ray, the coupling ray theory will correctly converge to the anisotropic ray theory for high frequencies. For other choices of reference rays, the high-frequency limit of the coupling ray theory at fixed anisotropy is incorrect, although the differences may be negligible at the finite frequencies under consideration.

In the *anisotropic common-ray approximation*, the common reference ray is traced using the averaged Hamiltonian of both *S*-wave polarizations (BAKKER, 2002).

In the less accurate *isotropic common-ray approximation*, the reference ray is traced in the reference isotropic model. Moreover, the reference isotropic model may be selected in different ways, yielding quasi-isotropic approximations of various accuracy.

A.2. Quasi-isotropic Projection of the Polarization Vectors

The coupling ray theory solution (3) may be approximated by its projection

$$\tilde{u}_i = h_{iM} h_{mM} u_m \qquad (A.1)$$

onto the orthonormal reference polarization vectors h_{k1}, h_{k2}. This approximation may simplify the modification of existing isotropic ray tracing codes for the coupling ray theory. The error of this approximation is obvious and can be calculated simply.

A.3. Quasi-isotropic Approximation of the Christoffel Matrix

Denote the polarization vectors of the isotropic ray theory, or the reference polarization vectors in general, by h_{k1}, h_{k2} and h_{k3}. If the Christoffel matrix is approximated by its projections onto plane span$\{h_{j1}, h_{k2}\}$ and onto vector h_{l3},

$$\begin{aligned}
\tilde{\Gamma}_{jk} &= h_{jM} h_{mM} \Gamma_{mn} h_{nN} h_{kN} + h_{j3} h_{m3} \Gamma_{mn} h_{n3} h_{k3} \\
&= \Gamma_{jk} - (h_{jM} h_{k3} + h_{j3} h_{kM}) h_{mM} \Gamma_{mn} h_{n3} \ ,
\end{aligned} \qquad (A.2)$$

eigenvectors g_{k1} and g_{k2} become situated in plane span$\{h_{j1}, h_{k2}\}$ as in the zero-order quasi-isotropic approximation of PŠENČÍK (1998) and ČERVENÝ (2001). This approximation includes the approximation of Subappendix A.2.

A.4. Quasi-isotropic Perturbation of Travel Times

The linearized perturbation of equations (2) with respect to the density-normalized elastic parameters yields approximation

$$\frac{d\tau_1}{d\tau} \approx (\Gamma_{jk}^0 g_{j1} g_{k1})^{-\frac{1}{2}} - \frac{1}{2} (\Gamma_{jk} - \Gamma_{jk}^0) g_{j1} g_{k1} (\Gamma_{mn}^0 g_{m1} g_{n1})^{-\frac{3}{2}} \ . \qquad (A.3)$$

Assuming that

$$\Gamma^0_{jk}g_{j1}g_{k1} = 1 \ , \tag{A.4}$$

see Subappendix A.3, equation (A.3) reads

$$\frac{d\tau_1}{d\tau} \approx \frac{3}{2} - \frac{1}{2}\Gamma_{jk}g_{j1}g_{k1} \ , \tag{A.5}$$

as in the quasi-isotropic approximation of PŠENČÍK (1998) and ČERVENÝ (2001). Analogously for $d\tau_2/d\tau$. The quasi-isotropic perturbation of travel times leads to an erroneous time shift in synthetic seismograms but has a negligible impact on the amplitudes.

Appendix B: Derivation of Equations (2)

We present the derivation for $d\tau_1/d\tau$, equations for $d\tau_2/d\tau$ are analogous. The derivation is based on the assumption that equation

$$\tau_1 = \tau_1(\tau) \tag{B.1}$$

can be applied approximately also in the vicinity of the reference ray. Equation (B.1) means that the gradient of τ_1 is parallel to the gradient of τ in the vicinity of the reference ray. The Christoffel matrix corresponding to travel time τ_1 is then

$$\frac{\partial\tau_1}{\partial x_i}a_{ijkl}\frac{\partial\tau_1}{\partial x_l} = \left[\frac{d\tau_1}{d\tau}\right]^2 \Gamma_{jk}(\tau) \ , \tag{B.2}$$

and has the same eigenvectors as the reference Christoffel matrix $\Gamma_{jk}(\tau)$ defined by equation (1). The eigenvalue of Christoffel matrix (B.2) corresponding to the eigenvector $g_{i1}(\tau)$ has to be unity,

$$\left[\frac{d\tau_1}{d\tau}\right]^2 \Gamma_{jk}(\tau)g_{j1}(\tau)g_{k1}(\tau) = 1 \ , \tag{B.3}$$

which yields equations (2).

The alternative derivation of equations (2), based on the second-order travel-time perturbations, has been presented by KLIMEŠ (2002, eqs. 43 and 65).

Appendix C: Derivation of Equation (18)

Denote by e_{j1} the unit vector along the line of intersection of planes span$\{g^0_{j1}, g^0_{k2}\}$ and span$\{g_{j1}, g_{k2}\}$. Denote by e^0_{j2} and e_{j2} unit vectors perpendicular to e_{j1} in planes span$\{g^0_{j1}, g^0_{k2}\}$ and span$\{g_{j1}, g_{k2}\}$, respectively.

We assume that

(a) vectors g_{j1}^0, g_{j2}^0 are rotated about vector g_{j3}^0, perpendicular to plane span$\{g_{j1}^0, g_{k2}^0\}$, through angle φ_1 to reach vectors e_{j1}, e_{j2}^0,

$$(e_{j1}, e_{j2}^0, g_{j3}^0) = (g_{j1}^0, g_{j2}^0, g_{j3}^0) \begin{pmatrix} \cos\varphi_1 & -\sin\varphi_1 & 0 \\ \sin\varphi_1 & \cos\varphi_1 & 0 \\ 0 & 0 & 1 \end{pmatrix}, \tag{C.1}$$

(b) vectors e_{j2}^0, g_{j3}^0 are rotated about vector e_{j1} through angle α to reach vectors e_{j2}, g_{j3},

$$(e_{j1}, e_{j2}, g_{j3}) = (e_{j1}, e_{j2}^0, g_{j3}^0) \begin{pmatrix} 1 & 0 & 0 \\ 0 & \cos\alpha & -\sin\alpha \\ 0 & \sin\alpha & \cos\alpha \end{pmatrix}, \tag{C.2}$$

(c) vectors e_{j1}, e_{j2} are rotated about vector g_{j3}, perpendicular to plane span$\{g_{j1}, g_{k2}\}$, through angle φ_2 to reach vectors g_{j1}, g_{j2},

$$(g_{j1}, g_{j2}, g_{j3}) = (e_{j1}, e_{j2}, g_{j3}) \begin{pmatrix} \cos\varphi_2 & -\sin\varphi_2 & 0 \\ \sin\varphi_2 & \cos\varphi_2 & 0 \\ 0 & 0 & 1 \end{pmatrix}. \tag{C.3}$$

Then

$$\begin{pmatrix} g_{j1}g_{j1}^0 & g_{j2}g_{j1}^0 \\ g_{j1}g_{j2}^0 & g_{j2}g_{j2}^0 \end{pmatrix} = \begin{pmatrix} \cos\varphi_1 & -\sin\varphi_1 \\ \sin\varphi_1 & \cos\varphi_1 \end{pmatrix} \begin{pmatrix} 1 & 0 \\ 0 & \cos\alpha \end{pmatrix} \begin{pmatrix} \cos\varphi_2 & -\sin\varphi_2 \\ \sin\varphi_2 & \cos\varphi_2 \end{pmatrix}$$

$$= \begin{pmatrix} \cos\varphi_1\cos\varphi_2 - \sin\varphi_1\cos\alpha\sin\varphi_2 & -\cos\varphi_1\sin\varphi_2 - \sin\varphi_1\cos\alpha\cos\varphi_2 \\ \sin\varphi_1\cos\varphi_2 + \cos\varphi_1\cos\alpha\sin\varphi_2 & -\sin\varphi_1\sin\varphi_2 + \cos\varphi_1\cos\alpha\cos\varphi_2 \end{pmatrix}.$$

$$\tag{C.4}$$

We see that

$$g_{j1}g_{j2}^0 - g_{j2}g_{j1}^0 = (1 + \cos\alpha)\sin(\varphi_1 + \varphi_2) \tag{C.5}$$

and

$$g_{j1}g_{j1}^0 + g_{j2}g_{j2}^0 = (1 + \cos\alpha)\cos(\varphi_1 + \varphi_2), \tag{C.6}$$

which yields equation (18) for the sum

$$\Delta\varphi = \varphi_1 + \varphi_2 \tag{C.7}$$

of rotation angles.

Appendix D: Comparison with the Runge-Kutta Methods

Coupling equation (11) can be solved by a Runge-Kutta method if the first derivative $d\varphi/d\tau$ of rotation angle $\varphi(\tau)$ is smooth and finite, whereas the method proposed in this paper requires nothing but condition (37) and only outside the singular regions.

The numerical integration of coupling equation (11) by a Runge-Kutta method requires the derivative (5) of rotation angle $\varphi(\tau)$ to be calculated. The calculation of this derivative is much more involved than the calculation of angular increment (18), and fails in the singular regions where the Christoffel matrix has two equal eigenvalues.

To efficiently control the accuracy, the Runge-Kutta methods must be applied with a variable step along the reference ray, similarly as the method proposed in this paper. Unfortunately, the correct length of the integration step of the Runge-Kutta method cannot be determined reliably in advance. The variable step is also the reason why the predictor-corrector methods, designed primarily for a regular step, are not considered.

To keep the error of the fourth-order Runge-Kutta method with the integration step composed of two equal half-steps $\Delta\tau$ (RALSTON, 1965, eq. 5.6-48) below the maximum specified limit δ, the very roughly derived condition

$$\frac{\alpha^5}{\Delta\tau} \leq \frac{1}{\tau}\frac{45}{73}\delta \qquad (D.1)$$

should be satisfied. Here α is defined by equations (15), (18) and (12). Condition (D.1) is always more restrictive than condition (37). At high frequencies, condition (36) implies proportionality $\Delta\tau \sim \omega^{-\frac{1}{2}}$, whereas condition (D.1) implies proportionality $\Delta\tau \sim \omega^{-\frac{5}{4}}$. That is why the fourth-order Runge-Kutta method requires shorter integration steps than the method proposed in this paper at high frequencies, at which the two S-wave polarizations are split.

Runge-Kutta methods of lower orders are much less efficient than the fourth-order Runge-Kutta method for the numerical solution of the coupling equation at all frequencies.

Acknowledgements

The authors thank Ivan Pšenčík for his indispensible cooperation during the preparation of the paper and both anonymous reviewers for their suggestions which resulted in the considerable improvement of the manuscript.

This research was supported by the Grant Agency of the Czech Republic under Contract 205/01/0927, by the Grant Agency of the Charles University under Contract 229/2002/B-GEO/MFF, by the Ministry of Education of the Czech Republic within Research Project J13/98 113200004, and by the members of the consortium "Seismic Waves in Complex 3-D Structures" (see "http://sw3d.mff. cuni.cz").

REFERENCES

BAKKER, P.M. (2002), *Coupled Anisotropic Shear-wave Raytracing in Situations where Associated Slowness Sheets Are Almost Tangent*, Pure appl. geophys. *159*, 1403–1417.

BUCHA, V., BULANT, P., and KLIMEŠ, L. (eds.) (2000), *SW3D–CD–4*. In *Seismic Waves in Complex 3-D Structures, Report 10* (Dep. Geophys. Charles Univ., Prague) p. 227 (online at "http://sw3d.mff. cuni.cz").

BUCHA, V., and KLIMEŠ, L. (eds.) (1999), *SW3D–CD–3*. In *Seismic Waves in Complex 3-D Structures, Report 8* (Dep. Geophys. Charles Univ., Prague) p. 193 (online at "http://sw3d.mff.cuni.cz").

BULANT, P., KLIMEŠ, L., and PŠENČÍK, I. (2000), *Comparison of ray methods with the exact solution in the 1-D anisotropic "twisted crystal" model*. In *Expanded Abstracts of 70th Annual Meeting (Calgary)* (Soc. Explor. Geophysicists, Tulsa) pp. 2289–2292.

ČERVENÝ, V. (1972), *Seismic Rays and Ray Intensities in Inhomogeneous Anisotropic Media*, Geophys. J. R. astr. Soc. *29*, 1–13.

ČERVENÝ, V. (2001), *Seismic Ray Theory* (Cambridge Univ. Press, Cambridge).

COATES, R.T., and CHAPMAN, C.H. (1990), *Quasi-shear Wave Coupling in Weakly Anisotropic 3-D Media*, Geophys. J. int. *103*, 301–320.

KLIMEŠ, L. (ed.) (1998), *SW3D–CD–2*. In *Seismic Waves in Complex 3-D Structures, Report 7* (Dep. Geophys. Charles Univ., Prague) p. 405 (online at "http://sw3d.mff.cuni.cz").

KLIMEŠ, L. (2002), *Second-order and Higher-order Perturbations of Travel Time in Isotropic and Anisotropic Media*, Stud. geophys. geod. *46*, 213–248.

PŠENČÍK, I. (1998), *Green's Functions for Inhomogeneous Weakly Anisotropic Media*, Geophys. J. int. *135*, 279–288.

PŠENČÍK, I., and DELLINGER, J. (2001), *Quasi-shear Waves in Inhomogeneous Weakly Anisotropic Media by the Quasi-isotropic Approach: A Model Study*, Geophysics 66, 308–319.

RALSTON, A. (1965), *A First Course in Numerical Analysis* (McGraw-Hill, New York).

(Received October 10, 2000, revised March 21, 2001, accepted April 4, 2001)

Pure appl. geophys. 159 (2002) 1437–1445
0033–4553/02/081437–09 $ 1.50 + 0.20/0

© Birkhäuser Verlag, Basel, 2002

❚ Pure and Applied Geophysics

Perturbation of the Polarization Vectors in the Isotropic Ray Theory

LUDĚK KLIMEŠ[1]

Summary—The equations for the linear paraxial approximation of the polarization vectors and for the variation of the polarization vectors with a velocity perturbation are presented.

Key words: Polarization vectors, paraxial ray approximation, ray perturbation, isotropic ray theory, elastic waves.

Introduction

Paraxial ray methods have found many applications in the forward and inverse problems of wave propagation and are still rapidly developing. Refer to the book by ČERVENÝ (2001) for more details and references. Paraxial ray approximations may take advantage of the first and higher partial derivatives of travel time (e.g., KLIMEŠ, 1999), geometrical spreading, and polarization vectors.

The equations for the linear paraxial approximation of the polarization vectors and for the variation of the polarization vectors with a velocity perturbation were derived by COATES and CHAPMAN (1990). Here the equations are derived in more detail, and the final equations presented make it more obvious which numerical quadratures along the central ray are required.

Notation, Polarization Vectors and Coordinates

In the case of component notation, the capital-letter indices take values $K, L, \ldots = 1, 2$; the lower-case indices take values $k, l, \ldots = 1, 2, 3$. The Einstein summation convention is used with respect to repeated subscripts.

We denote by H_{iK} two mutually perpendicular unit vectors, perpendicular to the ray (polarization vectors in the case of an S wave). We require them not to rotate about the ray (POPOV and PŠENČÍK, 1978a,b),

[1] Department of Geophysics, Charles University, Ke Karlovu 3, 121 16 Praha 2, Czech Republic. E-mail: klimes@seis.karlov.mff.cuni.cz

$$H_{iK} \frac{\partial H_{iL}}{\partial s} = 0 \, , \tag{1}$$

where s is the arclength along the ray. Similarly, we denote by H_{i3},

$$H_{i3} = p_i v \, , \tag{2}$$

the unit vector tangent to the ray (polarization vector in the case of a P wave).

We introduce three coordinate systems: (a) Cartesian coordinates x_i. (b) Ray coordinates γ_i, where γ_1, γ_2 are the take-off ray parameters and γ_3 is an independent variable along rays (e.g., travel time τ, arclength s, $\sigma = \int v \, ds$, or another parameter). (c) Ray-centered coordinates q_i connected with their central ray, where q_I are Cartesian coordinates in plane $q_3 = constant$ perpendicular to the central ray, and q_3 is the arclength s along the central ray. We choose coordinates q_I in such a way that

$$H_{ik} = \frac{\partial x_i}{\partial q_k} \tag{3}$$

at the central ray. Since $\partial x_i / \partial q_K$ is constant in the plane perpendicular to the central ray,

$$\frac{\partial^2 x_i}{\partial q_K \partial q_L} = 0 \, . \tag{4}$$

For a simple 2-D sketch of the coordinate systems refer to Figure 1.

Geometrical Spreading Matrix and the Travel-time Derivatives

We denote by Q_{im} the matrix of geometrical spreading,

$$Q_{im} = \frac{\partial q_i}{\partial \gamma_m} \, , \tag{5}$$

and by P_{im} the matrix

$$P_{im} = \frac{\partial^2 \tau}{\partial \gamma_m \partial q_i} = \frac{\partial}{\partial \gamma_m} \left(p_j \frac{\partial x_j}{\partial q_i} \right) = \frac{\partial p_j}{\partial \gamma_m} \frac{\partial x_j}{\partial q_i} + p_j \frac{\partial q_l}{\partial \gamma_m} \frac{\partial^2 x_j}{\partial q_l \partial q_i} \, . \tag{6}$$

Removing zero derivatives (4), equation (6) reads

$$P_{im} = \frac{\partial p_j}{\partial \gamma_m} \frac{\partial x_j}{\partial q_i} + p_j \frac{\partial q_3}{\partial \gamma_m} \frac{\partial^2 x_j}{\partial q_3 \partial q_i} + \delta_{i3} p_j \frac{\partial q_L}{\partial \gamma_m} \frac{\partial^2 x_j}{\partial q_3 \partial q_L} \, , \tag{7}$$

where δ_{ij} is the Kronecker delta (components of the identity matrix). At the central ray, relation (7) yields

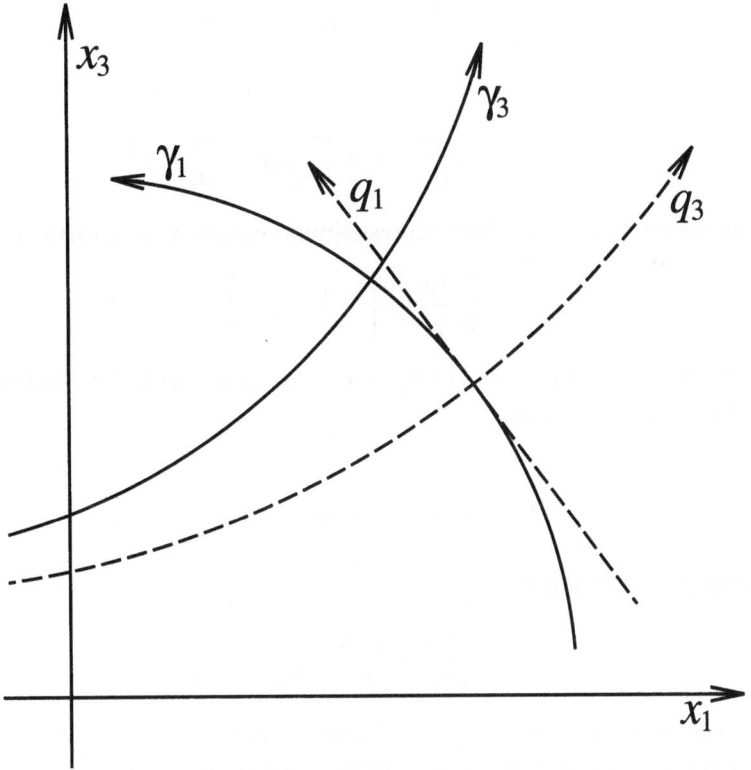

Figure 1
A simple 2-D sketch of (a) Cartesian coordinates x_i, (b) ray coordinates γ_i, (c) ray-centered coordinates q_i.

$$P_{im} = \frac{\partial p_j}{\partial \gamma_m} H_{ji} + p_j \frac{\partial H_{ji}}{\partial q_3} Q_{3m} + \delta_{i3}\, p_j \frac{\partial H_{jL}}{\partial q_3} Q_{Lm} \ . \tag{8}$$

Matrices Q_{im} and P_{im} are related as

$$P_{im} = M_{ij} Q_{jm} \ , \tag{9}$$

where

$$M_{ij} = \frac{\partial^2 \tau}{\partial q_i \partial q_j} \tag{10}$$

is the matrix of the second travel-time derivatives in the ray-centered coordinates.

Equation for the Polarization Vectors

The very simple form of the equation to trace the polarization vectors along rays reads (COATES and CHAPMAN, 1990, equation C3)

$$\frac{\partial H_{ik}}{\partial \gamma_3} = W_{ij} H_{jk} \quad, \tag{11}$$

where

$$W_{ij} = \frac{\partial p_i v}{\partial \gamma_3} p_j v - \frac{\partial p_j v}{\partial \gamma_3} p_i v = \frac{\partial p_i}{\partial \gamma_3} p_j v^2 - \frac{\partial p_j}{\partial \gamma_3} p_i v^2 \quad. \tag{12}$$

For the sake of conciseness, we shall express similar equations in a form analogous to

$$W_{ij} = \left[\frac{\partial p_i v}{\partial \gamma_3} p_j v\right] - \left[\cdots_{i \leftrightarrow j}\right] \quad. \tag{13}$$

Note that the derivative with respect to γ_3 in (11) is really a partial derivative, because it is applied for γ_1, γ_2 constant.

Paraxial Polarization Vectors

We are interested in the derivatives

$$\frac{\partial H_{ik}}{\partial q_n} = \frac{\partial H_{ik}}{\partial \gamma_m} Q_{mn}^{-1} \tag{14}$$

of the polarization vectors in ray-centered coordinates.

Equation (11) together with the ray tracing equations yields

$$\frac{\partial H_{ik}}{\partial q_3} = \left(\frac{\partial p_i}{\partial q_3} p_j v^2 - \frac{\partial p_j}{\partial q_3} p_i v^2\right) H_{jk} = \left(-\frac{\partial v}{\partial x_i} p_j + \frac{\partial v}{\partial x_j} p_i\right) H_{jk} \quad, \tag{15}$$

where $q_3 = s$ is the arclength along the ray. Since H_{i3} is a unit vector perpendicular to the wavefront,

$$\frac{\partial H_{i3}}{\partial q_N} = H_{iL} M_{LN} v \quad, \tag{16}$$

where M_{LN} are the second travel-time derivatives in the ray-centered coordinates, see (10). Unit vectors H_{i1}, H_{i2} are mutually perpendicular and both of them are perpendicular to H_{i3}. Their derivatives can thus be expressed as

$$\frac{\partial H_{iK}}{\partial q_N} = \epsilon_{KL} H_{iL} \Omega_N - H_{i3} M_{KN} v \quad, \tag{17}$$

where $\epsilon_{11} = \epsilon_{22} = 0$, $\epsilon_{12} = -\epsilon_{21} = 1$ and

$$\Omega_N = \frac{\partial H_{i1}}{\partial q_N} H_{i2} = -\frac{\partial H_{i2}}{\partial q_N} H_{i1} \ . \tag{18}$$

Note that equation (17) is equivalent to equation (C9) of COATES and CHAPMAN (1990). We now have to find the equations for Ω_1 and Ω_2.

Equations for the Paraxial Polarization Vectors

The differentiation of (11) with respect to ray coordinates γ_m yields

$$\frac{\partial}{\partial \gamma_3} \frac{\partial H_{ik}}{\partial \gamma_m} = W_{ij} \frac{\partial H_{jk}}{\partial \gamma_m} + \frac{\partial W_{ij}}{\partial \gamma_m} H_{jk} \ . \tag{19}$$

Since H_{ik} is a unitary 3×3 matrix satisfying equation (11), the solution of (19) may be expressed in the form

$$\frac{\partial H_{nl}}{\partial \gamma_m}(\gamma_3) = H_{nk}(\gamma_3) \left[H_{ki}^{-1}(\gamma_3^0) \frac{\partial H_{il}}{\partial \gamma_m}(\gamma_3^0) + \int_{\gamma_3^0}^{\gamma_3} d\gamma_3 H_{ki}^{-1} \frac{\partial W_{ij}}{\partial \gamma_m} H_{jl} \right]$$

$$= H_{nk}(\gamma_3) \left[H_{ik}(\gamma_3^0) \frac{\partial H_{il}}{\partial \gamma_m}(\gamma_3^0) + \int_{\gamma_3^0}^{\gamma_3} d\gamma_3 H_{ik} \frac{\partial W_{ij}}{\partial \gamma_m} H_{jl} \right] \ . \tag{20}$$

Note that equations (19) and (20) are identical to equations (C5) and (C6) of COATES and CHAPMAN (1990), who introduced 3×3 propagator matrix $E_{ni}(\gamma_3, \gamma_3^0) = H_{nk}(\gamma_3) H_{ki}^{-1}(\gamma_3^0)$ by equation (C4).

We shall now consider the integral on the right-hand side of this relation,

$$\int d\gamma_3 H_{ik} \frac{\partial W_{ij}}{\partial \gamma_m} H_{jl} = \left[\int d\gamma_3 H_{ik} H_{jl} \frac{\partial}{\partial \gamma_m} \left(\frac{\partial p_i v}{\partial \gamma_3} p_j v \right) \right] - \left[\dots_{\ k \leftrightarrow l} \right]$$

$$= \left[\int d\gamma_3 H_{ik} H_{jl} \left(p_j v \frac{\partial}{\partial \gamma_3} \frac{\partial p_i v}{\partial \gamma_m} + \frac{\partial p_i v}{\partial \gamma_3} \frac{\partial p_j v}{\partial \gamma_m} \right) \right] - \left[\dots_{\ k \leftrightarrow l} \right] \ , \tag{21}$$

see (13). Since $p_i v = H_{i3}$, see (2), and $H_{i3} \frac{\partial p_i v}{\partial \gamma_m} = 0$,

$$\int d\gamma_3 H_{ik} \frac{\partial W_{ij}}{\partial \gamma_m} H_{jl}$$

$$= \left[\int d\gamma_3 \left(H_{ik} \delta_{3l} \frac{\partial}{\partial \gamma_3} \frac{\partial p_i v}{\partial \gamma_m} + H_{iK} \delta_{Kk} \frac{\partial p_i v}{\partial \gamma_3} \frac{\partial p_j v}{\partial \gamma_m} H_{jL} \delta_{Ll} \right) \right] - \left[\dots_{\ k \leftrightarrow l} \right] \ . \tag{22}$$

Since (22) is skew (antisymmetric) in indices k and l,

$$\int \mathrm{d}\gamma_3 H_{ik} \frac{\partial W_{ij}}{\partial \gamma_m} H_{jl}$$

$$= \left[\int \mathrm{d}\gamma_3 \left(H_{iK}\delta_{Kk}\delta_{3l} \frac{\partial}{\partial \gamma_3} \frac{\partial p_i v}{\partial \gamma_m} + H_{iK}\delta_{Kk} \frac{\partial p_i v}{\partial \gamma_3} \frac{\partial p_j v}{\partial \gamma_m} H_{jL}\delta_{Ll} \right) \right] - \left[\cdots_{k \leftrightarrow l} \right] . \quad (23)$$

Since $\frac{\partial H_{iK}}{\partial \gamma_n} \frac{\partial p_i v}{\partial \gamma_m} = 0$ and $H_{iK} \frac{\partial p_i v}{\partial \gamma_3} = -v^{-1} \frac{\partial s}{\partial \gamma_3} \frac{\partial v}{\partial q_K}$,

$$\int \mathrm{d}\gamma_3 H_{ik} \frac{\partial W_{ij}}{\partial \gamma_m} H_{jl}$$

$$= \left[\int \mathrm{d}\gamma_3 \left(\delta_{Kk}\delta_{3l} \frac{\partial}{\partial \gamma_3} \left(\frac{\partial p_i v}{\partial \gamma_m} H_{iK} \right) - \delta_{Kk}v^{-1} \frac{\partial s}{\partial \gamma_3} V_K \frac{\partial p_j v}{\partial \gamma_m} H_{jL}\delta_{Ll} \right) \right] - \left[\cdots_{k \leftrightarrow l} \right] , \quad (24)$$

which reads

$$\int \mathrm{d}\gamma_3 H_{ik} \frac{\partial W_{ij}}{\partial \gamma_m} H_{jl}$$

$$= \left[\delta_{Kk}\delta_{3l}v \frac{\partial p_i}{\partial \gamma_m} H_{iK} - \delta_{Kk}\delta_{Ll} \int \mathrm{d}s\, V_K \frac{\partial p_i}{\partial \gamma_m} H_{iL} \right] - \left[\cdots_{k \leftrightarrow l} \right] . \quad (25)$$

Here

$$V_k = \frac{\partial v}{\partial x_i} H_{ik} \quad (26)$$

is the velocity gradient in the ray-centered coordinate system at the central ray. Insertion of (15) into (8) yields

$$P_{km} = \frac{\partial p_j}{\partial \gamma_m} H_{jk} - 2v^{-2}\delta_{k3}V_3 Q_{3m} + v^{-2}V_k Q_{3m} + v^{-2}\delta_{k3}V_l Q_{lm} . \quad (27)$$

Considering (27), relation (25) takes the form

$$\int \mathrm{d}\gamma_3 H_{ik} \frac{\partial W_{ij}}{\partial \gamma_m} H_{jl}$$

$$= \left[\delta_{Kk}\delta_{3l}v\left(P_{Km} - v^{-2}V_K Q_{3m}\right) - \delta_{Kk}\delta_{Ll} \int \mathrm{d}s\, V_K P_{Lm} \right] - \left[\cdots_{k \leftrightarrow l} \right]$$

$$= \left(\delta_{Kk}\delta_{3l} - \delta_{3k}\delta_{Kl}\right)\left(vP_{Km} - v^{-1}V_K Q_{3m}\right) - \delta_{Kk}\delta_{Ll}\epsilon_{KL} \int \mathrm{d}s\, V_I \epsilon_{IJ} P_{Jm} , \quad (28)$$

suitable to convert the right-hand side of equation (20).

Equations (14) and (20) yield

$$H_{iK}(\gamma_3) \frac{\partial H_{iL}}{\partial q_n}(\gamma_3) = \left[H_{iK}(\gamma_3^0) \frac{\partial H_{iL}}{\partial \gamma_m}(\gamma_3^0) + \int_{\gamma_3^0}^{\gamma_3} \mathrm{d}\gamma_3 H_{iK} \frac{\partial W_{ij}}{\partial \gamma_m} H_{jL} \right] Q_{mn}^{-1}(\gamma_3) . \quad (29)$$

Inserting (28) into (29), we arrive at

$$H_{iK}(\gamma_3)\frac{\partial H_{iL}}{\partial q_n}(\gamma_3) = \left[H_{iK}(\gamma_3^0)\frac{\partial H_{iL}}{\partial \gamma_m}(\gamma_3^0) - \epsilon_{KL}\int\limits_{s(\gamma_3^0)}^{s(\gamma_3)} ds\, V_I\epsilon_{IJ}P_{Jm} \right] Q_{mn}^{-1}(\gamma_3) \ . \qquad (30)$$

Inserting this into (18) and applying (14) for $\gamma_3 = \gamma_3^0$, we obtain this expression for Ω_N:

$$\Omega_N(\gamma_3) = \left[H_{i2}(\gamma_3^0)\frac{\partial H_{i1}}{\partial q_n}(\gamma_3^0)Q_{nm}(\gamma_3^0) + \int\limits_{s(\gamma_3^0)}^{s(\gamma_3)} ds\, V_I\epsilon_{IJ}P_{Jm} \right] Q_{mN}^{-1}(\gamma_3) \ . \qquad (31)$$

Using (1) and

$$Q_{J3} = 0 \ , \quad P_{J3} = 0 \ , \qquad (32)$$

which follows from (5) and (7), equation (31) reads

$$\Omega_N(\gamma_3) = \left[H_{i2}(\gamma_3^0)\frac{\partial H_{i1}}{\partial q_J}(\gamma_3^0)Q_{JM}(\gamma_3^0) + \int\limits_{s(\gamma_3^0)}^{s(\gamma_3)} ds\, V_I\epsilon_{IJ}P_{JM} \right] Q_{MN}^{-1}(\gamma_3) \ , \qquad (33)$$

which is equivalent to equation (C10) of COATES and CHAPMAN (1990). Here Q_{MN} and P_{MN} are the standard 2×2 paraxial ray tracing matrices in the ray-centered coordinate system. In order to evaluate the paraxial changes (17) of the S-wave polarization vectors, we need to compute two integrals

$$\int\limits_{s(\gamma_3^0)}^{s(\gamma_3)} ds(V_1 P_{2M} - V_2 P_{1M}) \qquad (34)$$

along the ray, whereas derivatives (15) and (16) of the polarization vectors require no quadrature. The numerical quadrature can be calculated after ray tracing, along rays stored together with the polarization vectors and the paraxial ray propagator matrix in disk files (ČERVENÝ et al., 1988, section 7.26).

Variation of the Polarization Vectors with a Velocity Perturbation

We denote by δA the variation of quantity A with respect to any of the model parameters. During the velocity perturbations, we keep the coordinate systems fixed. The perturbations (i.e., the variations with respect to the model parameters) then have the properties of partial derivatives, and commute with the partial derivatives with respect to the coordinates. Equations (19)–(25) thus remain valid if we replace partial derivative $\partial/\partial \gamma_m$ by variation δ with respect to a model parameter. With this substitution, equations (20) and (25) yield

$$\delta H_{nl}(\gamma_3) = H_{nk}(\gamma_3)\left[H_{ik}(\gamma_3^0)\delta H_{il}(\gamma_3^0) + (\delta_{Kk}\delta_{3l} - \delta_{3k}\delta_{Kl})\left[v(\gamma_3)H_{iK}(\gamma_3)\delta p_i(\gamma_3)\right.\right.$$

$$\left.\left. - v(\gamma_3^0)H_{iK}(\gamma_3^0)\delta p_i(\gamma_3^0)\right] - (\delta_{Kk}\delta_{Ll} - \delta_{Lk}\delta_{Kl})\int_{s(\gamma_3^0)}^{s(\gamma_3)} ds\, V_K H_{iL}\delta p_i\right] . \qquad (35)$$

This equation may be rearranged to read

$$\delta H_{nl}(\gamma_3) = H_{nk}(\gamma_3)\left[\delta_{Kk}\delta_{Ll}H_{iK}(\gamma_3^0)\delta H_{iL}(\gamma_3^0) + (\delta_{Kk}\delta_{3l} - \delta_{3k}\delta_{Kl})\right.$$

$$\left. \times v(\gamma_3)H_{iK}(\gamma_3)\delta p_i(\gamma_3) - \epsilon_{KL}\delta_{Kk}\delta_{Ll}\int_{s(\gamma_3^0)}^{s(\gamma_3)} ds\, V_I\epsilon_{IJ}H_{iJ}\delta p_i\right] . \qquad (36)$$

In order to evaluate the perturbation of the S-wave polarization vectors, we need to compute the integral

$$\int_{s(\gamma_3^0)}^{s(\gamma_3)} ds\, (V_1\delta P_2 - V_2\delta P_1) \qquad (37)$$

along the ray, whereas the perturbation of the P-wave polarization vector requires no quadrature. Here

$$\delta P_k = H_{ik}\delta p_i \qquad (38)$$

is the slowness vector perturbation in ray-centered coordinates.

Conclusions

Equations (15), (16) and (17) with (33) may be used to calculate the first derivatives of the polarization vectors. Two numerical quadratures (34) are required to calculate the derivatives of the S-wave polarization vectors in planes perpendicular to the ray.

Equation (36) may be used to calculate the variations of the polarization vectors with velocity perturbations. Equation (36) takes a very simple form for the variation of the P-wave polarization vector or for the ray-tangent component of the variation. In addition to the quadratures for the variation of the central ray (FARRA and MADARIAGA, 1987; KLIMEŠ, 2002), one numerical quadrature (37) is required for the ray-normal components of the variation of the S-wave polarization vectors per each perturbation.

Acknowledgements

This paper was initiated by the fruitful discussions with Ivan Pšenčík. The first version of the manuscript was written at the Institute of Geology and Geotechnics of the Czechoslovak Academy of Sciences in Prague. It was revised and converted into TEX during the author's visiting fellowship at the Geological Survey of Canada in Ottawa, Ontario. The final revision of the paper has been supported by the Grant Agency of the Czech Republic under Contract 205/01/0927, by the Ministry of Education of the Czech Republic within Research Project J13/98 113200004, and by the members of the consortium "Seismic Waves in Complex 3-D Structures" (see "http://sw3d.mff.cuni.cz").

REFERENCES

ČERVENÝ, V., *Seismic Ray Theory* (Cambridge Univ. Press, Cambridge 2001).
ČERVENÝ, V., KLIMEŠ, L., and PŠENČÍK, I., *Complete seismic-ray tracing in three-dimensional structures.* In *Seismological Algorithms* (ed. Doornbos, D.J.) (Academic Press, New York 1988) pp. 89–168.
COATES, R. T., and CHAPMAN, C. H. (1990), *Ray Perturbation Theory and the Born Approximation*, Geophys. J. int. *100*, 379–392.
FARRA, V., and MADARIAGA, R. (1987), *Seismic Waveform Modeling in Heterogeneous Media by Ray Perturbation Theory*, J. geophys. Res. *92B*, 2697–2712.
KLIMEŠ, L. (1999), *Calculation of the third and higher travel-time derivatives in isotropic and anisotropic media.* In *Expanded Abstracts of 69th Annual Meeting (Houston)* (Soc. Explor. Geophysicists, Tulsa) pp. 1751–1754.
KLIMEŠ, L. (2002), *Second-order and Higher-order Perturbations of Travel Time in Isotropic and Anisotropic Media*, Stud. geophys. geod. *46*, 213–248.
POPOV, M. M., and PŠENČÍK, I. (1978a), *Ray Amplitudes in Inhomogeneous Media with Curved Interfaces*, Travaux Instit. Géophys. Acad. Tchécosl. Sci. No.454, Geofys. Sborník *24*, 111–129 (Academia, Praha).
POPOV, M. M., and PŠENČÍK, I. (1978b), *Computation of Ray Amplitudes in Inhomogeneous Media with Curved Interfaces*, Stud. geophys. geod. *22*, 248–258.

(Received October 10, 2000, revised March 21, 2001, accepted March 28, 2001)

 To access this journal online:
http://www.birkhauser.ch

Pure appl. geophys. 159 (2002) 1447–1464
0033–4553/02/081447–18 $ 1.50 + 0.20/0

❘ Pure and Applied Geophysics

Synthetic Seismograms and Wide-angle Seismic Attributes from the Gaussian Beam and Reflectivity Methods for Models with Interfaces and Velocity Gradients

Robert L. Nowack[1] and Stephen M. Stacy[1]

Abstract—The effects of interfaces and velocity gradients on wide-angle seismic attributes are investigated using synthetic seismograms. The seismic attributes considered include envelope amplitude, pulse instantaneous frequency, and arrival time of selected phases. For models with interfaces and homogeneous layers, head waves can propagate which have lower amplitudes, as well as frequency content, compared to the direct arrivals. For media with interfaces and velocity gradients, higher amplitude diving waves and interference waves can also occur. The Gaussian beam and reflectivity methods are used to compute synthetic seismograms for simple models with interfaces and gradients. From the results of these methods, seismic attributes are obtained and compared. It was found that both methods were able to simulate wide-angle seismic attributes for the simple models considered. The advantage of using the Gaussian beam method for seismic modeling and inversion is that it is fast and also asymptotically valid for laterally varying media.

Key words: Wide-angle seismic attributes, synthetic seismograms, seismic refraction.

Introduction

In this paper synthetic seismograms are used to investigate the effects of interfaces and velocity gradients on wide-angle seismic attributes. The seismic attributes considered include envelope amplitude, pulse instantaneous frequency, as well as arrival time of selected phases. These attributes can be used for seismic inversion for elastic and anelastic structure and provide an alternative to more complete wavefield inversions. For example, Nowack and Matheney (1997a) and Matheney *et al.* (1997) performed inversions using observed seismic attributes for smoothly varying upper crustal structure.

One of the important aspects of seismology is the interplay between smoothly varying media and interfaces in the earth. Although many material discontinuities occur in the earth, smoothly varying, homogenized earth models separated by a relatively small number of interfaces can often be used. Here we will investigate the

---

[1] Department of Earth and Atmospheric Sciences, Purdue University, West Lafayette, IN 47907. E-mail: nowack@purdue.edu

effects of large contrast interfaces on wide-angle seismic attributes. For models with interfaces between homogeneous layers, head waves can propagate which have lower amplitudes, as well as frequency content, compared to direct arrivals. However, higher amplitude refracted waves are often observed resulting from velocity gradients. For these types of media, a number of body-wave phases can propagate including diving waves, reflected waves, and interference waves.

The Gaussian beam and reflectivity methods are used to compute synthetic seismograms in simple models with interfaces and velocity gradients, and from these, seismic attributes are obtained and compared. The reflectivity method is a full-wave method which assumes layered earth models. The Gaussian beam method is an asymptotic method for high-frequency seismic waves. The method is valid at caustics, and when wide beam parameters are used, can simulate head-wave phases. With a careful selection of parameters, both methods were found to be able to simulate seismic attributes for the models considered. The Gaussian beam method has the advantage of being fast, as well as being asymptotically valid in laterally varying media.

Refraction effects on seismic attributes

The first example we investigate is for a layer over a half-space for which a head wave can propagate. For this case, if the direct wave has a spectrum of $S(\omega)$, then the head-wave branch will have a spectrum proportional to $(i/\omega)S(\omega)$ (AKI and RICHARDS, 1980). In the time-domain, this corresponds to a smoother, integrated waveform compared to the direct arrival. Using seismic attributes derived from synthetic seismograms, NOWACK and MATHENEY (1997b) found that the time-domain instantaneous frequency of the pure head wave is also lowered with respect to the direct wave. For distances greater than the critical distance, the head-wave amplitude decays as $r^{-1/2}L^{-3/2}$ where r is the distance and L is the distance traveled in the lower medium. For distances much greater than the interface depth, the head-wave amplitude approximately decays as r^{-2} which is lower than that of the direct wave amplitude.

Figure 1 illustrates the effects of a layer over a half-space on the frequency content of the first arrivals. Figure 1A shows a velocity model with an interface at 25 km with a velocity contrast from 6 to 8 km/s. Synthetic seismograms for this model were computed using the reflectivity method (FUCHS and MÜLLER, 1971; KIND, 1978; MÜLLER, 1985). Two models are considered: first a model with low attenuation ($Q = 2000$), and then a model with a constant Q of 150. Figure 1B depicts the pulse instantaneous frequencies obtained using the method of MATHENEY and NOWACK (1995) for the low attenuation case (squares) and the $Q = 150$ case (triangles). The gap in distance near 130 km is where the attributes are not computed due to interference near the cross-over distance. The refracted arrivals occur for

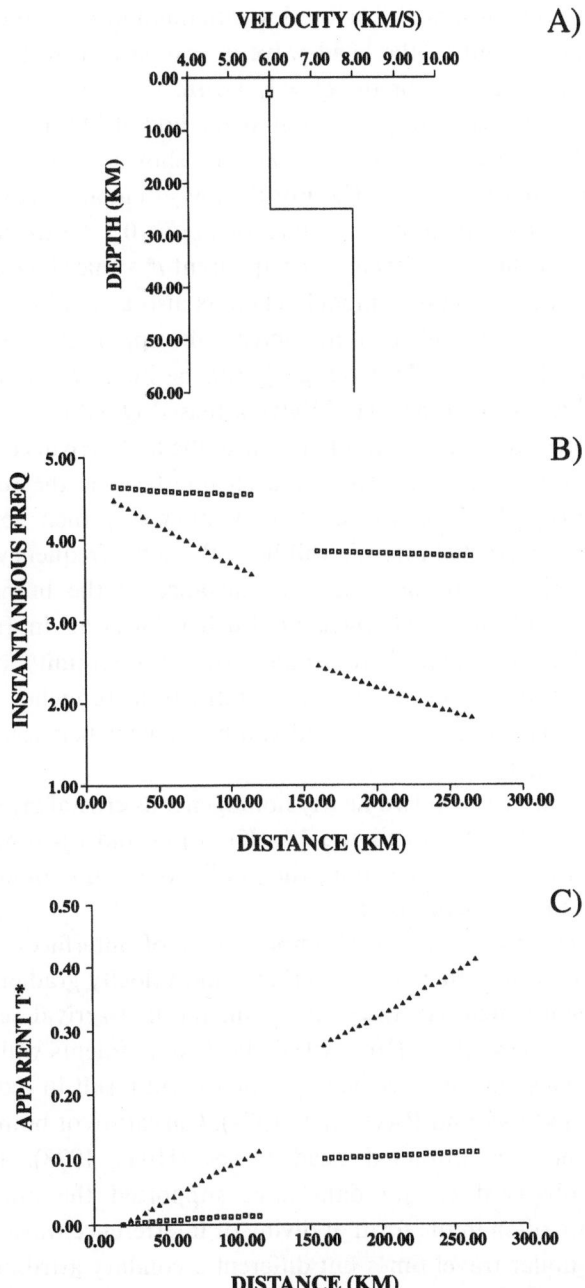

Figure 1

A) A layer over a half-space with an interface at 25 km and a velocity increase from 6 to 8 km/s. B) The computed pulse instantaneous frequencies for the first arrivals for a low attenuation case (squares) and a $Q = 150$ case (triangles). The gap in distance is near the cross-over distance for the branches where attributes for the crossing phases were not computed. C) The apparent t^* values from the instantaneous frequencies in B) are shown using the matching method of MATHENEY and NOWACK (1995).

distances exceeding this distance. For the low attenuation case, the refracted pulse frequency lowers as a result of the head wave in comparison to the direct wave. A similar frequency effect is seen for the $Q = 150$ case.

Using the instantaneous frequency matching method of MATHENEY and NOWACK (1995), apparent t^* values can be estimated, and are shown in Figure 1C for a source pulse with a center frequency of 4.77 Hz. For the low attenuation case (squares) there is a step increase in the apparent t^* values of about 0.1 between the direct and refracted arrivals. For the $Q = 150$ case, the apparent t^* values increase linearly with distance related to the attenuation model. There is also a step increase between the direct and refracted arrivals which would increase the apparent t^* estimate by about 38% at a distance of 150 km. An average Q can be inferred from the travel-time divided by t^*, and at this distance would give a biased Q value.

For a model with two layers over a half-space, the first arrivals consist of a direct wave and two head-wave phases, one for each interface. If the reference pulse is taken along a refracted phase for a shallow interface, then seismic attributes extracted for the two refracted arrivals will have the same frequency characteristics. Thus frequency attributes at the cross-over distance of the branches between a shallow and deeper interface will have no further lowering in frequency. If an attenuation model is included, there would now be continuity of the changing frequency characteristics between the two refraction branches (NOWACK and MATHENEY, 1997b). However, there would still be changes between the direct wave and the refracted branches.

An example of a model with two nearly homogeneous crustal layers and an upper mantle lid is given by IASP91 in Figure 2A (KENNETT and ENGDAHL, 1991). Even considering a spherical earth correction, the gradients for this model are relatively low, and head-wave effects can occur.

The previous examples were for simple cases of interfaces between nearly homogeneous layers. For models with interfaces and velocity gradients, a number of waves can propagate which can have effects on the first-arrival seismic attributes (ČERVENÝ and RAVINDRA, 1971; HILL, 1971). Positive gradients will result in diving and interference waves. Negative velocity gradients will result in reduced amplitude diffracted waves (ČERVENÝ and RAVINDRA, 1971). Curvature of boundaries will also result in interference or diffracted head waves (HILL, 1973). Refracted wave amplitudes from observed crustal data have supported the interpretation that refracted arrivals are often in the form of diving or interference waves instead of pure head waves, with similar travel times but different secondary attributes (BRAILE and SMITH, 1975).

HILL (1971) showed that by using the amplitudes of the refracted wave phases, the effects of velocity and attenuation structure cannot be easily separated. BRAILE (1977), however, inferred by reflectivity modeling that the use of travel times and amplitudes of crustal phases, in addition to P_n phases, allows for some resolution of velocity and attenuation structure for models with interfaces. Nonetheless, even in

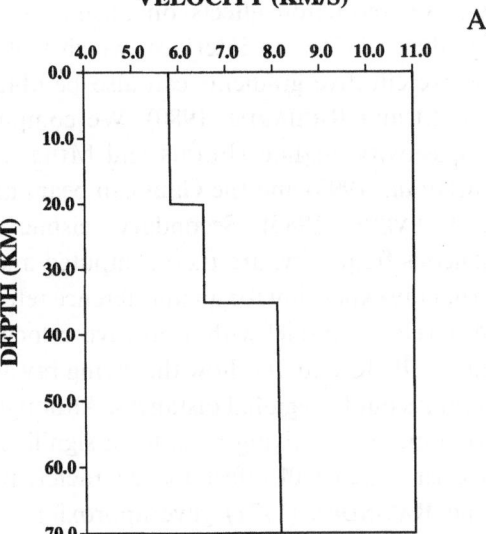

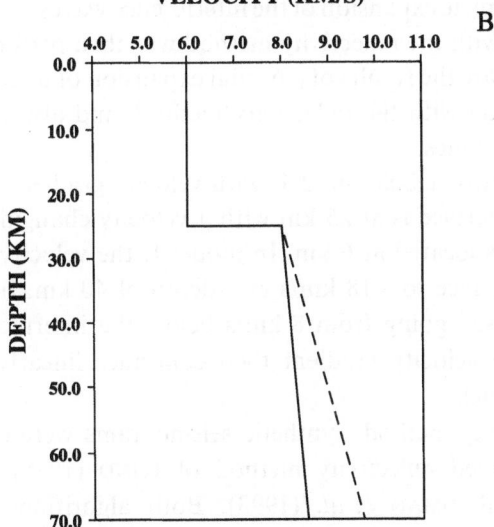

Figure 2

A) The velocity model IASP91 showing the crust and upper mantle lid (KENNETT and ENGDAHL; 1991). B) Velocity models 1 and 2 with two different velocity gradients below an interface (solid and dashed lines, respectively).

spherically symmetric earth models, attenuation estimates using P_n can be biased by refraction effects (SERENO and GIVEN, 1990).

Here we will investigate refraction effects on seismic attributes using simple models with velocity gradients below an interface, which will result in diving and interference waves. Positive effective gradients can also be obtained by flattening a spherical earth model (AKI and RICHARDS, 1980). We compute synthetic seismograms using both the reflectivity method (FUCHS and MÜLLER, 1971; KIND, 1978; MÜLLER, 1985; RUDMAN et al., 1993) and the Gaussian beam method (POPOV, 1982; ČERVENÝ et al., 1982; ČERVENÝ, 1985). Secondary seismic attributes, including amplitude and instantaneous frequency, are then computed and compared.

An example of a partial ray expansion for an interference refracted wave is given in Figure 3. In Figure 3A a velocity model with a positive velocity gradient below an interface is shown. Figures 3B, 3C and 3D show the diving ray as well as the first and second subinterface multiples out to regional distances. Although refracted waves are often modeled for travel times using diving rays, for a significant distance range an interference wave can exist which will affect the characteristics of the secondary attributes. ČERVENÝ and RAVINDRA (1971) gave approximate expressions for the interference refracted waves, and also at what distances successive waves separate from the interference wave. The separation of individual phases depends on the velocity gradient, as well as the pulse frequency. Finite frequency interference effects were studied by CORMIER and RICHARDS (1977) for the inner core boundary using a full wave theory based on a complete expansion of the interference waves. A comparison between this full wave theory with the reflectivity method was then performed by CHOY et al. (1980). Here we compare the results of a partial expansion of interface waves using the Gaussian beam method with the reflectivity method, and also compare the derived secondary seismic attributes.

Figure 2B shows two velocity models with velocity gradients below an interface. In each model, the interface is at 25 km with a velocity changing from 6 to 8 km/s. An explosive source is located at 6 km. In model 1, the velocity gradient goes from 8 km/s below the interface to 8.18 km/s at a depth of 40 km. In model 2, a steeper velocity gradient is used going from 8 km/s below the interface to 8.57 km/s at a depth of 40 km. The velocity gradient then continues linearly to greater depths, depending on the model.

Using the reflectivity method, synthetic seismograms were computed using two algorithms, the modified reflectivity method of KIND (1978) and the reflectivity algorithm given by RUDMAN et al. (1993). Both algorithms were found to be sensitive to the layer thickness used to approximate the velocity gradient. With the integration parameters used, the algorithm of RUDMAN et al. (1993) was found to be more time consuming but gave somewhat better results for the direct and reflected phases in the upper homogeneous layer. The reflectivity synthetics shown in Figures 4 and 5 were obtained using the algorithm described by RUDMAN et al. (1993). For the reflectivity synthetics using model 1, a layer thickness of 0.1875 km

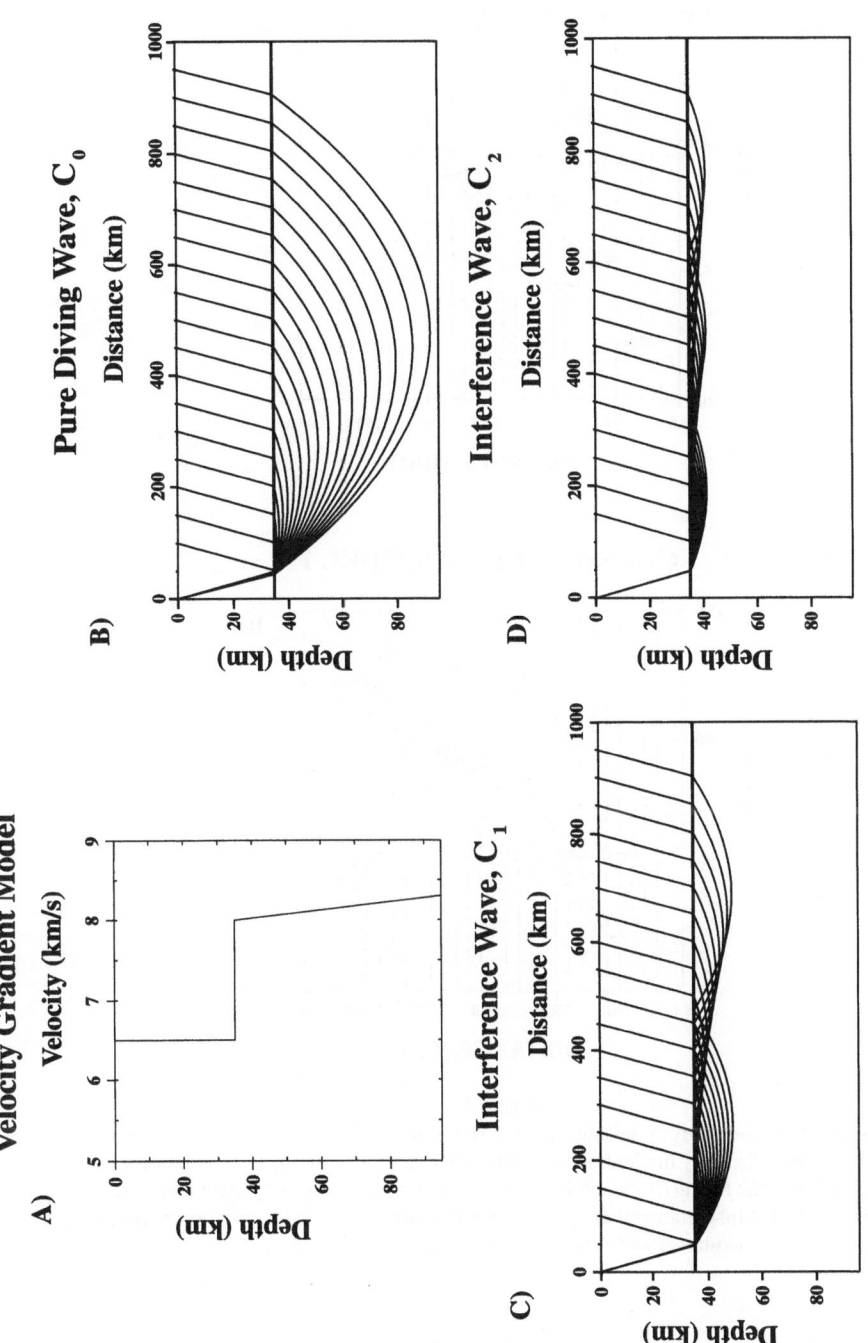

Figure 3

An example of several rays of a partial ray expansion for the interference wave out to regional distances for a positive velocity gradient below an interface. A) A velocity model showing a linear velocity gradient below an interface. B) shows the diving ray, C) the first subinterface multiple, and D) the second sub-interface multiple.

REFLECTIVITY MODEL 1

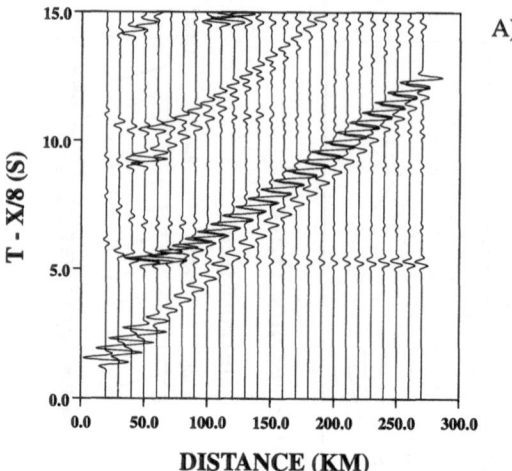

A)

GAUSSIAN BEAM MODEL 1

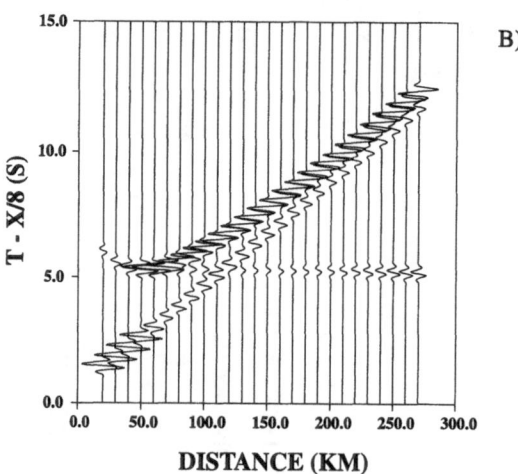

B)

Figure 4

Synthetic seismograms and derived seismic attributes for model 1 shown in Figure 2. A) Synthetic seismograms using the reflectivity method. B) Synthetic seismograms using the Gaussian beam method. C) Envelope amplitudes of the first arrivals obtained from reflectivity modeling (squares) and Gaussian beam modeling (plus signs). D) Instantaneous frequencies for first arrivals obtained from reflectivity modeling (squares) and Gaussian beam modeling (plus signs).

was used to approximate the velocity gradient down to 55 km. In Figure 4D, a finer layer thickness of 0.1 km and the double precision algorithm of KIND (1978) were used for the comparison of the pulse instantaneous frequencies. For model 2, a layer

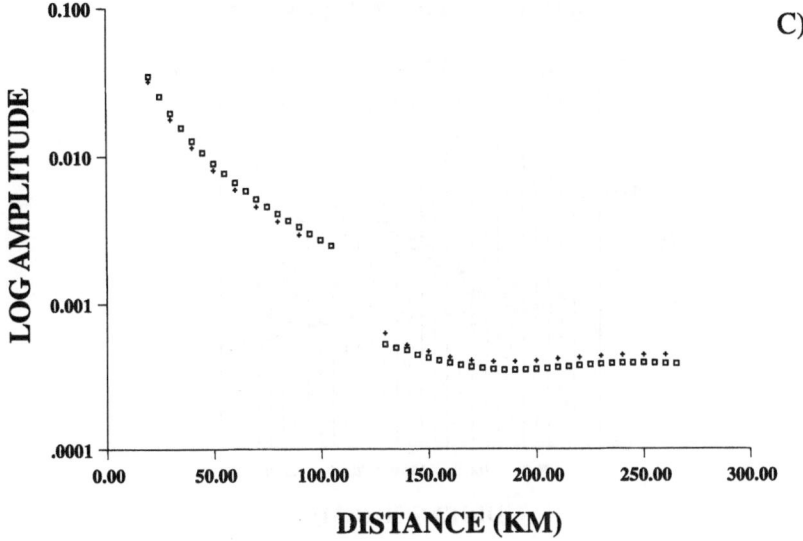

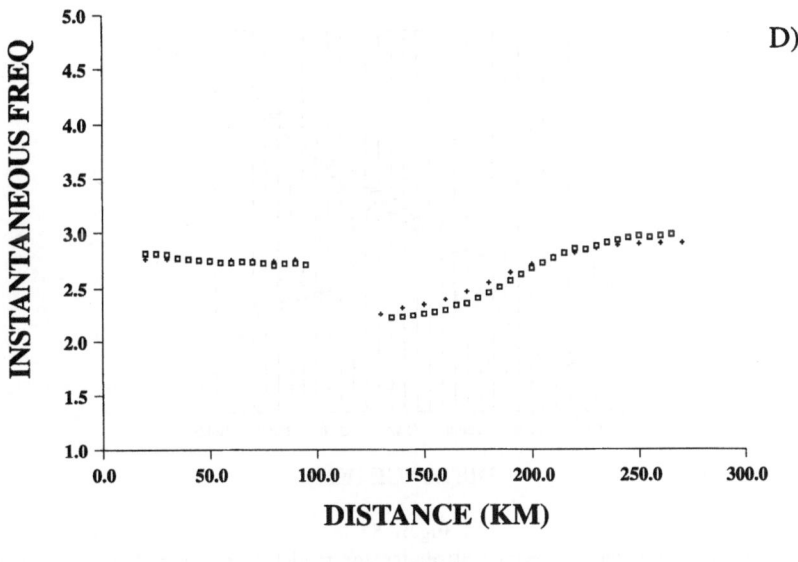

Figure 4C,D

thickness of 0.34375 km was used to a depth of 80 km, since for this case the diving ray penetrates to greater depths. The reflectivity synthetics were then used to compute secondary seismic attributes which are compared with the results from the Gaussian beam method.

The Gaussian beam method is an asymptotic method for the computation of high frequency seismic waves, and was proposed by POPOV (1982) and initially applied by

REFLECTIVITY MODEL 2

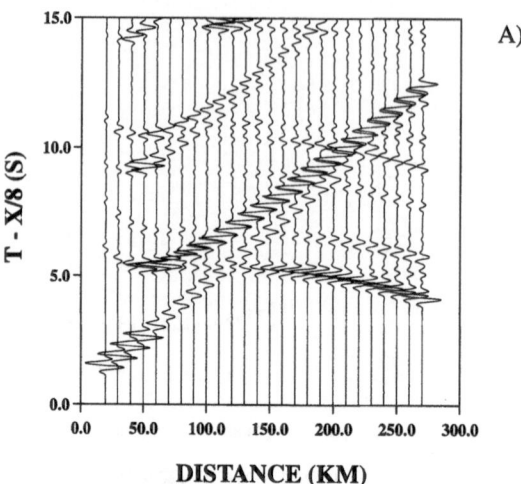

A)

GAUSSIAN BEAM MODEL 2

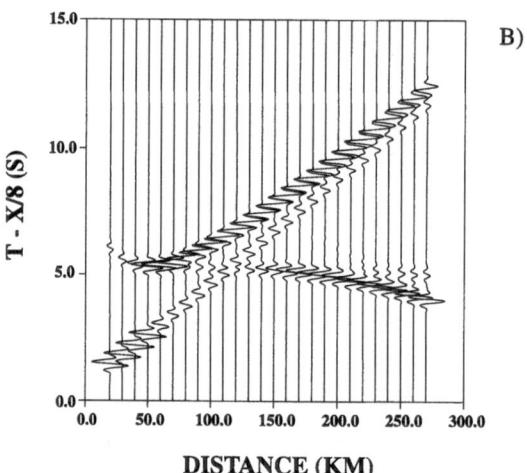

B)

Figure 5

Synthetic seismograms and derived seismic attributes for model 2 shown in Figure 2. A) Synthetic seismograms using the reflectivity method. B) Synthetic seismograms using the Gaussian beam method. C) Envelope amplitudes of the first arrivals obtained from reflectivity modeling (squares) and Gaussian beam modeling (plus signs). D) Instantaneous frequencies for first arrivals obtained from reflectivity modeling (squares) and Gaussian beam modeling (plus signs).

ČERVENÝ *et al.* (1982). It has an advantage over the standard ray method of providing finite results at caustics. The method involves the summation of individual paraxial Gaussian beams along ray trajectories as

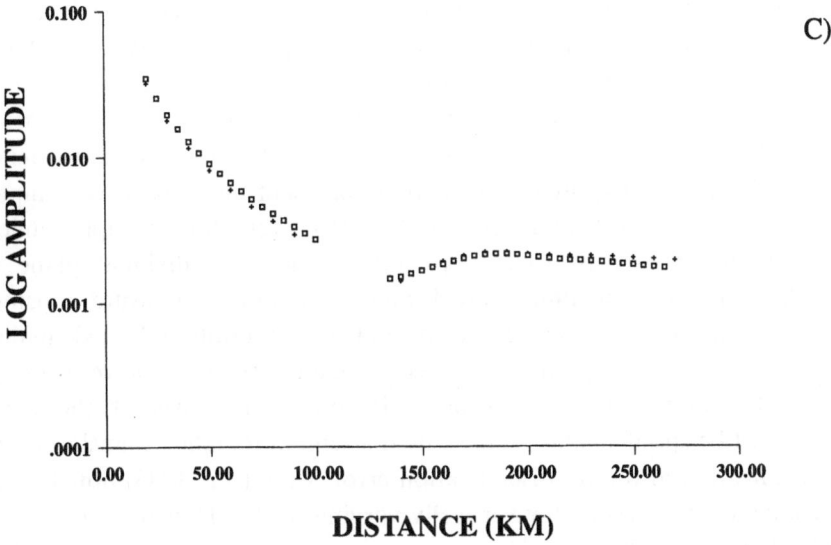

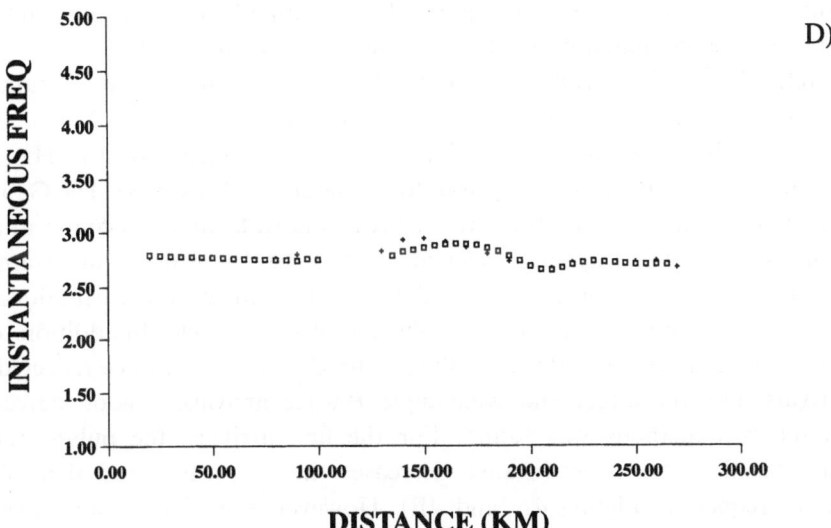

Figure 5C,D

$$\vec{u}(\vec{x}, \omega) = \iint\limits_{D} \Phi(\gamma_I)\vec{u}^{GB}_{\gamma_I}(\vec{x}, \omega, M(s_b))\, d^2\gamma \ ,$$

where $\vec{u}^{GB}_{\gamma_I}$ are the individual beam solutions, $\Phi(\gamma_I)$ are the weighting functions, and $M(s_b)$ are the beam parameters for a specified position s_b along the ray. The ray parameters γ_I $(I = 1, 2)$ are used to specify the central ray for each beam along the initial wavefront, and the domain D depends on the type of source to be decomposed into beams. A review of the Gaussian beam method is given by ČERVENÝ (1985), and a recent overview is presented by NOWACK (2002).

For the case of the interference wave, the specification of beam parameters is complicated by different requirements of the solution. For the wide-angle reflected phase, broad beams are required to accurately represent the head-wave component of the solution (NOWACK and AKI, 1984). However, for multiple underside reflections, both caustics, as well as pseudo-caustics of individual plane wave components, can result and finite sized beams are required to ensure nonsingular solutions. For both the reflected waves and the multiply reflected underside reflections, the beams are specified by effective plane waves at the receiver which provide stable summations. In addition, the imaginary parts of the complex curvatures must be specified and we have used a semi-automatic expansion resulting in broad beams that also limit discretization error. ČERVENÝ (1985) noted that this choice produces better results for vertically varying media. However, for the wide-angle reflections, somewhat broader beams were used to ensure that the head-wave contribution was obtained. Finally, the Gaussian beam method was run in ray mode for the direct wave since this wave is regular. The resulting Gaussian beam solution is then obtained as a combination of the individual wave components.

The reflectivity seismograms for model 1 are shown in Figure 4A, and the seismograms using the Gaussian beam method are shown in Figure 4B. The source pulse was given by a Gabor wavelet with a dominant frequency of 2.77 Hz and a gamma value of 3.75 which specifies the width of the spectral envelope. The Gaussian beam calculations include only the direct arrival, the wide-angle reflection and the interference wave made of up to 8 underside reflections. The wide beam parameters used for the wide-angle reflections also allow for a head wave contribution. The reflectivity method, in contrast, computes the complete wavefield. In addition to the direct, reflected and refracted P-waves, the results also include surface reflected and later arrivals. For the direct and wide-angle P-wave arrivals, a good agreement between the two methods was found. For the first arrivals, the pulses initially broaden after the cross-over distance decreases the frequency, and then slowly increase in frequency (Figures 4A and 4B). However, the diving wave does not separate from the interference waves for these distances.

The results for model 2 are shown in Figure 5 where the reflectivity results are given in Figure 5A and the Gaussian beam results in Figure 5B. For this model, the diving wave separates from the interference wave at about 220 km. For greater distances, the first arrival is the diving wave which is a geometric arrival. The separation distances for multiply reflected underside reflections from the interference wave package are given by ČERVENÝ and RAVINDRA (1971). Although a greater layer thickness was used for the reflectivity modeling in this case because of the greater penetration of the diving wave, there is still a good agreement between the reflectivity and Gaussian beam synthetics for the compressional waves modeled (Figures 5A and 5B).

In Figures 4C and 5C, the envelope amplitudes for the first arrival synthetics are shown for models 1 and 2. For both cases, the analytic signals are computed and from these the envelope amplitudes are obtained following the procedure of MATHENEY and NOWACK (1995). For the given first arrival, the peak of the envelope amplitude is then obtained. For each model, the squares represent the peak envelope amplitudes from the reflectivity modeling and the plus signs are from the Gaussian beam modeling. For all cases, a large Q_p value of 2000 was used to emphasize only the structural effects on the amplitudes. A gap in the amplitudes near the cross-over distance results from the picking algorithm that discards traces if there are overlapping arrivals, such as at the cross-over distance.

In Figure 4C, the refracted wave amplitudes for model 1 from the reflectivity modeling (squares) are slightly lower than the amplitudes from the Gaussian beam modeling (plus signs). Nonetheless, the overall shape of the curve is very similar for the two methods. For the refracted arrivals, the envelope amplitudes are lower after the cross-over distance, and then are constant or very slightly increasing resulting from wave interference.

The envelope amplitudes for model 2 are shown in Figure 5C, and again are very similar between the two approaches. However, the reflectivity amplitudes decay slightly more with increasing distance for the refracted arrivals. After the cross-over distance, the refracted amplitudes are higher than for the previous case since the diving waves are now more significant at shorter distance ranges resulting from the steeper gradient. The amplitudes of the refracted waves increase slightly with distance until a distance of about 180 km, and then begin to decrease as the diving wave begins to separate. For both models 1 and 2, the incorporation of attenuation into the models would have the effect of further decreasing the amplitude curves with distance.

The frequency content of a specific arrival is characterized by the signal's frequency spectrum. However, the signal must be windowed to avoid secondary arrivals which may overlap with the primary arrival. Several methods have been used to represent the average frequency content of a signal in terms of a single measure. These have included measures of pulse broadening (TONN, 1989), the peak or centroid of the signal spectrum (QUAN and HARRIS, 1997) and the pulse instantaneous frequency (MATHENEY and NOWACK, 1995). Here we use the pulse instantaneous frequency as a measure of a representative pulse frequency. An advantage of using the pulse instantaneous frequency is that it is a localized time-domain measurement. However, an averaging window is usually applied for stability. Also, a slight damping of the estimate has been applied when the envelope amplitude is low (MATHENEY and NOWACK, 1995). In our case, the instantaneous frequency estimates are taken at the peak of the envelope of the first arrival. A disadvantage of instantaneous frequency or other single measures of the frequency content is that it cannot easily separate the effects of overlapping signals. For this, a sonogram approach based on either the short-time Fourier transform or a continuous wavelet transform is required (DAUBECHIES, 1992).

GB MOD1 SPECTRA WITH IF, CENT AND PEAK

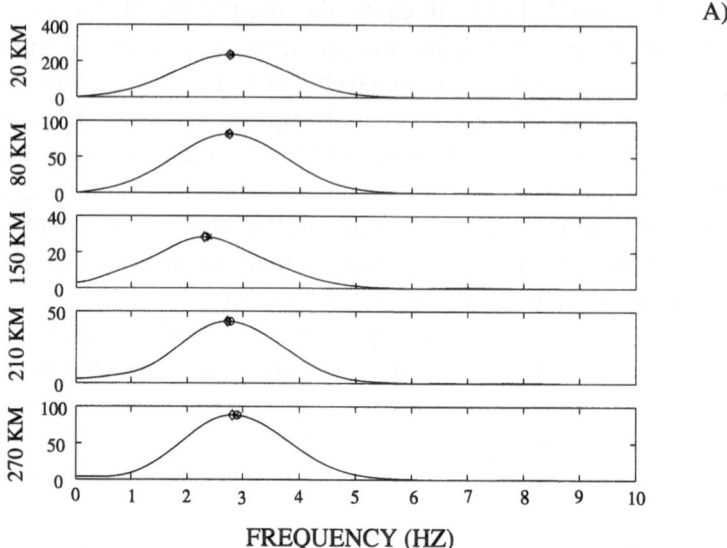

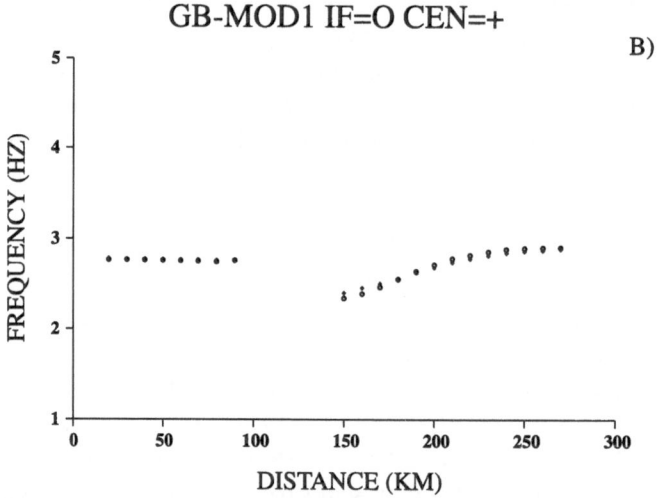

Figure 6

A) Spectra of first arrivals at several distances using Gaussian beam method showing the peak spectral amplitude (diamonds), the centroid frequency (crosses) and the instantaneous frequency (circles). B) Centroid frequencies (plus signs) and instantaneous frequencies (circles) are displayed for first arrivals with distance from Gaussian beam modeling.

Figure 4D shows the pulse instantaneous frequencies for model 1 obtained from the reflectivity calculations (squares) and the Gaussian beam method (plus signs). The gap in the estimates at the cross-over distance is where seismic attributes were not computed because of crossing phases. The Gaussian beam estimates required large beam parameters for the wide-angle reflections in order to obtain the broadening effects of the refracted arrival after the cross-over distance for this model. Also, the instantaneous frequencies resulting from the reflectivity computations were found to be sensitive to the layer thickness used to approximate the gradient. For the comparison in Figure 4D, the double precision reflectivity algorithm of KIND (1978) was used with a reduced layer thickness of 0.1 km for the first 15 km below the interface.

At distances less than 100 km, the direct wave instantaneous frequencies are shown, and for distances greater than 125 km, the refracted wave instantaneous frequencies are shown. After the cross-over distance the instantaneous frequencies initially lower and then increase to values greater than those of the direct wave. This was initially surprising until an early description of this effect was found in ČERVENÝ and RAVINDRA (1971) for an asymptotic interference wave for a layer over a gradient. In their Figure 6.6, the spectra of the refracted waves are displayed at different distances. It shows that the peaks of the spectra initially lower after the critical distance compared to the incident pulse, then increase to values greater than the incident pulse, and finally lower to that of the incident wave after the separation of the diving wave. This is similar to the results for instantaneous frequencies found here for distances less than the separation of the diving wave.

To verify that the pulse instantaneous frequency estimates are representative of this frequency effect, the spectra of the first arrivals from the Gaussian beam modeling are shown in Figure 6A. In this figure, the crosses are the instantaneous frequencies, the circles are the centroid frequencies weighted by the power spectrum and the diamonds are the peaks of the spectra. In Figure 6B, a comparison of the weighted centroid frequencies and the pulse instantaneous frequencies at different distances are given, and the estimates are seen to compare reasonably well.

To understand why the instantaneous frequencies and the centroid frequencies are similar, the power theorem can be used (BRACEWELL, 1986), where

$$\int s_1(t)s_2^*(t)\,dt = \int s_1(\omega)s_2^*(\omega)\frac{d\omega}{2\pi}\ .$$

If the analytic signal is written as $y(t) = a(t)e^{i\phi(t)}$, then $a(t)$ is the envelope amplitude, $\phi(t)$ is the instantaneous phase and $f(t) = d\phi/dt(t)$ is the instantaneous frequency, where the spectrum of the analytic signal is zero for negative frequencies. Letting $s_1(t) = dy(t)/dt = \dot{a}(t)e^{i\phi(t)} + if(t)y(t)$ with Fourier transform $s_1(\omega) = i\omega y(\omega)$, $s_2(t) = y(t)$ with Fourier transform $s_2(\omega) = y(\omega)$, and then equating imaginary parts in the power theorem above gives

$$\frac{\int f(t)a^2(t)\,dt}{\int a^2(t)\,dt} = \frac{\int \omega y^2(\omega)\frac{d\omega}{2\pi}}{\int y^2(\omega)\frac{d\omega}{2\pi}}\ .$$

Thus, as the windowed average of the instantaneous frequency weighted by the squared envelope becomes larger, the estimate approaches the centroid of the power spectrum, where the frequency integrals are over the positive frequencies. For the example in Figure 6, the instantaneous frequencies are weighted over 11 points by the squared envelope. This suggests that the weighted instantaneous frequency is a useful local measure of the characteristic frequency of a signal, particularly if taken at the peak amplitude of a signal and with the averaging window chosen appropriately.

Figure 5D shows the instantaneous frequency estimates for the synthetic first arrivals obtained from model 2. This model has a steeper gradient than model 1 and results in the diving wave going to greater depths. As a result, the diving wave separates from the interference package at shorter distances as seen in Figures 5A and 5B. In Figure 5D, the plus symbols are for the Gaussian beam results and the squares are for the reflectivity results. The gap in the values at the cross-over distance is where attribute values were not estimated because of crossing phases. For distances exceeding the cross-over distance, the instantaneous frequency now initially increases to greater values than those of the direct wave, then slightly decreases and finally flattens with the separation of the diving wave.

The overall characteristics between the two synthetic results are similar, with the exception of the somewhat lower values of the pulse instantaneous frequencies for the reflectivity results just after the cross-over distance. This also occurred for model 1 when a larger layer thickness was used for the reflectivity modeling. As a result, the reflectivity results are slightly underestimated for this distance range. Nonetheless, the overall shape of the curves is similar between the two approaches and illustrates how the frequency characteristics of the refracted arrival change with distance out to the separation of the diving wave for this model.

Finally, models with interfaces and negative velocity gradients have been described by ČERVENÝ and RAVINDRA (1971) and result in diffracted head waves that have reduced amplitudes even compared to pure head waves. It has been suggested based on petrologic arguments that negative velocity gradients below the Moho should occur in many regions of high heat flow. However, this is contrary to the general observation that P_n arrivals are found worldwide. TITTGEMEYER et al. (1999) proposed that scattering from small-scale features superposed on negative velocity gradients can also generate P_n arrivals. It is as yet unclear if these small-scale features can be homogenized to larger scale, effective velocity and attenuation parameters for the modeling of wide-angle seismic data.

Conclusions

In this paper, several examples of refracted arrivals from simple layered models have been used to illustrate refraction effects of interfaces and velocity gradients on first arrival seismic attributes. The reflectivity and Gaussian beam methods have been used to compute synthetic seismograms, and from these seismic attributes obtained and compared. Using wide beam parameters for the reflected phases in the Gaussian beam method, and a small layer thickness for the reflectivity modeling, a good agreement was found for the seismic attributes obtained from the two methods. Thus, seismogram modeling can be used to simulate refraction effects on wide-angle seismic attributes. These effects can then be incorporated into the modeling and inversion for elastic and anelastic structure. The advantage of using the Gaussian beam method for the modeling of seismic attributes is that it is fast, and also asymptotically valid in laterally varying media.

Acknowledgements

The authors would like to acknowledge Prof. V. ČERVENÝ for his insightful work on ray and wave methods over the years. We also thank L. W. Braile, A. J. Rudman and M. Sen for providing reflectivity algorithms used in this paper.

REFERENCES

AKI, K. and RICHARDS, P. G., *Quantitative Seismology Theory and Methods* (W. H. Freeman and Co., San Francisco, 1980).

BRACEWELL, R. N., *The Fourier Transform and its Applications* (McGraw-Hill, New York, 1986).

BRAILE, L. W., *Interpretation of crustal velocity gradients and Q structure using amplitude-corrected seismic refraction profiles*. In *The Earth's Crust* (ed. J.G. Heacock) (Am. Geophys. Un. Monograph 20, 1977), pp. 427–439.

BRAILE, L. W. and SMITH, R. B. (1975), *Guide to the Interpretation of Crustal Refraction Profiles*, Geophys. J.R. astr. Soc. *40*, 145–176.

ČERVENÝ, V., POPOV, M. M., and PŠENČÍK, I. (1982), *Computation of Wavefields in Inhomogeneous Media — Gaussian Beam Approach*, Geophys. J. R. astr. Soc. *70*, 109–128.

ČERVENÝ, V. (1985), *Gaussian Beam Synthetic Seismograms*, J. Geophys. *58*, 44–72.

ČERVENÝ, V. and RAVINDRA, R., *Theory of Seismic Head Waves* (University of Toronto Press, Toronto, 1971).

CHOY, G. L., CORMIER, V. F., KIND, R., MÜLLER, G., and RICHARDS, P. G. (1980), *A Comparison of Synthetic Seismograms of Core Phases Generated by the Full Wave Theory and by the Reflectivity Method*, Geophys. J. R. astr. Soc. *61*, 21–39.

CORMIER, V. F. and RICHARDS. P. G. (1977), *Full Wave Theory Applied to a Discontinuous Velocity Increase: The Inner Core Boundary*, J. Geophys. *43*, 3–31.

DAUBECHIES, I., *Ten Lectures on Wavelets* (SIAM, Philadelphia, 1992).

FUCHS, K. and MÜLLER, G. (1971), *Computation of Synthetic Seismograms with the Reflectivity Method and Comparison with Observations*, Geophys. J.R. astr. Soc. *23*, 417–433.

HILL, D. (1971), *Velocity Gradients and Anelasticity from Crustal Body-wave Amplitudes*, J. Geophys. Res. *76*, 3309–3325.

HILL, D. (1973), *Critically Refracted Waves in a Spherically Symmetric Radially Heterogeneous Earth Model*, Geophys. J.R. astr. Soc. *34*, 149–177.

KENNETT, R. L. N. and ENGDAHL, E. R. (1991), *Traveltimes for Global Earthquake Location and Phase Identification*, Geophys. J. Int. *105*, 429–465.

KIND R. (1978), *The Reflectivity Method for a Buried Source*, J. Geophys. *44*, 603–612.

MATHENEY, M. P. and NOWACK, R. L. (1995), *Seismic Attenuation Values Obtained from Instantaneous-frequency Matching and Spectral Ratios*, Geophys. J. Int. *123*, 1–15.

MATHENEY, M. P., NOWACK, R. L. and TREHU, A. M. (1997), *Seismic Attribute Inversion for Velocity and Attenuation Structure*, J. Geophys. Res. *102*, 9949–9960.

MÜLLER, G, 1985, *The Reflectivity Method: A Tutorial*, J. Geophys. *58*, 153–174.

NOWACK, R. L. (2002), *Calculation of Synthetic Seismograms with Gaussian Beams*, Pure appl. geophys., in press.

NOWACK, R. L. and AKI, K (1984), *The Two-dimensional Gaussian Beam Synthetic Method: Testing and Application*, J. Geophys. Res. *89*, 7797–7819.

NOWACK, R. L. and MATHENEY, M. P. (1997a), *Inversion of Seismic Attributes for Velocity and Attenuation Structure*, Geophys. J. Int. *128*, 689–700.

NOWACK, R. L. and MATHENEY, M. P. (1997b), *Extraction of Seismic Attributes from Wide-angle Synthetic Data Derived with Models with Interfaces*, Trans. Am. Geophys. Un. EOS *78*.

QUAN, Y. and HARRIS, J. M. (1997), *Seismic Attenuation Tomography Using the Frequency Shift Method*, Geophysics *62*, 895–905.

POPOV M. M. (1982), *A New Method of Computing Wavefields in the High-frequency Approximation*, Wave Motion *4*, 85–97.

RUDMAN, A. J., MALLICK, S., FRAZER, L. N., and BROMIRSKI, P. (1993), *Workstation Computation of Synthetic Seismograms for Vertical and Horizontal Profiles: A Full Wavefield Response for a Two-dimensional Layered Half-space*, Computers and Geosciences *19*, 447–474.

SERENO, T. J. and GIVEN, J. W. (1990), P_n *Attenuation for a Spherically Symmetric Earth Model*, Geophys. Res. Lett. *17*, 1141–1144.

TITTGEMEYER, M., WENZEL, F. and FUCHS, K. (1999), *On the Nature of* P_n, Am. Geophys. Un. EOS *80*, F711.

TONN, R. (1991), *Comparison of Seven Methods for the Computation of Q*, Phys. Earth Planet. Int. *55*, 259–268.

(Received September 29, 2000, revised March 3, 2001, accepted March 30, 2001)

To access this journal online:
http://www.birkhauser.ch

Pure appl. geophys. 159 (2002) 1465–1485
0033–4553/02/081465–21 $ 1.50 + 0.20/0

© Birkhäuser Verlag, Basel, 2002

▌Pure and Applied Geophysics

Lyapunov Exponents for 2-D Ray Tracing
Without Interfaces

LUDĚK KLIMEŠ[1]

Summary—The Lyapunov exponents quantify the exponential divergence of rays asymptotically, along infinitely long rays. The Lyapunov exponent for a finite 2-D ray and the average Lyapunov exponents for a set of finite 2-D rays and for a 2-D velocity model are introduced. The equations for the estimation of the average Lyapunov exponents in a given smooth 2-D velocity model without interfaces are proposed and illustrated by a numerical example. The equations allow the average exponential divergence of rays and exponential growth of the number of travel-time branches in the velocity model to be estimated prior to ray tracing.

Key words: Velocity models, travel times, ray tracing, paraxial rays, deterministic chaos.

1. Introduction

If heterogeneities in a velocity model (macro model) exceed a certain degree, the average geometrical spreading exponentially increases with the length of the rays and, in consequence, the average number of travel times (i.e., the average number of rays intersecting at the same point) exponentially increases with distance from the source. This behaviour of rays strictly limits the possibility of calculating two-point rays and travel times in overly complex models because:

(a) The geometrical spreading is so large that two-point rays cannot be found within the numerical accuracy. Similarly, the ray tubes cannot be sufficiently narrow for travel-time interpolation.

(b) The number of two-point travel times at each point is so large that the travel times cannot be calculated within reasonable computational time and costs.

(c) The number of two-point travel times at each point is so large that they can hardly be useful for any application, independently of the applicability of the ray theory which is not considered here.

[1] Department of Geophysics, Charles University, Ke Karlovu 3, 121 16 Praha 2, Czech Republic.
E-mail: klimes@seis.karlov.mff.cuni.cz

It is thus of principal interest to quantify the exponential divergence of neighbouring rays with respect to the complexity of the model, and to formulate explicit criteria enabling models suitable for ray tracing to be constructed.

The exponential divergence of rays is quantified by the Lyapunov exponents (LYAPUNOV, 1949; OSELEDEC, 1968). The aim of this paper is to introduce and estimate the *average Lyapunov exponent*, describing the average spreading of ray tubes and average number of travel times, in smooth 2-D models without interfaces.

Note that the average Lyapunov exponent and the average frequency of caustic points along rays (see the Appendix) are two different characteristics of ray chaos.

2. Paraxial-ray Propagator Matrix

In this section we still consider a 3-D space, however the form of the quantities and equations in 2-D (or N-D) space is obvious.

Let us denote by $\mathbf{w} = (x^1, x^2, x^3, p_1, p_2, p_3)^T$ the phase-space coordinates, x^1, x^2, x^3 being the spatial coordinates and p_1, p_2, p_3 being the slowness-vector components. The ray tracing equations may be expressed in the form of

$$\frac{d\mathbf{w}_\alpha}{d\vartheta} = \Sigma_{\alpha\beta} \frac{\partial H}{\partial w^\beta} \quad , \tag{1}$$

where

$$\Sigma = \begin{pmatrix} \mathbf{0} & \mathbf{1} \\ -\mathbf{1} & \mathbf{0} \end{pmatrix} \quad , \tag{2}$$

with $\mathbf{0}$ and $\mathbf{1}$ being the 3×3 (in 3-D space) zero and identity matrices, respectively. Here parameter ϑ along a ray is determined by the form of Hamiltonian $H = H(w_\alpha)$. For more details on Hamiltonian ray tracing and corresponding references refer to the book by ČERVENÝ (2001).

The paraxial approximation of deviation $\delta\mathbf{w}(\vartheta)$ between the phase-space coordinates of the points of two infinitesimally close rays is

$$\delta\mathbf{w}(\vartheta) = \mathbf{\Pi}(\vartheta, \vartheta_0) \, \delta\mathbf{w}(\vartheta_0) \quad , \tag{3}$$

where $\delta\mathbf{w}(\vartheta_0)$ is the initial value of the deviation. Here $\mathbf{\Pi}$ is the *paraxial-ray propagator matrix*,

$$\Pi_{\alpha\beta}(\vartheta, \vartheta_0) = \frac{\partial w^\alpha(\vartheta)}{\partial w^\beta(\vartheta_0)} \quad . \tag{4}$$

The derivative of the paraxial-ray propagator matrix along a ray is given by the *dynamic ray tracing* equation

$$\frac{d\mathbf{\Pi}(\vartheta, \vartheta_0)}{d\vartheta} = \mathbf{\Sigma}\mathbf{H}(\vartheta)\mathbf{\Pi}(\vartheta, \vartheta_0) \ , \tag{5}$$

where

$$H_{\alpha\beta} = \frac{\partial^2 H}{\partial w^\alpha \partial w^\beta} \tag{6}$$

are the second phase-space partial derivatives of the Hamiltonian (ČERVENÝ, 1972). Dynamic ray tracing equation (5) directly follows from ray tracing equations (1) differentiated with respect to the initial conditions.

3. Lyapunov Exponents

The Lyapunov exponents may be defined in several ways (LYAPUNOV, 1949; OSELEDEC, 1968; KATOK, 1980; MATYSKA, 1999). Some definitions rely on unspecified norm $\| \bullet \|$ in phase space, which may be chosen arbitrarily. Although the phase-space norm does not affect the values of the Lyapunov exponents defined asymptotically for infinitely long rays, it may considerably affect the estimated values of the Lyapunov exponents along finite rays in finite models.

The estimates of the Lyapunov exponents based on the *characteristic values* of the paraxial-ray propagator matrix are not affected by the free parameters corresponding to the norm in the phase space. On the other hand, the characteristic values oscillate along rays which makes the estimation of the Lyapunov exponents difficult. The oscillations of the characteristic values are caused by the rotation of ray tubes in the phase space.

Let us denote by $\mu_1, \mu_2, \ldots, \mu_{2N}$ the *characteristic values* of the $2N \times 2N$ (in N-D space) propagator matrix $\mathbf{\Pi}$, i.e., the solutions of the characteristic equation

$$\det[\mathbf{\Pi}(\vartheta, \vartheta_0) - \mu(\vartheta, \vartheta_0)\mathbf{1}] = 0 \ , \tag{7}$$

sorted according to their absolute values,

$$|\mu_1| \geq |\mu_2| \geq \cdots \geq |\mu_{2N}| \ . \tag{8}$$

The complex-valued characteristic values, with the same absolute value, are assumed to be sorted according to their argument $-\pi < \arg \mu \leq \pi$. The characteristic values are also sometimes called eigenvalues. We prefer the term characteristic values because the corresponding eigenvectors usually do not exist.

The *positive Lyapunov exponents* along a ray parametrized by monotonically increasing parameter σ may be defined as

$$\lambda_k = \limsup_{\vartheta \to +\infty} \frac{\ln |\mu_k(\vartheta, \vartheta_0)|}{\sigma(\vartheta) - \sigma(\vartheta_0)}, \quad k = 1, 2, \ldots, N , \qquad (9)$$

see MATYSKA (1999). The Lyapunov exponents are thus defined with respect to a particular monotonic parameter $\sigma = \sigma(\vartheta)$ which may or may not differ from parameter ϑ, determined by the form of the Hamiltonian.

Because of the symplectic property of the paraxial-ray propagator matrix, its inverse $\boldsymbol{\Pi}^{-1}$ has the same set of characteristic values as $\boldsymbol{\Pi}$. That is why the characteristic values of all Hamiltonian systems form reciprocal pairs $\mu_1 \mu_{2N} = 1$, $\mu_2 \mu_{2N-1} = 1, \ldots, \mu_N \mu_{N+1} = 1$. Each positive Lyapunov exponent of ray tracing (as of other Hamiltonian systems) is thus accompanied by a negative Lyapunov exponent of the same absolute value. It is thus sufficient to study the positive Lyapunov exponents for ray tracing.

4. Ray-centred Coordinates

Assume that the Hamiltonian is a homogeneous function of the second degree with respect to the slowness vector. Then the corresponding parameter along a ray is the travel time, $\vartheta = \tau$. In the ray-centred coordinates, $\mathbf{w}^{(q)} = (q^1, q^2, q^3, p_1^{(q)}, p_2^{(q)}, p_3^{(q)})^{\mathrm{T}}$, the second phase-space derivatives of the Hamiltonian take the form

$$\mathbf{H} = \begin{pmatrix} \mathbf{H}_{11} & 0 & \mathbf{H}_{12} & 0 \\ 0\ 0 & 0 & 0\ 0 & 0 \\ \mathbf{H}_{21} & 0 & \mathbf{H}_{22} & 0 \\ 0\ 0 & 0 & 0\ 0 & 1 \end{pmatrix}, \qquad (10)$$

and the paraxial-ray propagator matrix takes the form

$$\boldsymbol{\Pi} = \begin{pmatrix} \mathbf{Q}_1 & 0 & \mathbf{Q}_2 & 0 \\ 0\ 0 & 1 & 0\ 0 & \tau \\ \mathbf{P}_1 & 0 & \mathbf{P}_2 & 0 \\ 0\ 0 & 0 & 0\ 0 & 1 \end{pmatrix}, \qquad (11)$$

see KLIMEŠ (1994). Two characteristic values are thus unity, $\mu_3 = 1$ and $\mu_4 = 1$, with the corresponding Lyapunov exponents identical to zero, $\lambda_3 = 0$ and $\lambda_4 = 0$.

Two positive Lyapunov exponents λ_1 and λ_2 correspond to 4×4 dynamic ray tracing in the ray-centred coordinates, with matrices

$$\mathbf{H} = \begin{pmatrix} \mathbf{H}_{11} & \mathbf{H}_{12} \\ \mathbf{H}_{21} & \mathbf{H}_{22} \end{pmatrix}, \quad \boldsymbol{\Pi} = \begin{pmatrix} \mathbf{Q}_1 & \mathbf{Q}_2 \\ \mathbf{P}_1 & \mathbf{P}_2 \end{pmatrix}. \qquad (12)$$

The ray-centred coordinates may be chosen in such a way that $\mathbf{H}_{12} = \mathbf{0}$ and $\mathbf{H}_{21} = \mathbf{0}$ in (12),

$$\mathbf{H} = \begin{pmatrix} \mathbf{H}_{11} & \mathbf{0} \\ \mathbf{0} & \mathbf{H}_{22} \end{pmatrix} \tag{13}$$

(KLIMEŠ, 1994). If we denote $\mathbf{H}_{11} = \mathbf{V}$ and $\mathbf{H}_{22} = \mathbf{G}$ in (13),

$$\mathbf{V} = \begin{pmatrix} \frac{\partial^2 H}{\partial q^1 \partial q^1} & \frac{\partial^2 H}{\partial q^1 \partial q^2} \\ \frac{\partial^2 H}{\partial q^1 \partial q^2} & \frac{\partial^2 H}{\partial q^2 \partial q^2} \end{pmatrix}, \quad \mathbf{G} = \begin{pmatrix} \frac{\partial^2 H}{\partial p_1^{(q)} \partial p_1^{(q)}} & \frac{\partial^2 H}{\partial p_1^{(q)} \partial p_2^{(q)}} \\ \frac{\partial^2 H}{\partial p_1^{(q)} \partial p_2^{(q)}} & \frac{\partial^2 H}{\partial p_2^{(q)} \partial p_2^{(q)}} \end{pmatrix}, \tag{14}$$

the dynamic ray tracing equation takes the form

$$\frac{d}{d\tau} \begin{pmatrix} \mathbf{Q}_1 & \mathbf{Q}_2 \\ \mathbf{P}_1 & \mathbf{P}_2 \end{pmatrix} = \begin{pmatrix} \mathbf{0} & \mathbf{G} \\ -\mathbf{V} & \mathbf{0} \end{pmatrix} \begin{pmatrix} \mathbf{Q}_1 & \mathbf{Q}_2 \\ \mathbf{P}_1 & \mathbf{P}_2 \end{pmatrix}. \tag{15}$$

5. Positive Lyapunov Exponent in 2-D

5.1. Lyapunov Exponent of a Finite Ray

We decompose the 2×2 paraxial-ray propagator matrix $\mathbf{\Pi}$ into the 2×2 identity matrix $\mathbf{1}$ and the Pauli matrices

$$\boldsymbol{\sigma}_1 = \begin{pmatrix} 0 & 1 \\ 1 & 0 \end{pmatrix}, \quad \boldsymbol{\sigma}_2 = \begin{pmatrix} 0 & -i \\ i & 0 \end{pmatrix}, \quad \boldsymbol{\sigma}_3 = \begin{pmatrix} 1 & 0 \\ 0 & -1 \end{pmatrix}, \tag{16}$$

which satisfy relations

$$\boldsymbol{\sigma}_1 \boldsymbol{\sigma}_2 = -\boldsymbol{\sigma}_2 \boldsymbol{\sigma}_1 = i\boldsymbol{\sigma}_3, \quad \boldsymbol{\sigma}_2 \boldsymbol{\sigma}_3 = -\boldsymbol{\sigma}_3 \boldsymbol{\sigma}_2 = i\boldsymbol{\sigma}_1, \quad \boldsymbol{\sigma}_3 \boldsymbol{\sigma}_1 = -\boldsymbol{\sigma}_1 \boldsymbol{\sigma}_3 = i\boldsymbol{\sigma}_2 \tag{17}$$

and

$$\boldsymbol{\sigma}_1 \boldsymbol{\sigma}_1 = \boldsymbol{\sigma}_2 \boldsymbol{\sigma}_2 = \boldsymbol{\sigma}_3 \boldsymbol{\sigma}_3 = \mathbf{1}. \tag{18}$$

The decomposition of the real-valued matrix $\mathbf{\Pi}$ reads

$$\mathbf{\Pi} = \Pi_0 \mathbf{1} + \Pi_1 \boldsymbol{\sigma}_1 + \Pi_2 i\boldsymbol{\sigma}_2 + \Pi_3 \boldsymbol{\sigma}_3, \tag{19}$$

where Π_0, Π_1, Π_2 and Π_3 are real-valued coefficients. Note that coefficients Π_1 and Π_2 are just formal. A reasonable physical meaning and units correspond to the off-diagonal terms $\Pi_1 + \Pi_2$ and $\Pi_1 - \Pi_2$ of matrix $\mathbf{\Pi}$, not just the terms Π_1 or Π_2 on their own. Since $\det \mathbf{\Pi} = 1$, the coefficients satisfy equation

$$\Pi_0^2 - (\Pi_1 + \Pi_2)(\Pi_1 - \Pi_2) - \Pi_3^2 = 1. \tag{20}$$

Since $\det \mathbf{\Pi} = 1$, the greater characteristic value of matrix $\mathbf{\Pi}$ is

$$\mu_1 = |\Pi_0| + \sqrt{\Pi_0^2 - 1}. \tag{21}$$

Since the characteristic value oscillates along the ray, we shall look for a different "norm" which could substitute the characteristic value. We require the "norm" to be independent of free parameters (e.g., of the metric in the phase space) and of the length and time units. It is not acceptable to have a different measure of ray divergence for the same ray expressed once in kilometres and then in feet. Both the off-diagonal components of Π are dependent on the units (kilometres, feet), only their product is independent. However, since $\det \Pi = 1$, the product of the off-diagonal components contains no additional information to the information carried by the diagonal components. The "norm" can thus depend only on the diagonal components of Π, i.e., on coefficients Π_0 and Π_3. Characteristic value (21) is of little use at the points along a ray where Π_0 is small with respect to the other coefficients. We thus replace characteristic value (21) by "norm"

$$M = \sqrt{\Pi_0^2 + \Pi_3^2} + \sqrt{\Pi_0^2 + \Pi_3^2 - 1} \; , \tag{22}$$

which is also good for small Π_0 and large Π_3. If both Π_0 and Π_3 are small, product $(\Pi_1 - \Pi_2)(\Pi_1 + \Pi_2)$ of the off-diagonal components is small, too, because of (20). "Norm" (22) can thus reasonably reflect the exponential ray divergence along the whole ray. We hope that "norm" (22) enables Lyapunov exponents

$$\lambda(\tau, \tau_0) = \frac{L(\tau, \tau_0)}{\sigma(\tau) - \sigma(\tau_0)} \tag{23}$$

to be defined for rays of finite lengths, and to replace the limes superior in (9) by a simple limit,

$$\lambda = \lim_{\tau \to +\infty} \lambda(\tau, \tau_0) \; . \tag{24}$$

Here we have denoted

$$L(\tau, \tau_0) = \ln[M(\tau, \tau_0)] \tag{25}$$

the function characterizing propagator matrix $\Pi(\tau, \tau_0)$. An example of the dependence of $L(\tau, \tau_0)$ on τ for the rays shot from a point source is shown in Figure 7. The magnitude of the variation of the Lyapunov exponents from ray to ray is considerable and the Lyapunov exponent for a single ray obviously cannot be linked to the overall properties of the velocity model.

Fortunately, equation (23) allows the average Lyapunov exponent

$$\bar{\lambda} = \frac{\sum_{\text{ray}} L_{\text{ray}}(\tau, \tau_0)}{\sum_{\text{ray}} [\sigma_{\text{ray}}(\tau) - \sigma_{\text{ray}}(\tau_0)]} \tag{26}$$

over all rays to be introduced. This average Lyapunov exponent expresses the global properties of the velocity model. However, for a single source, it is still dependent on the source geometry and position with respect to the model boundaries. The average Lyapunov exponent over various sources should describe the global properties of the velocity model and should be dependent on the model boundaries only if the statistical properties of the model are anisotropic.

5.2. Approximation of the Paraxial-ray Propagator Matrix

In 2-D, dynamic ray tracing equation (15) simplifies to equation

$$\frac{d\Pi}{d\tau} = \begin{pmatrix} 0 & G \\ -V & 0 \end{pmatrix} \Pi \ . \tag{27}$$

The infinitesimal propagator matrix for small travel time increment $\delta\tau$ is

$$\Pi_{\delta\tau} = 1 + \begin{pmatrix} 0 & G \\ -V & 0 \end{pmatrix} \delta\tau + O(\delta\tau^2) \ . \tag{28}$$

In *defocusing regions*, where $VG < 0$, equation (28) reads

$$\Pi_{\delta\tau} = \begin{pmatrix} |\frac{V}{G}|^{-\frac{1}{4}} & 0 \\ 0 & |\frac{V}{G}|^{\frac{1}{4}} \end{pmatrix} \left(1 + \sigma_1 \sqrt{-VG}\ \delta\tau \right) \begin{pmatrix} |\frac{V}{G}|^{\frac{1}{4}} & 0 \\ 0 & |\frac{V}{G}|^{-\frac{1}{4}} \end{pmatrix} + O(\delta\tau^2) \ , \tag{29}$$

whereas in *focusing regions*, where $VG > 0$, equation (28) reads

$$\Pi_{\delta\tau} = \begin{pmatrix} |\frac{V}{G}|^{-\frac{1}{4}} & 0 \\ 0 & |\frac{V}{G}|^{\frac{1}{4}} \end{pmatrix} \left(1 + i\sigma_2 \sqrt{VG}\ \delta\tau \right) \begin{pmatrix} |\frac{V}{G}|^{\frac{1}{4}} & 0 \\ 0 & |\frac{V}{G}|^{-\frac{1}{4}} \end{pmatrix} + O(\delta\tau^2) \ . \tag{30}$$

Note that

$$\begin{pmatrix} (|\frac{V}{G} + \delta(\frac{V}{G})|^{\frac{1}{4}} & 0 \\ 0 & |\frac{V}{G} + \delta(\frac{V}{G})|^{-\frac{1}{4}} \end{pmatrix} \begin{pmatrix} |\frac{V}{G}|^{-\frac{1}{4}} & 0 \\ 0 & |\frac{V}{G}|^{\frac{1}{4}} \end{pmatrix} = \begin{pmatrix} [1 + (\frac{V}{G})^{-1}\delta(\frac{V}{G})]^{\frac{1}{4}} & 0 \\ 0 & [1 + (\frac{V}{G})^{-1}\delta(\frac{V}{G})]^{-\frac{1}{4}} \end{pmatrix}$$

$$= 1 + \sigma_3 \frac{1}{4} \frac{\delta(VG^{-1})}{VG^{-1}} + O(\delta\tau^2) \ . \tag{31}$$

We now introduce three parameters:

$$\Lambda = \int \sqrt{\text{neg}(VG)}\, d\tau \ , \tag{32}$$

$$\Phi = \int \sqrt{\text{pos}(VG)}\, d\tau \tag{33}$$

and

$$\Psi = \tfrac{1}{4}\ln(|VG^{-1}|) \ . \tag{34}$$

Here

$$\mathrm{pos}(x) = \max(x, 0) \ , \quad \mathrm{neg}(x) = -\min(x, 0)$$

denote the positive and negative parts of x, respectively. Parameter Λ accounts for the exponential spreading of the ray tube in phase space, and parameter Φ describes how the ray tube is twisted in phase space, see the Appendix.

The product of infinitesimal propagator matrices (29) is then

$$\mathbf{\Pi}_{\Delta\Lambda} = \cdots (1 + \boldsymbol{\sigma}_1 \delta\Lambda)(1 + \boldsymbol{\sigma}_3 \delta\Psi)(1 + \boldsymbol{\sigma}_1 \delta\Lambda)(1 + \boldsymbol{\sigma}_3 \delta\Psi) \cdots + O(\delta\tau^2) \tag{35}$$

and the product of infinitesimal propagator matrices (30) is

$$\mathbf{\Pi}_{\Delta\Phi} = \cdots (1 + i\boldsymbol{\sigma}_2 \delta\Phi)(1 + \boldsymbol{\sigma}_3 \delta\Psi)(1 + i\boldsymbol{\sigma}_2 \delta\Phi)(1 + \boldsymbol{\sigma}_3 \delta\Psi) \cdots + O(\delta\tau^2) \ . \tag{36}$$

Note that the values of $\delta\Lambda$, $\delta\Phi$ and $\delta\Psi$ in (35) and (36) vary along the considered segment of the ray. If the segment of the ray extends across the whole defocusing or focusing region, i.e., if $V = 0$ at the endpoints of the segment, we assume that the distribution of parameter Ψ along the segment is roughly symmetric and that it does not influence matrix products (35) and (36) considerably. Neglecting the multiplicands with Ψ, (35) and (36) simplify to

$$\mathbf{\Pi}_{\Delta\Lambda} \approx [\mathbf{1}\cosh(\Delta\Lambda) + \boldsymbol{\sigma}_1 \sinh(\Delta\Lambda)] \tag{37}$$

and

$$\mathbf{\Pi}_{\Delta\Phi} \approx [\mathbf{1}\cos(\Delta\Phi) + i\boldsymbol{\sigma}_2 \sin(\Delta\Phi)] \ , \tag{38}$$

respectively. The propagator matrix along the ray can be composed of the propagator matrices corresponding to the individual defocusing segments $\mathbf{\Pi}_{\Delta\Lambda}$ and focusing segments $\mathbf{\Pi}_{\Delta\Phi}$,

$$\mathbf{\Pi} = \mathbf{\Pi}_{\Delta\Phi_N} \mathbf{\Pi}_{\Delta\Lambda_N} \cdots \mathbf{\Pi}_{\Delta\Phi_2} \mathbf{\Pi}_{\Delta\Lambda_2} \mathbf{\Pi}_{\Delta\Phi_1} \mathbf{\Pi}_{\Delta\Lambda_1} \mathbf{\Pi}_{\Delta\Phi_0} \ , \tag{39}$$

i.e.,

$$\mathbf{\Pi} \approx [\mathbf{1}\cos(\Delta\Phi_N) + i\boldsymbol{\sigma}_2 \sin(\Delta\Phi_N)][\mathbf{1}\cosh(\Delta\Lambda_N) + \boldsymbol{\sigma}_1 \sinh(\Delta\Lambda_N)]$$

$$\times \cdots \times [\mathbf{1}\cos(\Delta\Phi_2) + i\boldsymbol{\sigma}_2 \sin(\Delta\Phi_2)][\mathbf{1}\cosh(\Delta\Lambda_2) + \boldsymbol{\sigma}_1 \sinh(\Delta\Lambda_2)]$$

$$\times [\mathbf{1}\cos(\Delta\Phi_1) + i\boldsymbol{\sigma}_2 \sin(\Delta\Phi_1)][\mathbf{1}\cosh(\Delta\Lambda_1) + \boldsymbol{\sigma}_1 \sinh(\Delta\Lambda_1)]$$

$$\times [\mathbf{1}\cos(\Delta\Phi_0) + i\boldsymbol{\sigma}_2 \sin(\Delta\Phi_0)] \ . \tag{40}$$

Taking only the terms with cosines for $\Delta\Phi_n$, $n = 1, 2, \ldots, N - 1$, we arrive at the zero-order approximation of the right-hand side of (40),

$$\mathbf{\Pi}^{(0)} = \prod_{n=1}^{N-1} \cos(\Delta\Phi_n) \left[\mathbf{1} \cosh\left(\sum_{n=1}^{N} \Delta\Lambda_n \right) \cos(\Delta\Phi_N + \Delta\Phi_0) \right.$$

$$\left. + \boldsymbol{\sigma}_3 \sinh\left(\sum_{n=1}^{N} \Delta\Lambda_n \right) \sin(\Delta\Phi_N - \Delta\Phi_0) \right] + \boldsymbol{\sigma}_1 \cdots + \boldsymbol{\sigma}_2 \cdots \quad . \qquad (41)$$

The correction with one cosine term replaced by the sine term for $\Delta\Phi_J$ is

$$\mathbf{\Pi}^{(J)} = \frac{\sin(\Delta\Phi_J)}{\cos(\Delta\Phi_J)} \prod_{n=1}^{N-1} \cos(\Delta\Phi_n)$$

$$\times \left[-\mathbf{1} \cosh\left(\sum_{n=J}^{N} \Delta\Lambda_n - \sum_{n=1}^{J-1} \Delta\Lambda_n \right) \sin(\Delta\Phi_N + \Delta\Phi_0) \right.$$

$$\left. + \boldsymbol{\sigma}_3 \sinh\left(\sum_{n=J}^{N} \Delta\Lambda_n - \sum_{n=1}^{J-1} \Delta\Lambda_n \right) \cos(\Delta\Phi_N - \Delta\Phi_0) \right] + \boldsymbol{\sigma}_1 \cdots + \boldsymbol{\sigma}_2 \cdots \quad . \qquad (42)$$

The correction with two cosine terms replaced by the sine terms for $\Delta\Phi_J$ and $\Delta\Phi_K$ is

$$\mathbf{\Pi}^{(JK)} = \frac{\sin(\Delta\Phi_J)\sin(\Delta\Phi_K)}{\cos(\Delta\Phi_J)\cos(\Delta\Phi_K)} \prod_{n=1}^{N-1} \cos(\Delta\Phi_n)$$

$$\times \left[-\mathbf{1} \cosh\left(\sum_{n=K}^{N} \Delta\Lambda_n - \sum_{n=J}^{K-1} \Delta\Lambda_n + \sum_{n=1}^{J-1} \Delta\Lambda_n \right) \cos(\Delta\Phi_N + \Delta\Phi_0) \right.$$

$$\left. - \boldsymbol{\sigma}_3 \sinh\left(\sum_{n=K}^{N} \Delta\Lambda_n - \sum_{n=J}^{K-1} \Delta\Lambda_n + \sum_{n=1}^{J-1} \Delta\Lambda_n \right) \sin(\Delta\Phi_N - \Delta\Phi_0) \right]$$

$$+ \boldsymbol{\sigma}_1 \cdots + \boldsymbol{\sigma}_2 \cdots \quad . \qquad (43)$$

Further corrections $\mathbf{\Pi}^{(JKL)}$, $\mathbf{\Pi}^{(JKLM)}$, etc., may be calculated in a similar way.

Since corrections (42) are smaller than zero-order approximation (41) roughly by the factors $\exp[-2\min(\sum_{n=J}^{N} \Delta\Lambda_n, \sum_{n=1}^{J-1} \Delta\Lambda_n)]$, and corrections (43) are smaller roughly by the factors $\exp(-2\sum_{n=J}^{K-1} \Delta\Lambda_n)$, we approximate the propagator matrix by the zero-order approximation only,

$$\mathbf{\Pi}(\tau, \tau_0) \approx \mathbf{\Pi}^{(0)}(\tau, \tau_0) \quad . \qquad (44)$$

Note that (44) is a very bad approximation of the propagator matrix at a single point of a single ray, but is a reasonably good estimate of the "average" propagator matrix.

5.3. Approximation of the Lyapunov Exponent

Inserting coefficients Π_0 and Π_3 from (41) into (22) and (25), we approximately obtain

$$L \approx \sum_{n=1}^{N} \Delta\Lambda_n + \sum_{n=1}^{N-1} \ln|\cos(\Delta\Phi_n)| + \frac{1}{2}\ln[\cos^2(\Delta\Phi_0 + \Delta\Phi_N) + \sin^2(\Delta\Phi_0 - \Delta\Phi_N)] \ .$$

$$(45)$$

The mean value of $\ln|\cos(\Delta\Phi_n)|$ over all values of angle $\Delta\Phi_n$ is

$$\langle\ln|\cos(\Delta\Phi_n)|\rangle = -\ln 2 \tag{46}$$

and the standard deviation of $\ln|\cos(\Delta\Phi_n)|$ from the mean value is

$$\sqrt{\langle[\ln|\cos(\Delta\Phi_n)| - \langle\ln|\cos(\Delta\Phi_n)|\rangle]^2\rangle} = \frac{\pi}{\sqrt{12}} \ . \tag{47}$$

Because of the rough approximation (38) of argument $\Delta\Phi_n$ of the cosine, the cosine is considerably inaccurate for Φ_n comparable with $\pi/2$ and greater. We thus replace $\ln|\cos(\Delta\Phi_n)|$ for $\Delta\Phi_n > \pi/3$ with its mean value,

$$\Delta\Phi_n \longrightarrow \min\left(\Delta\Phi_n, \frac{\pi}{3}\right), \quad n = 1, 2, \ldots, N - 1 \ . \tag{48}$$

Similarly, the mean value of the last addend in (45) over all angles $\Delta\Phi_0$ and $\Delta\Phi_N$ is

$$\langle\frac{1}{2}\ln[\cos^2(\Delta\Phi_0 + \Delta\Phi_N) + \sin^2(\Delta\Phi_0 - \Delta\Phi_N)]\rangle$$

$$= \langle\frac{1}{2}\ln[1 - \sin(2\Delta\Phi_0)\sin(2\Delta\Phi_N)]\rangle = \langle\frac{1}{4}\ln[1 - \sin^2(2\Delta\Phi_0)\sin^2(2\Delta\Phi_N)]\rangle$$

$$= \langle\frac{1}{2}\ln\{\frac{1}{2}[1 + \cos^2(2\Delta\Phi_N)]\}\rangle = \langle\ln|\cos(\Delta\Phi_N)|\rangle = -\ln 2 \ . \tag{49}$$

The mean value is reached if both $\Delta\Phi_0 = \pi/6$ and $\Delta\Phi_N = \pi/6$. We thus replace

$$\Delta\Phi_n \longrightarrow \min\left(\Delta\Phi_n, \frac{\pi}{6}\right), \quad n = 0, N \ . \tag{50}$$

The final approximate estimate of function L in definition (26) is

$$L \approx \sum_{n=1}^{N} \Delta\Lambda_n + \sum_{n=1}^{N-1} \ln\left|\cos\left[\min\left(\Delta\Phi_n, \frac{\pi}{3}\right)\right]\right|$$

$$+ \frac{1}{2}\ln\left\{1 - \sin\left[2\min\left(\Delta\Phi_0, \frac{\pi}{6}\right)\right]\sin\left[2\min\left(\Delta\Phi_N, \frac{\pi}{6}\right)\right]\right\} , \tag{51}$$

where $\Delta\Lambda_n$, $n = 1, 2, \ldots, N$, correspond to the defocusing ray segments $(VG < 0)$ and $\Delta\Phi_n$, $n = 0, 1, 2, \ldots, N$, correspond to the focusing ray segments $(VG > 0)$, $\Delta\Phi_0$ and $\Delta\Phi_N$ may become zero if the ray starts or terminates in a defocusing region.

6. Isotropic Medium

In an isotropic medium with Hamiltonian $H = v^2 \mathbf{p}^T \mathbf{p}$, the second phase-space derivatives of the Hamiltonian in ray-centred coordinates q_1, q_2 are

$$\mathbf{V} = v^{-1} \begin{pmatrix} \frac{\partial^2 v}{\partial q^1 \partial q^1} & \frac{\partial^2 v}{\partial q^1 \partial q^2} \\ \frac{\partial^2 v}{\partial q^1 \partial q^2} & \frac{\partial^2 v}{\partial q^2 \partial q^2} \end{pmatrix} \tag{52}$$

and

$$\mathbf{G} = v^2 \mathbf{1} . \tag{53}$$

In 2-D, parameters (32) and (33) read

$$\Lambda = \int \sqrt{\mathrm{neg}\left(v \frac{\partial^2 v}{\partial q \partial q} \right)} \, d\tau \tag{54}$$

and

$$\Phi = \int \sqrt{\mathrm{pos}\left(v \frac{\partial^2 v}{\partial q \partial q} \right)} \, d\tau , \tag{55}$$

where $\frac{\partial^2 v}{\partial q \partial q}$ is the second derivative of the velocity with respect to the in-plane ray-centred coordinate q, perpendicular to the ray.

7. Average Lyapunov Exponent for the Model

Our estimate of the Lyapunov exponent of a single finite ray depends on the position and direction of the ray. Let us now average the Lyapunov exponent over the whole model volume.

We cover the model by a dense system of parallel straight lines. For each direction of the lines, we calculate the average Lyapunov exponent (26) using equation (51), similarly as for a system of rays. We then average the calculated directional Lyapunov exponent over the directions, applying a selected directional weighting function. For instance, the weighting function may have the shape of a model box, with the origin at the centre of the box or at the mean position of the intended point sources. The weighting function may also correspond to the probability of the ray directions estimated for ray tracing in the model. Refer to Figure 6 for examples of the directional dependence of the average Lyapunov exponent and of the directional weighting function.

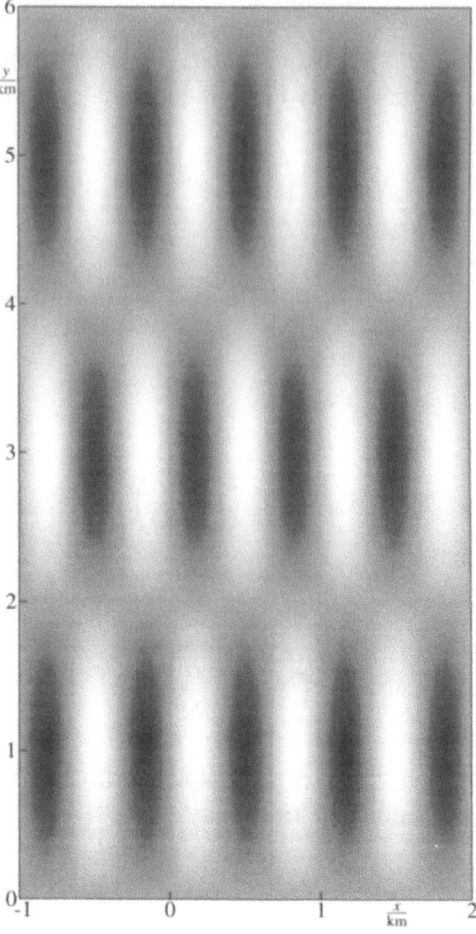

Figure 1

The model is formed by the homogeneous background of velocity 1.0 km s^{-1}, perturbed by a stretched bi-sine egg-box of amplitude 0.2 km s^{-1}. The zones situated close to the four corners are focusing low-velocity regions.

8. Numerical Example

The model designed by Jean-David Benamou is formed by the homogeneous background of velocity 1.0 km s^{-1}, perturbed by a stretched bi-sine egg-box of amplitude 0.2 km s^{-1}, see Figure 1. The horizontal dimension of the model box is 3 km, vertical 6 km. There are focusing low-velocity regions close to the four corners. The average travel times in seconds closely correspond to the average lengths in kilometres in this model.

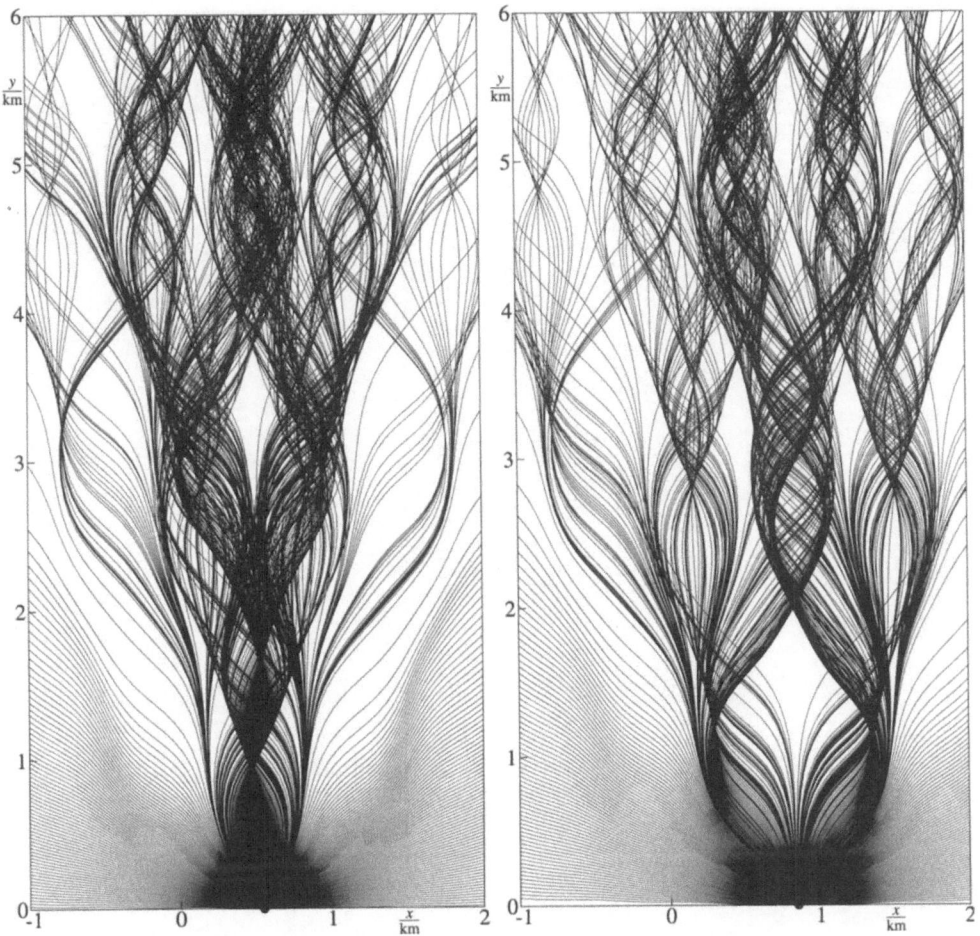

Figure 2
Rays corresponding to the point source situated at the bottom of the model box, 1.55 km from the left-hand corner.

Figure 3
Rays corresponding to the point source situated at the bottom of the model box, 1.85 km from the left-hand corner.

Some rays, shot from the point source situated at the bottom of the model box, 1.55 km from the left-hand corner, are shown in Figure 2. Figure 3 shows the rays shot from the point source shifted to 1.85 km from the left-hand corner. Although the model is highly regular and periodic, the ray paths are quite irregular. If we divide the model into the smallest equal cells, rays enter and leave the individual cells at and in quite different positions and directions, 0% of rays being periodic. The rays traced in this model can thus serve as an example of the chaotic behaviour of rays in general

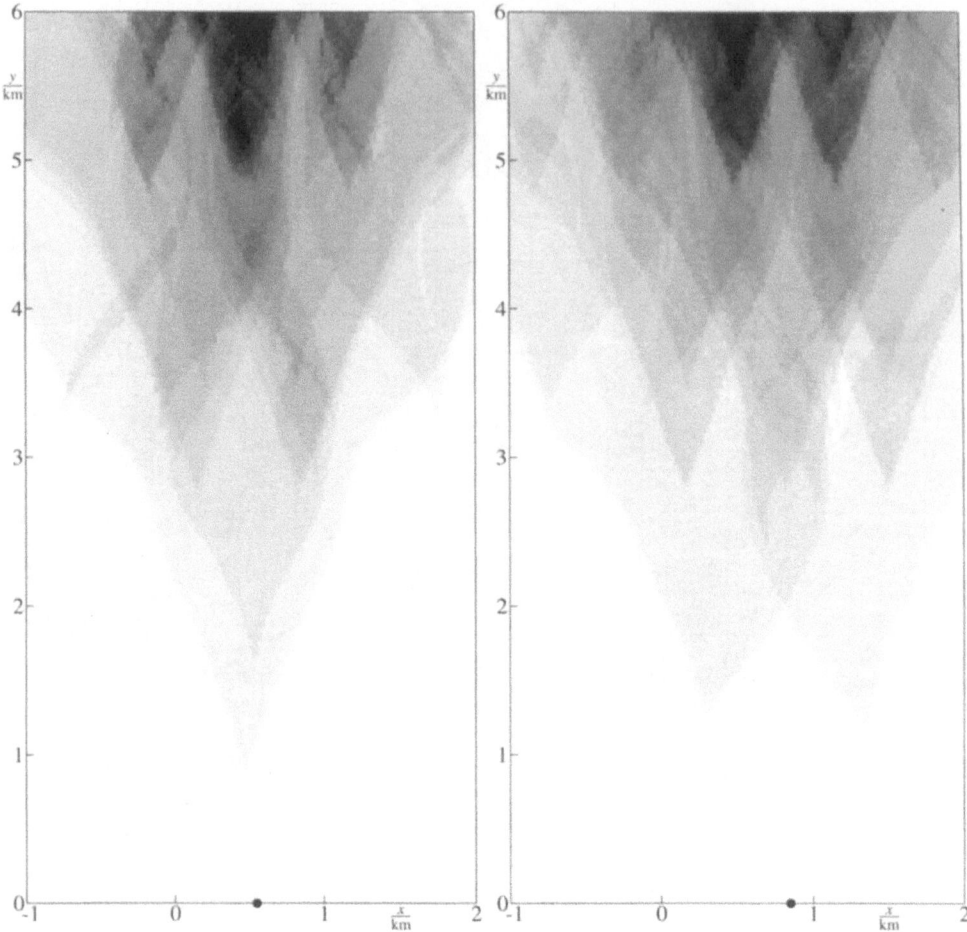

Figure 4

The numbers of travel times corresponding to the point source situated at the bottom of the model box, 1.55 km from the left-hand corner. The maximum number of found travel times is 49.

Figure 5

The numbers of travel times corresponding to the point source situated at the bottom of the model box, 1.85 km from the left-hand corner. The maximum number of found travel times is 59.

heterogeneous 2-D models, keeping in mind that the statistical properties of this model are strongly anisotropic.

Figure 4 displays the numbers of travel times corresponding to the point source situated at the bottom of the model box, 1.55 km from the left-hand corner. The travel times are calculated in a grid of 121×241 points covering the model box by means of interpolation within ray cells (BULANT, 1997, 1999a, b, c, 2000; BULANT and KLIMEŠ, 1999). The maximum number of found travel times is 49. Figure 5 shows the numbers of travel times for the point source shifted

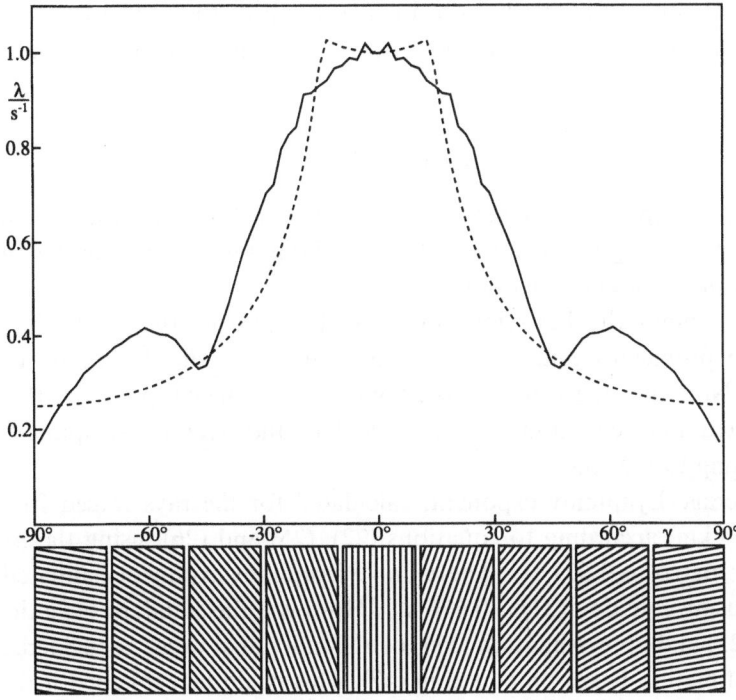

Figure 6

The angular dependence of the average Lyapunov exponents for the model [*solid line*] and the selected directional weighting function [*dashed line*]. The extent of the horizontal axis is 180 degrees, with the vertical direction at its centre, as schematically illustrated below the diagram. The average Lyapunov exponents vary between 0.170 s^{-1} and 1.019 s^{-1}. The selected directional weighting function corresponds to the model box with the origin at the centre of the bottom edge. We emphasize that the solid line describes the velocity heterogeneities whereas the dashed line characterizes the shape of the model box.

to 1.85 km from the left-hand corner. The maximum number of found travel times is 59.

Ninety directions with an angular increment of 2 degrees have been chosen to estimate the average Lyapunov exponent for the model. For each direction, the model has been covered by 45 equally spaced straight lines. The directional Lyapunov exponent according to equations (23) and (51) has been numerically calculated along the straight lines, with a step corresponding to 45 steps along the longest line for the direction. The average Lyapunov exponent for each direction has been calculated according to equation (26), with respect to the travel time, $\sigma(\tau) = \tau$. Since the statistical properties of the model are strongly anisotropic, the average Lyapunov exponents vary between 0.170 s^{-1} and 1.019 s^{-1} for different directions, see the solid line in Figure 6. The selected directional weighting function corresponds to the model box with the origin at the centre of the bottom edge, see the dashed line in Figure 6. This directional weighting function is suitable for the point sources situated at the bottom or at the top of the model box. The average Lyapunov exponent for the model, calculated with this directional weighting function, is

$$\bar{\lambda}_{\text{model}} = 0.698 \text{ s}^{-1} \ . \tag{56}$$

The average Lyapunov exponent does not noticeably vary with the horizontal translation of the origin of the directional weighting function within the middle third of the horizontal model dimension.

Figure 7 shows the behaviour of logarithm (25) of the "norm" (22) of the paraxial-ray propagator matrix along the individual rays shot from the second source (1.85 km). Note that the values of the logarithms at the endpoints of the rays have been inserted into equation (26) to calculate the average Lyapunov exponent corresponding to the source.

The average Lyapunov exponent, calculated for the rays traced from the first source (1.55 km) according to equations (22), (25) and (26), using the paraxial-ray propagator matrix, is 0.679 s^{-1}. The average Lyapunov exponent estimated along the same rays using equation (51) is 0.676 s^{-1}. However, the standard deviation between equation (25) and approximation (51) for the individual rays is considerably greater than the estimation based on equation (47). This could be explained if the standard deviations of approximations (37), (38) or (44) were comparable with the standard deviation (47) of approximations (48) and (50).

The average Lyapunov exponent calculated for the rays traced from the second source (1.85 km) according to equations (22), (25) and (26), using the paraxial-ray propagator matrix, is 0.613 s^{-1}. The average Lyapunov exponent estimated along the same rays using equation (51) is 0.697 s^{-1}.

Figure 8 displays the natural logarithms of the average and maximum numbers of travel times along the individual horizontal grid lines of Figures 4 and 5. The horizontal axis is the distance of the grid line from the bottom of the model box in kilometres, and serves as a rough approximation of the travel time in seconds. The slope of the straight solid lines is given by the average Lyapunov exponent (56) for the model.

The numerical example demonstrates the good correspondence between the average logarithms of the numbers of ray-theory travel times, the average Lyapunov exponents for the rays and the estimate of the average Lyapunov exponent for the model.

9. Conclusions

The proposed estimation of the average Lyapunov exponent for the model enables the suitability of a 2-D velocity model, without interfaces, for ray tracing to

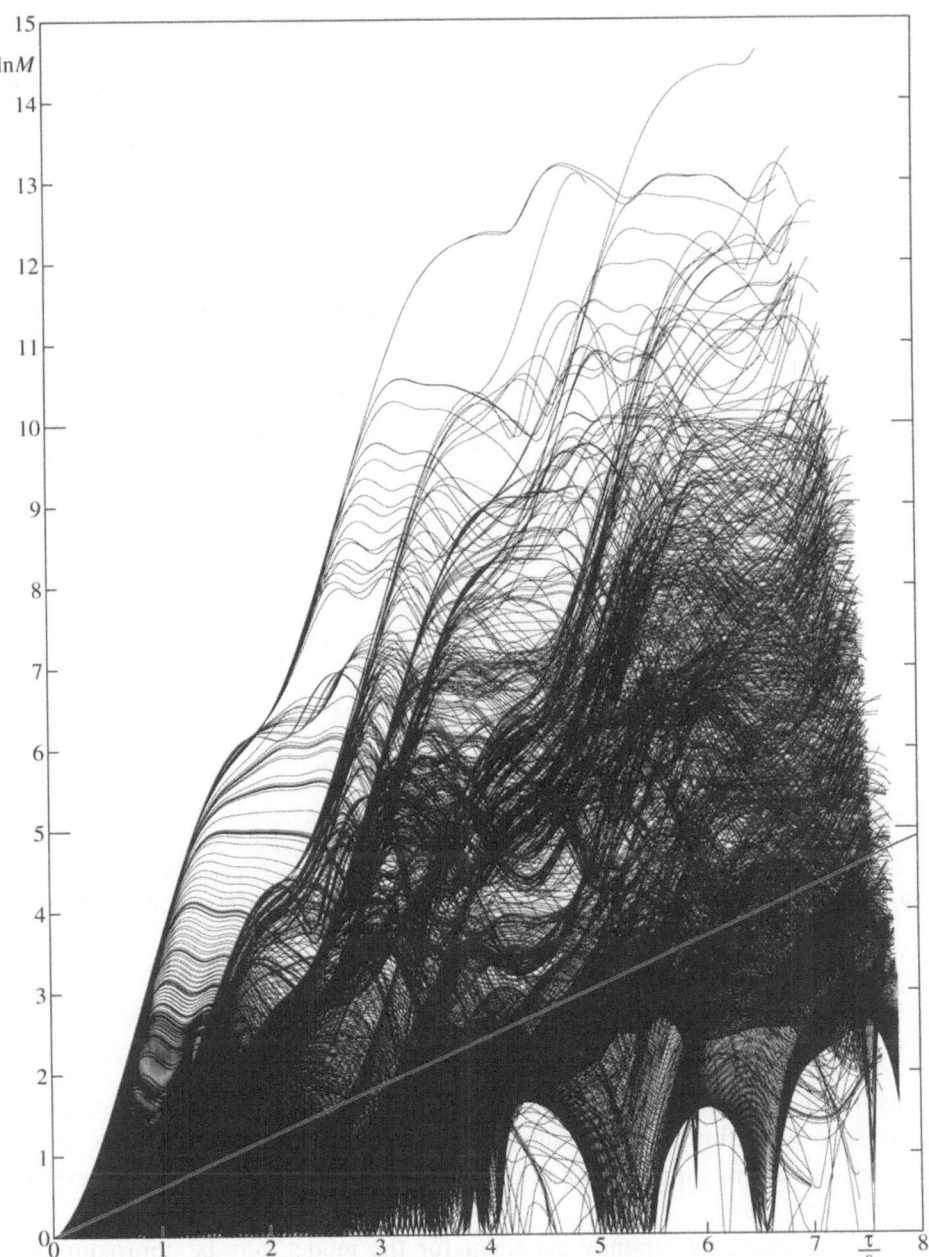

Figure 7

The behaviour of logarithm (25) of the "norm" (22) of the paraxial-ray propagator matrix along the individual rays shot from the source of Figure 5. The horizontal axis represents the travel time and extends from 0 s to 8 s. The vertical axis represents the natural logarithm and extends from 0 to 15. The straight line corresponds to the average Lyapunov exponent of 0.613 s^{-1} for this set of rays.

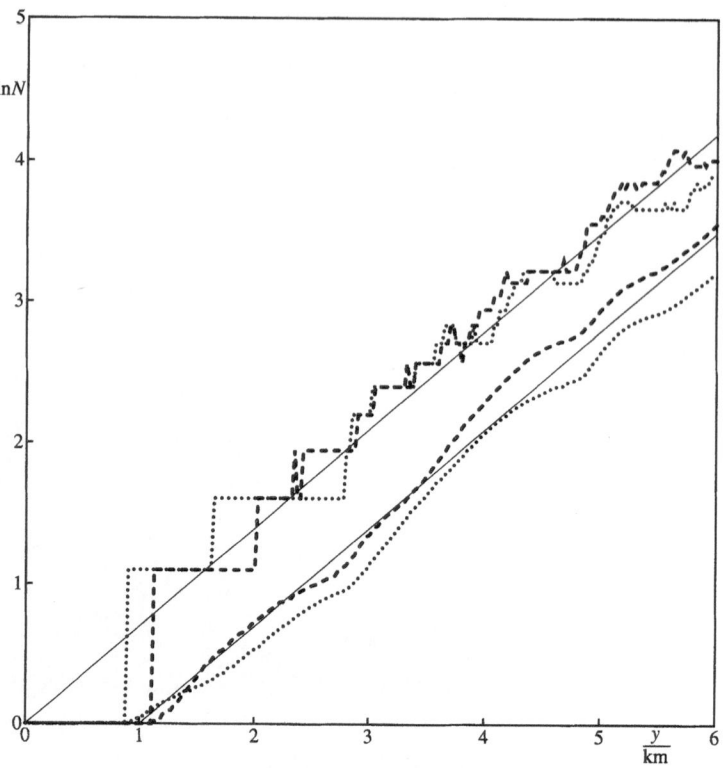

Figure 8

The natural logarithms of the average and maximum numbers of travel times along the individual horizontal grid lines of Figure 4 [*bold dotted lines*] and Figure 5 [*bold dashed lines*]. The horizontal axis represents the distance of the grid line from the bottom of the model box in kilometres, and serves as a rough approximation of the travel time in seconds. The slope of the *thin solid lines* is given by the average Lyapunov exponent (56) for the model.

be quantified. The numerical example demonstrates a good correspondence between the average logarithms of the numbers of ray-theory travel times, the average Lyapunov exponents for the rays and the estimate of the average Lyapunov exponent for the model.

Since the average Lyapunov exponent for the model may be approximated in terms of the Sobolev norm composed of the second velocity derivatives (KLIMEŠ, 2000), the results of this paper may be used in the construction of velocity models optimized for the calculation of ray-theory Green functions (BULANT, 2002; ŽÁČEK, 2002).

The theoretical results should be further generalized to 3-D models and to models with structural interfaces.

Appendix

Note on the Number of Caustics

The number of travel times and the number of caustics are not related in a simple way, because the number of travel times remains unchanged when the ray touches a caustic, whereas the number of travel times may be either increased by 2 or decreased by 2 when the ray crosses a caustic. The average Lyapunov exponent, characterizing the exponential increment of the number of travel times in dependence on the length of the rays, and the average frequency of caustic points along rays are thus two different characteristics of ray chaos. The caustic point is understood to be the point at which the ray touches the caustic.

Let us express, without derivation, the conjecture that the average angular velocity $\bar{\varphi}$ of twisting ray tubes in phase space may be approximated by

$$\bar{\varphi} \approx \frac{\sum\limits_{\text{ray}} \Phi_{\text{ray}}(\tau, \tau_0)}{\sum\limits_{\text{ray}} [\sigma_{\text{ray}}(\tau) - \sigma_{\text{ray}}(\tau_0)]} \ , \tag{A.1}$$

where parameter Φ along the individual rays is defined by (26). This conjecture is in accordance with the example of WHITE *et al.* (1988) but has not yet been further numerically tested.

The harmonically averaged distance between two consecutive caustic points along a ray, measured in terms of parameter σ, is

$$\bar{\sigma}_{\text{caustic}} = \frac{\pi}{\bar{\varphi}} \ . \tag{A.2}$$

At this distance, the average number of travel times increases by the factor

$$\bar{N}_{\text{caustic}} = \exp\left(\frac{\bar{\lambda}}{\bar{\varphi}} \pi\right) \ . \tag{A.3}$$

We can conjecture from equations (26), (51) and (A.1) that

$$0 \leq \bar{\lambda} < \bar{\varphi} \ , \tag{A.4}$$

depending on the kind of heterogeneities in the velocity model. Consequently, the average increment $\bar{N}_{\text{caustic}}$ of travel times between two consecutive caustic points depends on the kind of heterogeneities in the velocity model, varying within interval

$$1 \leq \bar{N}_{\text{caustic}} < \exp \pi = 23.141 \ . \tag{A.5}$$

Acknowledgements

This research has been supported by the Grant Agency of the Czech Republic under Contracts 205/01/0927 and 205/01/D097, by the Grant Agency of the Charles University under Contract 237/2001/B-GEO/MFF, by the Ministry of Education of the Czech Republic within Research Project J13/98 113200004, and by the members of the consortium "Seismic Waves in Complex 3-D Structures" (see "http:// sw3d.mff.cuni.cz").

REFERENCES

BULANT, P. (1997), *Calculation of multivalued ray-theory travel times at nodes of 3-D grids*. In *Seismic Waves in Complex 3-D Structures, Report 6* (Dep. Geophys. Charles Univ., Prague) pp. 71–74 (online at "http://sw3d.mff.cuni.cz").

BULANT, P. (1999a), *Two-point Ray Tracing and Controlled Initial-Value Ray Tracing in 3-D Heterogeneous Block Structures*, J. seism. Explor. *8*, 57–75.

BULANT, P. (1999b), *Calculation of multivalued ray-theory travel times in 3D — Interpolation within individual ray cells*. In *Extended Abstracts of 61th EAGE Conference (Helsinki)* (Eur. Assoc. Geoscientists and Engr., Houten) P081.

BULANT, P. (1999c), *Prismatic ray cells versus tetrahedra — numerical tests of interpolation methods*. In *Seismic Waves in Complex 3-D Structures, Report 8* (Dep. Geophys. Charles Univ., Prague) pp. 69–70 (online at "http://sw3d.mff.cuni.cz").

BULANT, P. (2000), *Is the bicubic travel-time interpolation time consuming?* In *Seismic Waves in Complex 3-D Structures, Report 10* (Dep. Geophys. Charles Univ., Prague) pp. 113–114 (online at "http:// sw3d.mff.cuni.cz").

BULANT, P. (2002), *Sobolev Scalar Products in the Construction of Velocity Models — Application to Model Hess and to SEG/EAGE Salt Model*, Pure appl. geophys. *159*, 1487–1506.

BULANT, P. and KLIMEŠ, L. (1999), *Interpolation of Ray Theory Travel Times Within Ray Cells*, Geophys. J. int. *139*, 273–282.

ČERVENÝ, V. (1972), *Seismic Rays and Ray Intensities in Inhomogeneous Anisotropic Media*, Geophys. J. R. astr. Soc. *29*, 1–13.

ČERVENÝ, V. (2001), *Seismic Ray Theory* (Cambridge Univ. Press, Cambridge).

KATOK, S. R. (1980), *The estimation from above for the topological entropy of a diffeomorphism*. In *Global Theory of Dynamical Systems* (eds. Netecki, Z. and Robinson, C.) (Lecture Notes in Mathematics, Vol. 819, Springer, Berlin, Heidelberg, New York) pp. 258–264.

KLIMEŠ, L. (1994), *Transformations for Dynamic Ray Tracing in Anisotropic Media*, Wave Motion *20*, 261–272.

KLIMEŠ, L. (2000), *Sobolev scalar products in the construction of velocity models*. In *Seismic Waves in Complex 3-D Structures, Report 10* (Dep. Geophys. Charles Univ., Prague) pp. 15–40 (online at "http:// sw3d.mff.cuni.cz").

LYAPUNOV, A. M. (1949), *Problème Général de la Stabilité du Mouvement* (Annals of Mathematical Studies, Vol. 17, Princeton Univ. Press).

MATYSKA, C. (1999), *Note on chaos in ray propagation*. In *Seismic Waves in Complex 3-D Structures, Report 8* (Dep. Geophys. Charles Univ., Prague) pp. 71–82 (online at "http://sw3d.mff.cuni.cz").

OSELEDEC, V. I. (1968), *A Multiplicative Ergodic Theorem: Lyapunov Characteristic Numbers for Dynamical Systems*, Trans. Moscow Math. Soc. *19*, 179–210 in Russian, 197–231 in English translation.

WHITE, B. S., NAIR, B. and BAYLISS, A. (1988), *Random Rays and Seismic Amplitude Anomalies*, Geophysics *53*, 903–907.

ŽÁČEK, K. (2002), *Smoothing the Marmousi Model*, Pure appl. geophys. *159*, 1507–1526.

(Received October 10, 2000, revised March 21, 2001, accepted April 5, 2001)

 To access this journal online:
http://www.birkhauser.ch

Pure appl. geophys. 159 (2002) 1487–1506
0033–4553/02/081487–20 $ 1.50 + 0.20/0

Pure and Applied Geophysics

Sobolev Scalar Products in the Construction of Velocity Models: Application to Model Hess and to SEG/EAGE Salt Model

PETR BULANT[1]

Summary—The minimization of the Sobolev norm during linearized inversion of given data allows control of the model parameters unresolved by the data being fitted. Even if a reasonably looking model can be obtained without minimizing the Sobolev norm, it may be too rough for some computational methods. We may construct models optimally smooth for given computational methods by minimizing the corresponding Sobolev norm during the inversion. Probably the smoothest models are required by the ray methods. The efficiency of ray tracing can be evaluated in terms of the "average Lyapunov exponent" for the model. The "average Lyapunov exponent" may be approximated by the square root of the corresponding Sobolev norm of the model, which allows models most suited for ray tracing to be constructed.

Key words: Model specification, smoothing, inversion, ray methods, Sobolev norm, Lyapunov exponent.

1. Introduction

The construction of velocity models is a complex task which covers the application of many different model building tools, including smoothing. We would like to show that one of the useful tools in the velocity model-building process is the application of Sobolev scalar products.

In this paper we shall describe the technique to take the given data, form an objective function, and build a continuous B-spline parameterized velocity (or slowness) model using the method of minimizing the selected objective function. As this is an inverse problem, we shall speak of the "inversion of the given data", or shortly "inversion". The data may be either measured travel times, values of material parameters and coordinates of points at interfaces obtained from wells, or gridded material parameters and triangulated interfaces obtained from any model-building process. In principle, we treat the construction of continuous velocity models from existing gridded velocities in the same way as the linearized travel-time inversion. We would like to show that including the minimization of the corresponding Sobolev

[1] Department of Geophysics, Charles University, Ke Karlovu 3, 121 16 Praha 2, Czech Republic.
E-mail: bulant@seis.karlov.mff.cuni.cz

norm into the inversion of the given data allows us to construct models optimal for the selected computational method.

The Sobolev scalar product of two functions is a linear combination of the L_2 Lebesgue scalar products of the zero, first, second or higher derivatives of the functions. The linear combination should preserve the properties required of scalar products (symmetry and positiveness). The Sobolev norm of two functions is the square root of the Sobolev scalar product of the functions.

The minimization of the Sobolev norm during linearized inversion of given data enables control of the model parameters unresolved by the data being fitted. For example, if we are fitting discrete values of a material parameter with a smooth function, the properties of interpolation between the discrete values may be controlled by means of minimizing the square of the selected Sobolev norm of the function. Similarly, if we are fitting discrete values of a structural interface, the behavior of the interpolated function between the discrete points, and also outside the region covered by the data for the interface, may be controlled by the given Sobolev norm of the function.

Even if a reasonably looking model can be obtained without minimizing the Sobolev norm, it may be too rough for some computational methods. The errors of many computational methods may be approximated by a function dependent on the Sobolev norm of functions describing the model. We may thus construct models optimally smooth for given computational methods.

Probably the smoothest models are required by the ray methods. In complex models, the number of arrivals increases rapidly with travel time. The initial-value ray tracing in such models is still possible, however, the solution of boundary-value problems, e.g., two-point ray tracing, is slow and expensive due to the large number of arrivals at each receiver. We thus consider the number of arrivals to be one of the criteria of applicability of the ray methods. In complex models, the behavior of rays becomes chaotic and the geometrical spreading, number of arrivals and density of caustic surfaces increase exponentially with travel time. The exponential increment may be described quantitatively in terms of the "average Lyapunov exponent" for the model. The "average Lyapunov exponent" may be approximated by the square root of the corresponding Sobolev norm of the model. This enables quantitative criteria on models, so as to be suitable for ray tracing, to be determined in terms of Sobolev scalar products. The criteria then allow the optimum models of geological structures for ray tracing to be constructed either by smoothing existing rough models, or by inversion of seismic data.

The construction of a model for ray methods is a typical example of using the Sobolev norm for smoothing. The task of smoothing a given rough model may be solved either as a direct problem (GRUBB and WALDEN, 1995; VERSTEEG, 1991) by, e.g., convolution or wavenumber filtering, or as an inverse problem (PRETLOVÁ, 1976, 1985) by fitting the values in the rough model with the new smooth and sparsely parameterized model. Smoothing the given model by inversion is advantageous not only because we treat the problem of model smoothing in the same way as inversion of measured data, but also because we may include the minimization of the suitable Sobolev norm into the inversion, and thus ensure the feasibility of ray tracing in the smooth model.

2. Sobolev Scalar Products and Sobolev Norms

We define the Sobolev scalar product (see, e.g., TARANTOLA, 1987) of functions f and g as a linear combination of the L_2 Lebesgue scalar products of the zero, first, second or higher derivatives of the functions:

$$((f,g)) = \left[\int dx\right]^{-1} \sum_{ij...} \sum_{ab...} \int dx \, b_{ij...|ab...} \, f^*_{,ij...}(\mathbf{x}) \, g_{,ab...}(\mathbf{x}) \, , \tag{1}$$

where $\int dx$ is the definite integral over the model volume, and $b_{ij...|ab...}$ are real-valued coefficients of the linear combination. Symbol $f_{,i}$ denotes the partial derivative of function f with respect to the i-th coordinate axis. The order of the considered partial derivative $f_{,ij...}$ equals the number of indices $ij...$. The indices take the values from 1 to 3 in 3-D, 1 to 2 in 2-D and 1 in 1-D. The linear combination should preserve the properties required of scalar products (symmetry and positiveness). The asterisk denoting complex-conjugacy is included just for completeness here; we consider real-valued functions and real-valued coefficients $b_{ij...|ab...}$ in this paper.

The Sobolev norm of functions f and g is defined as the square root of the Sobolev scalar product of the functions:

$$\|f,g\| = \sqrt{((f,g))} \, . \tag{2}$$

Note that if the linear combination defining the Sobolev scalar product does not contain the Lebesgue scalar product of the zero derivatives of the functions, the Sobolev scalar product of a constant function with any function is zero. In this case, the Sobolev scalar product is not positive-definite, but only positive-semidefinite, and the Sobolev norm is not a norm, but only a seminorm. However, we may factorize the considered linear space of functions with respect to the space of functions keeping the Sobolev scalar product identically zero. In factor space, the Sobolev scalar product becomes positive-definite and the corresponding Sobolev seminorm becomes a norm. We will thus always speak of the "Sobolev norm", even in the situations in which "Sobolev seminorm" would be more appropriate.

3. Sobolev Norm in the Objective Function for Inversion of Seismic Data

There are two main reasons for including the Sobolev norm in the inversion. Firstly, the Sobolev norm regularizes ill-conditioned inversions, which enables us to control the behavior of the model functions in regions not illuminated by the data. The Sobolev norm is thus suitable for fitting discrete values with a velocity model. A typical example of such application is the fitting of structural interfaces for which the data are available only in some parts of the model. Secondly, the errors of many computational methods may be approximated by a function dependent on the corresponding Sobolev norm, which enables us to construct seismic models optimally smooth for the particular methods. For example, if we construct a model for full-wave finite differences, we may minimize the Sobolev norm approximating the propagation-velocity error due to the grid dispersion.

The Sobolev norm may, in general, be composed of the zero, first, second or higher derivatives of the functions describing the model. The model is very often described by cubic, bicubic or tricubic splines. Since the third homogeneous partial derivatives of the splines are discontinuous, the Sobolev scalar product composed of the fourth and higher derivatives could be infinite, and thus should be avoided. Simultaneously, minimization of the (zero derivatives of the) model functions usually has no meaning. We thus usually consider only Sobolev norms composed of the first, second and third derivatives. Since studies of the third derivatives have not yet been finished, we will concentrate on the first and second derivatives in this paper.

3.1. Fitting Discrete Values and Regularizing an Ill-conditioned Inversion

Assume that we are fitting discrete values u^α at points $\mathbf{x}^\alpha$ inside the model volume by minimizing objective function

$$y = \sum_\alpha [\tilde{u}(\mathbf{x}^\alpha) - u^\alpha]^2 [\Delta u^\alpha]^{-2} + S^2((\tilde{u}, \tilde{u})) \ , \tag{3}$$

where $\tilde{u}(\mathbf{x})$ is the model being constructed, Δu^α is the given standard deviation of value u^α given at point $\mathbf{x} = \mathbf{x}^\alpha$, and S is a manually chosen free parameter of the objective function which acts as a weighting factor of the Sobolev norm $((\tilde{u}, \tilde{u}))$ of the model. For infinitely small weighting factor $S \to 0$, term $\sum_\alpha [\tilde{u}(\mathbf{x}^\alpha) - u^\alpha]^2 [\Delta u^\alpha]^{-2}$ in (3) is dominant, and the limit form

$$\tilde{\tilde{u}}(\mathbf{x}) = \lim_{S \to 0} \tilde{u}(\mathbf{x}) \tag{4}$$

of the solution corresponds (KLIMEŠ, 2000a) to the minimization of objective function

$$y = ((\tilde{\tilde{u}}, \tilde{\tilde{u}})) \tag{5}$$

with boundary conditions

$$\tilde{\tilde{u}}(\mathbf{x}^\alpha) = u^\alpha \ . \tag{6}$$

For larger values of S, the influence of the minimization of the Sobolev norm increases, and the constructed model $\tilde{u}(\mathbf{x})$ fits the given values u^α only approximately.

The properties of interpolation between discrete values by means of minimizing the squares of different Sobolev norms has been investigated by KLIMEŠ (2000a). He showed that the minimization of the first derivatives yields piecewise linear interpolation in 1-D. In 2-D and 3-D the interpolation tends to produce a constant function with singular spikes at given points $\mathbf{x}^{\alpha}$. The minimization of the second derivatives yields cubic splines in 1-D, in 2-D smooth interpolation with continuous first derivatives, and in 3-D smooth interpolation with continuous first derivatives between the given points, but with conical peaks at the given points. We can thus use the Sobolev norm composed of second derivatives for fitting discrete values.

Assume that we are fitting a structural interface with a smooth function. In such a case the data for the function are available only in that part of the model where the interface is located. The behavior of the function in other parts of the model is not given by the data, see, e.g., Figure 5. We might control the function by adding new artificial points to the data for the interface, however this manual intervention would probably slightly distort the modelled interface. It is considerably more convenient to control the function by increasing the weighting factor S of the Sobolev norm. A small increment of S will not influence the function in the region with the data, but will regularize its behavior in the remaining parts of the model.

Another typical example of strongly ill-conditioned inversion is fitting a structural interface with an implicit function. Assume that the discrete points at the interface are given and that the interface cannot be fitted with an explicit function, e.g., such as $x_3 = w(x_1, x_2)$, but has to be described as the zero isosurface of function $f(x_1, x_2, x_3)$. We then minimize the deviations of function $f(x_1, x_2, x_3)$ from zero at the given points. Without additional *a priori* requirements, we would thus obtain an entirely zero function. One of the agreeable solutions is to fix the functional value at a specified point and to control the function in other than the given and the specified points by minimizing a suitable Sobolev norm. The specified point should be sufficiently distant from the interface. If the surface is closed, it may be located inside the block limited by the surface.

3.2. Constructing Seismic Models of Optimal Smoothness for a Selected Numerical Method

The Sobolev norm is usually included in the inversion of seismic data if the numerical method to be applied to the model requires a smooth model. The Sobolev norm included in the objective function acts as a low-pass wavenumber filter (KLIMEŠ, 2000a).

We assume the objective function for the inversion of slowness or velocity $u(\mathbf{x})$ in the form of

$$y = \left[\int d\mathbf{x}\right]^{-1} \int d\mathbf{x}\tilde{C}(\mathbf{x}, \mathbf{x}) + S^2 \|\tilde{u}\|^2 \, , \qquad (7)$$

where

$$\tilde{C}(\mathbf{x}, \mathbf{x}') = \langle [\tilde{u}(\mathbf{x}) - u(\mathbf{x})][\tilde{u}(\mathbf{x}') - u(\mathbf{x}')] \rangle \tag{8}$$

is the covariance function describing the deviation of model $\tilde{u}(\mathbf{x})$ from geological structure $u(\mathbf{x})$ (KLIMEŠ, 2002a), and S is the weighting factor of the Sobolev norm $\|\tilde{u}\|$ of the functions describing the model.

The first summand in (7) is the mean squared difference between the model and the geological structure. If possible, we choose the Sobolev norm $\|\tilde{u}\|$ so that $S\|\tilde{u}\|$ approximate the r.m.s. difference of the results of the numerical method under consideration from the exact solution in the model, expressed in units of $u(\mathbf{x})$. Objective function (7) then gains the clear physical meaning of the squared deviation of the model from the geological structure plus the squared deviation of the exact from the numerical results in the model. Note that by "exact solution" we mean the results which we would obtain had we applied a hypothetical exact numerical method in the model.

For example, if the inaccuracy of the numerical method under consideration can be approximately characterized by the local relative travel-time standard deviation $\rho\tau_{\text{method}}(\mathbf{x})$ describing the travel-time difference between the results of the numerical method under consideration from the exact results in the model, and, furthermore, if the square $[\rho\tau_{\text{method}}(\mathbf{x})\tilde{u}(\mathbf{x})]^2$ of the deviation bilinearly depends on the derivatives of the interpolated parameter (slowness or velocity),

$$[\rho\tau_{\text{method}}(\mathbf{x})\tilde{u}(\mathbf{x})]^2 \approx S^2 b_{ij\ldots|ab\ldots}\tilde{u}_{,ij\ldots}(\mathbf{x})\tilde{u}_{,ab\ldots}(\mathbf{x}) \ , \tag{9}$$

the Sobolev scalar product to be used in objective function (7) is defined by coefficients $b_{ij\ldots|ab\ldots}$ obtained from approximation (9):

$$\|\tilde{u}\|^2 = \left[\int d\mathbf{x}\right]^{-1} \sum_{ij\ldots} \sum_{ab\ldots} \int d\mathbf{x} \, b_{ij\ldots|ab\ldots}\tilde{u}_{,ij\ldots}(\mathbf{x})\tilde{u}_{,ab\ldots}(\mathbf{x}) \ . \tag{10}$$

The selection of the Sobolev scalar products for particular numerical methods has been demonstrated by KLIMEŠ (2000a) on examples of centred finite differences of the second order, network shortest-path ray tracing (KLIMEŠ and KVASNIČKA, 1994), 2-D first-order travel-time tracing (PODVIN and LECOMTE, 1991), and 2-D second-order grid travel-time tracing (KLIMEŠ, 1996).

3.3. Constructing Seismic Models for Ray Tracing and Kirchhoff Migrations

The feasibility and efficiency of ray tracing imposes relatively strong constraints on the smoothness of the velocity or slowness models. Unfortunately, we still have no quantitative criteria of applicability and accuracy of ray methods or their extensions. The situation is thus more complicated than in the case of the methods mentioned in Section 3.2. On the other hand, the numerical efficiency of ray tracing can be evaluated in terms of the "average Lyapunov exponent", introduced by KLIMEŠ (2000c, 2002b) for 2-D models without structural interfaces.

In complex models, the behavior of rays becomes chaotic and the geometrical spreading, number of arrivals and density of caustic surfaces increase exponentially with travel time. The exponential increment may be quantitatively described in terms of the "average Lyapunov exponent" for the model. The "average Lyapunov exponent" may roughly be approximated by the square root of the Sobolev norm composed of the second partial derivatives of the functions describing the slowness or velocity field in the model (KLIMEŠ, 2000a). This enables us to determine the quantitative criteria on models, so as to be suitable for ray tracing, in terms of Sobolev scalar products. We choose the maximum "average Lyapunov exponent" for the model, we estimate the maximum Sobolev norm of the model, and we estimate the value of weighting factor S. We then perform the inversion, compute the values of the "average Lyapunov exponent" and the Sobolev norm in the inverted model, re-estimate the value of S, and continue iteratively with the inversion until we obtain the model of the required smoothness. We may thus construct models in which ray tracing will be possible and numerically efficient. We remind the reader that we know nothing about the applicability and accuracy of ray tracing in the models.

For example, to construct a model suitable for ray-based migrations, KLIMEŠ (2000b) chose the value of the "average Lyapunov exponent" $\bar{\lambda}$ for the model as

$$\bar{\lambda} \leq \frac{2}{T_{\max}} , \tag{11}$$

where $T_{\max} \geq T_S + T_R$ is the maximum sum of travel times from the source and from the receiver. The product of the average numbers of arrivals from a source and a receiver then probably does not exceed 1.8, and the product of maximum numbers of arrivals from a source and a receiver then probably does not exceed 7.

If we choose coefficients

$$b_{ij|ab} = \tfrac{1}{3}\left(\delta_{ij}\delta_{ab} + \delta_{ia}\delta_{jb} + \delta_{ib}\delta_{ja}\right) \tag{12}$$

of the Sobolev norm according to KLIMEŠ (2000b), the Sobolev norm $\|u\|$ of the model satisfying equation (11) should obey equation (KLIMEŠ, 2000b)

$$\|u\| \leq \sqrt{\frac{8}{3}}\left[\frac{4 + 2K_{\mathrm{osc}}\ln 2}{T_{\max}}\right]^2 \frac{1}{\langle v\rangle^3} , \tag{13}$$

where $\langle v\rangle$ is the average velocity in the model, and K_{osc} is the average number of velocity oscillations (i.e., the average number of focusing and defocusing velocity zones) along rays of length corresponding to $T_{\max}$. δ_{ij} in equation (12) is the Kronecker delta.

4. Numerical Examples

Routines for application of the above described minimization of the Sobolev norm during the seismic inversion have been coded and incorporated into existing Fortran 77 package MODEL, which serves to specify heterogeneous seismic models composed of smooth blocks with arbitrary varying seismic parameters. For the description of the basic ideas of the model specification refer to ČERVENÝ et al. (1988), the latest version of the package may be found at http://sw3d.mff.cuni.cz. Package CRT, available at the same place, is used for ray tracing in the models. All the computations are carried out on a PC Pentium with 64 Mbytes of memory.

The memory requirements of the current coding of the method are $M(D + M)$ storage locations, where M is the number of model parameters, and D is the number of data points. The requirements create no problems for 2-D model Hess, but they disallow the use of all the data and achieving the minimum error in inversion of the 3-D SEG/EAGE Salt Model.

Smoothing 2-D model Hess is an example of smoothing a given rough model by generating the values in the rough model and fitting them with the smooth model. In the case of the SEG/EAGE 3-D Salt Model, only gridded velocities and triangulated interfaces are given. From the point of view of ray tracing, the gridded velocities and the points at the interfaces are not a model, because for ray tracing the values of material parameters must be known at each point of the model volume. Building the SEG/EAGE 3-D Salt Model is thus an example of constructing a 3-D model by inversion of given data.

For another example of using the Sobolev norms to construct velocity models, refer to ŽÁČEK (2002).

4.1. Smoothing 2-D Model Hess for Kirchhoff Migrations

4.1.1. Data and model parameterization

2-D model Hess is specified in the form of a file with input data for the MODEL package. Note that the dimensions of the model are given in feet, and feet are used during the calculation, however we use metres throughout this paper. The dimensions of the model are 12252.96 times 6400.8 m. Two interfaces in the model, limiting a salt body, are parameterized by B-splines with 64 spline points spaced at 60.96 m. Background velocities are parameterized by B-splines with 202 × 106 spline points with a spacing of 60.96 m × 60.96 m; the velocity inside the salt body is constant (see Figure 1).

We perform several tests of ray tracing in the original model. The behavior of the ray field seems to be too complicated with too many caustics, especially in the low velocity channel near the surface of the model (see Figures 2 to 4). We thus decide to smooth the model by means of inversion of the values at the spline points with minimization of the Sobolev norms of the model.

In Figure 1, we can see that the horizontal variation of the velocity field is less pronounced than the vertical variation, and that the vertical variation is greater in the shallow parts of the model than in its deeper parts. The interfaces are quite smooth, but the edges are sharp. According to these observations, we manually choose a new parameterization of the model. Let us emphasize that the parameterization is based on the author's opinion, and is expected to be optimized during the inversion. The parameterization is as follows: for the interfaces we take an irregular B-spline grid of 28 points with 26 points regularly spaced at 152.4 m in the region of the salt body and 2 points at the sides of the model. The two points enable us to control the behavior of the functions describing the interfaces in the parts of the model not covered by the data for the interfaces. For the background velocity we take a horizontally regular grid of 17 points spaced at 762 m, and a vertically irregular grid of 31 points spaced at 60.96 m at the top to 243.84 m at the bottom.

4.1.2. Choosing the maximum Lyapunov exponent

To estimate the maximum sum of travel times from a source and a receiver, we compute the average slowness of the model $\langle u \rangle = 0.395705 \ 10^{-3}$ s m^{-1}. If we consider the maximum sum of the ray paths to be about 15,240 m, we arrive at the estimation that the maximum sum of travel times from a source and a receiver is limited by $T_{max} = 6$ s. According to equation (11), we choose the maximum "average Lyapunov exponent" $\bar{\lambda}_{max}$ of the new model as

$$\bar{\lambda}_{max} = \frac{2}{T_{max}} = 0.333 \ \text{s}^{-1} \ . \tag{14}$$

4.1.3. Choosing the maximum Sobolev norm of the model

Analogously to equation (13) we choose the maximum Sobolev norm $\|v\|_{max}$ of the velocity field in the new model as

$$\|v\|_{max} = \sqrt{\frac{8}{3}} \left[\frac{4 + 2K_{osc} \ln 2}{T_{max}} \right]^2 \frac{1}{\langle v \rangle} = 8.53 \ 10^{-4} (\text{m s})^{-1} \ , \tag{15}$$

where $\langle v \rangle = 2461.18$ m s^{-1} is the average velocity of the background, and the average number K_{osc} of focusing and defocusing zones is roughly estimated to be 3 for the rays which pass the salt, and to be 1 for the other rays; we thus calculate with $K_{osc} = 2$.

In this example, we first try to apply the minimum reasonable smoothing of the interfaces. We thus do not prescribe the maximum Sobolev norm of the interfaces.

Figure 1   Figure 2

Figure 1 *P*-wave velocity in the original model (before smoothing). The background velocity varies from $1524\,\mathrm{m\,s^{-1}}$ at the top of the model to $2987.04\,\mathrm{m\,s^{-1}}$ at the bottom. The velocity of the salt body is constant and equals $4511.04\,\mathrm{m\,s^{-1}}$. The horizontal dimension of the model is 12252.96 m; its vertical dimension is 6400.8m.

Figure 2 The numbers of travel times and the rays computed for the refracted wave in the original model. The point source is located in the top left corner of the model. The yellow color corresponds to no travel time, green to 2 arrivals, cyan to 4, blue to 6, magenta to 8, red to 10 arrivals. The extents of the figure are the same as of Figure 1.

Figure 3   Figure 4

Figure 3 The numbers of travel times and the rays computed for the source located in the middle of the top of the model. The color scale is the same as in Figure 2. The extents of the figure are the same as of Figure 1.

Figure 4 The numbers of travel times and the rays computed for the source located in the top right corner of the model. The color scale is the same as in Figure 2. The extents of the figure are the same as of Figure 1.

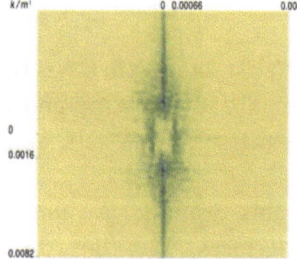

Figure 5   Figure 6

4.1.4. Inversion and smoothing

The objective function is taken in the form

$$y = \left(\frac{|x - x_0|}{\sigma_x}\right)^2 + (S_x\|x\|)^2 + \left(\frac{|v - v_0|}{\sigma_v}\right)^2 + (S_v\|v\|)^2 , \qquad (16)$$

where $|x - x_0|$ is the standard deviation of the position of the interfaces in the smooth model from the gridded interfaces in the original model, $\|x\|$ is the 2-D Sobolev norm of the curvature of the interfaces in the smoothed model, $|v - v_0|$ is the standard velocity deviation of the model from the original gridded velocities, and $\|v\|$ is the 2-D Sobolev norm of the velocity field in the smoothed model. S_x and S_v are numerical parameters which control the smoothing during the inversion. We use unit given velocity deviation $\sigma_v = 1\,\text{ft s}^{-1} = 0.3048\,\text{m s}^{-1}$ and unit given interface deviation $\sigma_x = 1\,\text{ft} = 0.3048\,\text{m}$ in the selected objective function.

The inversion of the interfaces and the inversion of the velocity field are mutually independent in this example. We can thus perform the inversion of the interfaces in the first step, and the inversion of the velocity in the second step.

4.1.4.1. Inversion of the interfaces

We choose $S_v = 0\,\text{m s}$, and perform the inversion without smoothing of the velocity, which we are not interested in at the moment.

In this example we try to apply the minimum reasonable smoothing of the interfaces. We thus start the inversion without smoothing of the interfaces, with $S_x = 0\,\text{m}$ (the interfaces are smoothed just by the projection onto the B-splines). We obtain oscillations of both surfaces, which produce a salt body other than that given by the data (see Figure 5). In the next steps of inversion we thus include the minimization of the curvature of the interfaces by increasing the value of S_x. When we reach approximately the value of $S_x = 762\,\text{m}$, the spurious salt body moves outside the model volume. We thus consider the value of $S_x = 762\,\text{m}$ as the value corresponding to the minimum reasonable smoothing of the interfaces. Whether the interfaces are sufficiently smooth will be tested by ray tracing.

We compute standard interfaces deviation $|x - x_0| = 1.707\,\text{m}$. This deviation causes a travel-time error of approximately

◄

Figure 5 Inversion of the interfaces without the Sobolev norm. The oscillations of the surfaces limiting the lenticular salt body produced another spurious salt body on the right-hand side of the model. The color scale and the extents of the figure are the same as of Figure 1.

Figure 6 Fourier spectrum of the velocity deviation of the smoothed model obtained with $S_v = 15240\,\text{m s}$. The extents of the figure correspond to wavenumbers k from $-0.0082\,(\text{m})^{-1}$ to $0.0082\,(\text{m})^{-1}$. The light rectangle in the middle of the figure shows the wavenumbers k of the velocity field which the model fits. We can see that the model fits the velocity variations of wavelength greater than approximately 762 m horizontally and 304.8 m vertically.

$$dT = |x - x_0| \left| \frac{1}{\langle v \rangle} - \frac{1}{v_{\text{salt}}} \right| = 3.15 \ 10^{-4} \, \text{s} \ , \tag{17}$$

where $\langle v \rangle = 2461.18 \, \text{m s}^{-1}$ is the average velocity of the background and $v_{\text{salt}} = 4511.04 \, \text{m s}^{-1}$ is the velocity of the salt.

4.1.4.2. Inversion of the velocity field

We have selected the value of $S_x = 762 \, \text{m}$, which ensures the proper position of the interfaces. We will retain this value during the velocity inversion.

(a) First iteration

We first perform the inversion with $S_v = 0 \, \text{m s}$ (i.e., without smoothing), and obtain the standard velocity deviation of the model

$$|v - v_0| = 4.249 \, \text{m s}^{-1} \ . \tag{18}$$

The Sobolev norm of the velocity field smoothed just by the projection onto the B-splines

$$\|v\| = 121.4 \ 10^{-4} \, (\text{m s})^{-1} \ , \tag{19}$$

is large, see equation (15).

(b) Second iteration

We choose

$$S_v = \frac{|v - v_0|}{\sigma_v \|v\|_{\text{max}}} \ , \tag{20}$$

as in the paper by KLIMEŠ (2000b, eq. 20). Then

$$S_v = 16346.5 \, \text{m s} \ , \tag{21}$$

and we enter the rounded value of

$$S_v = 50000 \, \text{ft s} = 15240 \, \text{m s} \tag{22}$$

for the second iteration. The resulting Sobolev norm

$$\|v\| = 9.51 \ 10^{-4} \, (\text{m s})^{-1} \tag{23}$$

of the velocity looks acceptable, see equation (15).

(c) Subsequent iterations

The presence of the structural interfaces and the existence of the low velocity channel at the top of the model have not been taken into account during the estimation of S_v. We thus choose several values of S_v for subsequent iterations in order to see the behavior of the model in dependence on S_v. After each inversion we perform several tests. For each model we compute the standard velocity deviation from the velocity in the original model, the value of "average Lyapunov exponent" $\bar{\lambda}$

Table 1

Results of the inversion of the velocity field in model Hess in dependence on weighting factor S_v of the Sobolev norm used

	$|v-v_0|/(\text{m s}^{-1})$	$||v||/((\text{m s})^{-1})$	$\lambda/(\text{s}^{-1})$	N_{max}
Desired values		$8.53\ 10^{-4}$	0.333	
Original model			0.423	11
$S_v = 0\text{ m s}$	4.249	$121.4\ 10^{-4}$	0.380	7
$S_v = 5334\text{ m s}$	4.630	$16.08\ 10^{-4}$	0.345	6
$S_v = 10668\text{ m s}$	5.307	$11.48\ 10^{-4}$	0.316	5
$S_v = 15240\text{ m s}$	5.864	$9.51\ 10^{-4}$	0.300	5
$S_v = 21336\text{ m s}$	6.529	$7.87\ 10^{-4}$	0.287	5

and the maximum number of arrivals N_{max} from the source located in the left top corner of the model. The results of the computations are in Table 1.

We can see that for $S_v = 0\text{ m s}$ we obtain a standard velocity deviation of about 4.25 m s^{-1}. The desired value of $\bar{\lambda} = 0.333\text{ s}^{-1}$ corresponds to a velocity deviation of about 5 m s^{-1}. Most of the velocity deviation is thus caused by the projection onto the B-splines, and we should try the inversion with a finer parameterization of the model.

A figure of the Fourier spectrum of the velocity deviation should help us to decide in which direction and how many times we should densify the parameterization of the model. We must keep in mind that the parameterization must not be denser than the data.

4.1.5. Finer B-spline parameterization of the model

We choose the new parameterization of the model as follows: for the interfaces we retain the irregular B-spline grid of 28 points; for the background velocity we take a horizontally regular grid of 26 points spaced at 487.68 m and a vertically irregular grid of 58 points spaced at 60.96 to 121.92 m.

We now have $56 + 1 + 26 \times 58 = 1565$ B-spline points, and the memory requirements of the inversion are 36180865 storage locations. This is too much for our computer and we thus cannot use all the data. We resample the data in the horizontal direction and use a data grid of 67 points spaced at 182.88 m horizontally and the original 106 points spaced at 60.96 m vertically.

Note that, with the new parameterization, we are not able to perform the inversion with $S_v = 0\text{ m s}$, because we now have more background velocity B-spline gridpoints in the salt body, and the inversion becomes ill-conditioned. We performed several tests of S_v with the new parameterization of the model (see Table 2).

We see that the standard velocity deviation caused by the projection onto the B-splines is now much smaller than for the previous model parameterization, and that the deviation increases as we increase S_v in order to meet the requirements imposed on the model smoothness.

Table 2

Results of the inversion of the velocity field in model Hess with finer B-spline parameterization of the model

	$|v-v_0|/(\text{m s}^{-1})$	$||v||/((\text{m s})^{-1})$	$\lambda/(\text{s}^{-1})$	$N_{\max}$
Desired values		$8.53 \ 10^{-4}$	0.333	
Original model			0.423	11
$S_v = 304.8$ m s	1.68	$40.3 \ 10^{-4}$	0.398	6
$S_v = 3048$ m s	2.22	$21.3 \ 10^{-4}$	0.365	5
$S_v = 5334$ m s	2.75	$17.1 \ 10^{-4}$	0.343	5
$S_v = 7620$ m s	3.26	$14.4 \ 10^{-4}$	0.327	5
$S_v = 10668$ m s	3.89	$12.1 \ 10^{-4}$	0.313	5

4.1.6. Numerical tests of the smoothed model

The model with the finer parameterization obtained after the inversion with values of $S_x = 762$ m and $S_v = 7620$ m s has been selected as the best model.

The standard relative velocity deviation from the velocity in the original model is 0.18%. The standard deviation of the interfaces is 1.707 m.

For the gridded *P*-wave velocity in the model, the gridded velocity deviation, the Fourier spectrum of the gridded velocity deviation, and the tests of ray tracing of a refracted wave, see Figures 7 to 12.

4.2. Smoothing the SEG/EAGE 3-D Salt Model for Kirchhoff Migrations

The SEG/EAGE 3-D Salt Model (AMINZADEH *et al.*, 1997) is specified in a form suitable for the GOCAD modeling software. Several interfaces in the model are stored in the form of triangulated surfaces, and the 3-D velocity cube is given. From the point of view of ray tracing, the gridded velocities and the points at the interfaces are not a model, because the values of the material parameters must be known at each point of the model volume for ray tracing. This example is thus principally different from the example of smoothing of the Hess model. On the other hand, the process of building the Salt model is technically the same as the process of smoothing the Hess model.

We choose the maximum Lyapunov exponent of the model being constructed, estimate the maximum Sobolev norm of the model and the starting value of weighting factor S_v, select the starting parameterization of the model, and choose the corresponding data grid. We take the data and build a 3-D model suitable for ray tracing by inversion. We then optimize the parameterization of the model so as to

▶

Figure 11 The numbers of travel times and the rays computed in the smoothed model for the source located in the middle of the top of the model. The color scale is the same as in Figure 2.
Figure 12 The numbers of travel times and the rays computed in the smoothed model for the source located in the top right corner of the model. The color scale is the same as in Figure 2.

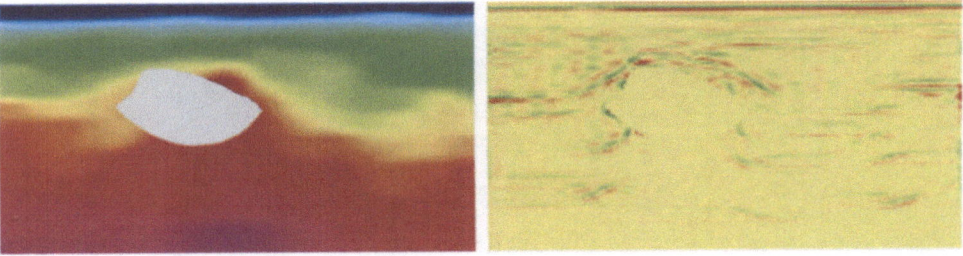

Figure 7   Figure 8

Figure 7 *P*-wave velocity in the smoothed model. The background velocity varies from $1524\,\mathrm{m\,s^{-1}}$ at the top of the model to $2987.04\,\mathrm{m\,s^{-1}}$ at the bottom. The velocity of the salt body is constant and equals $4511.04\,\mathrm{m\,s^{-1}}$. The horizontal dimension of the model is 12252.96 m; its vertical dimension is 6400.8 m.
Figure 8 Deviation of the gridded velocity in the smoothed model from the gridded velocity in the original model. The yellow color corresponds to zero deviation, red to negative and green to positive deviation. The maximum velocity deviation is $31.42\,\mathrm{m\,s^{-1}}$. The standard velocity deviation is $3.26\,\mathrm{m\,s^{-1}}$, the standard relative velocity deviation is 0.18%.

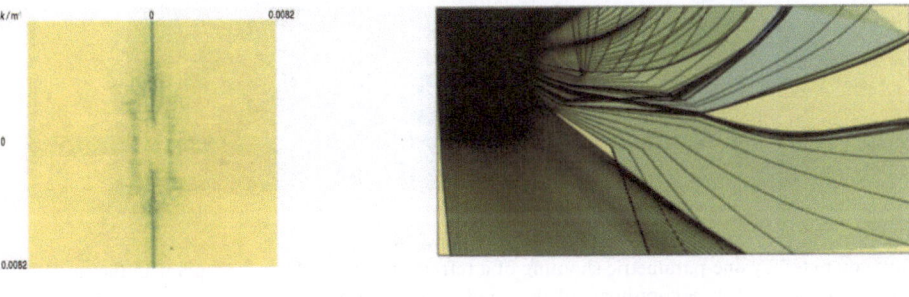

Figure 9   Figure 10

Figure 9 Fourier spectrum of the velocity deviation from Figure 8. The extents of the figure correspond to wavenumbers k from $-0.0082\,(\mathrm{m})^{-1}$ to $0.0082\,(\mathrm{m})^{-1}$.
Figure 10 The numbers of travel times and the rays computed for the refracted wave in the smoothed model. The source is located in the left top corner of the model. The color scale is the same as in Figure 2.

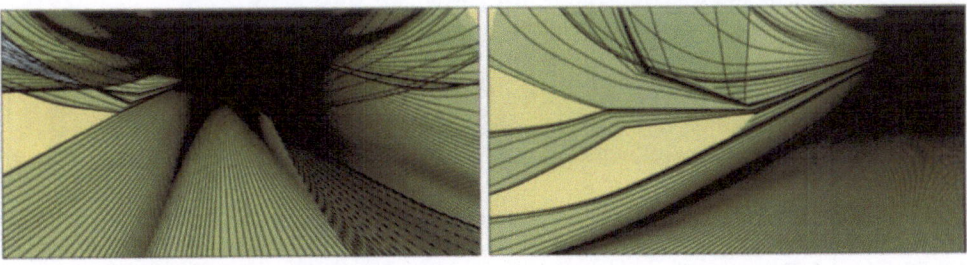

Figure 11   Figure 12

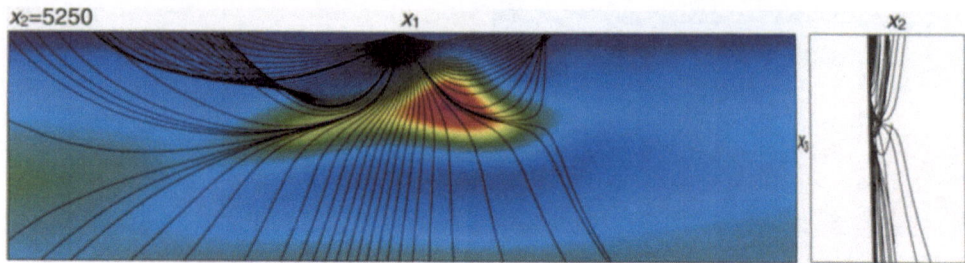

Figure 13

Rays computed by one-parametric shooting of a refracted wave in the $x_1 - x_3$ plane in the smooth version of the Salt Model. The rays are projected onto the $x_1 - x_3$ and onto the $x_2 - x_3$ planes. The coordinates of the point source are [7000, 5250, 0] m. The $x_1 - x_3$ plane is colored according to the P-wave velocity in the $x_1 - x_3$ section containing the point source. The color scale ranges from dark blue for $1500 \, \text{m s}^{-1}$ in the water to red for $4450 \, \text{m s}^{-1}$ in the salt body. The horizontal extents of both the $x_1 - x_3$ and $x_2 - x_3$ sections are 14000 m, the vertical extents of the sections are 4200 m. The reader is reminded that the standard relative deviation of the velocity in the model from the data is 8.22%.

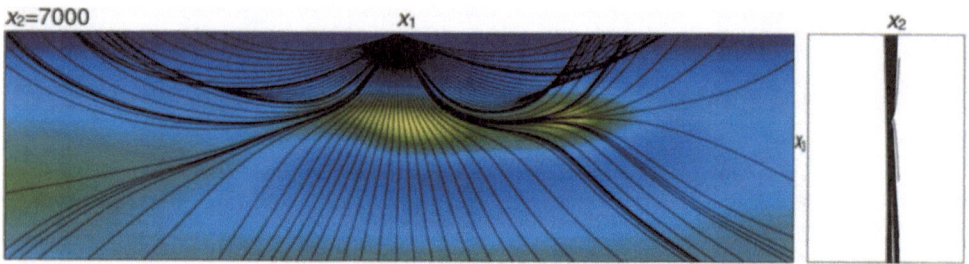

Figure 14

Rays computed by one-parametric shooting of a refracted wave in the $x_1 - x_3$ plane in the smooth version of the Salt Model. The x_2 coordinate of the point source and of the velocity section is now 7000 m. The color scale is the same as in Figure 13.

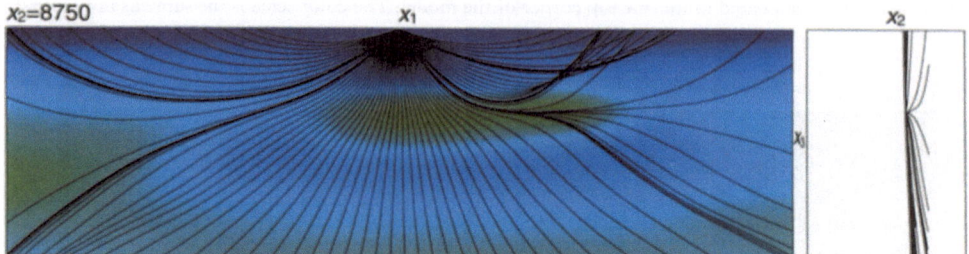

Figure 15

Rays computed by one-parametric shooting of a refracted wave in the $x_1 - x_3$ plane in the smooth version of the Salt Model. The x_2 coordinate of the point source and of the velocity section is 8750 m. The color scale is the same as in Figure 13.

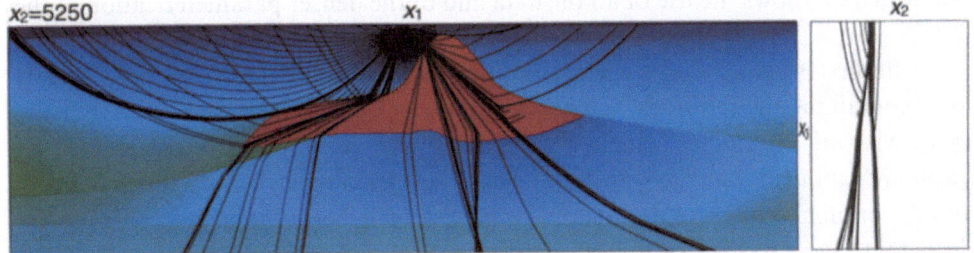

Figure 16

Rays computed by one-parametric shooting of a refracted wave in the $x_1 - x_3$ plane in the Salt Model, the version with interfaces. The x_2 coordinate of the point source and of the velocity section is 5250 m. The color scale is the same as in Figure 13. The reader is reminded that the standard relative deviation of the velocity in the model from the data is 5.24%.

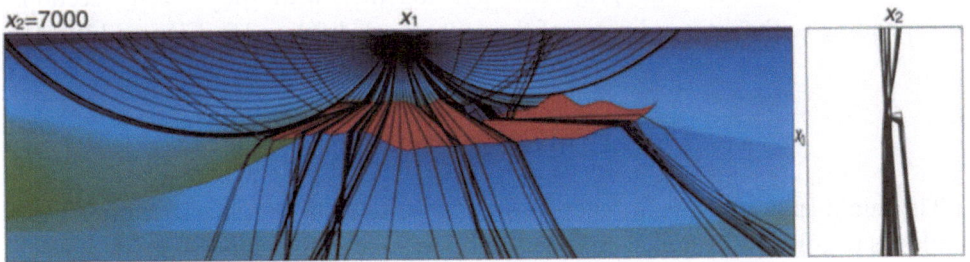

Figure 17

Rays computed by one-parametric shooting of a refracted wave in the $x_1 - x_3$ plane in the Salt Model, the version with interfaces. The x_2 coordinate of the point source and of the velocity section is 7000 m. The color scale is the same as in Figure 13.

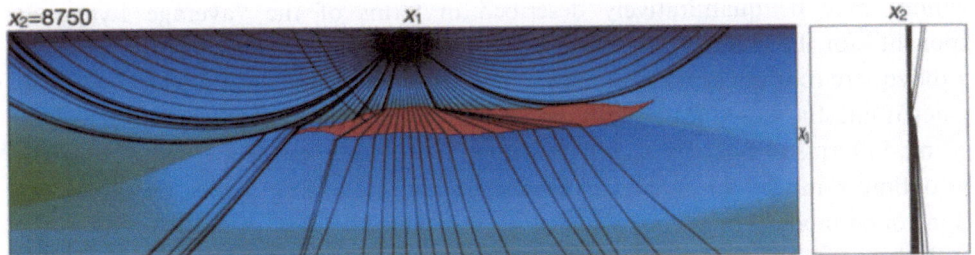

Figure 18

Rays computed by one-parametric shooting of a refracted wave in the $x_1 - x_3$ plane in the Salt Model, the version with interfaces. The x_2 coordinate of the point source and of the velocity section is 8750 m. The color scale is the same as in Figure 13.

obtain the smallest possible deviation of the model from the data. Finally, we optimize the weighting factor S_v in order to obtain the model of prescribed smoothness. Note that the generalization of the method from 2-D to 3-D is straightforward. The only problem is the limited memory of the computer being used, which disallows the use of all the data and of the denser parameterization of the model.

In the first step we construct a smooth model without interfaces, and then a model containing the salt body and some other selected interfaces. The best obtained model without interfaces, which satisfies our requirements with respect to the Lyapunov exponent, has the standard relative deviation of the velocity in the model from the original data cube $|(v - v_0)/v_0| = 8.22\%$. The best obtained model with the selected interfaces has the standard relative velocity deviation $|(v - v_0)/v_0| = 5.24\%$. For a more detailed description of the process of inversion refer to BULANT (2001).

We performed several tests of ray tracing in the models obtained after the inversions. As it is difficult to clearly display the results of 3-D ray tracing, we decided to shoot the rays only in the direction of the $x_1 - x_3$ plane (one-parametric shooting in a 3-D model). We then projected the computed rays onto the $x_1 - x_3$ plane and onto the $x_2 - x_3$ plane. The projected rays together with the velocity sections corresponding to the x_2 coordinates of the sources are displayed in Figures 13 to 18.

5. Conclusions

The minimization of the Sobolev norm of the model during linearized inversion enables us to control the model parameters unresolved by the data being fitted, while the parameters determined by the data remain almost unchanged. It enables us to minimize not only the difference between the data and the model, but, for many computational methods, also the difference between the exact solution and the results obtained by the considered numerical method to be used.

The smoothness of the velocity field of the models constructed for the ray methods may be quantitatively described in terms of the "average Lyapunov exponent" for the model. The "average Lyapunov exponent" may be approximated by the square root of the corresponding Sobolev norm of the model. This enables us to determine the quantitative criteria on the models, so as to be suitable for ray tracing, in terms of Sobolev scalar products. The criteria then allow us to construct the optimum models of geological structures for ray tracing, either by smoothing given rough models, or by inversion of seismic data.

Both numerical examples demonstrate the application of Sobolev norm to the regularization of the ill-conditioned inversion. The behavior of the model functions in the regions not illuminated by the data is controlled by the Sobolev norm.

The example of smoothing model Hess shows how the smoothing of the velocity field by including the Sobolev norm in the velocity inversion regularizes the chaotic behavior of rays, while the velocity field changes only slightly with the standard relative deviation of 1.8 per mille. Note that the standard relative velocity deviation is equivalent to the standard relative travel-time deviation for very short rays.

The example of building the SEG/EAGE Salt Model demonstrates the use of Sobolev norms in inverting 3-D data.

Acknowledgements

This research has been supported by the Grant Agency of the Czech Republic under Contracts 205/01/D097 and 205/01/0927, by the grant agency of the Charles University under Contract 237/2001/B-GEO/MFF, by the Ministry of Education of the Czech Republic within Research Project J13/98 113200004, and by the members of the consortium "Seismic Waves in Complex 3-D Structures," see http:// sw3d.mff.cuni.cz. The data for model Hess have been provided by Scott Morton of the Amerada Hess Corporation. The author expresses his thanks to Luděk Klimeš for his kind guidance of the work being presented in the paper, and to Ivan Pšenčík and both anonymous reviewers for their comments and suggestions resulting in substantial improvement of the paper.

REFERENCES

AMINZADEH, F., BRAC, J., and KUNZ, T., *3-D Salt and Overthrust Models, SEG/EAGE 3-D Modeling Series No.1.* (Soc. Explor. Geophysicists, Tulsa 1997).

BULANT, P. (2001), *Smoothing SEG/EAGE Salt Model for ray tracing using Sobolev scalar products*, Expanded Abstracts of 71st Annual Meeting (San Antonio), Errata, SP 5.9, Soc. Explor. Geophysicists, Tulsa.

ČERVENÝ, V., KLIMEŠ, L., and PŠENČÍK, I., *Complete seismic-ray tracing in three-dimensional structures*. In *Seismological Algorithms* (ed. Doornbos, D. J.) (Academic Press, New York 1988) pp. 89–168.

GRUBB, H. J. and WALDEN, A. T. (1995), *Smoothing Seismically Derived Velocities*, Geophys. Prosp. *43*, 1061–1082.

KLIMEŠ, L. and KVASNIČKA, M. (1994), *3-D Network Ray Tracing*, Geophys. J. Int. *116*, 726–738.

KLIMEŠ, L. (1996), *Grid Travel-time Tracing: Second-order Method for the First Arrivals in Smooth Media*, Pure appl. geophys. *148*, 539–563.

KLIMEŠ, L. (2000a) *Sobolev scalar products in the construction of velocity models*. In *Seismic Waves in Complex 3-D Structures*, Report 10 (Dep. Geophys., Charles Univ., Prague) pp. 15–40, online at http:// sw3d.mff.cuni.cz.

KLIMEŠ, L. (2000b) *Smoothing the Marmousi model for Gaussian-packet migrations*. In *Seismic Waves in Complex 3-D Structures*, Report 10 (Dep. Geophys., Charles Univ., Prague) pp. 63–74, online at http:// sw3d.mff.cuni.cz.

KLIMEŠ, L. (2000c), *Lyapunov exponents for 2-D ray tracing without interfaces*, Expanded Abstracts of 70th Annual Meeting (Calgary), 2293–2296, Soc. Explor. Geophysicists, Tulsa.

KLIMEŠ, L. (2002a), *Application of the Medium Covariance Functions to Travel-time Tomography*, Pure appl. geophys., this volume.

KLIMEŠ, L. (2002b), *Lyapunov Exponents for 2-D Ray Tracing Without Interfaces*, Pure appl. geophys., this issue.

PODVIN, P. and LECOMTE, I. (1991), *Finite Difference Computation of Traveltimes in Very Contrasted Velocity Models: A Massively Parallel Approach and its Associated Tools*, Geophys. J. Int. *105*, 271–284.

PRETLOVÁ, V. (1976), *Bicubic Spline Smoothing of Two-dimensional Geophysical Data*, Studia geophys. geod. *20*, 168–177.

PRETLOVÁ, V. (1985), *Bicubic Spline Smoothing of the Data Given at Points of a Rectangular Network*, Studia geophys. geod. *29*, 238–247.

TARANTOLA, A., *Inverse Problem Theory* (Elsevier, Amsterdam 1987).

VERSTEEG, R. J., *Analysis of the Problem of the Velocity Model Determination for Seismic Imaging* (Ph.D. thesis, University of Paris VII, Paris 1991).

ŽÁČEK, K. (2002), *Smoothing the Marmousi Model*, Pure appl. geophys., this issue.

(Received October 18, 2000, revised February 13, 2001, accepted May 17, 2001)

To access this journal online:
http://www.birkhauser.ch

Pure appl. geophys. 159 (2002) 1507–1526
0033–4553/02/081507–20 $ 1.50 + 0.20/0

❙ Pure and Applied Geophysics

Smoothing the Marmousi Model

KAREL ŽÁČEK[1]

Summary — The only way to make an excessively complex velocity model suitable for application of ray-based methods, such as the Gaussian beam or Gaussian packet methods, is to smooth it. We have smoothed the Marmousi model by choosing a coarser grid and by minimizing the second spatial derivatives of the slowness. This was done by minimizing the relevant Sobolev norm of slowness. We show that minimizing the relevant Sobolev norm of slowness is a suitable technique for preparing the optimum models for asymptotic ray theory methods. However, the price we pay for a model suitable for ray tracing is an increase of the difference between the smoothed and original model. Similarly, the estimated error in the travel time also increases due to the difference between the models. In smoothing the Marmousi model, we have found the estimated error of travel times at the verge of acceptability. Due to the low frequencies in the wavefield of the original Marmousi data set, we have found the Gaussian beams and Gaussian packets at the verge of applicability even in models sufficiently smoothed for ray tracing.

Key words: Velocity model, smoothing, asymptotic ray theory, Gaussian beams, Lyapunov exponent, Sobolev norm.

1. Introduction

The computation of rays is extremely sensitive to the smoothness of the model. In rough models, the behaviour of rays becomes chaotic and geometrical spreading and the number of arrivals increase with travel time rapidly (e.g., SMITH *et al.*, 1992; ABDULLAEV, 1993; TAPPERT and TANG, 1996; WITTE *et al.*, 1996; KEERS *et al.*, 1997). Moreover, a large number of two-point rays to each receiver makes calculation of two-point travel times slow and expensive. Often, two-point rays cannot be found within the numerical accuracy.

We need a reasonably smooth velocity model for a depth migration technique based on Gaussian packets. In the Gaussian packet method (e.g., KLIMEŠ, 1989), we do not need to find two-point rays, however a sufficiently dense set of rays must be calculated. Thus, the desired model should be suitable for ray tracing. Since we wish to keep the width of the Gaussian packets sufficiently small, the width of the

[1] Department of Geophysics, Charles University, Ke Karlovu 3, 121 16 Praha 2, Czech Republic, E-mail: zacek@karel.troja.mff.cuni.cz

Gaussian packet also depending on frequency, the model should be sufficiently smooth for the frequencies under consideration.

Various methods of smoothing the velocity model have been developed and published. The authors have been interested in determining the most suitable physical quantity to be smoothed (MÜLLER and SHAPIRO, 2000; GOLD et al., 2000), in finding a way to smooth the velocity model (GRUBB and WALDEN, 1995; VERSTEEG, 1991; BRAC and NGUYEN, 1990), or in studying the effects of smoothing on the wavefield (VERSTEEG, 1991, 1993).

An optimum way to smooth a complex velocity model for ray-based methods, which is presented in this paper, is to minimize the appropriate Sobolev norm of the velocity or slowness. We show that minimizing the Sobolev norm may be used for efficiently controlling the behaviour of rays in complex structures.

2. The Marmousi Model

Since we wish to use a "realistic" 2-D velocity model, we have decided to smooth the Marmousi model (VERSTEEG and GRAU, 1991; VERSTEEG, 1991, 1993). The Marmousi model, based on a real geological structure, is very complex, see Figure 1. The dimensions of the model are 9200 metres (length) by 3000 metres (depth). Values of velocity, which correspond to P waves, are defined at each gridpoint of the grid of cells of 4×4 metres. The grided values of velocity vary from 1500 ms^{-1} to 5500 ms^{-1}.

In the Marmousi model, the synthetic seismograms were computed by the finite difference method (VERSTEEG and Grau, 1991). We wish to use these seismograms as the "real data" for the migration. The length of the seismograms is 2.9 seconds with a sampling interval of 4 milliseconds. A trapezoidal frequency filter determined by frequencies of 0 Hz, 10 Hz, 35 Hz and 55 Hz has been applied to the data by the developers of the Marmousi data set.

3. Basic Ideas about the Desired Model

The desired smoothed model must fulfil two main and, unfortunately, contradictive requirements:
(a) to be in "good agreement" with the original Marmousi model, and
(b) to be "sufficiently smooth" for ray tracing and Gaussian packet computations.

Under the term "good agreement," we understand a slight difference between the smoothed and original model, expressed in terms of the standard L2 Lebesgue norm.

The meaning of the term "sufficiently smooth" is more complicated. In a complex model, the geometrical spreading and number of arrivals exponentially increase with increasing travel time. The exponential increment is controlled by the Lyapunov exponent (LYAPUNOV, 1949; MCCAULEY, 1993; ADDISON, 1997; KLIMEŠ, 2001). Since

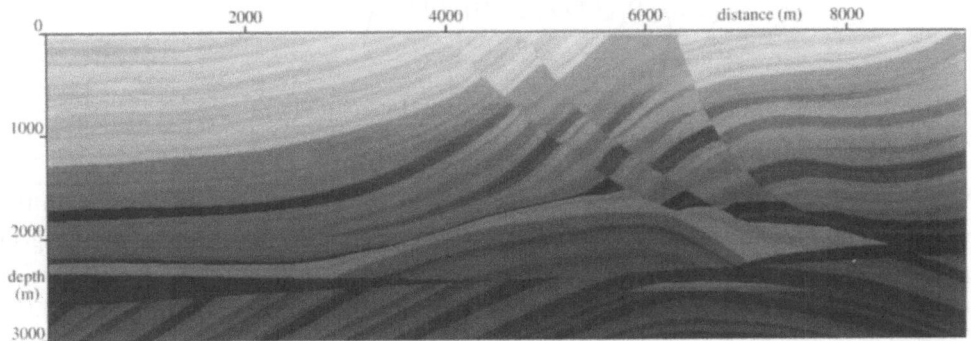

Figure 1
The Marmousi model.

the Lyapunov exponent depends on the second spatial derivatives of the velocity or slowness, the second derivatives should be minimized.

By minimizing the square of the Sobolev norm of slowness we may minimize the corresponding partial derivatives. The Sobolev scalar product is a linear combination of the L2 Lebesgue scalar products of the zero, first, second or higher partial derivatives (TARANTOLA, 1987).

The vague terms "good agreement" and "sufficiently smooth" cannot be easily quantified before a detailed study of the behaviour of rays and Gaussian packets in smoothed models is made.

The original Marmousi velocity model consists of discrete values of velocity at grid points of a regular, dense grid. In obtaining a smoothed model we

(a) choose a coarser data grid (which is a subgrid of the original grid) to reduce the amount of data to fit,

(b) arithmetically average the densely sampled slowness of the Marmousi model over cells centred at the grid points of the coarse data grid,

(c) choose a coarse B-spline model grid (which is a subgrid of the coarse data grid) and

(d) fit the averaged slowness values by the smoothed model.

We need to interpolate the discrete values of slowness on a coarse model grid for ray tracing. We have chosen bicubic B-splines as the interpolating functions, benefiting from the continuity of the second derivatives.

We summarize all types of grids being used in this paper. The first one is the original grid of the Marmousi model. The second is the coarser data grid constructed from the Marmousi model as explained above, which is used to fit the smoothed model. The third is the B-spline grid of the smoothed model.

4. Inversion

In order to find optimum parameters of the smoothed model, we minimize the objective function S defined by formula

$$S = \sum_{\text{GRID}} \left(\frac{u^{\text{D}}(\mathbf{x}^{\text{GRID}}) - u^{\text{M}}(\mathbf{x}^{\text{GRID}})}{\sigma^{\text{GRID}}} \right)^2$$

$$+ \left[\int \mathrm{d}^2 x \right]^{-1} \int b_{ijkl} \left(\frac{\partial^2 u^{\text{M}}(\mathbf{x})}{\partial x_i \partial x_j} \right) \left(\frac{\partial^2 u^{\text{M}}(\mathbf{x})}{\partial x_k \partial x_l} \right) \mathrm{d}^2 x \;, \tag{1}$$

where u^{D} is the value of slowness in the data grid, u^{M} is the value of slowness in the model being sought, $\mathbf{x} = (x_1, x_2)$, σ^{GRID} are the weighting parameters of the grid points, b_{ijkl} are the weighting coefficients of the Sobolev scalar product. Superscript GRID takes values $\text{GRID} = 1, 2, \ldots, N$, where N is the number of grid points of the coarser data grid with averaged values of slowness mentioned above. Subscripts take values $i, j, k, l = 1, 2$ in a 2-D model. Einstein summation over the pairs of identical indices is used. Integration is performed over the whole model.

We can express u^{M} as a linear combination of bicubic B-splines $B_\alpha(\mathbf{x})$

$$u^{\text{M}}(\mathbf{x}) = B_\alpha(\mathbf{x}) u_\alpha \;, \tag{2}$$

where u_α are the model parameters (values of slowness at grid points of the B-spline grid). Subscript α takes values $\alpha = 1, 2, \ldots, P$, where P is the number of model parameters. Consequently, P is the number of B-splines describing the smoothed model.

Equation (1) now reads

$$S = \sum_{\text{GRID}} \left(\frac{u^{\text{D}}(\mathbf{x}^{\text{GRID}}) - B_\alpha(\mathbf{x}^{\text{GRID}}) u_\alpha}{\sigma^{\text{GRID}}} \right)^2 + u_\alpha D_{\alpha\beta} u_\beta \;, \tag{3}$$

where

$$D_{\alpha\beta} = \left[\int \mathrm{d}^2 x \right]^{-1} \int b_{ijkl} \left(\frac{\partial^2 B_\alpha(\mathbf{x})}{\partial x_i \partial x_j} \right) \left(\frac{\partial^2 B_\beta(\mathbf{x})}{\partial x_k \partial x_l} \right) \mathrm{d}^2 x \;. \tag{4}$$

Since we do not know the coefficients b_{ijkl} which lead to the optimum model, the problem is not linear. Thus, parameters u_α cannot be determined analytically. Since we do not want to solve the nonlinear inverse problem numerically, we need to "linearize" formula (4). The linearization of (4) yields

$$D_{\alpha\beta} = s^2 D'_{\alpha\beta} \;, \tag{5}$$

$$D'_{\alpha\beta} = \left[\int \mathrm{d}^2 x \right]^{-1} \int b'_{ijkl} \left(\frac{\partial^2 B_\alpha(\mathbf{x})}{\partial x_i \partial x_j} \right) \left(\frac{\partial^2 B_\beta(\mathbf{x})}{\partial x_k \partial x_l} \right) \mathrm{d}^2 x \;, \tag{6}$$

where s is a free parameter and b'_{ijkl} are fixed coefficients of the Sobolev scalar product. The choice of coefficients b'_{ijkl} will be discussed in Section 6.

We can now rewrite equation (3) to read

$$S = [\mathbf{u}^{\text{D}} - \mathbf{B}\mathbf{u}]^{\text{T}} \mathbf{C}^{-1} [\mathbf{u}^{\text{D}} - \mathbf{B}\mathbf{u}] + s^2 \mathbf{u}^{\text{T}} \mathbf{D}' \mathbf{u} \;, \tag{7}$$

where **B** is defined as $B_{i\alpha} = B_\alpha(x_i)$, **D'** is a $P \times P$ matrix given by formula (6), **C** is a $N \times N$ diagonal matrix, composed of $(\sigma^{\mathrm{GRID}})^2$, see equation (1). N is the number of grid points.

The condition for the minimum of the objective function is

$$\frac{\partial S}{\partial u_\alpha} = 0 \ , \tag{8}$$

which yields

$$\mathbf{B}^{\mathrm{T}}\mathbf{C}^{-1}[\mathbf{B}\mathbf{u} - \mathbf{u}^{\mathrm{D}}] + s^2 \mathbf{D}'\mathbf{u} = 0 \ . \tag{9}$$

The resulting vector of the model parameters is

$$\mathbf{u} = [\mathbf{B}^{\mathrm{T}}\mathbf{C}^{-1}\mathbf{B} + s^2 \mathbf{D}']^{-1}\mathbf{B}^{\mathrm{T}}\mathbf{C}^{-1}\mathbf{u}^{\mathrm{D}} \ . \tag{10}$$

5. Criteria of Acceptability

In a complex 2-D model, the width of ray tube Q increases with increasing travel time τ approximately according to the asymptotic formula

$$Q \propto e^{\lambda\tau} \ , \tag{11}$$

where λ is the Lyapunov exponent corresponding to the ray (LYAPUNOV, 1949; OSELEDEC, 1968; KATOK, 1980).

The number of arrivals at each point of the model is an important indication as to whether the behaviour of rays is regular or chaotic. We wish the number of arrivals not to exceed, let us say, 10. In a finite model, the number of arrivals ν is proportional to the widths of the ray tubes. This is caused by the overlaping of the ray tubes, see Figure 5. As we wish to smooth the model for migration, the sum of travel times from source τ_{S} and receiver τ_{R} to a point of the model should be substituted for travel time τ in equation (11). Hence,

$$\nu \propto e^{\lambda(\tau_{\mathrm{S}}+\tau_{\mathrm{R}})} \ . \tag{12}$$

Consequently, the number of arrivals ν may be expressed as the product of the numbers of arrivals from source ν_{S} and receiver ν_{R}

$$\nu = \nu_{\mathrm{S}}\nu_{\mathrm{R}} \ . \tag{13}$$

For $\tau = \tau_{\mathrm{S}} + \tau_{\mathrm{R}} = 0$, we obtain $e^{\lambda\tau} = 1$. Since this corresponds to the number of arrivals in the nearest vicinity of the source (or of the receiver for the migration), we can alter equation (12) to read

$$\nu \approx e^{\lambda(\tau_{\mathrm{S}}+\tau_{\mathrm{R}})} \ . \tag{14}$$

We want to work with the "average Lyapunov exponent" $\hat{\lambda}$. The "average Lyapunov exponent" $\hat{\lambda}$ is the Lyapunov exponent averaged over a large set of rays (KLIMEŠ, 1999). The value of the "average Lyapunov exponent" may be one of the criteria of the smoothness of the model. Hence, we wish $e^{\hat{\lambda}\tau_{max}}$ not to exceed 10, τ_{max} being the maximum sum of travel times from the source and receiver to a point of the model. Since the sum of travel times from the source and receiver cannot exceed the length of the seismogram, $\tau_{max} = 2.9$ s may be used for estimating the optimum value of $\hat{\lambda}$. Thus, for the number of arrivals not exceeding 10, we obtain the optimum value of $\hat{\lambda}$ close to 0.8 s^{-1}.

The width of Gaussian packets should be kept small. For very wide packets the obtained wavefield would not be the correct solution of the equations being solved. Accordingly, the desired migrated section would be wrong. The maximum halfwidth should probably not be greater than the B-spline interval.

We mention that the width of the Gaussian beams or packets depends not only on the smoothness of the model, but also on the frequencies under consideration. From this point of view the model is not complex for Gaussian beams or packets by itself, but in relation to the frequency.

The relative root-mean-square (RMS) difference of slowness between the original and the smoothed model may be the criterion of "good agreement." The relative RMS difference of slowness corresponds approximately to the relative error of the travel time. This is an asymptotic relation valid for short rays. The relative error of the travel time may be smaller for longer rays.

By the term "error of travel time" we understand the difference between the real travel time in the original structure and the computed travel time in the smoothed model. Although we cannot determine the real travel time, we can estimate the error caused by the difference between the original and the smoothed model.

6. Choice of the Coefficients and of the Density of the Grids

We need to specify coefficients b'_{ijkl} and s, the matrix $\mathbf{C}$ and the density of the grids before the computation.

As we have no prior information, we choose $\sigma^{GRID} = \sqrt{N}$, where N is the number of values to be fitted. This makes the value of objective function S approximately independent of the number of gridpoints.

Coefficients b'_{ijkl} may be constructed as a completely symmetric tensor (e.g., BULANT, 2001). The 4×4 matrix $\mathbf{b}'$ is then defined by

$$b'_{ijkl} = \frac{d(d+2)}{3} \langle e_i e_j e_k e_l \rangle \ , \tag{15}$$

where $\mathbf{e}$ is a unit vector, $\langle \ldots \rangle$ indicates averaging over all directions of the unit vector, $d = 1$ in 1-D, $d = 2$ in 2-D and $d = 3$ in 3-D. We have introduced the formal scaling coefficient $1/3d(d+2)$ in order to make the respective coefficients b'_{ijkl} equal in 1-D, 2-D and 3-D.

The average of the unit vector over all directions can be calculated analytically. Generally in d-D for $d = 1$, 2 or 3, we may put

$$b'_{ijkl} = \frac{\delta_{ij}\delta_{kl} + \delta_{ik}\delta_{jl} + \delta_{il}\delta_{jk}}{3} \, , \tag{16}$$

where δ_{ij} is the Kronecker symbol. In 2-D,

$$b'_{1111} = b'_{2222} = 1,$$
$$b'_{1122} = b'_{1212} = b'_{1221} = b'_{2112} = b'_{2121} = b'_{2211} = \tfrac{1}{3} \text{ and}$$
$$b'_{1112} = b'_{1121} = b'_{1211} = b'_{1222} = b'_{2111} = b'_{2122} = b'_{2212} = b'_{2221} = 0 \, .$$

We must keep in mind that this is only our special choice of coefficients b'_{ijkl}, and that there are various other ways of constructing matrix $\mathbf{b}'$. For example, we can increase coefficient b'_{1111} and decrease coefficient b'_{2222} (or *vice versa*) and use this new matrix of coefficients for anisotropic smoothing.

The original grid of the Marmousi model consists of cells of 4×4 metres, which yields $751 \times 2301 = 1,728,051$ grid points. Three B-spline grids of cells of (a) 100×400 metres, (b) 200×230 metres and (c) 200×400 metres are studied. These grids consist of $P = 744$, 656 and 384 grid points, respectively.

Three data grids of cells of (a) 20×80 metres, (b) 40×40 metres and (c) 40×80 metres are used in the inversion. The values at the grid points are calculated by averaging the values of slowness in the Marmousi model, as described in Section 3. These grids consist of $N = 17516$, 17556 and 8816 grid points, respectively.

Finally, we need to choose the values of parameter s. We choose the initial value of parameter s for the linearized inversion as

$$s_{\text{init}} \approx \frac{|u - u_0|}{\sigma} \frac{1}{\|u\|_{\text{init}}} \, , \tag{17}$$

where $\|u\|_{\text{init}}$ is the initial value of the Sobolev norm of slowness, $|u - u_0|$ is the standard slowness deviation of the model, and σ is the given slowness deviation. We have made this rough estimate assuming that the first term on the right-hand side of equation (1) does not exceed dramatically the second term, or *vice versa*. According to equation (A.10), we can estimate the maximum value of the Sobolev norm of slowness as $\|u\|_{\text{init}} \approx \sqrt{8/3}(2\Lambda_{\text{init}})^2 u_A^3$, where u_A is the average slowness in the model, and Λ_{init} is the initial value of the "average Lyapunov exponent" without compensation for the focusing low-velocity zones, see equation (A.1). The standard slowness deviation of the model may be estimated by $|u - u_0| \approx \varepsilon u_A$, ε being the relative travel-time error. The given slowness deviation is determined by $\sigma = \sigma^{\text{GRID}} N^{-1/2}$. Hence, we can alter equation (17) as

$$s_{\text{init}} \approx \sqrt{\frac{3}{8}} \, \varepsilon (2 u_A \Lambda_{\text{init}})^{-2} \sqrt{N} (\sigma^{\text{GRID}})^{-1} \, . \tag{18}$$

Note that parameter s is proportional to $\hat{\lambda}^{-2}$, see equation (18). Thus, the n-fold decrement of $\hat{\lambda}$ requires an n^4-fold increment of the square of the Sobolev norm in the objective function in equation (1). Consequently, the decrement of $\hat{\lambda}$ increases the difference between the new and the original model. From this point of view $\hat{\lambda}$ should not be too small. We should keep $\hat{\lambda}$ close to the optimum value estimated above.

For $\varepsilon = 0.1$, $(u_A)^{-1} = 3000$ ms^{-1} and $\Lambda_{init} = 1.3$ s^{-1}, see equation (A.12), we obtain $s_{init} = 81529$. We study the values of s of (1) 0 m^2 (without the Sobolev norm included in the inversion), (2) $\sqrt{3/8} \cdot 10,000$ m$^2 \approx 6124$ m^2, (3) $\sqrt{3/8} \cdot 25,000$ m$^2 \approx 15,309$ m^2, (4) $\sqrt{3/8} \cdot 50,000$ m$^2 \approx 30,619$ m^2, (5) $\sqrt{3/8} \cdot 100,000$ m$^2 \approx 61,237$ m^2 and (6) $\sqrt{3/8} \cdot 225,000$ m$^2 \approx 137,784$ m^2.

7. Numerical Examples

We have calculated the smoothed models, the corresponding values of the relative RMS difference of slowness between the smoothed and original model, the angular dependence of the Lyapunov exponents, the values of the "average Lyapunov exponents," rays, numbers of arrivals and the halfwidths of Gaussian beams.

Figure 2 displays the models without the minimized Sobolev norm, smoothed just by the use of the coarse B-spline grid.

Figure 3 presents the smoothed models with the grid of cells of 200×400 metres and with the minimized Sobolev norm. The respective figures for the models with the grid of cells of 100×400 or 200×230 metres look similar and are not shown. At first glance we can see that the models with the values of parameter $s = 61,237$ m^2 and $s = 137,784$ m^2 do not show features of the original Marmousi model.

Figure 4 displays the angular dependence of the Lyapunov exponents and the values of the "average Lyapunov exponents" for smoothed models with the B-spline grid of cells of 200×400 metres. We can see that the model without the minimized Sobolev norm of slowness seems to be too rough, whereas the "average Lyapunov exponents" of the models with the minimized Sobolev norm are close to or less than our initial assumption of the optimum value. Unfortunately, the strong angular dependence of the Lyapunov exponents (and consequently the excessive maximum value of the Lyapunov exponent) indicates that even models with $\hat{\lambda}$ close to 1 (as the model with $s = 6124$ m^2) may still be too rough.

The synthetic seismograms, used for the migration in the Marmousi model, are computed with the length of 2.9 seconds. The streamer composed of hydrophone groups has been used for data acquisition for each shot. The farthest hydrophone was located 2575 metres from the watergun. Since the maximum travel time cannot exceed the length of the seismogram, let us assume a fixed travel time of 2.9 seconds. For an almost horizontal ray with the endpoint in the farthest hydrophone and for the velocity of 1500 ms^{-1} we obtain the farthest possible reflection point, at 4350 metres from the watergun and 900 metres from the farthest hydrophone. As 900

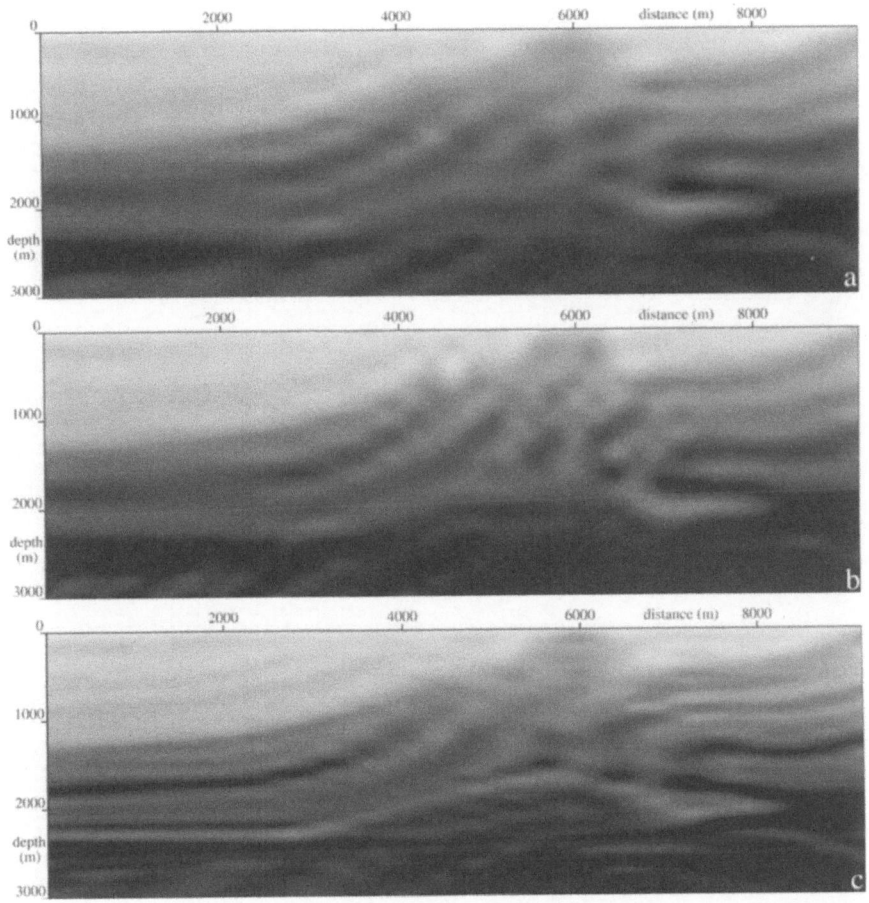

Figure 2

The smoothed models without minimized Sobolev norm of slowness ($s = 0$ m^2) for the B-spline grids of cells of (a) 200 × 400 metres, (b) 200 × 230 metres and (c) 100 × 400 metres.

metres corresponds to 0.6 seconds, we may estimate the maximum useful value of the travel time as 2.3 seconds. The rays have been calculated for this value of the maximum travel time.

Figure 5 displays rays computed in the smoothed models with a constant step in the take-off angles. We can see the dependence of the behaviour of rays on parameter s. We moved the source along the whole profile and tested ray tracing. The behaviour of rays was always of the same kind as in these illustrative figures. Models with $s = 0$ m^2 and $s = 6124$ m^2 seem to be unsuitable for ray methods due to the density of caustics. In Figure 5a we can see rays trapped in the low velocity channels.

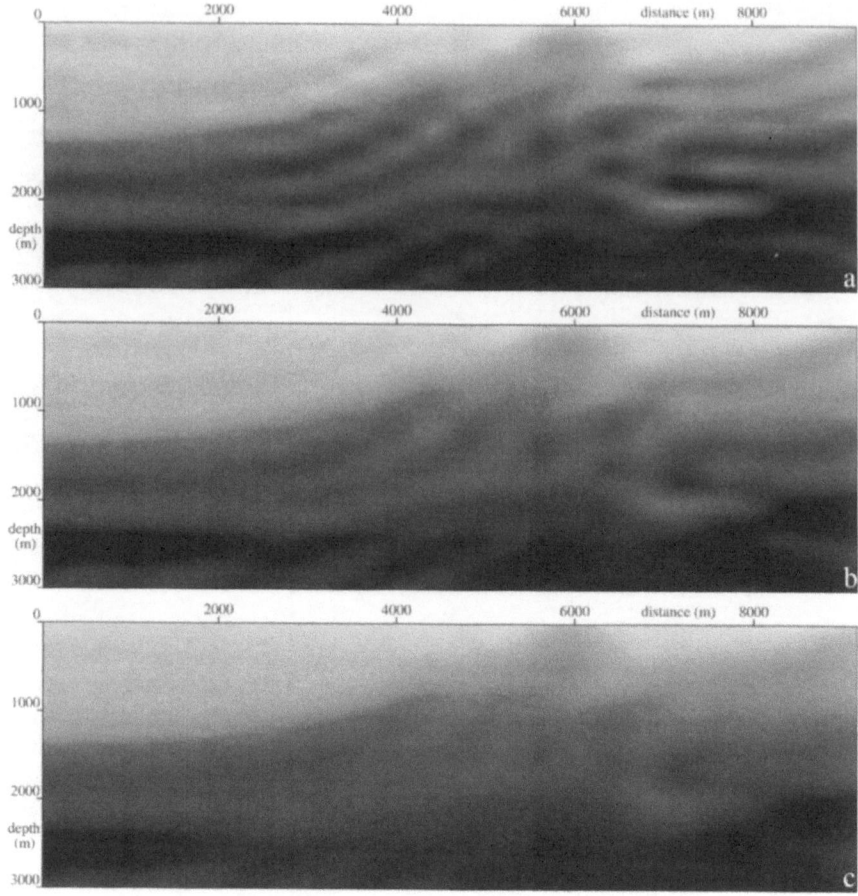

Figure 3
The smoothed models for the B-spline grid of cells of 200×400 metres and for values of parameter s of
(a) 0 m^2, (b) 6124 m^2, (c) 15,309 m^2, (d) 30,619 m^2, (e) 61,237 m^2 and (f) 137,784 m^2.

The maximum number of arrivals for the models with the B-spline grid of cells of 200×400 metres is

(a) 19 for the value of $s = 0$ m^2,
(b) 18 for the value of $s = 6124$ m^2,
(c) 7 for the value of $s = 15,309$ m^2,
(d) 5 for the value of $s = 30,619$ m^2,
(e) 3 for the value of $s = 61,237$ m^2 and
(f) 2 for the value of $s = 137,784$ m^2.

In the model without interfaces, the number of arrivals should be odd. Even numbers
for $s = 6124$ m^2 and $s = 137,784$ m^2 may be explained by the influence of the

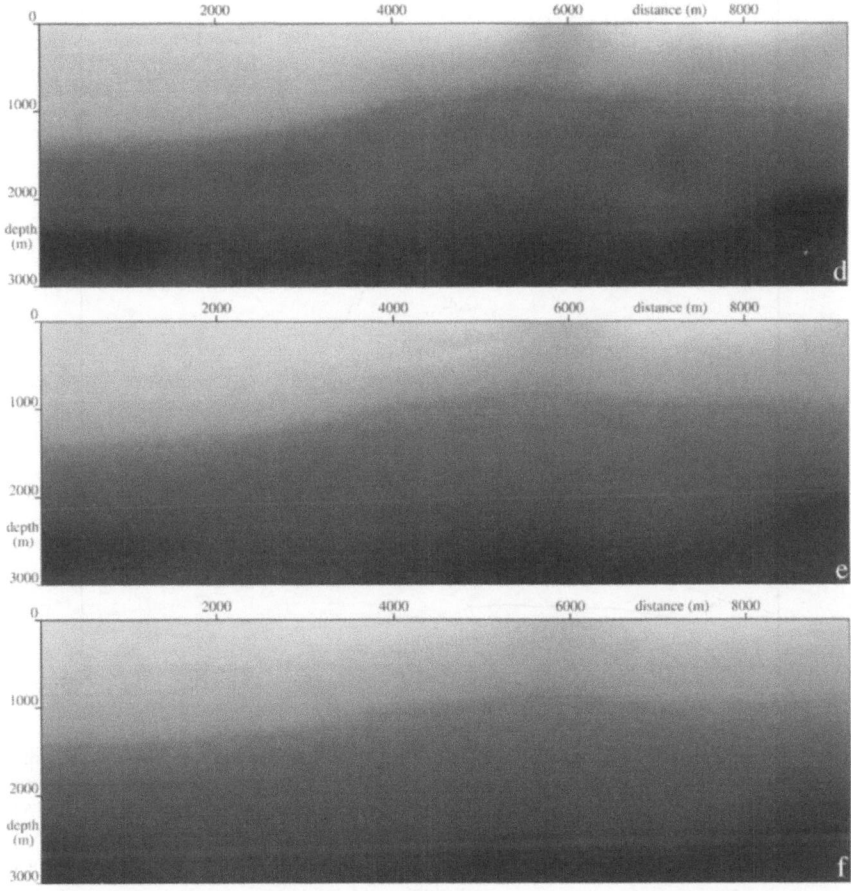

Figure 3d–f

borders of the model. Due to the requirements established above, the models with $s = 0 \text{ m}^2$ and $s = 6124 \text{ m}^2$ are probably not suitable for ray-based methods.

The relative RMS difference of slowness between all calculated models and the original Marmousi model is in Table 1. We can see that the price for a model suitable for ray tracing is a considerable increment of the relative RMS difference of slowness between the smoothed and original model, representing here the geological structure. This resulted from the complexity of the original Marmousi model. If the value of parameter s is larger, the relative RMS difference is the same for all the studied B-spline grids. Hence, the model with the B-spline grid of cells of 200×400 metres (with only 384 B-spline grid points) is probably the best choice.

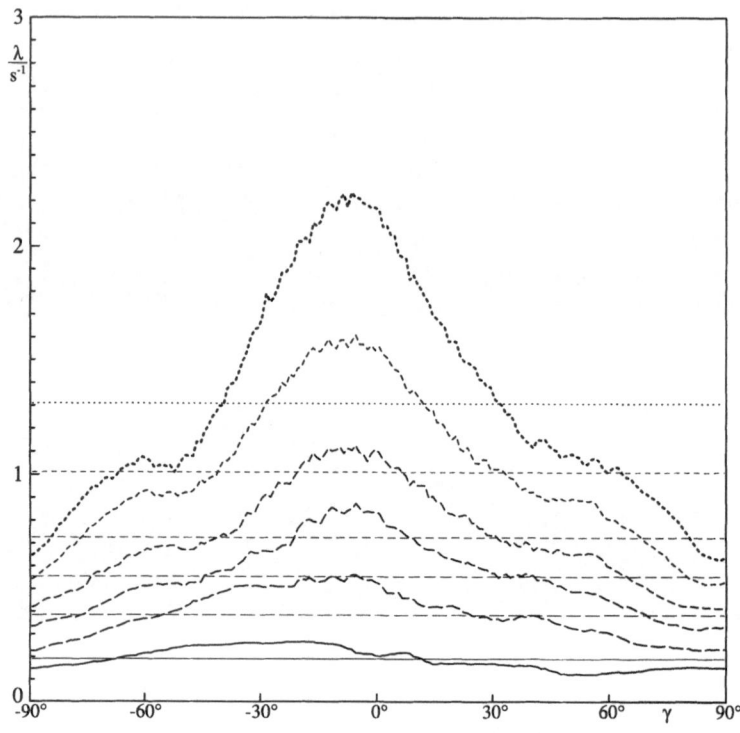

Figure 4

The angular dependence of the Lyapunov exponents for the models with the B-spline grid of cells of 200×400 metres. The left and right border corresponds to the vertical ray, the middle corresponds to the horizontal ray. The values of parameter s are (1) 0 m^2 for the dotted lines, (2) 6124 m^2, (3) $15,309$ m^2, (4) $30,619$ m^2 and (5) $61,237$ m^2 for the dashed lines and (6) $137,784$ m^2 for the solid lines. The thin horizontal lines correspond to the "average Lyapunov exponents," averaged over angles with a uniform weight.

In general, we believe it is useless to work with an overly dense B-spline grid, because the smoothed models with various densities of the model grid converge with increasing weight of the Sobolev norm.

Let us form a short summary of what we have accomplished. We have prepared the smoothed models. We have calculated the values of corresponding "directional" and "average Lyapunov exponents" and the values of the relative RMS difference between the smoothed and the original Marmousi model. Finally, we have studied the behaviour of rays in the smoothed models. That is, we already know how to smooth the Marmousi model for the computation of rays and travel times.

8. Effects of Smoothing on Gaussian Beams

Our primary objective was to prepare a suitable velocity model for Gaussian packet migration. Since the width of the Gaussian beam is equal to the width of the

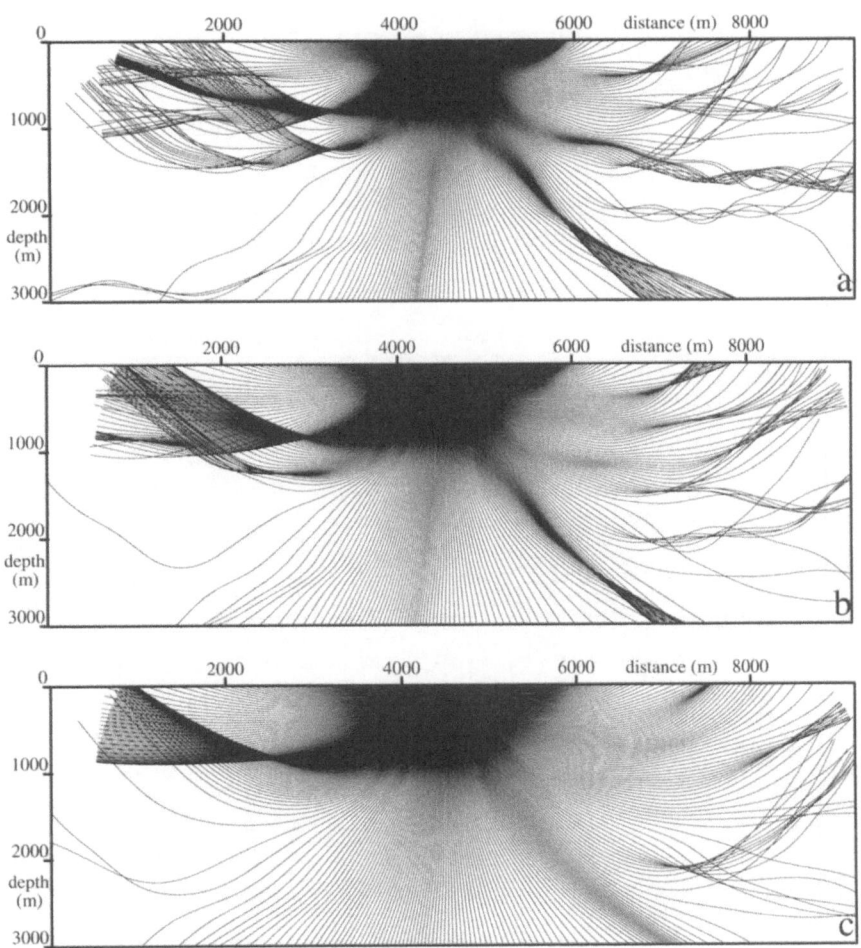

Figure 5
Rays in models with the B-spline grid of cells of 200×400 metres and with values of parameter s of
(a) 0 m^2, (b) 6124 m^2, (c) 15,309 m^2, (d) 30,619 m^2, (e) 61,237 m^2, and (f) 137,784 m^2. The maximum travel
time is of 2.3 seconds.

corresponding symmetric Gaussian packets, and the computation of the beams is
easier, we study the width of the Gaussian beams. In 2-D, the profile of the Gaussian
beam in a cross section orthogonal to the ray is controlled by the factor

$$\exp\left(i\pi f M q^2\right) , \tag{19}$$

where i is the imaginary unit, f is frequency, q is the ray-centred coordinate
orthogonal to the ray, and M is the second derivative of the complex-valued travel
time. The quadratic term in the Taylor expansion of the complex-valued travel time
of the Gaussian beam thus reads

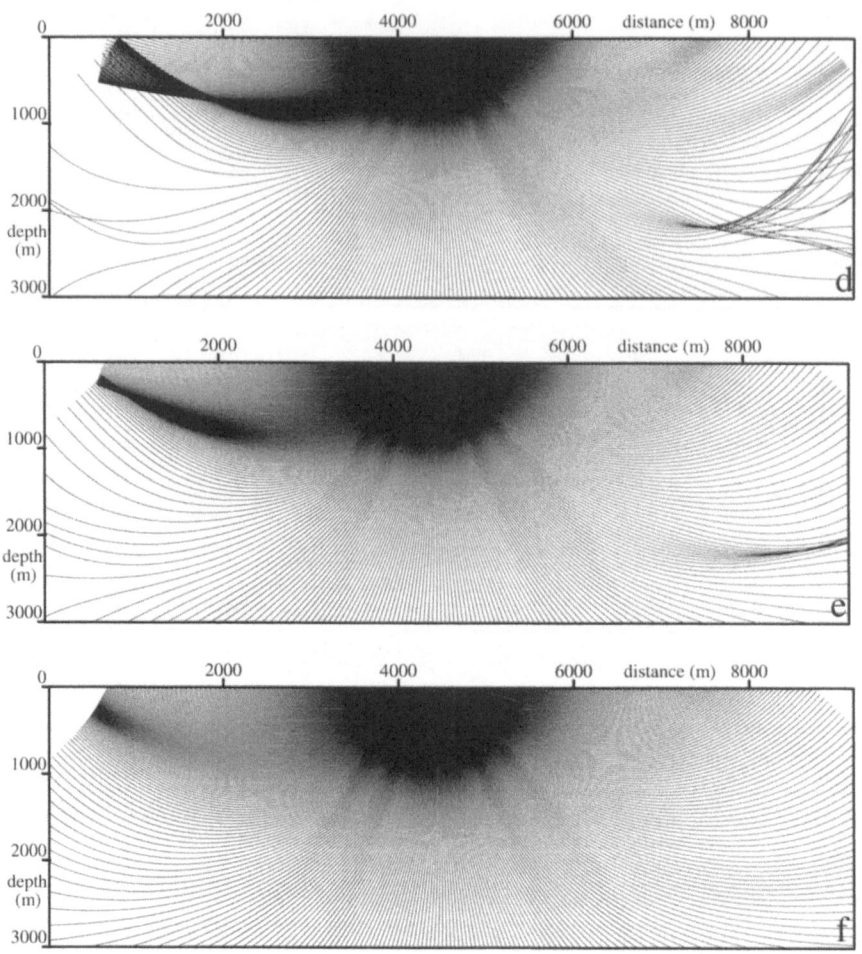

Figure 5d–f

$$\tfrac{1}{2} M q^2 \ . \tag{20}$$

The quadratic term in the Taylor expansion of the complex-valued travel time of the Gaussian beam along the surface is

$$\tfrac{1}{2}(G^R + iG^I)(x - x_0)^2 \ , \tag{21}$$

where $(x - x_0)$ is the distance from the initial point of the central ray of the Gaussian beam to the respective point on the surface and G^R and G^I are real-valued parameters determining M uniquely (KLIMEŠ, 1984).

We have calculated the standard halfwidths of Gaussian beams in various smoothed models for various initial values of parameters G^R and G^I. Standard halfwidth a of a Gaussian beam of cross section

$$\exp\left(-\frac{q^2}{2a^2}\right) , \tag{22}$$

multiplied by the square root of $(2\pi f)$, has been interpolated between the rays and displayed,

$$W = a\sqrt{2\pi f} . \tag{23}$$

The halfwidths of Gaussian beams calculated for the models with the B-spline grid of cells of 200×400 metres and with $s = 15,309$ m^2, $s = 30,619$ m^2, $s = 61,237$ m^2 and $s = 137,784$ m^2 are shown in Figure 6. These halfwidths have been calculated for the initial values of parameters $G^R = 0$ and $G^I = 0.250 \times 10^{-6}$. The models with lower values of s were excluded.

The gray-scale coded quantity W is displayed at the respective points along the central rays of the beams. The white colour corresponds to the Gaussian beam halfwidth of 0 metres for all frequencies. The black colour corresponds to the Gaussian beam halfwidths of 1010 metres and more for the frequency of 35 Hz, and of 1890 metres and more for the frequency of 10 Hz. Therefore, the black coloured regions of Figure 6 indicate too wide Gaussian beams for the frequencies under consideration.

We can see that the model with $s = 15,309$ m^2 is not suitable for Gaussian beams or packets. Especially if the position of the source is close to the middle of the profile, the Gaussian beams become wider too quickly. On the other hand, the models with $s = 61,237$ m^2 and $s = 137,784$ m^2 seem to be acceptable. Unfortunately, these models are smoothed to an extent which may jeopardize the migration. We hope that we will be able to use the model with $s = 30,619$ m^2 in the migration. We have studied the behaviour of Gaussian beams for various initial parameters G^R and G^I. We have realized that different initial values of these parameters are suitable for different positions of the source, or of the receiver in the migration. Consequently, we will try to develop a method to optimize the shapes of Gaussian beams or packets dependent on the position of the source, or of the receiver in the migration. This would allow the use of models not so smoothed.

We mention that even in models with a sufficiently small number of arrivals, the widths of Gaussian beams are at the verge of acceptability. This is due to the low frequencies under consideration.

Let us summarize that models with parameter s equal to or greater than $15,309$ m^2 seem to be suitable for ray tracing with the travel time of 2.3 seconds, see Figures 5 and 6. From this point of view, these models are sufficiently smooth. However, the low frequencies under consideration make the use of the Gaussian beam or packet method almost impossible. We probably should improve the applicability of the Gaussian packet method by using shapes of Gaussian packets optimized in dependence on the position of the source, or of the receiver in the migration.

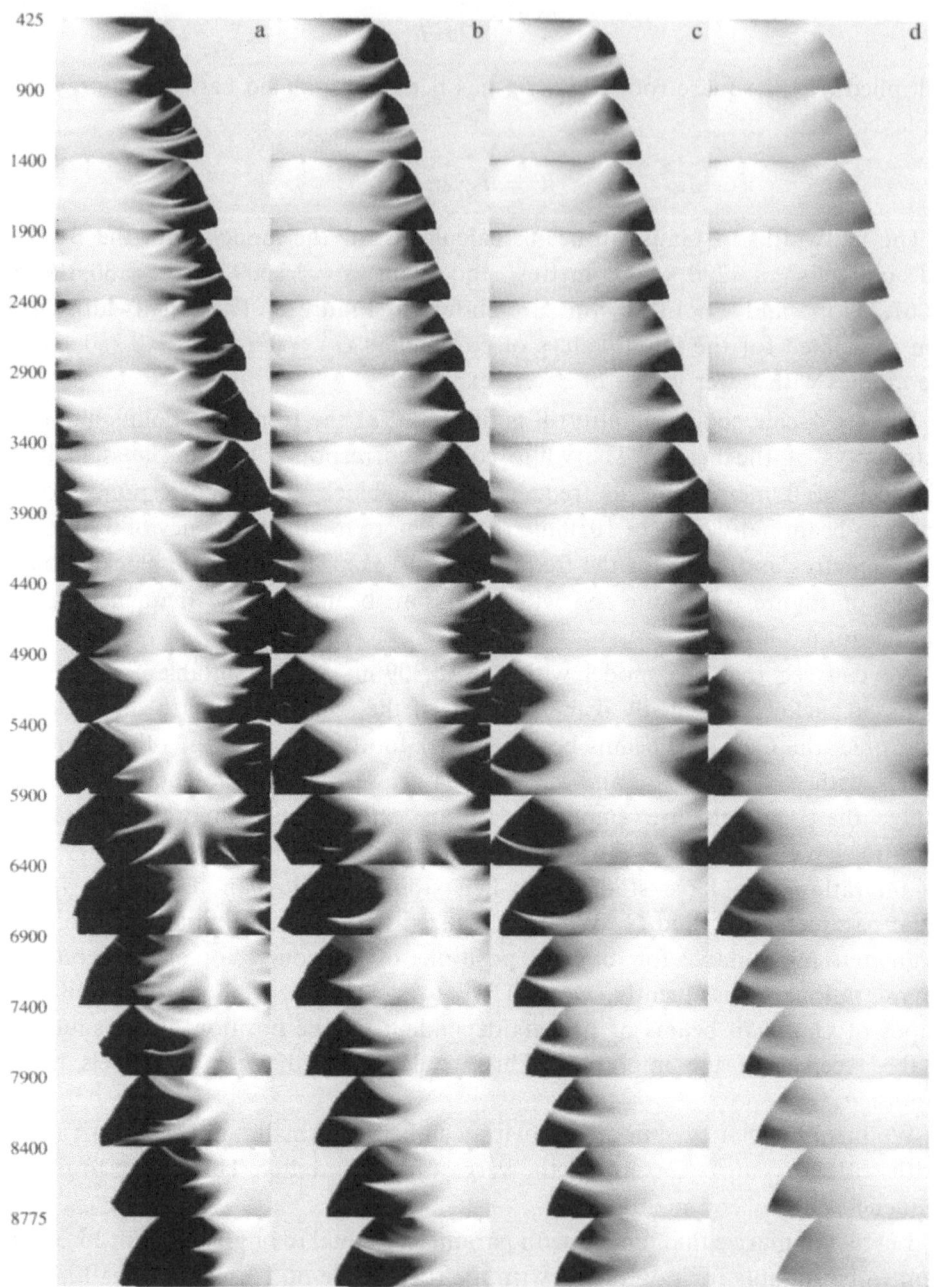

Figure 6
The halfwidths of Gaussian beams in smoothed models with the B-spline grid of cells of 200×400 metres. Columns correspond from left to right to the values of parameter s of (a) 15,309 m^2, (b) 30,619 m^2, (c) 61,237 m^2 and (d) 137,784 m^2. Rows correspond to various positions of the source (in metres).

Table 1

The relative RMS difference between the smoothed and the Marmousi model. Columns correspond to various B-spline grids, rows correspond to various values of parameter s

	100–400 m	200–230 m	200–400 m
$0 \ m^2$	8.3%	10.8%	11.1%
$6,124 \ m^2$	11.4%	11.9%	12.0%
$15,309 \ m^2$	13.2%	13.2%	13.3%
$30,619 \ m^2$	14.3%	14.3%	14.3%
$61,237 \ m^2$	15.1%	15.1%	15.1%
$137,784 \ m^2$	15.8%	15.8%	15.8%

9. Conclusions

The minimization of the relevant Sobolev norm of slowness is a powerful tool for preparing the optimum models for the asymptotic ray theory methods. As we have illustrated in numerical examples, it can be used for smoothing very complex models. However, the difference of slowness between the smoothed and the original model then increases rapidly. Also, the error of the travel time then increases.

We must keep in mind that there exists a natural relation between the complexity of the original model and the resulting difference between the sufficiently smoothed model and the original model. The more complex the original model, the more change it requires. Thus, the decision rests with the user, whether or not the model is too complex for smoothing. The required maximum error of travel time is then a key argument.

We have also demonstrated that even in models sufficiently smoothed for ray tracing, the Gaussian beams may still be too wide for the frequencies under consideration. In preparing a model for Gaussian beams or packets, we cannot judge solely from the number of arrivals and values of the "average Lyapunov exponents", whether the model is convenient. The widths of Gaussian beams or packets in relation to the frequency should be studied as well.

Acknowledgements

The author wishes to thank Luděk Klimeš for his kind guidance throughout the work on this paper, Professor V. Červený and Johana Brokešová for valuable comments and advice, and also Andreas Ehinger for providing the Marmousi model and data set. This research has been supported by the Ministry of Education of the Czech Republic within Research Project J13/98 113200004 and by the members of the consortium "Seismic Waves in Complex 3-D Structures" (see "http:/seis.karlov.mff.cuni.cz/consort/main.htm").

Appendix A

In evaluating a meaningful initial value of parameter s (see Section 6), we need to find some approximative relation between the Sobolev norm and the Lyapunov exponent.

According to KLIMEŠ (1999), the "average Lyapunov exponent" $\hat{\lambda}$ may be approximated by

$$\hat{\lambda} \approx \Lambda + \Delta\Phi \ , \tag{A.1}$$

where $\Delta\Phi$ is the decompensation for the low-velocity focusing zones. In 2-D, Λ is defined as

$$\Lambda = \left[\int v^{-1} \, \mathrm{d}^2x \right]^{-1} \int \sqrt{\mathrm{neg}(v_{,ij}\, e_i\, e_j)v^{-1}} \, \mathrm{d}^2x \ , \tag{A.2}$$

where $\mathrm{neg}(f) = 1/2(f - |f|)$ is the negative part of f, v is the velocity, $v_{,ij}$ is the second velocity derivative and $\mathbf{e}$ is a unit vector perpendicular to the ray.

We assume that the model is so smooth that the number of velocity oscillations, K_{osc}, along rays of length corresponding to τ_{max} is small,

$$K_{\mathrm{osc}} = \frac{\tau_{\mathrm{max}}}{\tau_{\mathrm{osc}}} \ , \tag{A.3}$$

where τ_{osc} is the average wavelength of the velocity oscillations in the smoothed model, expressed in travel-time units. This assumption allows for the approximation

$$\Delta\Phi \approx -\frac{\ln 2}{\tau_{\mathrm{osc}}} = -\frac{K_{\mathrm{osc}} \ln 2}{\tau_{\mathrm{max}}} \ . \tag{A.4}$$

Let us now perform several approximations to express Λ in terms of the Sobolev norm of slowness u in the model without interfaces,

$$\Lambda \approx \left[\int u \, \mathrm{d}^2x \right]^{-1} \int \sqrt{\mathrm{pos}(u_{,ij}\, e_i\, e_j)u^{-1}} \, \mathrm{d}^2x \ , \tag{A.5}$$

where $\mathrm{pos}(f) = 1/2(f + |f|)$ is the positive part of f,

$$\Lambda \approx \frac{1}{2} \left[\int u^{\frac{3}{2}} \, \mathrm{d}^2x \right]^{-1} \int \sqrt{|u_{,ij}\, e_i\, e_j|} \, \mathrm{d}^2x \ , \tag{A.6}$$

and

$$\Lambda \approx \frac{1}{2} u_{\mathrm{A}}^{-\frac{3}{2}} \left\{ \left[\int \mathrm{d}^2x \right]^{-1} \int (u_{,ij}\, e_i\, e_j)^2 \, \mathrm{d}^2x \right\}^{\frac{1}{4}} \ , \tag{A.7}$$

where

$$u_{\text{A}} = \left\{ \left[\int \text{d}^2x \right]^{-1} \int u^{\frac{3}{2}} \text{d}^2x \right\}^{\frac{2}{3}} .$$ (A.8)

Finally, we arrive at

$$\Lambda \approx \frac{1}{2} u_{\text{A}}^{-\frac{3}{2}} \left(\frac{3}{8} \right)^{\frac{1}{4}} \sqrt{\|u\|} ,$$ (A.9)

where $\|u\|$ is the Sobolev norm of slowness given by matrix $\mathbf{b}'$, see equation (17). This approximation may also be expressed as

$$\|u\| \approx \sqrt{\frac{8}{3}} \, u_{\text{A}}^3 (2\Lambda)^2 .$$ (A.10)

As we need to find some initial value of parameter s, we should estimate the respective value of the Sobolev norm $\|u\|_{\text{init}}$. Since we have already derived an approximative relation between $\|u\|$ and Λ, see equation (A.10), we need to find the value of Λ_{init}. We have decided to keep the number of arrivals less than 10, see Section 5. With a view to equations (14), (A.1) and (A.4),

$$\Lambda \leq \frac{\ln 10 + K_{\text{osc}} \ln 2}{\tau_{\text{max}}} .$$ (A.11)

Since we assume at least one shift of $- \ln 2$ for the source and one for the receiver, we assume $K_{\text{osc}} = 2$. For $\tau_{\text{max}} = 2.9$ s, we can put

$$\Lambda_{\text{init}} \approx 1.3 \, \text{s}^{-1} .$$ (A.12)

REFERENCES

ABDULLAEV, S.S., *Chaos and Dynamics of Rays in Waveguide Media* (Gordon and Breach, Amsterdam 1993).

ADDISON, P.S., *Fractals and Chaos: An Illustrated Course* (IOP Publishing Ltd, Bristol and Philadelphia 1997).

BRAC, J. and NGUYEN, L.L., *Modeling geological objects with splines*. In *PSI 1990 Annual Report* (Institut Francais du Petrole, Rueil Malmaison, France 1990).

BULANT, P. (2001), *Sobolev Scalar Products in the Construction of Velocity Models — Application to Model Hess and to SEG/EAGE Salt Model*, Pure appl. geophys. *159*, 1487–1506.

GOLD, N., SHAPIRO, S.A., BOJINSKY, S., and MÜLLER, T.M. (2000), *An Approach to Upscaling for Seismic Waves in Statistically Isotropic Heterogenous Elastic Media*, Geophysics 65, 1837–1850.

GRUBB, H.J. and WALDEN, A.T. (1995), *Smoothing Seismically Derived Velocities*, Geophys. Prosp. 43, 1061–1082.

KEERS, H., DAHLEN, F.A., and NOLET, G. (1997), *Chaotic Ray Behaviour in Regional Seismology*, Geophys. J. Int. *131*, 361–380.

KLIMEŠ, L. (1984), *Expansion of a High-frequency Time-harmonic Wavefield Given on an Initial Surface into Gaussian Beams*, Geophys. J. R. astr. Soc. *79*, 105–118.

KLIMEŠ, L. (1989), *Gaussian Packets in the Computation of Seismic Wavefields*, Geophys. J. Int. *99*, 421–433.

KLIMEŠ, L., *Lyapunov exponents for 2-D ray tracing without interfaces*. In *Seismic Waves in Complex 3-D Structures, Report 8* (Dep. Geophys., Charles Univ., Prague 1999) pp. 83–96.

KLIMEŠ, L. (2001), *Lyapunov Exponents for 2-D Ray Tracing Without Interfaces*, Pure appl. geophys. *159*, 1465–1485.

KATOK, S.R., *The estimation from above for the topological entropy of a diffeomorphism*. In *Global Theory of Dynamical Systems. Lecture Notes in Mathematics, vol. 819* (eds. Netecki, Z. and Robinson, C.) (Springer, Berlin, Heidelberg, New York 1980) pp. 258–264.

LYAPUNOV, A.M., *Problème Général de la Stabilité du Mouvement, Annals of Mathematical Studies, vol. 17* (Princeton University Press 1949).

MCCAULEY, J.L., *Chaos, Dynamics and Fractals: An Algorithmic Approach to Deterministic Chaos* (Cambridge University Press, Cambridge 1993).

MÜLLER, T.M. and SHAPIRO, S.A. (2000), *Most Probable Seismic Pulses in Single Realizations of Two- and Three-dimensional Random Media*, Geophys. J. Int. *144*, 83–95.

OSELEDEC, V.I. (1968), *A Multiplicative Ergodic Theorem: Lyapunov Characteristic Numbers for Dynamical Systems*, Trans. Moscow Math. Soc. *19*, 179–210 in Russian, 197–231 English translation.

SMITH, K.B., BROWN, M.G., and TAPPERT, F.D. (1992), *Ray Chaos in Underwater Acoustics*, J. Acoust. Soc. Am. *91*, 1939–1949.

TAPPERT, F.D. and TANG, X. (1996), *Ray Chaos and Eigenrays*, J. Acoust. Soc. Am. *99*, 185–195.

TARANTOLA, A., *Inverse Problem Theory* (Elsevier, Amsterdam 1987).

VERSTEEG, R.J. (1993), *Sensitivity of Prestack Depth Migration to the Velocity Model*, Geophysics *58*, 873–882.

VERSTEEG, R.J., *Analysis of the Problem of the Velocity Model Determination for Seismic Imaging* (Ph.D. Thesis, University of Paris VII 1991).

VERSTEEG, R.J. and GRAU, G. (eds.), *The Marmousi Experience* (Eur. Assoc. Explor. Geophysicists, Zeist 1991).

WITTE, O., ROTH, M., and MÜLLER, G. (1996), *Ray Tracing in Random Media*, Geophys. J. Int. *124*, 159–169.

(Received October 18, 2000, revised April 9, 2001, accepted May 15, 2001)

 To access this journal online:
http://www.birkhauser.ch

Pure appl. geophys. 159 (2002) 1527–1562
0033–4553/02/081527–36 $ 1.50 + 0.20/0

Converted *PS*-wave Reflection Coefficients in Weakly Anisotropic Media

PETR JÍLEK[1]

Abstract — I derive converted *PS*-wave reflection coefficients at a planar weak-contrast interface separating two weakly anisotropic half-spaces using first-order perturbation theory. The general expressions are further specified for the interface separating any of the two following media: isotropic, transversely isotropic with a vertical symmetry axis (VTI), transversely isotropic with a horizontal symmetry axis (HTI) and orthorhombic. Relatively simple forms of small-angle reflection coefficients are also obtained. The coefficients are expressed as functions of Thomsen-type medium parameters and incidence and azimuthal phase angles. Derived expressions, as well as their application, are more complicated than the corresponding expressions for *PP*-wave reflection coefficients. General characteristics and pitfalls are discussed. Numerical tests reveal a good agreement between exact and approximate coefficients for most models presented.

Key words: Reflection coefficients, converted waves, seismic anisotropy, perturbation theory, AVO analysis.

Introduction

In seismic exploration, amplitude variation with offset (AVO) and amplitude variation with azimuth (AVAZ) analyses are frequently-used methods for "direct detection" of hydrocarbons. Continual development of AVO and AVAZ techniques helps to improve estimates of such properties of hydrocarbon reservoirs as porosity, crack density, orientation of cracks and fluid content.

In AVO and AVAZ, the crucial quantity to be analyzed is the reflection coefficient at the target horizon (such as the top of a hydrocarbon reservoir) as a function of incidence angle and azimuth. The reflection and transmission coefficients are, however, generally complicated functions of medium parameters, especially in anisotropic media, where their analytic expressions may not exist. Therefore, simplified approximations of reflection/transmission coefficients are of great importance in practical application of AVO and AVAZ analyses.

[1] Center for Wave Phenomena, Colorado School of Mines, Golden, CO, 80401.
E-mail: pjilek@dix.mines.edu

Several approximations of reflection/transmission coefficients for isotropic media can be found, for example, in BORTFELD (1961), RICHARDS and FRASIER (1976), AKI and RICHARDS (1980), and SHUEY (1985). The derivation of such approximations is commonly based on the assumption of weak contrasts of P- and S-wave velocities and density across the interface under investigation.

For anisotropic models, derivation of the approximations is significantly more complicated and usually includes the additional assumption of weak anisotropy above and below the interface. The P-wave reflection coefficient for weakly anisotropic VTI media (transversely isotropic with a vertical symmetry axis) and small incidence angles was first introduced by BANIK (1987) and later extended for larger incidence angles by THOMSEN (1993). RÜGER (1996, 1997, 1998) refined THOMSEN's (1993) expression and derived the complete set of approximate reflection and transmission coefficients for VTI media. He further extended the derivation for symmetry planes of HTI (transversely isotropic with a horizontal symmetry axis) and orthorhombic media, provided that their vertical symmetry planes in the incidence and reflecting half-spaces are aligned. Also, he obtained the azimuthally-dependent P-wave reflection coefficient for an interface between two HTI media. Finally, PŠENČÍK and VAVRYČUK (1998) and VAVRYČUK and PŠENČÍK (1998) presented azimuthally-dependent weak-contrast, weak-anisotropy P-wave reflection/transmission coefficients valid for arbitrarily anisotropic media.

Recent development of acquisition techniques, such as ocean bottom cable (OBC) surveys, has resulted in collecting high-quality 3-D 3-C data, therefore a joint P- and S-waves analysis becomes realistic. In terms of AVO and AVAZ analyses, converted PS-wave reflection coefficients may provide important additional information (JIN, 1999; MILEY, 1999) with relatively little additional cost. For isotropic media, approximate forms of PS-wave reflection coefficients are described, for example, in DONATI (1998), LARSEN et al. (1999), ALVAREZ et al. (1999) and NEFEDKINA and BUZLUKOV (1999). An approximation of PS-wave reflection coefficients, derived for symmetry-plane reflection at an HTI/HTI interface, has been used by LI et al. (1996). RÜGER (1996) derived the approximate PS-wave reflection coefficients for VTI media and also for symmetry planes of HTI media. The most general approximations for converted PS-wave reflection coefficients probably have been published by VAVRYČUK (1999). Unfortunately, the results of Vavryčuk are not well suited for practical AVO analysis.

Here, I derive azimuthally-dependent weak-contrast, weak-anisotropy PS-wave reflection coefficients for a horizontal interface separating two arbitrarily anisotropic media. The resulting formulas are further discussed for the following half-spaces and their combinations: isotropic, VTI, HTI and orthorhombic. An arbitrary azimuth of vertical symmetry planes is allowed in the cases of HTI and orthorhombic media. The coefficients are written as functions of Thomsen-type medium parameters (THOMSEN, 1986; TSVANKIN, 1997a,b), and incidence and azimuthal angles. Relatively simple forms of small-incidence-angle reflection coefficients are also obtained. The derived approximations are consistent with the existing expressions derived by RÜGER (1996). I discuss some characteristics of the approximations, based on the comparison with the approximations previously derived for *PP* reflection coefficients, and test the accuracy of the obtained expressions on several azimuthally anisotropic models.

Analytic Development

To find a linearized form of the plane *PS*-wave reflection coefficients for arbitrarily anisotropic media, I follow the approach by VAVRYČUK and PŠENČÍK (1998) introduced for *PP*-wave reflection coefficients (Appendix A). Their derivation is, in principle, similar to that of BANIK (1987) and THOMSEN (1993). However, a different medium parameterization and more complicated machinery are necessary in order to account for arbitrary anisotropy.

Perturbation Approach

As in the case of exact reflection/transmission coefficients, the approximate forms result from evaluation of boundary conditions satisfied at the interface (i.e., the continuity of the displacement and traction vectors across the interface). Here, however, the boundary conditions are evaluated only approximately, assuming weak contrasts of elastic medium parameters across the interface, and weak anisotropy in both half-spaces.

As an auxiliary step, consider a homogeneous isotropic full-space separated by a fictitious planar interface into two half-spaces characterized by the same density ρ^0 and identical sets of density-normalized stiffness coefficients

$$a^0_{ijkl} = (\alpha^2 - 2\beta^2)\delta_{ij}\delta_{kl} + \beta^2(\delta_{ik}\delta_{jl} + \delta_{il}\delta_{jk}) \ , \tag{1}$$

where α and β represent the *P*- and *S*-wave velocities (for consistency, I use the same notation as VAVRYČUK and PŠENČÍK, 1998). I shall refer to such an isotropic space as the *background* medium. The elastic medium parameters of the true half-spaces under consideration can then be expressed in terms of perturbations from the background medium as

$$\begin{aligned} a^{(I)}_{ijkl} &= a^0_{ijkl} + \delta a^{(I)}_{ijkl}, \\ \rho^{(I)} &= \rho^0 + \delta\rho^{(I)} \ , \end{aligned} \tag{2}$$

where $I = 1, 2$ denote the incidence and reflecting half-spaces, respectively. Assuming that the perturbations $\delta a^{(I)}_{ijkl}$ and $\delta\rho^{(I)}$ for both $I = 1, 2$ are small, i.e.,

$$|\delta a^{(I)}_{ijkl}| \ll ||a^0_{ijkl}||, \quad |\delta\rho^{(I)}| \ll \rho^0 \ , \tag{3}$$

(for example, VAVRYČUK and PŠENČÍK, 1998, define the norm $||a^0_{ijkl}||$ as $||a^0_{ijkl}|| = \max|a^0_{ijkl}|$), it is possible to simplify the exact boundary conditions by linearizing them in δ quantities.

Notice that conditions (3) are equivalent to both the assumption of weak contrasts of the elastic parameters a_{ijkl} and density ρ across the interface, and the assumption of weak anisotropy in both half-spaces.

Slowness and Polarization Vectors

In order to linearize the exact boundary conditions at the interface, the slownesses and polarization vectors of the incident and all reflected and transmitted waves must be determined.

Consider a horizontal plane interface [the normal $n = (0, 0, 1)$], and an incident plane P-wave propagating in the half-space denoted as 1 and approaching the interface in the negative z direction; see Figure 1 for basic convention. For simplicity, the incidence plane, defined by the normal n and the slowness vector of the incident P-wave, coincides with the [x, z] plane of the reference Cartesian coordinate system (later, the formulas are generalized for an arbitrary azimuth, see Appendix B). In generally anisotropic media, three reflected and three transmitted waves, here denoted as P-wave (quasi compressional), and S_1- and S_2-waves (quasi shear), are generated at the interface.

Under the assumption of weak anisotropy, the slowness vectors can be written in terms of perturbations from the slowness vectors in the background medium, i.e.,

$$p^{(N)} \approx p^{0(N)} + \delta p^{(N)} \ , \tag{4}$$

where $N = 0, 1, 2, 3$ correspond to the incident P wave, and reflected S_1, S_2 and P waves, respectively, and $N = 4, 5, 6$ correspond to S_1-, S_2- and P- transmitted waves, respectively. The quantities $p^{0(N)}$ represent the slowness vectors in the isotropic background medium and satisfy the following relations:

$$\begin{aligned}
p^{0(0)} = p^{0(6)} &= (\sin i/\alpha, 0, \cos i/\alpha), \\
p^{0(1)} = p^{0(2)} &= (\sin i/\alpha, 0, -\cos j/\beta), \\
p^{0(3)} &= (\sin i/\alpha, 0, -\cos i/\alpha), \\
p^{0(4)} = p^{0(5)} &= (\sin i/\alpha, 0, \cos j/\beta) \ ,
\end{aligned} \tag{5}$$

where α and β are the isotropic background P- and S-wave velocities, as defined in equation (1), i is the P-wave incidence phase angle, and j is the S-wave reflection phase angle (Fig. 1). The quantities $\delta p^{(N)}$ in equation (4) are linear perturbations of the corresponding background slownesses due to the perturbations of medium parameters δa_{ijkl} defined in equation (2). Using the first-order perturbation theory (JECH and PŠENČÍK, 1989), VAVRYČUK and PŠENČÍK (1998) derived analytic

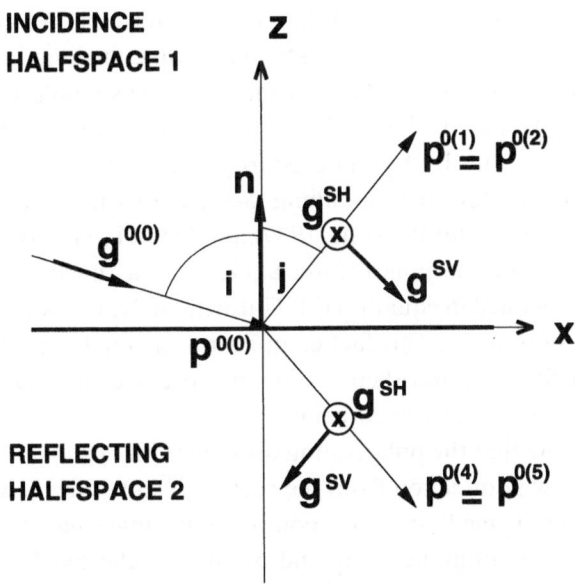

Figure 1
Convention for the zero-azimuth reflection/transmission (in the incidence plane [x, z]) in the background medium. The vector n is the normal to the interface, $p^{0(0)}$ is the background slowness vector of the incidence P wave, and $p^{0(1)} = p^{0(2)}$ and $p^{0(4)} = p^{0(5)}$ are the background slowness vectors of reflected and transmitted S waves, respectively. The angles i and j represent the P-wave incidence and the S-wave reflection/transmission phase angles, respectively. The vector $g^{0(0)}$ denotes the polarization vector of the incident P wave in the background medium, and g^{SV} and g^{SH} are the reflected SV- and SH-wave polarization vectors in the background medium (g^{SH} points away from the reader, parallel to the y axis).

expressions for the perturbations $\delta p^{(N)}$ as linear functions of the perturbations δa_{ijkl}; see also Appendix A.

Similarly to expressions for the slowness vectors (4), the polarization vectors can be written as

$$g^{(N)} \approx g^{0(N)} + \delta g^{(N)} \ , \tag{6}$$

where N has the same meaning as that in equation (4), $g^{0(N)}$ are the background slowness vectors, and $\delta g^{(N)}$ the corresponding linear perturbations derived by VAVRYČUK and PŠENČÍK (1998) (they can be found in Appendix A). The P-wave background polarization vectors are given as

$$g^{0(0)} = g^{0(6)} = (\sin i/\alpha, 0, \cos i),$$
$$g^{0(3)} = (\sin i/\alpha, 0, -\cos i) \ . \tag{7}$$

The S-wave polarization vectors in isotropic media usually are projected onto the incidence [x, z] plane (SV-wave polarization, here denoted as g^{SV}) and the direction perpendicular to the plane (SH-wave polarization denoted as g^{SH}); see Figure 1. JECH and PŠENČÍK (1989), however, showed that such a choice is no longer acceptable if the S-wave polarization vectors in a weakly anisotropic medium are to be found by means of small (linear) perturbations from the S-wave polarization vectors in the background isotropic medium. To minimize the perturbations, the chosen polarization vectors in the background isotropic medium must be rotated in the plane perpendicular to the corresponding S-wave slowness vector by a certain *polarization angle* Φ. JECH and PŠENČÍK (1989) and PŠENČÍK (1998) derived an explicit analytical expression for the polarization angle Φ as a function of the parameter perturbations δa_{ijkl} defined in equation (2). Unfortunately, the polarization angle Φ is not a linear function of δa_{ijkl}. This fact complicates the final approximations for *PS*-wave reflection coefficients, and their application becomes more involved than that for *PP*-wave reflections, as discussed below.

For now, I assume that the polarization angle Φ is known and is taken as positive in the counter-clockwise direction from the vector g^{SV} towards the vector g^{SH}. In the incidence background medium, the polarization angle Φ defines the "best" orientation of the two components S_1^0 and S_2^0 of the reflected S wave in the plane $[g^{SV}, g^{SH}]$. Similar to the relationships for the familiar SV and SH waves, the S_1^0 and S_2^0 components are mutually perpendicular; their polarization vectors are

$$g^{0(1)} = (-\cos j \cos \Phi, \sin \Phi, -\sin j \cos \Phi),$$
$$g^{0(2)} = (\cos j \sin \Phi, \cos \Phi, \sin j \sin \Phi) \ . \tag{8}$$

Clearly, if $\Phi = 0$ (as in isotropic media, see JECH and PŠENČÍK, 1989), S_1^0 reduces to the conventional SV wave (polarized in g^{SV} direction) and S_2^0 to the SH wave (polarized in g^{SH} direction). The background polarization vectors (8), together with the perturbations introduced in equation (6), define the reflected S_1 and S_2 waves in the weakly anisotropic medium.

Similarly, for the transmitted S waves,

$$g^{0(4)} = (\cos j \cos \Psi, \sin \Psi, -\sin j \cos \Psi),$$
$$g^{0(5)} = (-\cos j \sin \Psi, \cos \Psi, \sin j \sin \Psi) \ , \tag{9}$$

where Ψ is the corresponding polarization angle in the reflecting half-space.

Once the slowness and polarization vectors are specified, it is possible to linearize the boundary conditions at the interface and analytically evaluate all reflection and transmission coefficients for a given incidence wave. For detailed derivation of the linearized boundary conditions, see VAVRYČUK and PŠENČÍK (1998); their derivation is summarized in Appendix A.

General Forms of the Approximate PS-wave Reflection Coefficients

Appendices A and B describe how the linearized boundary conditions can be used to derive the approximations of the *PS*-wave reflection coefficients. In the first step, approximations are obtained for the incidence plane $[x, z]$ only (Appendix A); then they are generalized for an arbitrary azimuthal angle, denoted in the following as ψ (Appendix B). The final and most general formulas can be written as follows:

$$R_{PS_1} = R_{PSV} \cos \Phi + R_{PSH} \sin \Phi,$$
$$R_{PS_2} = -R_{PSV} \sin \Phi + R_{PSH} \cos \Phi .$$
(10)

Here, R_{PS_1} and R_{PS_2} are the approximate reflection coefficients of the PS_1- and PS_2-reflected waves, respectively, and the angle Φ is the polarization angle defined in equations (8). The terms R_{PSV} and R_{PSH} depend on the incidence angle i, the reflection angle j and the azimuthal angle ψ, and are linear functions of the contrasts

$$\Delta A_{ij} = A_{ij}^{(2)} - A_{ij}^{(1)} ,$$
(11)

where $A_{ij}^{(1)}$ and $A_{ij}^{(2)}$ represent the density normalized elastic medium parameters a_{klmn} of the incidence and reflecting half-spaces, respectively, written in the well-known Voigt convention. Explicit expressions for R_{PSV} and R_{PSH} are listed in Appendix B [equations (B-3)-(B-5)] and will be discussed in detail later.

It should be emphasized that, in the weak-anisotropy approximation, coefficients (10) determine the amplitudes of the PS_1- and PS_2-reflected waves polarized in proximity of the $g^{0(1)}$ and $g^{0(2)}$ directions, respectively [see equations (8)]. Thus, the true amplitudes $U^{(1)}$ and $U^{(2)}$ of the reflected PS_1 and PS_2 waves can be approximated as

$$U^{(1)} g^{(1)} \approx g^{0(1)} \cdot R_{PS_1} \cdot U^{(0)},$$
$$U^{(2)} g^{(2)} \approx g^{0(2)} \cdot R_{PS_2} \cdot U^{(0)} ,$$
(12)

where $g^{(1)}$ and $g^{(2)}$ are the corresponding true polarization vectors and $U^{(0)}$ is the amplitude of the incidence P wave. The perturbations of the true polarization vectors $\delta g^{(1)}$, $\delta g^{(1)}$ [equation (6)] are neglected in equation (12) since they contribute to the nonlinear (higher-order) terms only. This important concept will be used later to account for the polarization angle Φ.

Clearly, the coefficients (10) are not linear functions of the medium parameters A_{ij} due to the nonlinearity of the angle Φ. Therefore, the term "*linearized reflection coefficient*," commonly used for P-wave approximations, cannot be strictly applied to converted waves. Another important difference from the P-wave coefficient is that the coefficients (10) contain all 21 contrast parameters ΔA_{ij} (see Appendix B). As shown in VAVRYČUK and PŠENČÍK (1998), the approximations for R_{PP} contain only 13 contrasts of medium parameters, even in the most general case (triclinic symmetry).

Another interesting (but not surprising) observation, which can be made from equations (10) and (B-3)-(B-5), is that even a vertically incident P wave can generate a reflected S-wave at a horizontal interface. The quantities largely responsible for such a reflection, in the first-order approximation, are the elastic parameters ΔA_{34} and ΔA_{35}. In isotropic media, A_{34} and A_{35} vanish, and the conversion at normal incidence does not exist. This is also true for a horizontal interface between two anisotropic half-spaces with horizontal symmetry planes. Similarly, the contrasts in medium parameters mostly responsible for the "unusual" incidence-plane *P-SH* reflections (i.e., the reflected *S* wave has a nonzero displacement component perpendicular to the incidence plane) can be identified. For instance, for the incidence plane coinciding with the [x, z] plane of the reference coordinate system, nonzero contrasts ΔA_{14}, ΔA_{16}, ΔA_{34}, ΔA_{36}, ΔA_{45} and ΔA_{56} would generate such a reflection. It will be shown later that this reflection can occur, for example, at an isotropic/HTI or VTI/HTI interface.

R_{PS_1} and R_{PS_2} Coefficients for Orthorhombic Media

Here, the general equations (10) are specified for an interface between two orthorhombic media, using the Thomsen-type parameterization of TSVANKIN (1997b). Assuming orthorhombic (or higher) symmetry leads to a significant simplification of the general equations (10). The resulting expressions for orthorhombic media can then also be easily adapted for an interface separating any two of the following media: isotropic, VTI and HTI.

The first step is to choose the background velocities α and β and the density ρ^0. Following BANIK (1987) and others, the background values are defined as the arithmetic mean of the corresponding quantities in both half-spaces, i.e.,

$$
\begin{aligned}
\alpha &\equiv \bar{\alpha} \equiv \tfrac{1}{2}(\alpha^{(1)} + \alpha^{(2)}), \\
\beta &\equiv \bar{\beta} \equiv \tfrac{1}{2}(\beta^{(1)} + \beta^{(2)}), \\
\rho^0 &\equiv \bar{\rho} \equiv \tfrac{1}{2}(\rho^{(1)} + \rho^{(2)}) \ .
\end{aligned}
\tag{13}
$$

It is not immediately clear which velocities (in which direction) in (13) should be averaged for anisotropic media. It has been suggested (PŠENČÍK and MARTINS, 2000), however, that for small and moderate incidence angles, the choice of vertical velocities is appropriate. Small- and moderate-incidence-angle reflection coefficients are also most important for practical applications.

The choice of vertical velocity is still non-unique for *S* waves since anisotropic media have two vertically propagating *S* waves with different velocities. Numerical tests I carried out for the approximate R_{PP} coefficients suggest to choose the *S*-wave background velocity anywhere between the two vertical velocities of S_1 and S_2 waves. The resulting formulas differ slightly in accuracy, although these differences are often negligible, especially for small and moderate incidence angles. This conclusion is also in agreement with that of PŠENČÍK and MARTINS (2000). Therefore, I define the background velocities (13) as the average of the following velocities of the two half-spaces:

$$\alpha^{(I)} = \sqrt{A_{33}^{(I)}}, \quad \beta^{(I)} = \sqrt{A_{55}^{(I)}}, \quad I = 1, 2 \ . \tag{14}$$

The second important step in the derivation is the choice of medium parameterization. TSVANKIN (1997b) introduced an efficient way of parameterizing orthorhombic media by generalizing Thomsen's notation for VTI media (THOMSEN, 1986). An almost identical parameterization proved to be useful for the approximate reflection coefficients as well. Without changes I adopt the following anisotropy parameters (TSVANKIN, 1997b):

$$\epsilon_I^{(1)} \equiv \frac{A_{22}^{(I)} - A_{33}^{(I)}}{2A_{33}^{(I)}} \ , \tag{15}$$

$$\epsilon_I^{(2)} \equiv \frac{A_{11}^{(I)} - A_{33}^{(I)}}{2A_{33}^{(I)}} \ , \tag{16}$$

$$\gamma_I^{(S)} \equiv \frac{A_{44}^{(I)} - A_{55}^{(I)}}{2A_{55}^{(I)}} \ , \tag{17}$$

where $I = 1, 2$ represent the incidence and reflecting half-spaces, respectively. Further, I introduce three other parameters,

$$\tilde{\delta}_I^{(1)} \equiv \frac{A_{23}^{(I)} + 2A_{44}^{(I)} - A_{33}^{(I)}}{A_{33}^{(I)}} \ , \tag{18}$$

$$\tilde{\delta}_I^{(2)} \equiv \frac{A_{13}^{(I)} + 2A_{55}^{(I)} - A_{33}^{(I)}}{A_{33}^{(I)}} \ , \tag{19}$$

$$\tilde{\delta}_I^{(3)} \equiv \frac{A_{12}^{(I)} + 2A_{66}^{(I)} - A_{11}^{(I)}}{A_{11}^{(I)}} \ . \tag{20}$$

They differ from those in TSVANKIN (1997b) defined as

$$\delta_I^{(1)} \equiv \frac{\left(A_{23}^{(I)} + A_{44}^{(I)}\right)^2 - \left(A_{33}^{(I)} - A_{44}^{(I)}\right)^2}{2A_{33}^{(I)}\left(A_{33}^{(I)} - A_{44}^{(I)}\right)} \ , \tag{21}$$

$$\delta_I^{(2)} \equiv \frac{\left(A_{13}^{(I)} + A_{55}^{(I)}\right)^2 - \left(A_{33}^{(I)} - A_{55}^{(I)}\right)^2}{2A_{33}^{(I)}\left(A_{33}^{(I)} - A_{55}^{(I)}\right)} \ , \tag{22}$$

$$\delta_I^{(3)} \equiv \frac{\left(A_{12}^{(I)} + A_{66}^{(I)}\right)^2 - \left(A_{11}^{(I)} - A_{66}^{(I)}\right)^2}{2A_{11}^{(I)}\left(A_{11}^{(I)} - A_{66}^{(I)}\right)} \ . \tag{23}$$

SAYERS (1994) showed that the anisotropy parameters (18), (19) and (20) represent linearized forms of the parameters (21), (22) and (23), respectively, and, for weak anisotropy, the two sets of parameters are numerically close to one other. Although the derivation of R_{PS_1} and R_{PS_2} must be carried out using parameters (18)–(20), either set of parameters can be used in the final formulas without a significantly changing accuracy (PŠENČÍK and MARTINS, 2000). For strongly anisotropic media, where the two parameter sets would produce different results, the approximations are not sufficiently accurate anyway. It should be emphasized that the parameters (15)–(23) are defined in a local Cartesian coordinate system with the axes x_1, x_2 and x_3 lying within three symmetry planes of the orthorhombic medium under investigation. A generalized form of the parameterization (15)–(20) for arbitrary weakly anisotropic media has been introduced by MENSCH and RASOLOFOSAON (1997), and PŠENČÍK and GAJEWSKI (1998).

As shown below, the six anisotropy parameters (15)–(20) [together with the background parameters (13) and (14)] are sufficient for expressing the R_{PS_1} and R_{PS_2} coefficients, whereas orthorhombic media are fully described by seven anisotropy parameters (plus two vertical velocities). TSVANKIN's (1997b) original notation for orthorhombic media contains the parameters (15), (16), (21)–(23) and two other parameters γ_1 and γ_2. The parameter $\gamma^{(S)}$ [definition (17)] is then defined as an auxiliary parameter depending on γ_1 and γ_2. The approximations of R_{PS_1}, R_{PS_2}, however, are more sensitive to the difference $\gamma_1 - \gamma_2$ rather than to the parameters γ_1 and γ_2 separately. Thus, the number of parameters can be reduced from seven to six. TSVANKIN (1997b) shows that the parameter $\gamma^{(S)}$ reduces to the difference $\gamma_1 - \gamma_2$ in the weak-anisotropy approximation.

In the following, I assume that one of the symmetry planes of both orthorhombic half-spaces is horizontal. The x axis of the reference coordinate system introduced in Figure 1 lies within one of the vertical symmetry planes of the incidence orthorhombic half-space (the $[x_1, x_3]$ symmetry plane of the local coordinate system used in the definitions of the anisotropy parameters above). I also assume that the vertical symmetry planes of the reflecting orthorhombic half-space are rotated with respect to the vertical planes of the incidence orthorhombic half-space by an angle denoted as κ. Then, the following elastic parameters vanish:

$$A_{14}^{(1)} = A_{15}^{(1)} = A_{16}^{(1)} = A_{24}^{(1)} = A_{25}^{(1)} = A_{26}^{(1)} = A_{34}^{(1)}$$
$$= A_{35}^{(1)} = A_{36}^{(1)} = A_{45}^{(1)} = A_{46}^{(1)} = A_{56}^{(1)} = 0,$$
$$A_{14}^{(2)} = A_{15}^{(2)} = A_{24}^{(2)} = A_{25}^{(2)} = A_{34}^{(2)} = A_{35}^{(2)} = A_{46}^{(2)} = A_{56}^{(2)} = 0 \ . \tag{24}$$

Substituting relations (24), along with the definitions (13)–(20), into equations (B.3), (B.4) and (B.5), and further linearizing them in the anisotropy parameters (15)–(20), yields the final formulas for the PS-wave reflection coefficients

$$R_{PS_1} = R_{PSV} \cos \Phi + R_{PSH} \sin \Phi,$$
$$R_{PS_2} = -R_{PSV} \sin \Phi + R_{PSH} \cos \Phi . \tag{25}$$

Equations (25) are formally identical to general equations (10). Here, however,

$$R_{PSV} = V_1 \frac{\sin i}{\cos j} + V_2 \cos i \sin i + V_3 \frac{\sin^3 i}{\cos j} + V_4 \cos i \sin^3 i + V_5 \frac{\sin^5 i}{\cos j},$$

$$R_{PSH} = H_1 \sin i + H_2 \frac{\cos i \sin i}{\cos j} + H_3 \sin^3 i + H_4 \frac{\cos i \sin^3 i}{\cos j} , \tag{26}$$

and i, j are the incidence and reflection phase angles, respectively. Coefficients V_l and H_l are functions of the medium parameters [equations (13)–(20)], the azimuthal angle ψ and the angle κ. The explicit expressions for V_l and H_l are given in Appendix C [equations (C.1) and (C.2)]. Finally, the polarization angle Φ in equations (25) needs to be evaluated. The angle Φ can be computed analytically as long as all anisotropy parameters of the incidence medium are known. Equations for Φ are given in Appendix D. As discussed below, however, the polarization angle Φ can also be estimated directly from reflection data. This is important in practical applications since the knowledge of the medium parameters becomes unnecessary.

R_{PS_1} and R_{PS_2} Coefficients for Isotropic, VTI, and HTI Half-spaces

Since isotropic VTI and HTI media are special cases of orthorhombic media, the coefficients R_{PS_1} and R_{PS_2} derived in the previous section can be readily applied for an interface separating any two of such half-spaces. In order to do so, the anisotropy parameters in V_l and H_l [equations (26)] must be adapted for a specific model, as shown by TSVANKIN (1997b). The necessary anisotropy parameters used for VTI media are defined as

$$\epsilon \equiv \frac{A_{11} - A_{33}}{2A_{33}},$$

$$\delta \equiv \frac{(A_{13} + A_{55})^2 - (A_{33} - A_{55})^2}{2A_{33}(A_{33} - A_{55})} ,$$

and for HTI media with the symmetry axis in the x direction,

$$\epsilon^{(V)} \equiv \frac{A_{11} - A_{33}}{2A_{33}},$$

$$\delta^{(V)} \equiv \frac{(A_{13} + A_{55})^2 - (A_{33} - A_{55})^2}{2A_{33}(A_{33} - A_{55})}, \tag{27}$$

$$\gamma \equiv \frac{A_{44} - A_{66}}{2A_{66}} .$$

For a more detailed discussion of the anisotropy parameters for VTI and HTI media, see TSVANKIN (1996, 1997a) and THOMSEN (1986). For isotropic media, the anisotropy parameters (27) vanish in V_l, and $H_l = 0$ for all $l = 1, \ldots, 4$; see equations (C.1) and (C.2).

Table 1 shows the conversion key, relating anisotropy parameters for different anisotropic models. For example, if the upper medium is VTI, all parameters with the subscript 1 in equations (C.1) and (C.2) must be replaced by the corresponding VTI parameters listed in the third column of Table 1. Also, for isotropic VTI and HTI symmetry, the polarization angle Φ in equations (25) is simply determined from the expressions in Appendix D.

Discussion

Here, I compare expressions (25) for R_{PS_1} and R_{PS_2} with those obtained by RÜGER (1996) for the symmetry planes of HTI media. Then, I discuss fundamental differences of the R_{PS_1} and R_{PS_2} approximations from the approximations previously derived for R_{PP} coefficients. Finally, I describe the components R_{PSV} and R_{PSH} defined by equations (26).

Comparison of R_{PS_1} and R_{PS_2} with the Results of RÜGER (1996)

Expressions (10) and (25) are considerably more general that those obtained by RÜGER (1996). However, the derivation described in this paper is based on the same principle of linearization of the boundary conditions. Therefore, the resulting expressions presented here should be analogous to those obtained by RÜGER (1996) for the isotropy and symmetry-axis planes of HTI media (assuming that the symmetry axes of both HTI half-spaces are aligned). Indeed, expressions (25) specified for an HTI/HTI interface (using Table 1), for the azimuthal angles $\psi = 0°$ and $\psi = 90°$, and the symmetry-plane rotation angle $\kappa = 0$ [equations (C.1)–(C.2)],

Table 1

Conversion table of the anisotropy parameters (weak-anisotropy approximation) for VTI, HTI and orthorhombic media. VTI and HTI parameters are defined by equations (27). The symmetry axis of the HTI medium points in the x direction

Orthorhombic	HTI	VTI
$\epsilon^{(1)}$	0	ϵ
$\epsilon^{(2)}$	$\epsilon^{(V)}$	ϵ
$\gamma^{(S)}$	γ	0
$\tilde{\delta}^{(1)}$ or $\delta^{(1)}$	0	δ
$\tilde{\delta}^{(2)}$ or $\delta^{(2)}$	$\delta^{(V)}$	δ
$\tilde{\delta}^{(3)}$ or $\delta^{(3)}$	$\delta^{(V)} - 2\epsilon^{(V)}$	0

reduce to Rüger's expressions (see RÜGER, 1996, Appendix D). For the azimuths $\psi = 0°$ and $\psi = 90°$, the polarization angle Φ is equal to ψ (i.e., $0°$ and $90°$, respectively). A similar agreement has been also reported for the R_{PP} reflection coefficient by VAVRYČUK and PŠENČÍK (1998).

Strictly, the analogy is not exact, since expressions (25) include the linear parameters (18)–(20) rather than the nonlinear parameters (21)–(23) contained in Rüger's expressions. However, as discussed above, in weakly anisotropic media these two sets of parameters can be formally interchanged.

Another fundamental difference from the approach of RÜGER (1996) is that the general approach described here may provide a possibility to implement arbitrary isotropic background in equations (10) [see the discussion related to definitions (13) and (14)]. Having this choice is important for more complicated models, such as those with arbitrary orientation of the symmetry axis (for TI media) or symmetry planes (for orthorhombic media). Also, for such models it is more efficient to use a different parameterization from that defined by equations (15)–(20); see PŠENČÍK and MARTINS (2000).

R_{PS_1} and R_{PS_2} Coefficients Versus R_{PP} Coefficient

General equations (10), as well as more specific equations (25), indicate that practical application of R_{PS} reflection coefficients will be more problematic than that of R_{PP} coefficients.

The first apparent complication derives from the fact that equations (25) include both the incidence and reflection phase angles i and j, which are related by Snell's law in anisotropic media (this is not the case of the R_{PP} approximation in which the S-wave reflection angle j is eliminated during the derivation). This inconvenience, however, can be easily overcome by using the weak-contrast, weak-anisotropy assumption.

Since the factor $(\cos j)^{-1}$ in equations (26) is always multiplied by the contrasts in one of the medium parameters [contained in the corresponding coefficients V_1, V_3, V_5, and H_2, H_4, see equations (C.1)–(C.2)], it is possible to use purely isotropic Snell's law. The errors associated with such an approximation are of the second order, and thus can be neglected in the linearized approximation. Therefore,

$$\frac{1}{\cos j} \approx \frac{1}{\sqrt{1 - \frac{\bar{\beta}^2}{\bar{\alpha}^2}\sin^2 i}} \approx 1 + \frac{1}{2}\frac{\bar{\beta}^2}{\bar{\alpha}^2}\sin^2 i \ , \tag{28}$$

where $\bar{\alpha}$ and $\bar{\beta}$ are defined by equations (13) and (14). In the derivation of equation (28), an additional assumption of a small $\sin^2 i$ is used. Numerical tests show that such an assumption works well for a wide range of $\bar{\alpha}/\bar{\beta}$ ratios, and for incidence angles corresponding to the offset-to-depth ratios conventionally used in AVO. A similar approach to that used in equation (28) can be also used to express the coefficients R_{PS_1} and R_{PS_2} in terms of the reflection angle j (or horizontal slowness p_1, which may be convenient for some applications). Equation (28) can be directly substituted into either relations (B.3) or (26).

The second, more troublesome complication with R_{PS_1} and R_{PS_2} arises from the nonlinearity of the polarization angle Φ. For practical applications it would be desirable to eliminate the influence of the polarization angle Φ before further analysis.

Such an elimination is straightforward if the incidence medium is either isotropic or VTI, and $\cos \Phi = 1$ and $\sin \Phi = 0$. Also, for HTI media it is possible to use relations (D.6), if the azimuth (the angle between the incidence and symmetry-axis plane) is known. In principle, this should not represent a significant complication since the symmetry-axis plane (as well as isotropy plane) of HTI media can be identified by using the travel times and polarizations of S waves or the P-wave NMO ellipse (GRECHKA and TSVANKIN, 1998).

For an orthorhombic incidence half-space, however, the situation is more complicated. The general relations (D.1)–(D.5) can be applied only if all nine medium parameters of the half-space are known, which is not usually the case. Therefore, a different method must be used to account for the polarization angle Φ.

It is convenient to formally introduce the PS-wave reflection-coefficient vector as

$$\boldsymbol{R}_{PS} = (R_{PS_1}, R_{PS_2}) \ , \tag{29}$$

where R_{PS_1} and R_{PS_2} are defined by equations (25). As follows from equations (12), the vector (29) is specified in the coordinate system $\boldsymbol{g}^{0(1)} - \boldsymbol{g}^{0(2)}$ and approximates the exact reflection-coefficient vector $\boldsymbol{R}_{PS}^{ext} = (R_{PS_1}^{ext}, R_{PS_2}^{ext})$ specified in the coordinate system $\boldsymbol{g}^{(1)ext} - \boldsymbol{g}^{(2)ext}$ ($R_{PS_1}^{ext}$ and $R_{PS_2}^{ext}$ are the exact PS-wave reflection coefficients corresponding to the exact S-wave polarization vectors $\boldsymbol{g}^{(1)ext}$ and $\boldsymbol{g}^{(2)ext}$, respectively). The same reflection-coefficient vector (29) can also be projected onto $\boldsymbol{g}^{SV}$ and $\boldsymbol{g}^{SH}$ directions specified identically as in isotropic media (Fig. 1). It follows from equations (25) that such a projection results in the desired linear components R_{PSV} and R_{PSH}, i.e., $\boldsymbol{R}_{PS} = (R_{PSV}, R_{PSH})$. Since $\boldsymbol{R}_{PS}^{ext} \approx \boldsymbol{R}_{PS}$, the projection is, in the first-order approximation, equivalent to the projection of the exact reflection coefficient vector $\boldsymbol{R}_{PS}^{ext}$ extracted from field data onto the $\boldsymbol{g}^{SV}$ an $\boldsymbol{g}^{SH}$ directions.

The data projection described above is general, being applicable in any weakly anisotropic medium. For strong anisotropy, certainly, the projection will fail to produce satisfactory results, however equations (25) [as well as general equations (10)] are designed for only weak anisotropy. Note that the projection require's no additional operation with data; polarization analysis must be carried out in any case in order to properly recover the PS-wave amplitudes.

A similar approach can be used if only one of the R_{PS_1} and R_{PS_2} reflection coefficients is recovered from data. The directions of both S_1- and S_2-wave polarizations, however, still must be estimated. Subsequently, the polarization angle

Φ can be obtained directly as the angle between the $g^{(1)ext}$ and g^{SV} polarization vectors. Without the complementary PS-wave reflection coefficient, however, it is impossible to separate the R_{PSV} and R_{PSH} components. They must be analyzed together, using one of the relations (25).

Finally, we should highlight the problem of S-wave singular points in anisotropic media (slowness directions in which the slowness surfaces of the S_1 and S_2 waves intersect or touch each other, and both waves propagate with the same phase velocity). For an orthorhombic incidence half-space, approximations (25) generally suffer from numerical instability due to ambiguously determined polarization angle Φ at the S-wave singular points. For VTI and HTI incidence media, this instability is overcome by a formal separation of the S_1 and S_2 waves. Nevertheless, the approximations derived above should not be generally used near S-wave singularities since plane PS-wave reflection coefficients fail to represent amplitude signatures due to the complicated character of the wavefield in those regions.

R_{PSV} and R_{PSH} Components

In the following, let us assume that both reflection coefficients R_{PS_1} and R_{PS_2} can be recovered from seismic data, and thus the individual components R_{PSV} and R_{PSH} can be determined. Due to their linearity in medium parameters, the components R_{PSV} and R_{PSH} can then be used for AVO and AVAZ analyses similarly to R_{PP} reflection coefficients previously discussed in the literature. As can be seen from the corresponding equations, R_{PSV}, R_{PSH} and R_{PP} contain different linear combinations of the parameters of the medium under investigation (see equations (26), (C.1) and (C.2); for R_{PP} approximation, see VAVRYČUK and PŠENČÍK, 1998). Therefore, their joint linear inversion may provide a valuable tool for retrieving those parameters.

Equations (26), together with equations (C.1) and (C.2), show explicit forms of the components R_{PSV} and R_{PSH} for orthorhombic media. It follows from the equations that whereas the R_{PSV} component is a linear function of all anisotropy parameters (15)–(17) and (18)–(20) [or (21)–(23)] as well as isotropy contrasts $\Delta\rho/\bar{\rho}$ and $\Delta\beta/\bar{\beta}$, the other component, R_{PSH}, consists of the anisotropy parameters only. This is a consequence of the expected fact that there is only one nonzero reflection coefficient $R_{PS_1} = R_{PSV}$ in isotropic media [see equations (C.1)–(C.2), and equations (25) for $\Phi = 0°$]. A similar analysis reveals that only one nonzero coefficient $R_{PS_1} = R_{PSV}$ exists for isotropic/VTI, VTI/isotropic and VTI/VTI interfaces. All other interfaces generate both PS_1 and PS_2 reflections, although not necessarily for all azimuths. Notice that none of the approximations for PS_1 and PS_2 contains the P-wave velocity contrast $\Delta\alpha/\bar{\alpha}$, which always appears in the P-wave reflection coefficient.

Equations (26), (C.1), and (C.2) also clearly show how different medium parameters control R_{PSV} and R_{PSH} components in different ranges of the incidence and azimuthal angles. For example, small- and moderate-incidence-angle reflections are controlled mostly by the parameters $\Delta\rho/\bar{\rho}$, $\Delta\beta/\bar{\beta}$ and $\delta_1^{(1)}$, $\delta_2^{(1)}$, $\delta_1^{(2)}$, $\delta_2^{(2)}$, $\gamma_1^{(S)}$ and $\gamma_2^{(S)}$ [here, as well as in the following discussion, definitions (21)–(23) are used instead of (18)–(20)]. In contrast, large-incidence-angle reflections are also notably influenced by the parameters $\delta_1^{(3)}$, $\delta_2^{(3)}$ and $\epsilon_1^{(1)}$, $\epsilon_2^{(1)}$, $\epsilon_1^{(2)}$ and $\epsilon_2^{(2)}$. A similar conclusion is obtained for the R_{PP} approximations (RÜGER, 1997). If the symmetry planes of the incidence and reflecting orthorhombic half-spaces have close azimuths [small angle κ in equations (C.1) and (C.2)], then only the contrasts of the anisotropy parameters ($\Delta\delta^{(1)}$, $\Delta\delta^{(2)}$, $\Delta\delta^{(3)}$, $\Delta\gamma^{(S)}$, $\Delta\epsilon^{(1)}$, $\Delta\epsilon^{(2)}$) can be recovered.

Finally, as for the R_{PP} approximations, numerical accuracy of the R_{PSV} and R_{PSH} components generally decreases with increasing incidence angle i. Therefore, the medium parameters inverted using large-angle terms may contain major errors. Taking this into account, inversion for the parameters $\gamma^{(S)}$, $\delta^{(1)}$ and $\delta^{(2)}$ will certainly provide more reliable results than those for the parameters $\delta^{(3)}$, $\epsilon^{(1)}$ and $\epsilon^{(2)}$. Also, for weakly anisotropic half-spaces, the component R_{PSH} is expected to be small and thus less stable than the component R_{PSV}.

R_{PSV} and R_{PSH} Approximations for Small Incidence Angles

In practical AVO analysis one often uses reflection coefficients for relatively small incidence angles. Therefore, it is useful to simplify the components R_{PSV} and R_{PSH} from equations (26) by keeping only angular terms corresponding to small incidence angles.

Such a simplification is also justified by the fact that the approximations for all reflection coefficients generally lose accuracy at large incidence angles. This might be surprising since no assumption of small incidence angles has been used in the original derivation [an exception is equation (28), which is employed later in the derivation]. The loss of accuracy for large incidence angles has two main reasons. First, the terms associated with higher incidence angles contain increasing numbers of nonlinear factors (i.e., higher powers of the elastic parameters) and, eventually, linear factors may vanish from such terms completely. Thus, by neglecting all nonlinear factors we reduce the accuracy of the terms corresponding to higher incidence angles and, of course, that of the whole reflection coefficient. The second reason for losing the accuracy, applicable in anisotropic media only, is the choice of the background velocities $\bar{\alpha}$ and $\bar{\beta}$ [see discussion of equations (13) and (14)] that have been selected as the vertical velocities.

To obtain small-angle approximations of R_{PSV} and R_{PSH}, we neglect the second and higher powers of $\sin i$ in equations (26). Using expressions (C.1) and (C.2) and relation (28), we arrive at

$$R_{PSV} = \left\{ -\frac{1 + 2g}{2}\frac{\Delta\rho}{\bar{\rho}} - 2g\frac{\Delta\beta}{\bar{\beta}} + \frac{1}{2(1 + g)}\delta_2^{(2)}\cos^2(\psi - \kappa) \right.$$

$$+ \left[\frac{1}{2(1 + g)}\delta_2^{(1)} - 2g\gamma_2^{(S)} \right]\sin^2(\psi - \kappa) - \frac{1}{2(1 + g)}\delta_1^{(2)}\cos^2\psi$$

$$\left. - \left[\frac{1}{2(1 + g)}\delta_1^{(1)} - 2g\gamma_1^{(S)} \right]\sin^2\psi \right\}\sin i, \qquad (30)$$

$$R_{PSH} = \left\{ \left[\frac{1}{4(1 + g)}(\delta_2^{(2)} - \delta_2^{(1)}) + g\gamma_2^{(S)} \right]\sin 2(\psi - \kappa) \right.$$

$$\left. - \left[\frac{1}{4(1 + g)}(\delta_1^{(2)} - \delta_1^{(1)}) + g\gamma_1^{(S)} \right]\sin 2\psi \right\}\sin i ,$$

where g represents the background velocity ratio $g \equiv \bar{\beta}/\bar{\alpha}$.

Equations (30) work well up to incidence angles $15°$–$20°$. In order to improve the accuracy for larger angles, additional terms with the $\sin^3 i$ and $\cos i \sin^3 i$ factors would have to be included. Unfortunately, this would lead to a significant increase in complexity, as follows from equations (26), (C.1) and (C.2).

By analogy with the PP-wave reflection, equations (30) represent so-called PS-wave AVO gradients. Notice that the PS-wave AVO gradients are associated with the term $\sin i$, whereas the PP-wave AVO gradient corresponds to $\sin^2 i$. The PS-wave AVO gradients can be extracted from data by recovering the linear trends in plots $R_{PSV}(\sin i)$ and $R_{PSH}(\sin i)$. The AVO gradients can be used for fast rough estimates of the medium parameters as well as for a more sophisticated linear inversion of reflection coefficients. Joint inversion of the PP- and PS-wave AVO gradients can provide estimates of the parameters $\delta_{1,2}^{(1)}$, $\delta_{1,2}^{(2)}$ and $\gamma_{1,2}^{(S)}$ (assuming a nonzero angle κ). Such an inversion cannot be carried out using R_{PP} approximations only, since the corresponding expressions do not allow one to separate $\delta_{1,2}^{(1)}$ from $\gamma_{1,2}^{(S)}$ (see PŠENČÍK and MARTINS, 2000). The joint inversion of PP- and PS-wave AVO gradients for HTI media is also discussed in BAKULIN et al. (2000a).

Numerical Examples

I have tested the accuracy of the approximations (25) and (26) on several anisotropic models. Most of the models here represent fracture-induced anisotropic media, derived from the results of BAKULIN et al. (2000a,b).

For all models, I compute both exact R_{PSV} and R_{PSH} components, their approximations and the corresponding absolute errors as functions of incidence angle and azimuth (Figs. 2–4). The azimuth is measured from the x axis of the reference coordinate system as introduced in Figure 1, with the orientation of the x axis defined separately for each particular model. The exact R_{PSV} and R_{PSH} components are obtained by performing the projection of the exact R_{PS_1} and R_{PS_2} coefficients described above [see discussion related to equation (29)]. Thus, an error

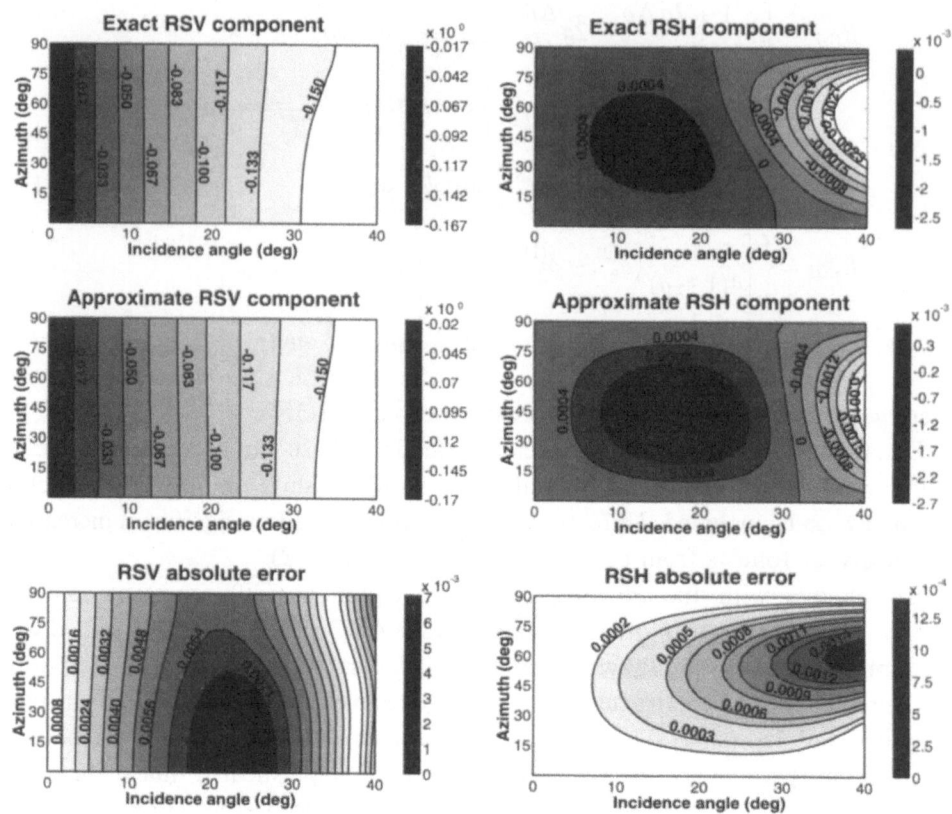

Figure 2

Components R_{PSV} (left) and R_{PSH} (right) of the PS-wave reflection coefficients for a VTI/HTI interface as functions of the incidence angle and azimuth: exact components (top), weak-contrast, weak-anisotropy approximations (middle) and absolute errors (bottom) defined as the difference between the exact and approximate components. Medium parameters are as follows: VTI incidence half-space: $\rho = 2.0 \text{g/cm}^3$, $V_{P0} = 2.9 \text{km/s}$, $V_{S0} = 1.5 \text{km/s}$, $\epsilon = 0.2$, $\delta = 0.1$, $\gamma = 0.1$; HTI reflecting half-space: $\rho = 2.2 \text{g/cm}^3$, $V_{P0} = 3.3 \text{km/s}$, $V_{44} = 1.8 \text{km/s}$, $\epsilon^{(V)} = -0.13$, $\delta^{(V)} = -0.14$ and $\gamma^{(V)} = -0.053$; $V_{44} = \sqrt{A_{44}}$.

associated with such a projection influences the final results. This may simulate the processing applied to field data in order to obtain the R_{PSV} and R_{PSH} components, which can be subsequently used in AVO analysis. Approximate R_{PSV} and R_{PSH} components are computed by using equations (26), (C.1), (C.2) and Table 1. The absolute error is determined as the absolute value of the difference between the exact and approximate coefficients.

Figure 2 shows the R_{PSV} (left) and R_{PSH} (right) components computed for an VTI/ HTI interface. The VTI incidence half-space represents a typical shale layer overlying an isotropic medium with vertical cracks resulting in HTI symmetry. In this model, the cracks are gas-filled and a moderate crack density $e = 5\%$ is assumed (the crack density e is defined as [number of cracks per unit volume] × [mean cubed diameter], for more detail, see BAKULIN *et al.*, 2000a). The medium parameters for both half-spaces can be found in the caption of Figure 2. The x axis of the reference coordinate system is horizontal and coincides with the symmetry axis of the HTI medium (perpendicular to the cracks).

The R_{PSV} component in Figure 2 is well represented by the approximate equations. The shape of the approximate component (i.e., its variation with the incidence and azimuthal angles) is close to that of the exact coefficient. The absolute error is small for the whole range of angles in the plot. However, because the R_{PSV} component increases with the incidence angle from zero at normal incidence, any estimates of the coefficient at small incidence angles will contain large relative errors. Likely, such a small-incidence-angle reflection event will be strongly distorted by noise. The shape of the R_{PSH} component is also approximated quite well, nonetheless the absolute errors are comparable to the magnitude of the coefficient itself. This should not be surprising since this component is small (note that the R_{PSH} component is always zero for isotropic media) and thus large relative errors can be expected. Therefore, if R_{PSH} is detected in field data, equations (26) and (C.2) can be used, primarily, for qualitative analysis rather than for quantitative estimates.

Figure 3 corresponds to an interface between two orthorhombic media with vertical symmetry planes rotated by 30°. Such a model can represent, for example, a vertical crack system above another one with a different azimuth embedded in an isotropic host rock. Each of the systems consists of two perpendicular crack sets with different crack densities. In the incidence half-space, the crack density of the cracks parallel to the $[x_2, x_3]$ plane of the local coordinate system (one of the symmetry planes) is $e_1 = 3\%$, and the perpendicular crack set (parallel to the other symmetry plane) has the crack density $e_2 = 5\%$ (for more details, see BAKULIN *et al.*, 2000b). The cracks are dry and embedded in the isotropic rock with the P- and S-wave velocities 3.6 km/s and 2.3 km/s, respectively. In the reflecting half-space, the crack sets are interchanged ($e_1 = 5\%$, $e_2 = 3\%$) and the isotropic host rock is characterized by the P- and S-wave velocities 4.17 km/s and 2.52 km/s, respectively. Additionally, the entire crack system of the reflecting half-space is rotated with respect to the crack system of the incidence half-space by the angle $\kappa = 30°$. The resulting medium parameters of both half-spaces are provided in the caption of Figure 3. The reference coordinate system is attached to the incidence orthorhombic medium, with the horizontal x axis perpendicular to the $[x_2, x_3]$ symmetry plane. This model represents the most general configuration that can be treated by equations (26), (C.1) and (C.2).

Figure 3 (left) depicts good accuracy of the approximation for the R_{PSV} component. The absolute error generally increases with azimuth but is small for all incidence angles plotted (especially for azimuths smaller than 60°). Slight differences in the azimuthal variation of the exact and approximate R_{PSV} components are associated with a small first-order term responsible for the azimuthal variation of the R_{PSV} approximation . In such a case, higher-order (neglected) terms become comparable to the first-order term and thus their influence is more significant. The

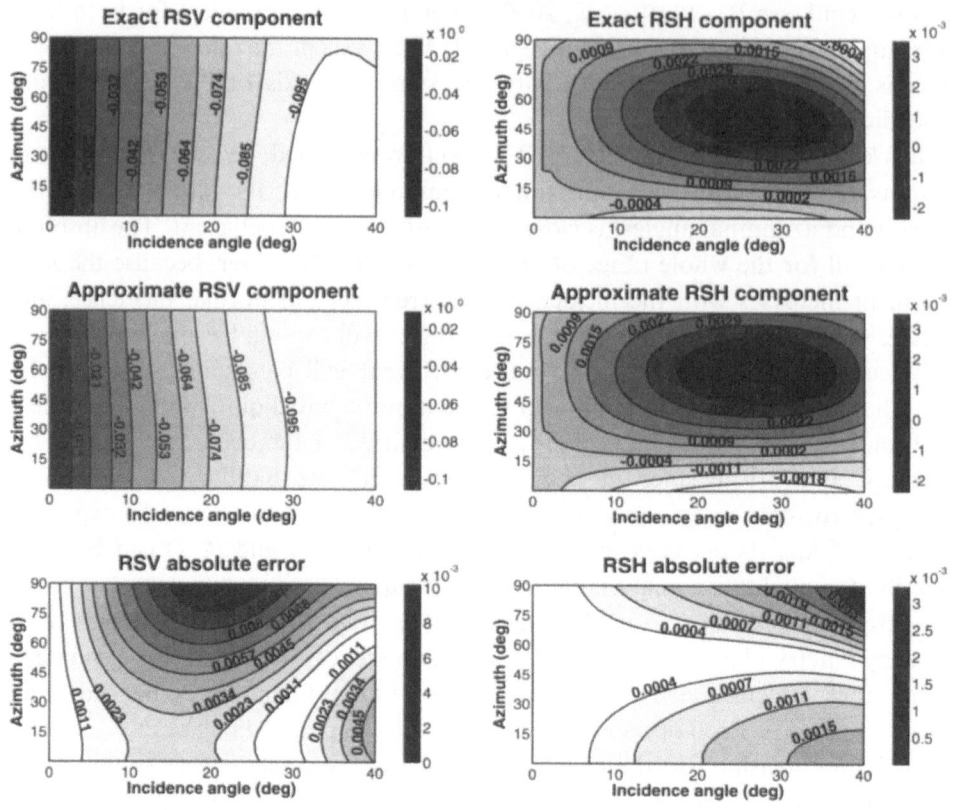

Figure 3

Components R_{PSV} (left) and R_{PSH} (right) of the PS-wave reflection coefficients for an orthorhombic/orthorhombic interface as functions of the incidence angle and azimuth: exact components (top), weak-contrast, weak-anisotropy approximations (middle) and absolute errors (bottom) defined as the difference between the exact and approximate components. The vertical symmetry planes of the orthorhombic half-spaces differ by the angle $\kappa = 30°$. Medium parameters are as follows: incidence medium: $\rho = 2.1 \text{g/cm}^3$, $V_{33} = 3.57 \text{km/s}$, $V_{44} = 2.16 \text{km/s}$, $\epsilon^{(1)} = -0.14$, $\delta^{(1)} = -0.14$, $\gamma^{(1)} = -0.06$, $\epsilon^{(2)} = -0.082$, $\delta^{(2)} = -0.082$, $\gamma^{(2)} = -0.035$, $\delta^{(3)} = -0.058$; reflecting medium: $\rho = 2.3 \text{g/cm}^3$, $V_{33} = 4.17 \text{km/s}$, $V_{44} = 2.52 \text{km/s}$, $\epsilon^{(1)} = -0.082$, $\delta^{(1)} = -0.082$, $\gamma^{(1)} = -0.035$, $\epsilon^{(2)} = -0.14$, $\delta^{(2)} = -0.14$, $\gamma^{(2)} = -0.06$, $\delta^{(3)} = 0.05$; $V_{33} = \sqrt{A_{33}}$ and $V_{44} = \sqrt{A_{44}}$.

shape of the R_{PSH} approximation (Fig. 3 right) is also sufficiently accurate, at least up to the incidence angle of 30°. Notice that the R_{PSH} component does not vanish for azimuths 0° and 90° as in the previous example. This is due to the misalignment of the vertical symmetry planes of the incidence and reflecting half-spaces. Figure 3 suggests that expressions (26), (C.1) and (C.2) can work reasonably well even for complicated models.

The final model (Fig. 4; medium parameters are provided in the caption) is derived from the previous one by increasing the strength of the anisotropy in both half-spaces (by increasing the crack densities to more than 10%). As expected, the accuracy of the R_{PSV} component has decreased relative to that for the previous model. Moreover, the azimuthal variation of R_{PSV} is no longer approximated adequately. A similar failure can be observed for the R_{PSH} component. Note that the exact R_{PSV} and R_{PSH} components have been obtained by the projection of the exact reflection-vector $\boldsymbol{R}_{PS}^{ext}$ onto the $\boldsymbol{g}^{SV}$ and $\boldsymbol{g}^{SH}$ directions: an operation that becomes inaccurate for a strongly anisotropic incidence medium.

Extensive numerical tests indicate that the accuracy of the R_{PSV} and R_{PSH} approximations generally decreases for lower anisotropic symmetries. However, the tests also indicate that approximations may still work acceptably for relatively

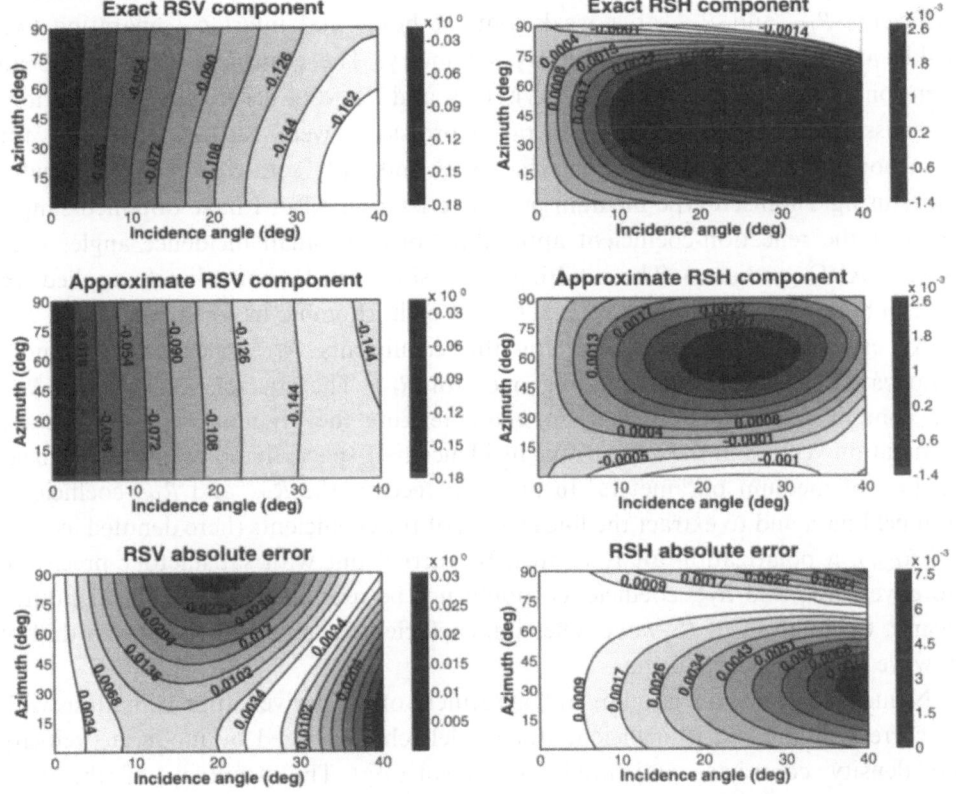

Figure 4

Components R_{PSV} (left) and R_{PSH} (right) of the *PS*-wave reflection coefficients for a strong-contrast, strong-anisotropy orthorhombic/orthorhombic interface as a function of the incidence angle and azimuth: exact components (top), weak-contrast, weak-anisotropy approximations (middle) and absolute errors (bottom) defined as the difference between the exact and approximate components. The vertical symmetry planes of the orthorhombic half-spaces differ by the angle $\kappa = 30°$. Medium parameters are as follows: incidence medium: $\rho = 2.0\text{g/cm}^3$, $V_{33} = 3.4\text{km/s}$, $V_{44} = 1.8\text{km/s}$, $\epsilon^{(1)} = -0.27$, $\delta^{(1)} = -0.28$, $\gamma^{(1)} = -0.1$, $\epsilon^{(2)} = -0.19$, $\delta^{(2)} = -0.20$, $\gamma^{(2)} = -0.07$, $\delta^{(3)} = -0.11$; reflecting medium: $\rho = 2.2\text{g/cm}^3$, $V_{33} = 4.2\text{km/s}$, $V_{44} = 2.4\text{km/s}$, $\epsilon^{(1)} = -0.18$, $\delta^{(1)} = -0.20$, $\gamma^{(1)} = -0.08$, $\epsilon^{(2)} = -0.24$, $\delta^{(2)} = -0.25$, $\gamma^{(2)} = -0.1$, $\delta^{(3)} = 0.06$; $V_{33} = \sqrt{A_{33}}$ and $V_{44} = \sqrt{A_{44}}$.

strongly anisotropic models with relatively strong velocity and density contrasts. Also, the accuracy of the approximate R_{PSV} component is usually comparable to that of the R_{PP} approximation tested previously in the literature. The accuracy of the R_{PSH} component, however, may be significantly lower for reasons mentioned above. In most cases, it likely will be impossible to extract the R_{PSH} component from field data due to a low signal-to-noise ratio.

Summary

Here, I presented the first-order approximations for converted-wave reflection coefficients R_{PS_1} and R_{PS_2} at a weak-contrast horizontal interface separating two weakly anisotropic media of arbitrary symmetry. The approach used here is an extension of the one described by VAVRYČUK and PŠENČÍK (1998) for PP-reflection/ transmission coefficients. The general expressions were written explicitly for an orthorhombic/orthorhombic interface with mutually rotated vertical symmetry planes using Thomsen-type medium parameterization. Also, I have obtained simple forms of the reflection-coefficient approximations for small incidence angles (i.e., PS-wave AVO gradients). The resulting expressions can be immediately applied for any combination of isotropic, VTI, HTI and orthorhombic half-spaces.

As expected, the expressions for the coefficients R_{PS_1} and R_{PS_2} are more complicated than the corresponding ones for R_{PP}. The coefficients R_{PS_1} and R_{PS_2} also contain a polarization angle Φ, characterizing the orientation of the S-wave polarization vectors in the anisotropic incidence half-space; Φ represents a nonlinear function of medium parameters. In order to recover the R_{PS_1} and R_{PS_2} coefficients from field data and to extract the linear parts of the coefficients (here denoted as R_{PSV} and R_{PSH}), a polarization analysis must be carried out with satisfactory precision. Moreover, R_{PS_1} and R_{PS_2} coefficients should not be used at S-wave singular points. Clearly, the analysis of PS-wave reflection coefficients is more involved than that of PP-wave reflection coefficients.

Numerical tests show good overall agreement of the derived approximations with the corresponding exact coefficients for models characterized by moderate velocity and density contrasts, and moderate anisotropy. The accuracy of the R_{PSV} component is usually comparable to that of the R_{PP} approximations. However, whereas the R_{PP} approximations are mostly applicable for small and moderate incidence angles, the R_{PSV} approximations frequently work better for relatively large incidence angles. Conversely, small-incidence-angle PS reflections are weak in general and thus likely distorted by noise.

The R_{PSH} component is a direct indicator of azimuthal anisotropy (R_{PSH} is zero for isotropic and VTI models). Its azimuthal variation is more significant than that of R_{PP} or R_{SV}. This fact can be used to obtain additional information regarding the medium (e.g., the directions of symmetry planes), if this component can be estimated from the data. Unfortunately, R_{PSH} is usually small and is likely to be distorted by noise. Nevertheless, for high-quality data it may still be possible to find local extrema of the R_{PSH} component as a function of azimuth and use them to constrain the model.

Azimuthal variation of both R_{PSV} and R_{PSH} components is well reproduced by the approximations for most models used here. However, decreasing accuracy can be generally expected for lower anisotropic symmetries. Extensive numerical tests nevertheless also reveal that both R_{PSV} and R_{PSH} approximations may still work acceptably for models with stronger anisotropy (up to a certain strength of the model anisotropy, the stronger anisotropy can produce more stable results).

Although I mainly discuss the PS-wave reflection coefficients for orthorhombic media with a horizontal symmetry plane, the equations derived in this paper can be used to analyze the R_{PS_1} and R_{PS_2} coefficients for an interface separating arbitrarily anisotropic half-spaces. The next logical step is to specify the general equations for TI half-spaces with arbitrarily tilted symmetry axes, and for orthorhombic half-spaces with no horizontal symmetry planes. However, the final expressions would be more complicated due to the presence of additional angles (for each half-space, two more angles for TI media and three more for orthorhombic media). This problem could be partially alleviated if a different parameterization of anisotropic media was adopted, such as the one introduced by PŠENČÍK and GAJEWSKI (1998).

Together with the R_{PP} coefficient, R_{PS_1} and R_{PS_2} coefficients may provide a tool for increasing the stability and reliability of AVO inversion. Such an inversion, however, will require high quality multiazimuth 3-D 3-C data.

Acknowledgments

I thank Ilya Tsvankin and Vladimir Grechka for their valuable advice and their insights they shared with me during this research. I am also grateful to Ilya Tsvankin, Vladimir Grechka and Ken Larner for their helpful suggestions in improving the manuscript. My special thanks to Ivan Pšenčík for numerous stimulating discussions, which are greatly appreciated. I wish to gratefully appreciate the sponsors of the Consortium Project at the Center for Wave Phenomena for partial support of this project.

Appendix A:
Linearized Boundary Conditions and Approximate PS-wave Reflection Coefficients
Within the Incidence Plane (Azimuth $\phi = 0$)

Here, I follow the derivation by VAVRYČUK and PŠENČÍK (1998) for the PP-wave reflection coefficient, applying it to the PS-wave reflection. For consistency, I use the same notation as that of these authors.

Exact Solution of the Reflection/Transmission Problem

Consider a planar interface with the normal n separating two homogeneous arbitrarily anisotropic half-spaces with the densities $\rho^{(1)}$, $\rho^{(2)}$ and the density-normalized stiffness tensor elements $a_{ijkl}^{(1)}$ and $a_{ijkl}^{(2)}$. A harmonic plane wave incident from the half-space denoted as 1 generates three reflected and three transmitted waves P, S_1 and S_2. The displacement vector of each wave can be written as

$$u^{(N)}(x,t) = U^{(N)} g^{(N)} \exp[-i\omega(t - p^{(N)} \cdot x)] \; , \tag{A.1}$$

where $N = 0$ corresponds to the incidence wave, $N = 1, 2, 3$ correspond to the reflected S_1, S_2 and P waves and $N = 4, 5, 6$ correspond to the transmitted S_1, S_2 and P waves, respectively. $U^{(N)}$ denotes the scalar amplitude (generally a complex number), $g^{(N)}$ represents the unit polarization vector of the wave N, $p^{(N)}$ is the corresponding slowness vector and ω is the circular frequency. The traction vectors associated with the displacement vectors (A.1) are

$$T^{(N)}(x,t) = i\omega U^{(N)} X^{(N)} \; , \tag{A.2}$$

where $X^{(N)}$ is so-called *amplitude normalized* traction vector,

$$
\begin{aligned}
X_i^{(N)} &= \rho^{(1)} a_{ijkl}^{(1)} n_j g_k^{(N)} p_l^{(N)}, \quad N = 0, 1, 2, 3, \\
X_i^{(N)} &= \rho^{(2)} a_{ijkl}^{(2)} n_j g_k^{(N)} p_l^{(N)}, \quad N = 4, 5, 6 \; .
\end{aligned}
\tag{A.3}
$$

In equation (A.3), Einstein summation convention applies.

For a welded contact between the half-spaces, displacement and traction vector fields must be continuous across the interface. These boundary conditions can be written in a compact matrix form as follows:

$$\hat{C} \cdot U = B \; . \tag{A.4}$$

The vector

$$B = -(g_1^{(0)}, g_2^{(0)}, g_3^{(0)}, X_1^{(0)}, X_2^{(0)}, X_3^{(0)})^{\mathrm{T}} \tag{A.5}$$

corresponds to the incidence wave of amplitude $U^{(0)} = 1$ (the superscript T denotes the transpose). Also, the 6 × 6 matrix

$$
\hat{C} =
\begin{bmatrix}
g_1^{(1)} & g_1^{(2)} & g_1^{(3)} & -g_1^{(4)} & -g_1^{(5)} & -g_1^{(6)} \\
g_2^{(1)} & g_2^{(2)} & g_2^{(3)} & -g_2^{(4)} & -g_2^{(5)} & -g_2^{(6)} \\
g_3^{(1)} & g_3^{(2)} & g_3^{(3)} & -g_3^{(4)} & -g_3^{(5)} & -g_3^{(6)} \\
X_1^{(1)} & X_1^{(2)} & X_1^{(3)} & -X_1^{(4)} & -X_1^{(5)} & -X_1^{(6)} \\
X_2^{(1)} & X_2^{(2)} & X_2^{(3)} & -X_2^{(4)} & -X_2^{(5)} & -X_2^{(6)} \\
X_3^{(1)} & X_3^{(2)} & X_3^{(3)} & -X_3^{(4)} & -X_3^{(5)} & -X_3^{(6)}
\end{bmatrix}
\tag{A.6}
$$

contains quantities constrained by the incidence wave. The polarization vectors $\mathbf{g}^{(N)}$ and slowness vectors $\mathbf{p}^{(N)}$, which are needed for evaluation of the amplitude-normalized traction vectors $X_i^{(N)}$ [equation (A.3)] and the matrix $\hat{\mathbf{C}}$, must be determined by solving the Christoffel equation,

$$(a_{ijkl}p_i^{(N)}p_l^{(N)} - \delta_{jk})g_j^{(N)} = 0 \ . \tag{A.7}$$

In general, system (A.7) must be solved numerically. Finally, the vector

$$\mathbf{U} = (R_{S_1}, R_{S_2}, R_P, T_{S_1}, T_{S_2}, T_P)^{\mathrm{T}} \tag{A.8}$$

contains the reflection/transmission coefficients to be found. This vector must be determined by solving matrix equation (A.4), together with Christoffel equation (A.7).

First-order Perturbation of the Exact Solution

Using perturbations of slowness and polarization vectors $\delta\mathbf{p}^{(N)}$ and $\delta\mathbf{g}^{(N)}$ introduced in equations (4) and (6), one can linearize the traction vectors $X^{(N)}$ defined by equations (A.3) and, consequently, vectors $\mathbf{B}$, $\mathbf{U}$ and matrix $\hat{\mathbf{C}}$ [see equations (A.5), (A.8) and (A.6)] as

$$\begin{aligned}
X^{(N)} &= X^{0(N)} + \delta X^{(N)}, \\
\hat{\mathbf{C}} &= \hat{\mathbf{C}}^0 + \delta\hat{\mathbf{C}}, \\
\mathbf{B} &= \mathbf{B}^0 + \delta\mathbf{B}, \\
\mathbf{U} &= \mathbf{U}^0 + \delta\mathbf{U} \ .
\end{aligned} \tag{A.9}$$

Substituting equations (A.9) into (A.4) and keeping only linear terms, we arrive at the final matrix equation for the perturbed vector of the reflection/transmission coefficients:

$$\delta\mathbf{U} = (\hat{\mathbf{C}}^0)^{-1}(\delta\mathbf{B} - \delta\hat{\mathbf{C}} \cdot \mathbf{U}^0) \ . \tag{A.10}$$

The background vector $\mathbf{B}^0$ does not appear in equation (A.10) because $\hat{\mathbf{C}}^0 \cdot \mathbf{U}^0 = \mathbf{B}^0$. Since the background medium is a uniform full space, a fictitious interface in such a medium generates only one transmitted wave of the same type as the incidence wave. Consequently, the perturbed reflection/transmission vector $\delta\mathbf{U}$ contains all desired reflection coefficients as perturbations of the vanishing reflection coefficients in the background medium. Equation (A.10) can be used to obtain the first-order reflection/transmission coefficients for all possible types of incidence waves and arbitrarily anisotropic media.

Approximate PS-wave Reflection Coefficients within the Incidence Plane

A specific type of incidence wave constrains all the quantities on the right-hand side of equation (A.10) needed to complete the derivation. For *P*-wave incidence, the background reflection/transmission vector U^0 is given by

$$U^0 = (0, 0, 0, 0, 0, 1)^T \ . \qquad (A.11)$$

This fact considerably simplifies general equation (A.10), since only the 6-th column of the perturbation matrix $\delta\hat{C}$ must be evaluated. Thus, equation (A.10) takes its final form

$$
\begin{aligned}
\delta U = (\hat{C}^0)^{-1} &\big(\delta g_1^{(6)} - \delta g_1^{(0)}, \delta g_2^{(6)} - \delta g_2^{(0)}, \delta g_3^{(6)} - \delta g_3^{(0)}, \\
&\delta X_1^{(6)} - \delta X_1^{(0)}, \delta X_2^{(6)} - \delta X_2^{(0)}, \delta X_3^{(6)} - \delta X_3^{(0)} \big)^T \ .
\end{aligned} \qquad (A.12)
$$

As shown in VAVRYČUK and PŠENČÍK (1998), the inverted matrix $(\hat{C}^0)^{-1}$ is given as

$$
(\hat{C}^0)^{-1} = \begin{pmatrix}
-\dfrac{\beta^2 p_1^0 Y \cos\Phi}{Z_S} & \dfrac{\sin\Phi}{2} & -\beta p_1^0 \cos\Phi & \dfrac{\cos\Phi}{2\rho^0\beta} & -\dfrac{\beta p_1^0 \sin\Phi}{Z_S} & \dfrac{\beta^2 (p_1^0)^2 \cos\Phi}{Z_S} \\[2mm]
\dfrac{\beta^2 p_1^0 Y \sin\Phi}{Z_S} & \dfrac{\cos\Phi}{2} & \beta p_1^0 \sin\Phi & -\dfrac{\sin\Phi}{2\rho^0\beta} & -\dfrac{\beta p_1^0 \cos\Phi}{Z_S} & -\dfrac{\beta^2 (p_1^0)^2 \sin\Phi}{Z_S} \\[2mm]
\dfrac{\beta^2 p_1^0}{\alpha} & 0 & -\dfrac{\beta^2 p_1^0 Y}{Z_P} & -\dfrac{\beta^2 (p_1^0)^2}{Z_P} & 0 & \dfrac{1}{2\rho^0\alpha} \\[2mm]
-\dfrac{\beta^2 p_1^0 Y \cos\Psi}{Z_S} & -\dfrac{\sin\Psi}{2} & \beta p_1^0 \cos\Psi & -\dfrac{\cos\Psi}{2\rho^0\beta} & -\dfrac{\beta p_1^0 \sin\Psi}{Z_S} & \dfrac{\beta^2 (p_1^0)^2 \cos\Psi}{Z_S} \\[2mm]
\dfrac{\beta^2 p_1^0 Y \sin\Psi}{Z_S} & -\dfrac{\cos\Psi}{2} & -\beta p_1^0 \sin\Psi & \dfrac{\sin\Psi}{2\rho^0\beta} & -\dfrac{\beta p_1^0 \cos\Psi}{Z_S} & -\dfrac{\beta^2 (p_1^0)^2 \sin\Psi}{Z_S} \\[2mm]
-\dfrac{\beta^2 p_1^0}{\alpha} & 0 & -\dfrac{\beta^2 p_1^0 Y}{Z_P} & -\dfrac{\beta^2 (p_1^0)^2}{Z_P} & 0 & -\dfrac{1}{2\rho^0\alpha}
\end{pmatrix} ,
$$

$$(A.13)$$

where Z_P, Z_S and Y are defined as

$$
\begin{aligned}
Z_P &\equiv 2\alpha\rho^0 \beta^2 p_1^0 p_3^{0P}, \\
Z_S &\equiv 2\rho^0 \beta^3 p_1^0 p_3^{0S}, \\
Y &\equiv \rho^0 (1 - 2\beta^2 (p_1^0)^2) \ .
\end{aligned}
$$

Based on the derivation of JECH and PŠENČÍK (1989) and VAVRYČUK and PŠENČÍK (1998), the perturbations of the polarization vectors in equation (A.12) can be expressed as

$$
\delta g_i^{(6)} - \delta g_i^{(0)} = (\delta c^{(6)} - \delta c^{(0)}) \left[p_i^{0(0)} - \frac{n_i}{\alpha^2 (n_k p_k^{0(0)})} \right] + (\delta G_i^{(6)} - \delta G_i^{(0)}) \ , \qquad (A.14)
$$

where *i* represents the *i*-th vector component. Here, $p^{0(0)}$ is the slowness vector of the incidence *P* wave in the background medium [equation (5)], n is the normal to the interface and α is the *P*-wave background velocity. $\delta c^{(N)}$ is the deviation of the phase velocity (its magnitude) of the incidence (N = 0) or transmitted (N = 6) *P* wave in the weakly anisotropic medium from the corresponding phase velocity in the isotropic background. By a linear perturbation of the Christoffel equation, JECH and PŠENČÍK (1989) obtained

$$\delta c^{(0)} = \tfrac{1}{2}\alpha\delta a_{ijkl}^{(1)}p_i^{0(0)}g_j^{0(0)}g_k^{0(0)}p_l^{0(0)},$$
$$\delta c^{(6)} = \tfrac{1}{2}\alpha\delta a_{ijkl}^{(2)}p_i^{0(0)}g_j^{0(0)}g_k^{0(0)}p_l^{0(0)},$$
(A.15)

where $g_i^{0(0)}$ is given in relations (7) and $\delta a_{ijkl}^{(I)}$ by relation (2). It immediately follows that

$$\delta c^{(6)} - \delta c^{(0)} = \tfrac{1}{2}\alpha\Delta a_{ijkl}p_i^{0(0)}g_j^{0(0)}g_k^{0(0)}p_l^{0(0)},$$
(A.16)

where

$$\Delta a_{ijkl} \equiv \delta a_{ijkl}^{(2)} - \delta a_{ijkl}^{(1)} = a_{ijkl}^{(2)} - a_{ijkl}^{(1)}$$
(A.17)

are the contrasts of the elastic parameters across the interface.

The vectors $\delta G_i^{(N)}$ in (A-14) describe the deviations of the polarization vectors in the weakly anisotropic medium from the corresponding polarization vectors in the isotropic background. In a fashion similar to that in the previous paragraph (using results of JECH and PŠENČÍK, 1989), it is possible to write

$$\delta G_m^{(6)} - \delta G_m^{(0)} = \frac{\alpha}{\alpha^2 - \beta^2}\Delta a_{ijkl}p_i^{0(0)}g_j^{0(0)}g_k^{0(0)}(\delta_{lm} - g_l^{0(0)}g_m^{0(0)}),$$
(A.18)

where δ_{lm} is Kronecker's delta. Notice that equation (A.14) describes the perturbations of the polarization vectors from the background due to both the elastic parameter contrasts across the interface and the anisotropy.

Finally, from equations (A.3), taking into account that $\boldsymbol{n} = (0, 0, 1)$, the difference of the first-order perturbations of amplitude-normalized traction vectors needed in equation (A.12) can be obtained as

$$\delta X_i^{(6)} - \delta X_i^{(0)} = \Delta\rho a_{i3kl}^0 g_k^{0(0)}p_l^{0(0)} + \rho^0\Delta a_{i3kl}g_k^{0(0)}p_l^{0(0)}$$
$$+ \rho^0 a_{i3kl}^0(\delta g_k^{(6)} - \delta g_k^{(0)})p_l^{0(0)} + \rho^0 a_{i3kl}^0 g_k^{0(0)}(\delta p_l^{(6)} - \delta p_l^{(0)}).$$
(A.19)

The quantities a_{i3kl}^0 and ρ^0 are defined in equations (2), and Δa_{i3kl} is given by (A.17). Similarly, $\Delta\rho \equiv \rho^{(2)} - \rho^{(1)}$ is the density contrast across the interface. The difference $\delta g_k^{(6)} - \delta g_k^{(0)}$ is specified by relation (A.14). The term $\delta p_l^{(6)} - \delta p_l^{(0)}$ has a physical meaning analogous to that of $\delta g_l^{(6)} - \delta g_l^{(0)}$; however, it is now applied to the slowness vector. Using expressions in JECH and PŠENČÍK (1989) and VAVRYČUK and PŠENČÍK (1998) yields the relation

$$\delta p_i^{(6)} - \delta p_i^{(0)} = -\frac{\delta c^{(6)} - \delta c^{(0)}}{\alpha^3 p_3^{0(0)}},$$
(A.20)

where all quantities are specified above. By substituting equations (A.13), (A.14) and (A.16)–(A.20) into equation (A.12), we can find the approximate P-wave incidence reflection/transmission coefficients. As follows from the convention above, the approximation of the R_{PS_1} reflection coefficient is obtained by using the first row of the matrix $(\hat{\mathbf{C}}^0)^{-1}$ in equation (A.12). The result can be formally written as

$$R_{PS_1} = f\left(\alpha, \beta, \rho^0, i, j, \Phi, \Delta A_{11}, \Delta A_{13}, \Delta A_{14}, \right.$$
$$\Delta A_{15}, \Delta A_{16}, \Delta A_{33}, \Delta A_{34}, \Delta A_{35}, \qquad \text{(A.21)}$$
$$\left. \Delta A_{36}, \Delta A_{45}, \Delta A_{55}, \Delta A_{56}\right),$$

where $\Delta A_{ij} = A_{ij}^{(2)} - A_{ij}^{(1)}$ are contrasts in density normalize medium parameters a_{klmn} written in the Voigt convention. The second coefficient R_{PS_2} can be derived using the second row of the matrix $(\hat{\mathbf{C}}^0)^{-1}$. However, it can be easily obtained from the coefficient R_{PS_1} by the substitution $\cos\Phi \rightarrow -\sin\Phi$ and $\sin\Phi \rightarrow \cos\Phi$, since this is the only difference between rows 1 and 2 of the matrix $(\hat{\mathbf{C}}^0)^{-1}$.

Appendix B:
General Explicit Expressions for the Reflection Coefficients R_{PS_1} and R_{PS_2}

To generalize the reflection coefficient R_{PS_1} (A.21) for an arbitrary azimuth ψ of the incidence plane, the medium parameters ΔA_{ij} must be expressed in terms of $\Delta A'_{ij}$ defined in the reference coordinate system. The tensor rotation is controlled by the following matrix of directional cosines:

$$\hat{\mathbf{R}} = \begin{pmatrix} \cos\psi & -\sin\psi & 0 \\ \sin\psi & \cos\psi & 0 \\ 0 & 0 & 1 \end{pmatrix}. \qquad \text{(B.1)}$$

Applying matrix (B.1) to the elements $\Delta A'_{ij}$ yields

$$\Delta A_{11} = \Delta A'_{11} \cos^4\psi + 2\Delta A'_{12} \cos^2\psi \sin^2\psi + 4\Delta A'_{16} \cos^3\psi \sin\psi$$
$$+ \Delta A'_{22} \sin^4\psi + 4\Delta A'_{26} \cos\psi \sin^3\psi + 4\Delta A'_{66} \cos^2\psi \sin^2\psi,$$

$$\Delta A_{13} = \Delta A'_{13} \cos^2\psi + \Delta A'_{23} \sin^2\psi + 2\Delta A'_{36} \cos\psi \sin\psi,$$

$$\Delta A_{14} = \Delta A'_{14} \cos^3 - \Delta A'_{15} \cos^2\psi \sin\psi + \Delta A'_{24} \cos\psi \sin^2\psi - \Delta A'_{25} \sin^3\psi$$
$$+ 2\Delta A'_{46} \cos^2\psi \sin\psi - 2\Delta A'_{56} \cos\psi \sin^2\psi,$$

$$\Delta A_{15} = \Delta A'_{14} \cos^2\psi \sin\psi + \Delta A'_{15} \cos^3\psi + \Delta A'_{24} \sin^3\psi + \Delta A'_{25} \cos\psi \sin^2\psi$$
$$+ 2\Delta A'_{46} \cos\psi \sin^2\psi + 2\Delta A'_{56} \cos^2\psi \sin\psi,$$

$$\Delta A_{16} = -\Delta A'_{11} \cos^3\psi \sin\psi + \Delta A'_{12}(\cos^3\psi \sin\psi - \cos\psi \sin^3\psi)$$
$$+ \Delta A'_{16}(\cos^4\psi - 3\cos^2\psi \sin^2\psi) + \Delta A'_{26}(3\cos^2\psi \sin^2\psi - \sin^4\psi)$$
$$+ \Delta A'_{66}(2\cos^3\psi \sin\psi - 2\cos\psi \sin^3\psi),$$

$$\Delta A_{33} = \Delta A'_{33},$$

$$\Delta A_{34} = \Delta A'_{34}\cos\psi - \Delta A'_{35}\sin\psi,$$

$$\Delta A_{35} = \Delta A'_{34}\sin\psi + \Delta A'_{35}\cos\psi,$$

$$\Delta A_{36} = -\Delta A'_{13}\cos\psi\sin\psi + \Delta A'_{23}\cos\psi\sin\psi + \Delta A'_{36}(\cos^2\psi - \sin^2\psi),$$

$$\Delta A_{45} = \Delta A'_{44}\cos\psi\sin\psi + \Delta A'_{45}(\cos^2\psi - \sin^2\psi) - \Delta A'_{55}\cos\psi\sin\psi,$$

$$\Delta A_{55} = \Delta A'_{44}\sin^2\psi\sin\psi + 2\Delta A'_{45}\cos\psi\sin\psi + \Delta A'_{55}\cos^2\psi,$$

$$\Delta A_{56} = -\Delta A'_{14}\cos\psi\sin^2\psi - \Delta A'_{15}\cos^2\psi\sin\psi + \Delta A'_{24}\cos\psi\sin^2\psi + \Delta A'_{25}\cos^2\psi\sin\psi$$

$$+ \Delta A'_{46}(\cos^2\psi\sin\psi - \sin^3\psi) + \Delta A'_{56}(\cos^3\psi - \cos\psi\sin^2\psi) . \tag{B.2}$$

Relations (B.2) and (A.21) are used to derive the approximations (10) for the azimuthally dependent reflection coefficients R_{PS_1} and R_{PS_2} valid for arbitrarily anisotropic half-spaces. The coefficients contain R_{PSV} and R_{PSH} components given by

$$R_{PSV} = v_1 + v_2\frac{\cos i}{\cos j} + v_3\frac{\sin i}{\cos j} + v_4\cos i\sin i + v_5\sin^2 i + v_6\frac{\cos i\sin^2 i}{\cos j}$$

$$+ v_7\frac{\sin^3 i}{\cos j} + v_8\cos i\sin^3 i + v_9\sin^4 i + v_{10}\frac{\cos i\sin^4 i}{\cos j} + v_{11}\frac{\sin^5 i}{\cos j},$$

$$R_{PSH} = h_1\frac{1}{\cos j} + h_2\cos i + h_3\sin i + h_4\frac{\cos i\sin i}{\cos j} + h_5\frac{\sin^2 i}{\cos j}$$

$$+ h_6\cos i\sin^2 i + h_7\sin^3 i + h_8\frac{\cos i\sin^3 i}{\cos j} + h_9\frac{\sin^4 i}{\cos j}, \tag{B.3}$$

where the coefficients v_i and h_i are as follows:

$$v_1 = \frac{\alpha}{2\beta(\alpha^2 - \beta^2)}[\Delta A_{35}\cos\psi + \Delta A_{34}\sin\psi],$$

$$v_2 = \frac{1}{2(\beta^2 - \alpha^2)}[\Delta A_{35}\cos\psi + \Delta A_{34}\sin\psi],$$

$$v_3 = -\frac{1}{2}\frac{\Delta\rho}{\rho^0} + \frac{1}{2(\alpha^2 - \beta^2)}[-\Delta A_{33} + (\Delta A_{13} + 2\Delta A_{55})\cos^2\psi$$

$$+ (\Delta A_{23} + 2\Delta A_{44})\sin^2\psi + (\Delta A_{36} + 2\Delta A_{45})\sin 2\psi],$$

$$v_4 = -\frac{\beta}{\alpha}\frac{\Delta\rho}{\rho^0} - \frac{1}{2\alpha\beta(\alpha^2 - \beta^2)}[-\beta^2\Delta A_{33} + (\beta^2\Delta A_{13} + 2\alpha^2\Delta A_{55})\cos^2\psi$$

$$+ (\beta^2\Delta A_{23} + 2\alpha^2\Delta A_{44})\sin^2\psi + (\beta^2\Delta A_{36} + 2\alpha^2\Delta A_{45})\sin 2\psi],$$

$$v_5 = \frac{1}{2\alpha\beta(\alpha^2 - \beta^2)}[-(\alpha^2 + 4\beta^2)(\Delta A_{35} \cos \psi + \Delta A_{34} \sin \psi)$$
$$+ (\alpha^2 + 2\beta^2)(\Delta A_{15} \cos^3 \psi + (\Delta A_{14} + 2\Delta A_{56}) \cos^2 \psi \sin \psi$$
$$+ (\Delta A_{25} + 2\Delta A_{46}) \cos \psi \sin^2 \psi + \Delta A_{24} \sin^3 \psi)],$$

$$v_6 = \frac{1}{2\alpha^2(\alpha^2 - \beta^2)}[(2\beta^2 + 3\alpha^2)\Delta A_{35} \cos \psi + (2\beta^2 + 3\alpha^2)\Delta A_{34} \sin \psi$$
$$- 3\alpha^2(\Delta A_{15} \cos^3 \psi + (\Delta A_{14} + 2\Delta A_{56}) \cos^2 \psi \sin \psi$$
$$+ (\Delta A_{25} + 2\Delta A_{46}) \cos \psi \sin^2 \psi + \Delta A_{24} \sin^3 \psi)],$$

$$v_7 = \frac{\beta^2}{\alpha^2}\frac{\Delta \rho}{\rho^0} + \frac{1}{2\alpha^2(\alpha^2 - \beta^2)}[(\beta^2 + \alpha^2)\Delta A_{33}$$
$$- (\beta^2(\Delta A_{13} + 4\Delta A_{55}) + 2\alpha^2(\Delta A_{13} + \Delta A_{55})) \cos^2 \psi$$
$$- (\beta^2(\Delta A_{23} + 4\Delta A_{44}) + 2\alpha^2(\Delta A_{23} + \Delta A_{44})) \sin^2 \psi$$
$$- (\beta^2(\Delta A_{36} + 4\Delta A_{45}) + 2\alpha^2(\Delta A_{36} + \Delta A_{45})) \sin 2\psi$$
$$+ \alpha^2 \Delta A_{11} \cos^4 \psi + \alpha^2 \Delta A_{22} \sin^4 \psi + 4\alpha^2 \Delta A_{16} \cos^3 \psi \sin \psi$$
$$+ 4\alpha^2 \Delta A_{26} \cos \psi \sin^3 \psi + 2\alpha^2(\Delta A_{12} + 2\Delta A_{66}) \sin^2 \psi \cos^2 \psi],$$

$$v_8 = \frac{-\beta}{2\alpha(\alpha^2 - \beta^2)}[\Delta A_{33} - 2(\Delta A_{13} + 2\Delta A_{55}) \cos^2 \psi$$
$$- 2(\Delta A_{23} + 2\Delta A_{44}) \sin^2 \psi - 2(\Delta A_{36} + 2\Delta A_{45}) \sin 2\psi$$
$$+ \Delta A_{11} \cos^4 \psi + \Delta A_{22} \sin^4 \psi + 4\Delta A_{16} \cos^3 \psi \sin \psi + 4\Delta A_{26} \cos \psi \sin^3 \psi$$
$$+ 2(\Delta A_{12} + 2\Delta A_{66}) \sin^2 \psi \cos^2 \psi],$$

$$v_9 = \frac{2\beta}{\alpha(\alpha^2 - \beta^2)}[\Delta A_{35} \cos \psi + \Delta A_{34} \sin \psi - \Delta A_{15} \cos^3 \psi - \Delta A_{24} \sin^3 \psi$$
$$- (\Delta A_{14} + 2\Delta A_{56}) \cos^2 \psi \sin \psi - (\Delta A_{25} + 2\Delta A_{46}) \cos \psi \sin^2 \psi],$$

$$v_{10} = \frac{2\beta^2}{\alpha^2(\alpha^2 - \beta^2)}[-\Delta A_{35} \cos \psi - \Delta A_{34} \sin \psi + \Delta A_{15} \cos^3 \psi + \Delta A_{24} \sin^3 \psi$$
$$+ (\Delta A_{14} + 2\Delta A_{56}) \cos^2 \psi \sin \psi + (\Delta A_{25} + 2\Delta A_{46}) \cos \psi \sin^2 \psi],$$

$$v_{11} = \frac{\beta^2}{2\alpha^2(\alpha^2 - \beta^2)}[\Delta A_{33} - 2(\Delta A_{13} + 2\Delta A_{55}) \cos^2 \psi - 2(\Delta A_{23} + 2\Delta A_{44}) \sin^2 \psi$$
$$- 2(\Delta A_{36} + 2\Delta A_{45}) \sin 2\psi + \Delta A_{11} \cos^4 \psi + \Delta A_{22} \sin^4 \psi + 4\Delta A_{16} \cos^3 \psi \sin \psi$$
$$+ 4\Delta A_{26} \cos \psi \sin^3 \psi + 2(\Delta A_{12} + 2\Delta A_{66}) \sin^2 \psi \cos^2 \psi] ,$$

$$(B.4)$$

and

$$h_1 = \frac{\alpha}{2\beta(\beta^2 - \alpha^2)}[\Delta A_{34}\cos\psi - \Delta A_{35}\sin\psi],$$

$$h_2 = \frac{1}{2(\alpha^2 - \beta^2)}[\Delta A_{34}\cos\psi - \Delta A_{35}\sin\psi],$$

$$h_3 = \frac{1}{4(\alpha^2 - \beta^2)}[-2(\Delta A_{36} + 2\Delta A_{45})\cos 2\psi$$
$$+ (\Delta A_{13} - \Delta A_{23} - 2\Delta A_{44} + 2\Delta A_{55})\sin 2\psi],$$

$$h_4 = \frac{1}{4\alpha\beta(\alpha^2 - \beta^2)}[2(\beta^2\Delta A_{36} + 2\alpha^2\Delta A_{45})\cos 2\psi$$
$$+ (\beta^2(-\Delta A_{13} + \Delta A_{23}) + 2\alpha^2(\Delta A_{44} - \Delta A_{55}))\sin 2\psi],$$

$$h_5 = \frac{1}{2\alpha\beta(\alpha^2 - \beta^2)}[(\alpha^2 + \beta^2)\Delta A_{34}\cos\psi - (\alpha^2 + \beta^2)\Delta A_{35}\sin\psi$$
$$- (\alpha^2\Delta A_{14} + 2\beta^2\Delta A_{56})\cos^3\psi + (\alpha^2\Delta A_{25} + 2\beta^2\Delta A_{46})\sin^3\psi + (\alpha^2(\Delta A_{15} - 2\Delta A_{46})$$
$$+ 2\beta^2(\Delta A_{15} - \Delta A_{25} - \Delta A_{46}))\cos^2\psi\sin\psi + (2\beta^2\Delta A_{14} - \alpha^2\Delta A_{24} - 2\beta^2\Delta A_{24}$$
$$+ 2\alpha^2\Delta A_{56} + 2\beta^2\Delta A_{56})\cos\psi\sin^2\psi],$$

$$h_6 = \frac{1}{8(\alpha^2 - \beta^2)}[(\Delta A_{14} + 3\Delta A_{24} - 4\Delta A_{34} + 2\Delta A_{56})\cos\psi$$
$$+ 2(-3\Delta A_{15} + \Delta A_{25} + 2\Delta A_{35} + 2\Delta A_{46})\sin\psi$$
$$- (3\Delta A_{15} - 3\Delta A_{25} - 6\Delta A_{46})\cos 2\psi\sin\psi + 3(\Delta A_{14} - \Delta A_{24} + 2\Delta A_{56})\cos 3\psi],$$

$$h_7 = \frac{1}{8(\alpha^2 - \beta^2)}[-2(\Delta A_{16} + \Delta A_{26} - 2\Delta A_{36} - 4\Delta A_{45})\cos 2\psi$$
$$+ (\Delta A_{11} - 2\Delta A_{13} - \Delta A_{22} + 2\Delta A_{23} + 4\Delta A_{44} - 4\Delta A_{55})\sin 2\psi$$
$$- 2(\Delta A_{16} - \Delta A_{26})\cos 4\psi + \tfrac{1}{2}(\Delta A_{11} - 2\Delta A_{12} + \Delta A_{22} - 4\Delta A_{66})\sin 4\psi],$$

$$h_8 = \frac{\beta}{8\alpha(\alpha^2 - \beta^2)}[2(\Delta A_{16} + \Delta A_{26} - 2\Delta A_{36} - 4\Delta A_{45})\cos 2\psi$$
$$+ (-\Delta A_{11} + 2\Delta A_{13} + \Delta A_{22} - 2\Delta A_{23} - 4\Delta A_{44} + 4\Delta A_{55})\sin 2\psi$$
$$+ 2(\Delta A_{16} - \Delta A_{26})\cos 4\psi - \tfrac{1}{2}(\Delta A_{11} - 2\Delta A_{12} + \Delta A_{22} - 4\Delta A_{66})\sin 4\psi],$$

$$h_9 = \frac{\beta}{8\alpha(\alpha^2 - \beta^2)}[(\Delta A_{14} + 3\Delta A_{24} - 4\Delta A_{34} + 2\Delta A_{56})\cos\psi$$
$$+ 2(-3\Delta A_{15} + \Delta A_{25} + 2\Delta A_{35} + 2\Delta A_{46})\sin\psi$$
$$- (3\Delta A_{15} - 3\Delta A_{25} - 6\Delta A_{46})\cos 2\psi\sin\psi + 3(\Delta A_{14} - \Delta A_{24} + 2\Delta A_{56})\cos 3\psi] .$$

$$\text{(B.5)}$$

Here, i and j denote the incidence and reflection phase angles, respectively, and ψ denotes the azimuthal angle. The quantities A_{ij} are defined above, and α, β and ρ^0 are the background medium parameters (i.e., P- and S-wave velocities and density, respectively). Again, $\Delta x = x^{(2)} - x^{(1)}$ denotes the contrast of a general parameter x across the interface.

Appendix C:
Approximations R_{PS_1} and R_{PS_2} for Orthorhombic Media

In the text, the general approximations R_{PS_1} and R_{PS_2} [equations (10)] are specified for orthorhombic media. Resulting equations (25) contain R_{PSV} and R_{PSH} components (26), which consist of linear combinations of the V_j ($j = 1, \ldots, 5$) and H_k ($k = 1, \ldots, 4$) terms, respectively. The V_j terms read

$$
V_1 = -\frac{1}{2}\frac{\Delta\rho}{\bar{\rho}} + \frac{\bar{\alpha}^2}{2(\bar{\alpha}^2 - \bar{\beta}^2)}\left[\tilde{\delta}_2^{(2)}\cos^2(\psi - \kappa) + \tilde{\delta}_2^{(1)}\sin^2(\psi - \kappa)\right.
$$
$$
\left. - \tilde{\delta}_1^{(2)}\cos^2\psi - \tilde{\delta}_1^{(1)}\sin^2\psi\right],
$$

$$
V_2 = -\frac{\bar{\beta}}{\bar{\alpha}}\frac{\Delta\rho}{\bar{\rho}} - 2\frac{\bar{\beta}}{\bar{\alpha}}\frac{\Delta\beta}{\bar{\beta}} - \frac{\bar{\alpha}\bar{\beta}}{2(\bar{\alpha}^2 - \bar{\beta}^2)}\left[\tilde{\delta}_2^{(2)}\cos^2(\psi - \kappa) + \tilde{\delta}_2^{(1)}\sin^2(\psi - \kappa)\right.
$$
$$
\left. -\tilde{\delta}_1^{(2)}\cos^2\psi - \tilde{\delta}_1^{(1)}\sin^2\psi\right] - 2\frac{\bar{\beta}}{\bar{\alpha}}\left[\gamma_2^{(S)}\sin^2(\psi - \kappa) - \gamma_1^{(S)}\sin^2\psi\right],
$$

$$
V_3 = \frac{\bar{\beta}^2}{\bar{\alpha}^2}\frac{\Delta\rho}{\bar{\rho}} + 2\frac{\bar{\beta}^2}{\bar{\alpha}^2}\frac{\Delta\beta}{\bar{\beta}} + \frac{\bar{\beta}^2}{2(\bar{\alpha}^2 - \bar{\beta}^2)}\left[-\tilde{\delta}_2^{(2)}\cos^2(\psi - \kappa) - \tilde{\delta}_2^{(1)}\sin^2(\psi - \kappa)\right.
$$
$$
\left. +\tilde{\delta}_1^{(2)}\cos^2\psi + \tilde{\delta}_1^{(1)}\sin^2\psi\right] + \frac{\bar{\alpha}^2}{\bar{\alpha}^2 - \bar{\beta}^2}\left[\epsilon_2^{(2)}(\cos^4(\psi-\kappa) + 2\cos^2(\psi-\kappa)\sin^2(\psi-\kappa))\right.
$$
$$
+ \epsilon_2^{(1)}\sin^4(\psi - \kappa) - \epsilon_1^{(2)}(\cos^4\psi + 2\cos^2\psi\sin^2\psi) - \epsilon_1^{(1)}\sin^4\psi - \tilde{\delta}_2^{(1)}\sin^2(\psi - \kappa)
$$
$$
- \tilde{\delta}_2^{(2)}\cos^2(\psi - \kappa) + \tilde{\delta}_2^{(3)}\cos^2(\psi - \kappa)\sin^2(\psi - \kappa) + \tilde{\delta}_1^{(1)}\sin^2\psi + \tilde{\delta}_1^{(2)}\cos^2\psi
$$
$$
\left. - \tilde{\delta}_1^{(3)}\cos^2\psi\sin^2\psi\right] + 2\frac{\bar{\beta}^2}{\bar{\alpha}^2}\left[\gamma_2^{(S)}\sin^2(\psi - \kappa) - \gamma_1^{(S)}\sin^2\psi\right],
$$

$$
V_4 = -\frac{\bar{\alpha}\bar{\beta}}{\bar{\alpha}^2 - \bar{\beta}^2}\left[\epsilon_2^{(2)}(\cos^4(\psi - \kappa) + 2\cos^2(\psi - \kappa)\sin^2(\psi - \kappa)) + \epsilon_2^{(1)}\sin^4(\psi - \kappa)\right.
$$
$$
- \epsilon_1^{(2)}(\cos^4\psi + 2\cos^2\psi\sin^2\psi) - \epsilon_1^{(1)}\sin^4\psi - \tilde{\delta}_2^{(1)}\sin^2(\psi - \kappa) - \tilde{\delta}_2^{(2)}\cos^2(\psi - \kappa)
$$
$$
\left. + \tilde{\delta}_2^{(3)}\cos^2(\psi - \kappa)\sin^2(\psi - \kappa) + \tilde{\delta}_1^{(1)}\sin^2\psi + \tilde{\delta}_1^{(2)}\cos^2\psi - \tilde{\delta}_1^{(3)}\cos^2\psi\sin^2\psi\right],
$$

$$
V_5 = -\frac{\bar{\beta}^2}{\bar{\alpha}^2 - \bar{\beta}^2}\left[\epsilon_2^{(2)}(\cos^4(\psi - \kappa) + 2\cos^2(\psi - \kappa)\sin^2(\psi - \kappa))\right.
$$
$$
+ \epsilon_2^{(1)}\sin^4(\psi - \kappa) - \epsilon_1^{(2)}(\cos^4\psi + 2\cos^2\psi\sin^2\psi) - \epsilon_1^{(1)}\sin^4\psi - \tilde{\delta}_2^{(1)}\sin^2(\psi - \kappa)
$$
$$
- \tilde{\delta}_2^{(2)}\cos^2(\psi - \kappa) + \tilde{\delta}_2^{(3)}\cos^2(\psi - \kappa)\sin^2(\psi - \kappa) + \tilde{\delta}_1^{(1)}\sin^2\psi + \tilde{\delta}_1^{(2)}\cos^2\psi
$$
$$
\left. - \tilde{\delta}_1^{(3)}\cos^2\psi\sin^2\psi\right].
$$

$$
\text{(C.1)}
$$

The H_k terms are as follows:

$$H_1 = \frac{\bar{\alpha}^2}{4(\bar{\alpha}^2 - \bar{\beta}^2)} \left[(\tilde{\delta}_2^{(2)} - \tilde{\delta}_2^{(1)}) \sin 2(\psi - \kappa) + (\tilde{\delta}_1^{(1)} - \tilde{\delta}_1^{(2)}) \sin 2\psi \right],$$

$$H_2 = \frac{\bar{\alpha}\bar{\beta}}{4(\bar{\alpha}^2 - \bar{\beta}^2)} \left[(\tilde{\delta}_2^{(1)} - \tilde{\delta}_2^{(2)}) \sin 2(\psi - \kappa) + (\tilde{\delta}_1^{(2)} - \tilde{\delta}_1^{(1)}) \sin 2\psi \right]$$
$$+ \frac{\bar{\beta}}{\bar{\alpha}} \left[\gamma_2^{(S)} \sin 2(\psi - \kappa) - \gamma_1^{(S)} \sin 2\psi \right],$$

$$H_3 = \frac{\bar{\alpha}^2}{2(\bar{\alpha}^2 - \bar{\beta}^2)} \left[\left[\tilde{\delta}_2^{(1)} - \tilde{\delta}_2^{(2)} - \tilde{\delta}_2^{(3)} (\cos^2(\psi - \kappa) - \sin^2(\psi - \kappa)) \right. \right.$$
$$\left. + 2(\epsilon_2^{(2)} - \epsilon_2^{(1)}) \sin^2(\psi - \kappa) \right] \sin(\psi - \kappa) \cos(\psi - \kappa) + \left[-\tilde{\delta}_1^{(1)} + \tilde{\delta}_1^{(2)} \right.$$
$$\left. \left. + \tilde{\delta}_1^{(3)} (\cos^2 \psi - \sin^2 \psi) - 2(\epsilon_1^{(2)} - \epsilon_1^{(1)}) \sin^2 \psi \right] \sin \psi \cos \psi \right], \quad (C.2)$$

$$H_4 = -\frac{\bar{\alpha}\bar{\beta}}{2(\bar{\alpha}^2 - \bar{\beta}^2)} \left[\left[\tilde{\delta}_2^{(1)} - \tilde{\delta}_2^{(2)} - \tilde{\delta}_2^{(3)} (\cos^2(\psi - \kappa) - \sin^2(\psi - \kappa)) \right. \right.$$
$$\left. + 2(\epsilon_2^{(2)} - \epsilon_2^{(1)}) \sin^2(\psi - \kappa) \right] \sin(\psi - \kappa) \cos(\psi - \kappa) + \left[-\tilde{\delta}_1^{(1)} + \tilde{\delta}_1^{(2)} \right.$$
$$\left. \left. + \tilde{\delta}_1^{(3)} (\cos^2 \psi - \sin^2 \psi) - 2(\epsilon_1^{(2)} - \epsilon_1^{(1)}) \sin^2 \psi \right] \sin \psi \cos \psi \right].$$

See the text for definitions of the quantities involved. The reference coordinate system is defined as in Figure 1, with the horizontal x axis within $[x_1, x_3]$ symmetry plane of the incidence orthorhombic half-space.

Appendix D:
Polarization Angle Φ

Here, I evaluate theoretical expressions for $\cos \Phi$ and $\sin \Phi$ required in equations (25) for R_{PS_1}, R_{PS_2} coefficients.

JECH and PŠENČÍK (1989), and PŠENČÍK (1998) derived general expressions that can be specified for a particular anisotropic symmetry. Adopting the convention introduced in Figure 1 and equations (8), the polarization angle Φ is given for an orthorhombic incidence medium:

$$\cos \Phi = \left[\frac{1}{2} \left(1 + \frac{A}{D} \right) \right]^{1/2},$$
$$\sin \Phi = \frac{B}{|B|} \left[\frac{1}{2} \left(1 - \frac{A}{D} \right) \right]^{1/2}, \quad (D.1)$$

for $B \neq 0$, or

$$\cos \Phi = 1, \quad \sin \Phi = 0 \ (B = 0). \quad (D.2)$$

In (D.1),

$$D \equiv [A^2 + 4B^2]^{1/2} \tag{D.3}$$

and

$$
\begin{aligned}
A \equiv{} & 2A_{33}^{(1)}(\epsilon_1^{(2)}\cos^2\psi - \tilde{\delta}_1^{(2)})\cos^2\psi\cos^2 j\sin^2 j \\
& + 2A_{33}^{(1)}(\epsilon_1^{(1)}\sin^2\psi - \tilde{\delta}_1^{(1)})\sin^2\psi\cos^2 j\sin^2 j \\
& + 2A_{33}^{(1)}(1 + 2\epsilon_1^{(2)})\tilde{\delta}_1^{(3)}\cos^2\psi\sin^2\psi(\cos^2 j + 1)\sin^2 j \\
& + 4A_{33}^{(1)}\epsilon_1^{(2)}\cos^2\psi\sin^2\psi\cos^2 j\sin^2 j \\
& + 2A_{33}^{(1)}(\epsilon_1^{(2)} - \epsilon_1^{(1)})\cos^2\psi\sin^2\psi\sin^2 j \\
& - 2A_{55}^{(1)}\gamma_1^{(1)}\sin^2\psi\sin^2 j \\
& - 2A_{55}^{(1)}\gamma_1^{(S)}\cos^2(2\psi) \\
& - 2A_{55}^{(1)}(\gamma_1^{(1)} - \gamma_1^{(S)})\cos^2\psi\sin^2 j \ ,
\end{aligned}
\tag{D.4}
$$

$$
\begin{aligned}
B \equiv{} & A_{33}^{(1)}(1 + 2\epsilon_1^{(2)})\tilde{\delta}_1^{(3)}\cos\psi\sin\psi\cos 2\psi\cos j\sin^2 j \\
& + 2A_{33}^{(1)}(\epsilon_1^{(1)} - \epsilon_1^{(2)})\cos\psi\sin^3\psi\cos j\sin^2 j \\
& + A_{33}^{(1)}(\tilde{\delta}_1^{(2)} - \tilde{\delta}_1^{(1)})\cos\psi\sin\psi\cos j\sin^2 j \\
& + 2A_{55}^{(1)}\gamma_1^{(S)}\cos\psi\sin\psi\cos j \ ,
\end{aligned}
\tag{D.5}
$$

where $\gamma_1^{(1)} \equiv (A_{66}^{(1)} - A_{55}^{(1)})/2A_{55}^{(1)}$ (TSVANKIN, 1997b), and all the other quantities are defined in the text.

The functions $\cos\Phi$ and $\sin\Phi$ are singular ($D = 0$) for the directions of the reflected-wave slowness vector corresponding to S-wave singularities, where the S_1- and S_2-wave velocities are identical. For such cases, neither equations (D.1) nor the R_{PS_1} and R_{PS_2} approximations (25) can be used.

We can proceed exactly as above, if the incidence half-space is HTI. However, we can also use the fact that the polarizations of S waves propagating in HTI media are known. One of the waves is always polarized within the plane formed by the slowness vector and the horizontal symmetry axis, while the other is polarized in the isotropy (vertical) plane. These geometrical relationships make it possible to derive purely "geometrical" expressions for $\cos\Phi$ and $\sin\Phi$ that contain no medium parameters:

$$
\begin{aligned}
\cos\Phi &= \frac{\cos j\cos\psi}{\sqrt{1 - \sin^2 j\cos^2\psi}}, \\
\sin\Phi &= \frac{-\sin\psi}{\sqrt{1 - \sin^2 j\cos^2\psi}} .
\end{aligned}
\tag{D.6}
$$

As before, j and ψ denote the S-wave reflection phase angle and the azimuth of the incidence plane, respectively. Equations (D.6) are more stable than (D.1), with the only singular point for $j = 90°$ and $\psi = 0°$. Obviously, the reflection angle $j = 90°$ is far beyond the area of applicability of the approximations.

For a VTI incidence half-space, the situation is considerably simpler. One S wave is polarized in the incidence plane (SV) and the other is perpendicular (SH). Thus, it immediately follows that

$$\cos \Phi = 1, \quad \sin \Phi = 0 \ . \tag{D.7}$$

From equations (D.7) and (25), R_{PS_1} for VTI media represents the reflection coefficient of the P-SV wave (polarized in the incidence plane) and R_{PS_2} represents the reflection coefficient of the P-SH wave (polarized horizontally). Such a P-SH conversion is generated if the reflecting half-space has HTI, orthorhombic or lower symmetry [see equations (C.1) and (C.2)].

Equations (D.7) also hold for a purely isotropic incidence medium. In that case, R_{PS_1} and R_{PS_2} correspond to the SV and SH components of the S-wave, respectively; both components travel with the same velocity.

REFERENCES

AKI, K., and RICHARDS, P. G., *Quantitative Seismology: Theory and Methods* (W. N. Freeman & Co., San Francisco, 1980).

ALVAREZ, K., DONATI, M., ALDANA, M., *AVO analysis for converted waves*. In *69th Annual Internat. Mtg., Soc. Expl. Geophys., Expanded Abstracts* (SEG, 1999) pp. 876–879.

BAKULIN, A., GRECHKA, V., and TSVANKIN, I. (2000a), *Estimation of Fracture Parameters from Reflection Seismic Data, Part I: HTI Model due to a Single Fracture Set.*, Geophysics 65, 1788–1802.

BAKULIN, A., GRECHKA, V., and TSVANKIN, I. (2000b), *Estimation of Fracture Parameters from Reflection Seismic Data, Part II: Fractured Models with Orthorhombic Symmetry.*, Geophysics 65, 1803–1817.

BANIK, N. C. (1987), *An Effective Anisotropy Parameter in Transversely Isotropic Media.*, Geophysics 52, 1654–1664.

BORTFELD, R. (1961), *Approximation to the Reflection Coefficients of Plane Longitudinal and Transverse Waves.*, Geophys. Prospect. 9, 485–502.

DONATI, S. M., *Making AVO analysis for converted waves a practical issue*. In *68th Annual Internat. Mtg., Soc. Expl. Geophys., Expanded Abstracts* (SEG, 1998) pp. 2060–2063.

GRECHKA, V., and TSVANKIN, I. (1998), *3-D Description of Normal Moveout for Anisotropic Media.*, Geophysics 63, 1079–1092.

JECH, J., and PŠENČIK, I. (1989), *First-order Perturbation Method for Anisotropic Media.*, Geophys. J. Int. 99, 369–376.

JIN, S., *Characterizing reservoir by using jointly P- and S-wave AVO analyses*. In *69th Annual Internat. Mtg., Soc. Expl. Geophys., Expanded Abstracts* (SEG, 1999) pp. 687–690.

LARSEN, A. J., MARGRAVE, F.G., and LU, H-X., *AVO Analysis by simultaneous P-P and P-S weighted stacking applied to 3C-3D seismic data*. In *69th Annual Internat. Mtg., Soc. Expl. Geophys., Expanded Abstracts* (SEG, 1999) pp. 721–724.

LI, X-Y., KÜHNEL, T., and MACBETH, C., *Mixed mode AVO response in fractured media*. In *66th Annual Internat. Mtg., Soc. Expl. Geophys., Expanded Abstracts* (SEG, 1996), pp. 1822–1825.

MENSH, T., and RASOLOFOSAON, P. (1997), *Elastic-wave Velocities in Anisotropic Media of Arbitrary Symmetry – Generalization of Thomsen's Parameters ε, δ and γ.*, Geophys. J. Int. 128, 43–64.

MILEY, P. M., *Overpressure prediction using converted mode reflections from base salt*. In *69th Annual Internat. Mtg., Soc. Expl. Geophys., Expanded Abstracts* (SEG, 1999) pp. 880–883.

NEFEDKINA, T., and BUZLUKOV, V., *Seismic dynamic inversion using multiwave AVO data*. In *69th Annual Internat. Mtg., Soc. Expl. Geophys., Expanded Abstracts* (SEG, 1999), pp. 888–891.

PŠENČÍK, I. (1998), *Green's Functions for Inhomogeneous Weakly Anisotropic Media.*, Geophys. J. Int. *135*, 279–288.

PŠENČÍK, I., and GAJEWSKI, D. (1998), *Polarization, Phase Velocity and NMO Velocity of qP Waves in Arbitrary Weakly Anisotropic Media.*, Geophysics *63*, 1754–1766.

PŠENČÍK, I., and VAVRYČUK, V. (1998), *Weak Contrast PP Wave Displacement R/T Coefficients in Weakly Anisotropic Elastic Media*, Pure appl. geophys. *151*, 699–718.

PŠENČÍK, I. MARTINS, J. L. (2000), *Properties of Weak Contrast PP Reflection/Transmission Coefficients for Weakly Anisotropic Elastic Media.*, Studia Geophysica et Geodaetica, submitted.

RICHARDS, P. G., and FRASIER, C. W. (1976), *Scattering of Elastic Waves from Depth-dependent Inhomogeneities*. Geophysics *41*, 441–458.

RÜGER, A., *Reflection Coefficients and Azimuthal AVO Analysis in Anisotropic Media* (Ph.D. Thesis, Center for Wave Phenomena, Colorado School of Mines, Golden, Colorado, 1996).

RÜGER, A. (1997), *P-wave Reflection Coefficients for Transversely Isotropic Models with Vertical and Horizontal Axis of Symmetry*, Geophysics *62*, 713–722.

RÜGER, A. (1998), *Variation of P-wave Reflectivity with Offset and Azimuth in Anisotropic Media*, Geophysics *63*, 935–947.

SAYERS, C. M. (1994), *P-wave Propagation in Weakly Anisotropic Media*. Geophys. J. Int. *116*, 799–805.

SHUEY, R. T. (1985), *A Simplification of the Zoeppritz-equations*. Geophysics *50*, 609–614.

THOMSEN, L. (1986), *Weak Elastic Anisotropy*, Geophysics *51*, 1954–1966.

THOMSEN, L. (1993), *Weak anisotropic reflections*. In *Offset Dependent Reflectivity* (Castagna and Backus, eds.) (SEG, Tulsa) pp. 103–111.

TSVANKIN, I. (1996), *P-wave Signatures and Notation for Transversely Isotropic Media: An Overview*, Geophysics *61*, 467–483.

TSVANKIN, I. (1997a), *Reflection Moveout and Parameter Estimation for Horizontal Transverse Isotropy*, Geophysics *62*, 614–629.

TSVANKIN, I. (1997b), *Anisotropic Parameters and P-wave Velocity in Orthorhombic Media*. Geophysics *62*, 1292–1309.

VAVRYČUK, V. (1999), *Weak-contrast Reflection/Transmission Coefficients in Weakly Anisotropic Elastic Media: P-wave Incidence*, Geophys. J. Int. *138*, 553–562.

VAVRYČUK, V., and PŠENČÍK, I. (1998), *PP-wave Reflection Coefficients in Weakly Anisotropic Elastic Media*, Geophysics *63*, 2129–2141.

(Received November 11, 2000, revised/accepted March 9, 2001)

Pure appl. geophys. 159 (2002) 1563–1582
0033–4553/02/081563–20 $ 1.50 + 0.20/0

❙ Pure and Applied Geophysics

Three-dimensional Travel-time Calculation
Based on Fermat's Principle

VALENTIN MESHBEY,[1] EVGENY RAGOZA,[1] DAN KOSLOFF,[2,1]
UZI EGOZI,[1] and TAL WEXLER[1]

Abstract—We present a new travel-time calculation method based on Fermat's principle. In the method, travel times are recursively calculated on horizontal planes of increasing depth. For typical configurations of exploration geophysics, the distance between the planes is on the order of 100 m. The travel times on the first plane are calculated by connecting straight ray segments from the source point to the grid points on the plane and integrating the slowness along each segment. The times on the other planes are calculated by finding the minimum of the combination of the times on the plane above plus the additional time along segments connecting grid points on the two planes. The travel-time calculation method is designed for calculating either the first arrival times, or the time of the shortest travel path arrival. The method is extended to handle vertically transverse isotropic (VTI) media by an approach which increases the computing time only slightly. The algorithm is tested against synthetic examples for isotropic and VTI wave propagation.

Key words: Travel times, Fermat's principle, heterogeneous velocity fields, vertically transverse anisotropy (VTI).

Introduction

The calculation of travel times is an integral part of Kirchhoff depth migration. Considerable research effort has been devoted in recent years to devising new travel-time calculation algorithms which are fast and robust, and which can also handle multiple arrivals. A review of recent work on the subject can be found in LEIDENFROST *et al.*, (1999). The present work describes a new travel-time calculation algorithm, based on Fermat's principle.

The published travel-time calculation methods can be roughly divided into three categories namely, eikonal equation solvers, network methods, and ray-tracing based methods. Eikonal solvers, as their name implies, solve the eikonal equation numerically. These methods are usually fast and can also produce satisfactory results with very little smoothing of the velocities. However, eikonal solvers are

[1] Paradigm Geophysical, Gav Yam 3, 9 Shenkar Street, Herzliya B, Israel, 46120.
[2] Department of Geophysics, Tel Aviv University, Tel Aviv, Israel 69978, dan@seismo.tau.ac.il

usually limited to calculate only the time of the first arrival. Network methods calculate travel times to points on a spatial mesh. At a given stage, the grid points are divided into two groups, namely, points for which times have previously been calculated, and points for which times have not yet been obtained. At each step one new point from the second group is transferred to the first group. The selection of the point is based on the requirement that the travel times from it to one of its neighbors belonging to the first group, plus the calculated time of that neighbor point, be minimal amongst all candidate points for which times have not yet been calculated. Network methods are similar to eikonal solvers in accuracy and are able to handle unsmoothed velocities. Usually, they calculate only the first arrival.

There are many travel-time calculation methods based on ray-tracing. It is beyond the scope of this article to discuss all the work which has been done in this area. In general ray tracing based methods can handle multiple arrivals and can calculate additional parameters, as for example geometrical spreading and ray angles. However, ray-tracing methods are usually complicated and relatively slow, and require some smoothing of the velocity field, or explicit handling of the interfaces.

This work presents a new travel-time calculation method based on Fermat's principle. The method is designed especially for use in Kirchhoff prestack depth migration. Of the three categories listed, it resembles most the network methods. The travel times are calculated recursively on horizontal planes of increasing depth. After the calculation of the travel times on one horizontal plane, the travel time at a point on the next plane is calculated by connecting straight rays to it from the grid points on the plane above. The combination of the previously calculated time at the points on the plane above, plus the additional time increment along the ray connecting the points above to the point on the next plane is evaluated. The calculated travel-time is selected by finding the sum of times which produces a minimum.

The travel-time algorithm was designed for calculating the first arrival, but it can be modified to select the event with the shortest travel path length. The extension of the method to handle vertically transverse isotropic media (VTI) is quite simple.

In the following sections we first describe the travel-time calculation method for calculating the time of the first arrival in an isotropic medium. Subsequently the extension for calculating the shortest travel path is discussed. Following that the generalization of the method to handle VTI media is explained. The last sections present results of travel-time calculations in heterogeneous media, and in a VTI medium.

The Travel-time Algorithm for Calculating the First Arrival

Travel times from a specified surface location $(x_s, y_s, z_s = 0)$ to subsurface locations are calculated sequentially on horizontal planes of increasing depth. For a typical configuration in reflection seismics, the distance between the calculation planes is approximately 100 m and the maximum calculation depth is between 6–8 kilometers. The horizontal grid spacing on the calculation planes is on the order of 50 m. The subsurface velocities are represented on a velocity grid in which the horizontal sampling rate is equal to the horizontal cmp trace spacing, and the vertical sampling rate is equal to the depth step of the migration.

The travel times from the source point S to the first calculation plane (Fig. 1) are calculated by connecting straight lines from S to each of the grid points of that plane. The travel-time is obtained by a numerical integration of the slownesses along the ray segments. The use of straight ray segments, instead of the true ray paths, is in agreement with first-order perturbation using a homogeneous background,

$$t \approx t_0 + \Delta t = \int_\Omega 1/v_0 \, dl + \int_\Omega (1/v - 1/v_0) \, dl = \int_\Omega 1/v \, dl \ .$$

Ω describes the straight ray path in the homogeneous background medium with velocity v_0, while v is the real velocity of the medium. This approximation is valid when the velocity variation along the ray segments is small, which occurs when the distance between the calculation planes is not large.

After obtaining the times on the first plane, the travel times on the following planes are calculated recursively. Given the calculated travel times $t^n(x, y, z = z_n)$ on the n-th plane, the travel-time increments $dt(x, y, x', y')$ from the grid point $(x, y, z = z_n)$ to a point $(x', y', z = z_{n+1})$ on the $n + 1$ plane are calculated by integrating the slowness along the straight line segment connecting the two points. For fixed $(x', y', z = z_{n+1})$ the sum $\tau_{x'y'}(x, y) = t^n(x, y, z = z_n) + dt(x, y, x', y')$ defines a two-dimensional function of the (x, y) coordinates of the n-th plane (Fig. 2).

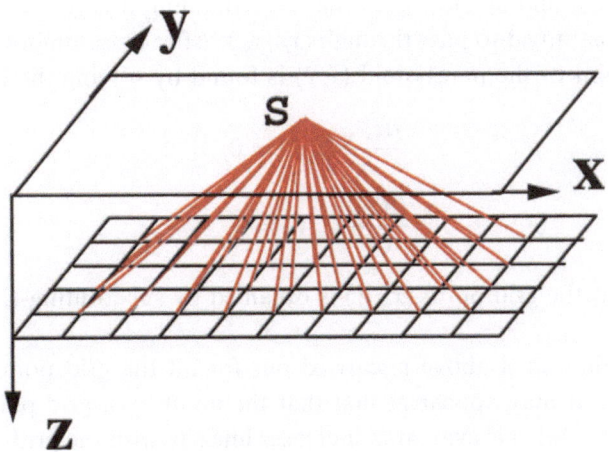

Figure 1
Travel-time calculation for the first calculation plane.

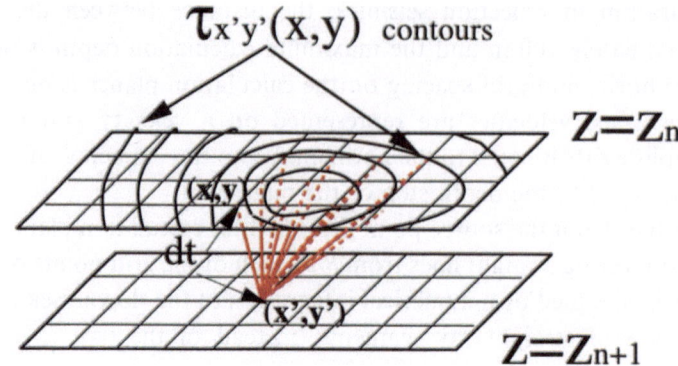

Figure 2
Calculation of the travel time on the $n+1$-th plane based on the calculated times on the n-th plane.

The time of the first arrival is given by,

$$t(x', y', z = z_{n+1}) = \min_{x,y} \tau_{x'y'}(x, y) \ . \tag{1}$$

In the numerical implementation, the minimum of $\tau_{x'y'}(x,y)$ is first found approximately by scanning its values on the grid points (x, y) of the n-th calculation plane. Denoting by (x^m, y^m) the grid point with the smallest value of $\tau_{x'y'}(x, y)$, the estimate of the time is improved with a least-squares paraboloid fit according to,

$$\tau_{x'y'}(x, y) = a_0 + a_1(x - x_m) + a_2(y - y_m) + a_3(x - x_m)^2$$
$$+ a_4(y - y_m)^2 + a_5(x - x_m)(y - y_m) \tag{2}$$

The six coefficients $\{a_k\}_{k=0}^{5}$ are determined by a least-squares fit of (2) to the values of $\tau_{x'y'}(x,y)$ at (x_m, y_m) and at the eight points surrounding it (Fig. 3) . In addition the paraboloid is constrained to pass through (x_m, y_m). After determining the coefficients, the minimum point of the paraboloid $(\hat{x}, \hat{y})$ is found by solving the linear equations

$$\frac{\partial \tau_{x'y'}}{\partial x}(\hat{x}, \hat{y}) = 0,$$

$$\frac{\partial \tau_{x'y'}}{\partial y}(\hat{x}, \hat{y}) = 0 \ .$$

The travel-time at the grid point (x', y') is obtained by substituting $(\hat{x}, \hat{y})$ for (x, y) in equation (2).

The process described above is carried out for all the grid points on the $n + 1$ calculation plane. It may appear at first that the number of grid points on the n-th plane for which $\tau_{x'y'}(x, y)$ is evaluated increases linearly with the grid size. However a search aperture which is centered one grid point away from the location of the calculated minimum of the previous grid point is used. A typical aperture size is 81

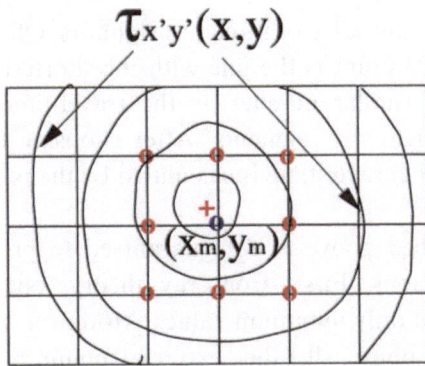

Figure 3
Finding the minimum travel time from travel-time values at grid points surrounding the grid point with the smallest time value.

points (9×9). In this manner, the amount of calculations per time point are independent of the problem size.

The Travel-time Algorithm for Calculating the Arrival with the Shortest Travel Path

Ideally Kirchhoff migration should use all ray arrivals for the imaging of the subsurface. From a practical viewpoint however this is very difficult because of the need to store separate travel-time tables for each arrival. Furthermore, the calculation of the migration weights for each arrival, depending on the type of migration, requires three to four additional tables. In terms of accuracy, the relative contribution of each arrival, which is proportional to the weights, is sensitive to the details of the velocity variation, which is usually not known with sufficient accuracy. The next best alternative would be to use the most energetic arrival. However, it has been shown that amplitude calculations are very sensitive to the degree of smoothing of the velocity field, and that very often the use of the most energetic arrival results in discontinuous images of the reflectors (NICHOLS et al., 1998).

Due to these difficulties, NICHOLS et al. (1998) suggested a different arrival selection criterion based on the shortest travel path. In many situations the shortest travel path coincides with the maximum energy path. However in regions of rapidly varying amplitudes the shortest path criterion is more robust. This section describes a modification of the Fermat method to select the shortest travel path arrival, instead of the first arrival.

The calculations of the shortest path travel times proceed in the same sequence as the first-arrival time calculation described in the previous section. The main difference is that the function $\tau_{x'y'}(x, y)$ is searched for all the local minima, instead of only the global minimum. A grid point of a local minimum in $\tau_{x',y'}$ value is a point with a value smaller than the value of its eight neighbors. Of all the points of minimal values found, the selected point is the one with the shortest travel path. The travel path is calculated in a similar manner as the travel-time by integration of the distances along the straight ray segments. After choosing the shortest travel path minimal point (x_m, y_m), the travel-time is calculated by the paraboloid fit described in the previous section.

The sequence described above is not guaranteed to produce the shortest path travel-time in all situations. First, from ray theory, one needs to consider all stationary values and not only minimum values. Moreover, after the selection of the shortest path minimum point, all other extremal points are not accounted for in subsequent calculations. The method, therefore, cannot track situations where the shortest path minimum on consecutive planes shifts from one local minimum to another. However, the main circumstance in which the shortest path criterion is useful is in imaging sediments close to steep salt flanks (Fig. 4). In such situations the first arrival is often the headwave traveling along the salt-sediment boundary. Conversely, the shortest travel path in this case coincides with the direct arrival which is a local minimum and does not shift in position. In this condition the selection of the shortest travel path arrival assures continuity of the imaged reflections when moving towards the salt boundaries.

The shortest path option in the Fermat method requires a more complicated search than the first arrival calculation. However, in imaging close to salt structures, the extra effort is justified. In later sections we present comparisons between first-arrival time calculation and shortest travel path times.

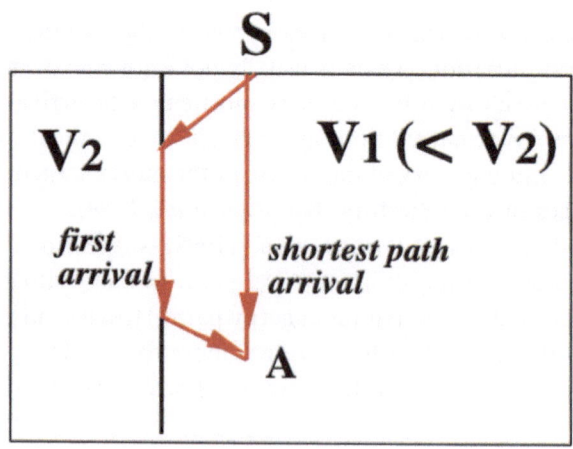

Figure 4
Direct arrival and refracted arrival in a structure with a steep vertical velocity contrast.

The Travel-time Algorithm for VTI Media

This section describes the generalization of the travel-time calculation method to VTI media. At first we review elements of VTI ray theory which are used in the travel-time algorithm.

The P-wave propagation in a VTI medium can be described by four parameters S_l, f, ϵ, δ, respectively. $S_l = 1/\alpha$ is the vertical P-wave slowness, with α the vertical P-wave velocity. $f = \sqrt{\beta/\alpha}$ where β is the vertical S-wave velocity. ϵ and δ are Thomsen parameters (THOMSEN, 1986).

For a ray propagating in a VTI medium, the position vector $\mathbf{x} = (x_1, x_2, x_3)$ and the slowness vector $\mathbf{p} = (p_1, p_2, p_3)$ satisfy the ray equations;

$$\frac{dx_i}{d\sigma} = \frac{\partial G}{\partial p_i}$$

$$\frac{dp_i}{d\sigma} = \frac{\partial G}{\partial x_i} \quad i = 1, \ldots, 3 \ , \tag{3}$$

where σ is a parameter which increases continuously along the ray,

$$G(\mathbf{p}) = -S_l^4 + K - L = 0 \ , \tag{4}$$

(ČERVENÝ; 1972, PAYTON, 1983; FARRA, 1989), with,

$$K = (1 + f)S_l^2(p_H^2 + p_z^2) + 2\epsilon S_l^2 p_H^2 \ ,$$

and,

$$L = \left((1 + 2\epsilon)p_H^2 + f p_z^2\right)(f p_H^2 + p_z^2) - \left(2\delta(1 - f) + (1 - f)^2\right)p_H^2 p_z^2 \ ,$$

where $p_H = (p_1^2 + p_2^2)^{1/2}$ and $p_z = p_3$.

The travel-time change along the ray is described by the equation

$$\frac{dt}{d\sigma} = \sum_{i=1}^{3} p_i \frac{dx_i}{d\sigma} \ . \tag{5}$$

Based on the first set of equations of (3), the unit vector pointing in the ray direction is given by,

$$n_i = \frac{(\nabla_\mathbf{p} G)_i}{\|\nabla_\mathbf{p} G\|_2} \ , \tag{6}$$

where $(\nabla_\mathbf{p} G)_i = \frac{\partial G}{\partial p_i}$, and $\|.\|_2$ denotes the l_2 norm of a vector.

The group slowness is defined by,

$$g_{sl} = \mathbf{p} \cdot \mathbf{n} = \sum_{i=1}^{3} p_i n_i \ .$$

Given a propagation distance increment *dl*, based on (5), the corresponding travel-time increment is equal to $dt = g_{sl} \, dl$. For a specified ray direction **n**, the calculation of the group slowness requires the value of the corresponding ray slowness vector **p**. The slowness is obtained by solving equation (6) numerically. The procedure is described in the appendix.

In the travel-time calculation, it is assumed that the subsurface structure consists of a finite number of closed volumes, each of which belongs to a formation with constant values of f, ϵ and δ, respectively. The same assumption was used in ČERVENÝ (1989) and was termed factorized anisotropy. The *P*-wave slowness S_l however, is allowed to vary in an arbitrary fashion. A formation identification grid is created whose integer values give the number of the volume to which each subsurface point belongs. The travel-time algorithm utilizes the fact that all the ray directions in the travel-time calculation method are known in advance (the ray segments are simply straight lines which connect grid points on two consecutive calculation planes). For each ray direction and each formation, a group slowness factor is calculated. This factor is equal to the group slowness for the formation values of f, ϵ, δ and for a vertical *P*-wave slowness of unity. The group factors are precalculated and stored in tables before the actual travel-time calculation.

During the travel-time calculation, the travel-time along a ray segment is first calculated isotropically using the vertical *P*-wave slowness. The result is then multiplied by the group slowness factor appropriate to the angle and the formation which the ray segment traverses. An average factor is calculated for segments which traverse more than one formation.

As the group slowness factors are precalculated, the extension of the Fermat method to VTI media requires very little additional computing time, compared to the isotropic case. The accuracy of the VTI travel times is comparable to that of the isotropic calculation. There are no restrictions on the degree of anisotropy, and for a ray segment traversing a homogeneous region the travel-time calculation is exact.

Example: Linear Velocity Variation

The first test of the travel-time algorithm was a comparison with the exact solution for a medium with a constant velocity gradient. The calculations used a grid with 100 points in the *x* and *y* horizontal directions respectively, and 200 points in the vertical *z* direction. The grid spacing was $dx = dy = 50$ m, and the vertical distance between the calculation planes was 100 m. The velocity variation in the medium was given by the function,

$$V(x, y, z) = V_0 + a_x(x - x_0) + a_y(y - y_0) + a_z z \ ,$$

with $V_0 = 2500$ m/sec, $x_0 = y_0 = 2500$ m, $a_x = 0.28 \, \text{sec}^{-1}$, $a_y = 0.08 \, \text{sec}^{-1}$, and $a_z = 1.25 \, \text{sec}^{-1}$. With these parameters the minimum and maximum velocities in the grid were 1600 m/sec, and 4883 m/sec respectively, with a velocity variation in all spatial directions.

Figure 5 presents calculated travel-time contours on a horizontal plane at a depth of 1500m. The asymmetry of the contours as a result of the lateral velocity gradients is apparent. Figure 6 presents a comparison between the numerical and the exact travel times along the line $y = 3750$ m on the same plane. In the scale of the plot the two travel times are almost indistinguishable. In all locations the difference between the travel times was less than 2 msec.

The conclusion from this test is that the travel-time algorithm is capable of producing accurate results with smooth velocity models, even when such models contain a strong velocity variation in all spatial directions.

Example: Vertical Interface

The second test of the Fermat algorithm was a structure containing two velocity regions which were separated by a vertical interface. The velocities in the two regions were 2000 m/sec and 4000 m/sec, respectively. The migrated section contained 200

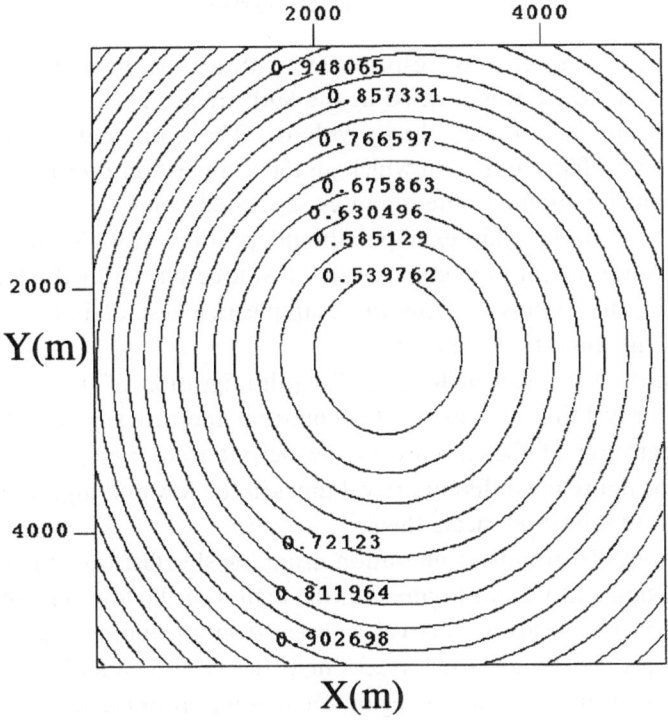

Figure 5
Linear velocity test: travel-time contours on a horizontal plane at a depth of 1500 m.

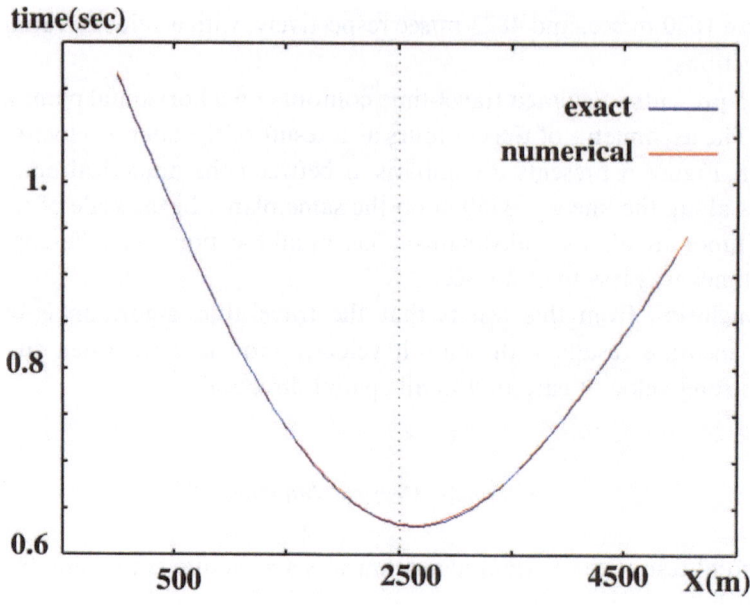

Figure 6
Linear velocity test: Comparison between numerical and analytical travel times in the x direction for $y = 3750\,\text{m}$, and $z = 1500\,\text{m}$.

traces in the x and y directions, respectively, with a trace increment of 25 m. The vertical velocity interface was located at the 80th trace. A small amount of lateral smoothing was applied to the velocity volume. The input data to the migration consisted of a zero offset volume in which all the samples were zero, except for those of trace 100 which contained nine impulses separated from each other by 0.25 sec. The migrated output for this example should produce impulse responses which closely resemble the travel-time contours for a source placed at the location of the trace containing the impulses. Thus the example is a test of both the travel-time algorithm and the migration algorithm.

Figure 7 shows the result of Kirchhoff depth migration of the input data where the first arrival travel times are used. The migration aperture size was 100 traces (this limited the migration of the impulse to the output trace range 50–150). One can clearly see in the figure the different arrival times in the low and high velocity regions, and the headwave which connects them.

Figure 8 shows the results of migration using the shortest travel path times. This figure differs significantly from Figure 7 in the region of the low velocity close to the vertical interface. The headwave arrival is absent from the figure which only contains the direct arrivals. There are a few traces near the vertical interface, for which the image shows discontinuities. In this region the minima from the refracted arrival and from the direct arrival are spatially close, and there are numerical difficulties to separate them on a discrete grid.

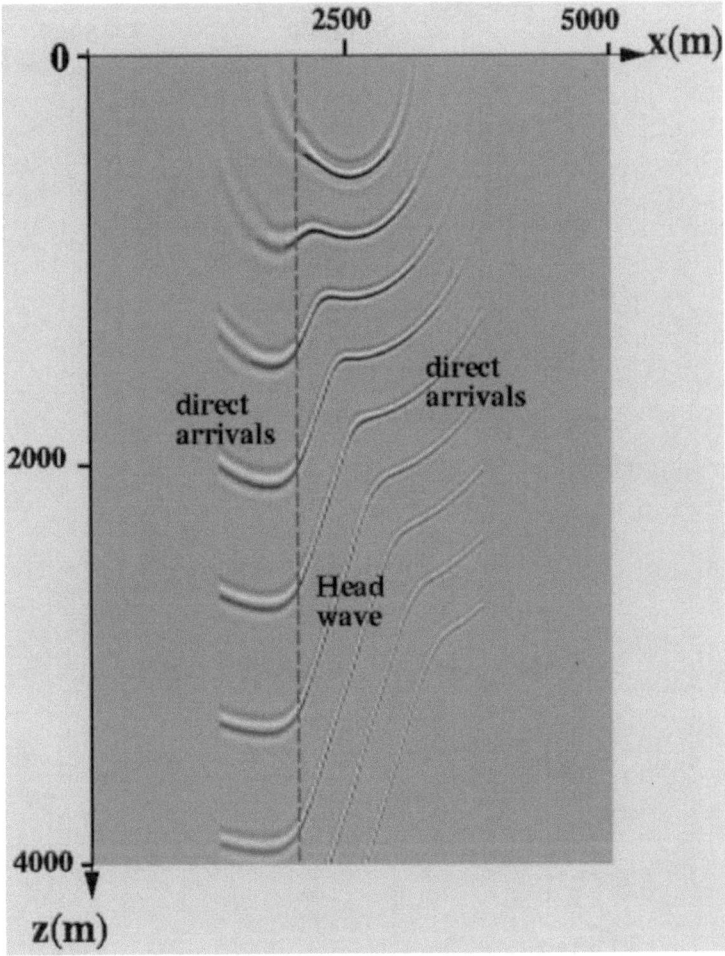

Figure 7
Vertical interface: First arrival impulse response.

This example has demonstrated that the Fermat method has been able to handle strong velocity contrasts, and that the shortest path option enabled the selection of the direct arrival instead of the headwave.

Example: The SEG Salt Model

The travel-time algorithm is used for imaging of the SEG salt model. The model contains a very rugose high velocity salt body embedded in a low velocity sedimentary medium. Because of the complicated salt boundary geometry, this example is a challenging test for Kirchhoff prestack depth migration.

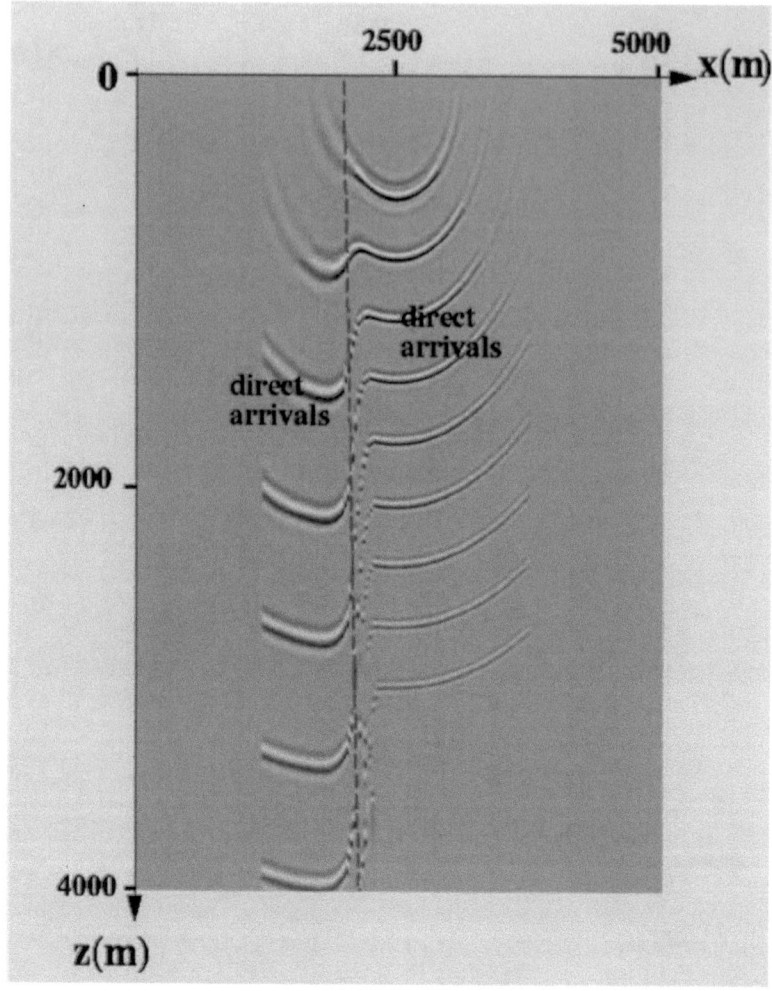

Figure 8
Vertical interface: Shortest travel path impulse response.

The model contains 670 cmp traces in the *x* direction (inlines), and 500 traces in the *y* direction (crosslines). The cmp trace spacing was 20 m in the inline direction and 20 m in the crossline direction. The maximum offset in the inline direction is 2680 m, and 140 m in the crossline direction. The migration used an aperture of 6000 m in both the inline and crossline directions.

The three-dimensional prestack depth migration was carried out on one vertical target plane along inline 360, where the velocity structure is most complicated. Figure 9 presents the velocity structure below this inline. The complexity of the velocity can be appreciated from Figure 10 which presents the velocity structure below crossline 260, which cuts inline 360 close to the highest point of the salt body.

Inline 360

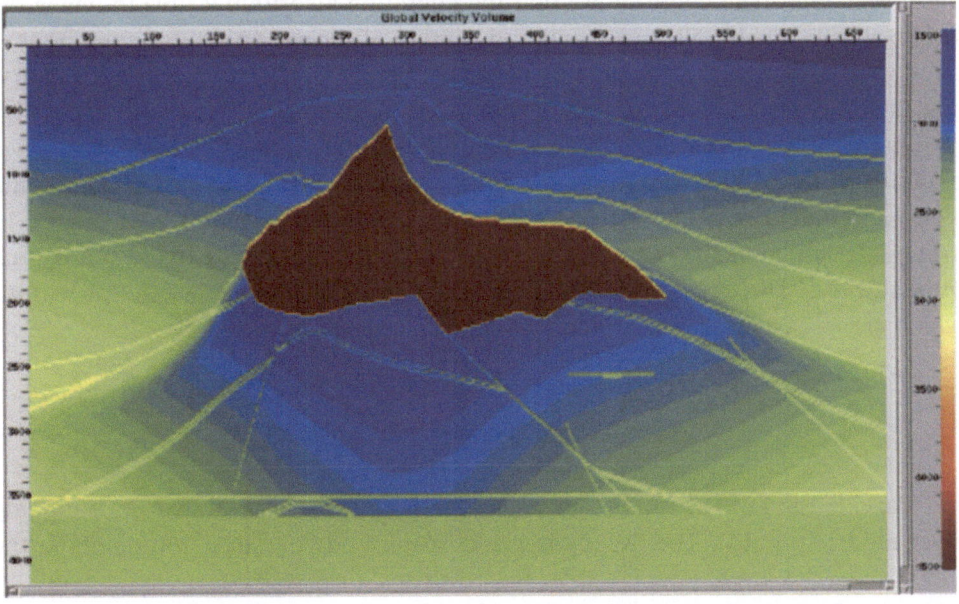

Figure 9
SEG salt model: Velocity section along inline 360.

Crossline 260

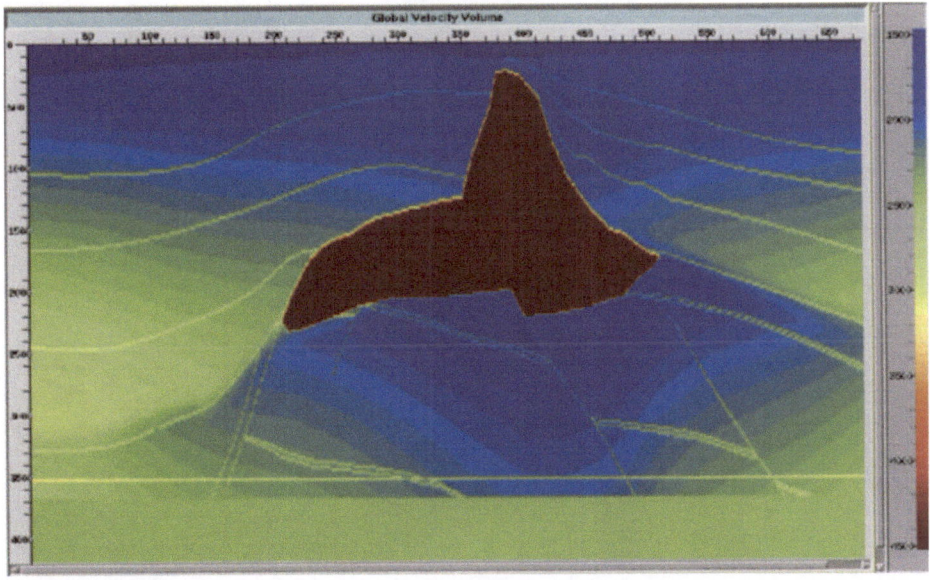

Figure 10
SEG salt model: Velocity section along crossline 260.

Figure 11 presents the migrated section of inline 360 where the first-arrival travel times were used. The figure shows that the migration algorithm is able to image the sediments above salt well, and also that the image of the right part of the salt body is quite good. However, the image of the left part of the salt body is less satisfactory, and in addition there are migration artifacts in the sedimentary region below salt.

Figure 12 presents the migrated section of the same inline using shortest travel path arrival times. This figure reveals a distinct improvement over the corresponding section generated with first arrival times. The noise which was present in Figure 11 around the left boundary of the salt has been eliminated, indicating that it was caused by headwaves. There is also improvement in the image quality below salt, for reasons which are less obvious.

Example: a Two-layer VTI Medium

The extension of the travel-time algorithm to VTI media is tested for a dipping layer model (Fig. 13). The travel time calculation grid contained 100 points in the

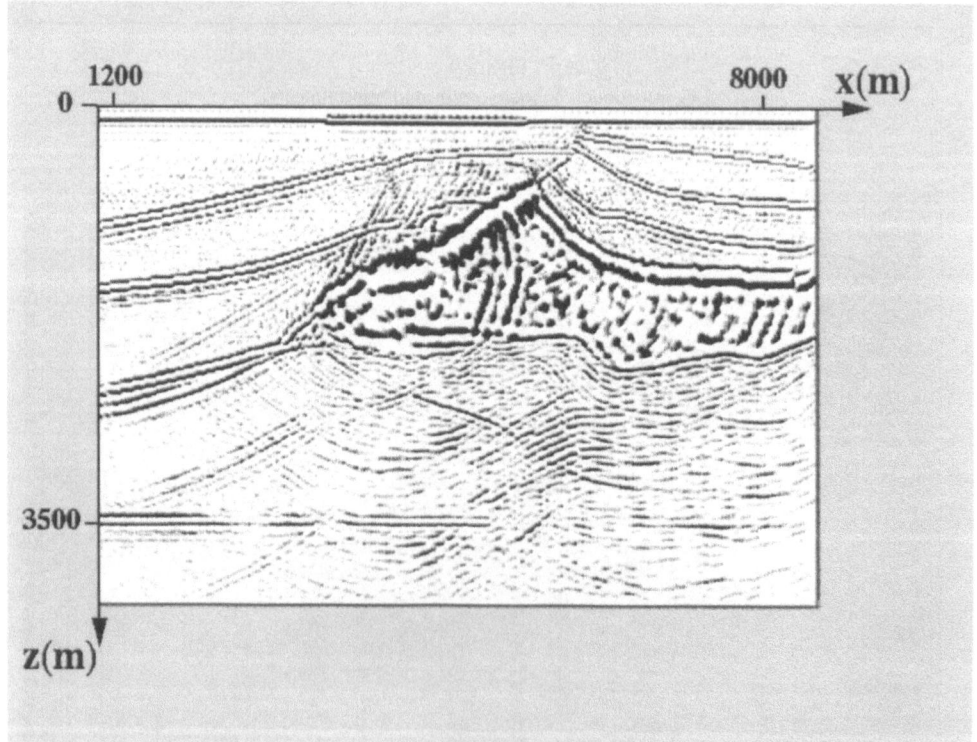

Figure 11
SEG salt model: Depth migrated section along inline 360 using first-arrival time tables.

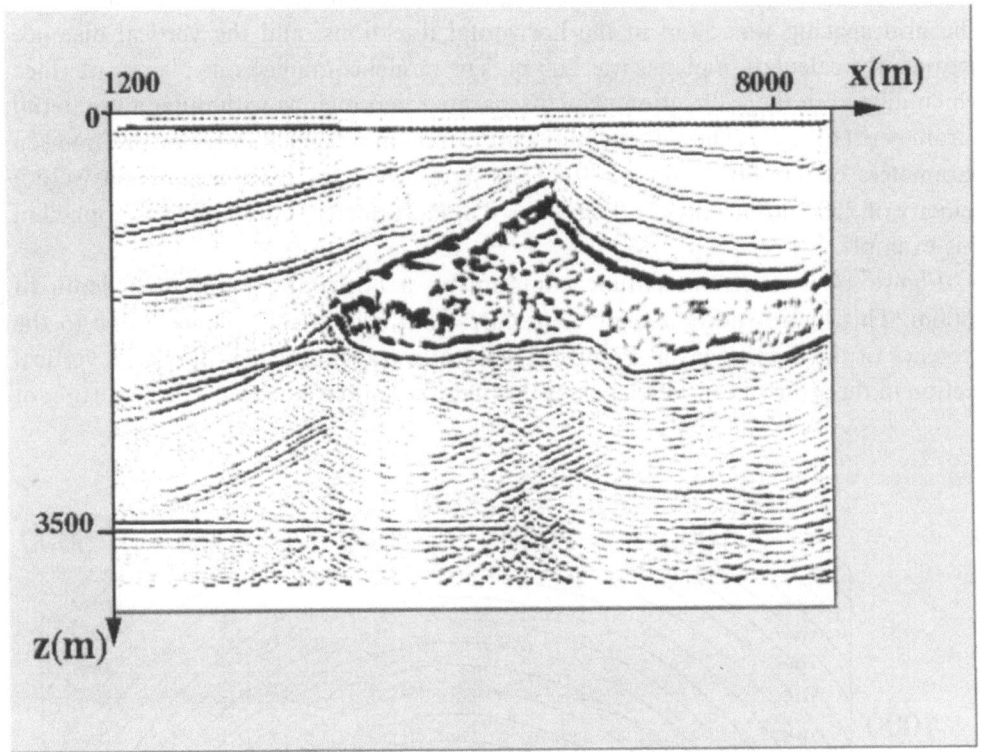

Figure 12
SEG salt model: Depth migrated section along inline 360 using the shortest path time tables.

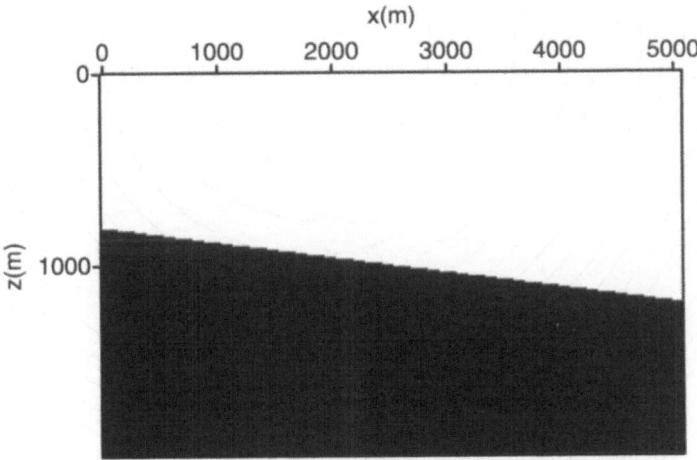

Figure 13
A cross section through the dipping layer model of the VTI test.

x and *y* horizontal directions respectively, and 200 points in the vertical *z* direction. The grid spacing was 50 m in the horizontal directions, and the vertical distance between the calculation planes was 200 m. The model contained one planar interface which dipped in the *x* direction, which separated two regions with different material parameters (Fig. 13). The velocity in the upper region was 2000 m/sec, with Thomsen parameters of $\epsilon = -0.2$, and $\delta = 0.2$. The parameters for the lower region were a velocity of 2500 m/sec and $\epsilon = 0$, and $\delta = 0.3$. No velocity smoothing was applied in this example.

Figure 14 presents travel-time contours on a horizontal plane at a depth of 1800m. This figure shows a slight departure from cylindrical symmetry due to the presence of the dipping layer. Figure 15 delineates travel-time contours on a vertical section in the *x* direction, which passes through the source location. The distortion of

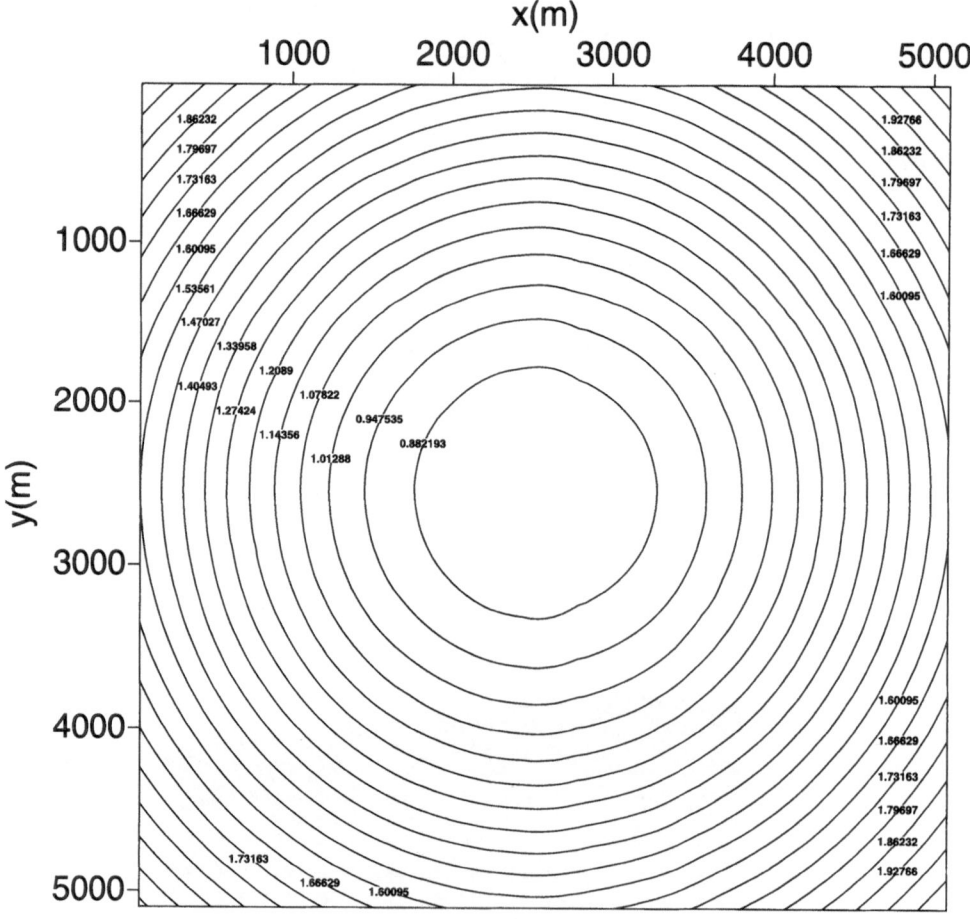

Figure 14
Travel-time contours for the dipping layer test on a horizontal plane at a depth of 1800 m.

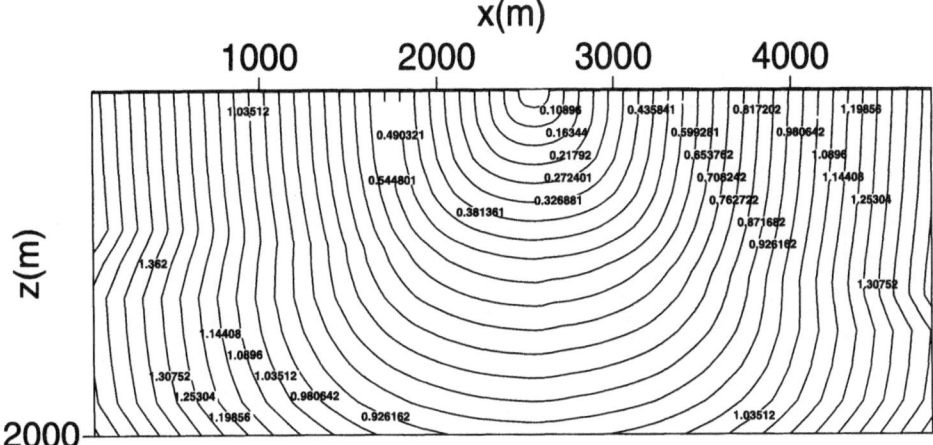

Figure 15

Dipping layer example: Travel-time contours on a vertical section in the x direction passing through the source location.

the contours due to the velocity change is apparent. Because of the strong anisotropy, the contours in the upper region of the model are very nonspherical.

The travel times from the algorithm are compared in Figure 16 (black color) to times produced by a 2-D wavefront reconstruction method (green color) at a depth of 1800, along a line in the x direction directly beneath the source location (for a vertical plane in the x direction containing the source point the problem becomes two-dimensional). The agreement between the travel-time curves in the figure is very good attesting to the validity of the VTI procedure of this study.

Conclusions

We have presented an algorithm based on Fermat's principle for calculating subsurface travel times. The method has been specifically designed for use in prestack Kirchhoff depth migration. Although the algorithm relies on a segmented representation of the geometrical ray paths, results have shown that it produces accurate and robust results. Because of the fact that a sparse calculation grid can be used, the method is comparable in speed to fast eikonal solvers.

The method has been extended to calculate the shortest travel path arrival times instead of the first arrivals. This extension allows the production of good images of sediments near steep salt flanks, where first arrival headwaves can pose problems to eikonal solvers. Migration results with the SEG salt model have confirmed this assertion. The main utility of the shortest travel path extension is for better imaging of steeply dipping interfaces. Thus, with a relatively simple method, we were able to obtain satisfactory results in models which can pose

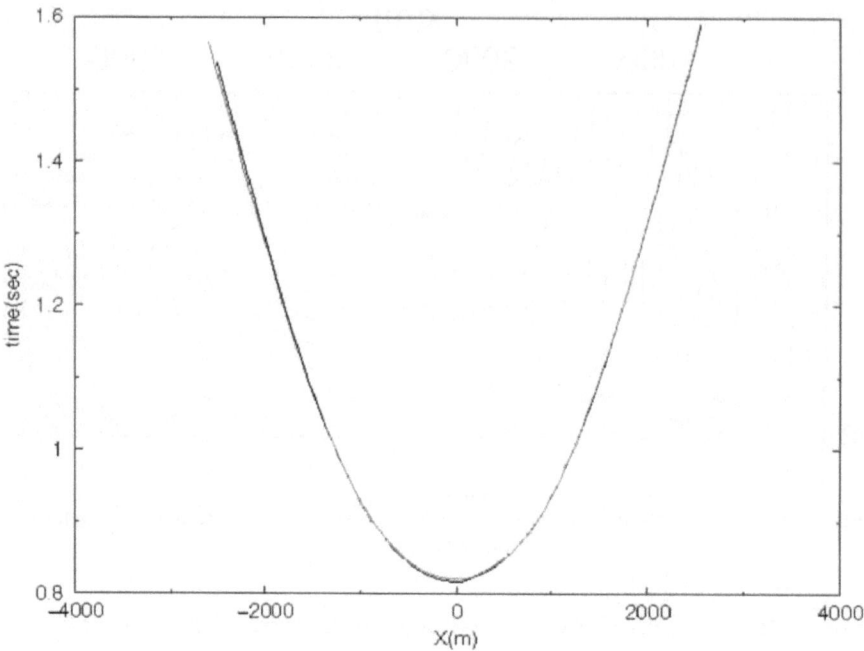

Figure 16
Dipping layer example: Comparison of calculated travel times along a horizontal line under the source at a depth of 1800 m, with travel times obtained with a 2-D wavefront reconstruction method.

problems to other fast travel-time calculation algorithms. There are however circumstances in which a more comprehensive travel-time selection criterion is needed, as for example, in imaging beneath a complex overburden. In such situations travel-time calculation methods based on ray tracing may be more appropriate. However those methods are considerably slower and usually require more velocity smoothing.

The extension of the Fermat method to handle VTI media is straightforward when a formation identification volume is used. The travel time along each ray segment is first calculated in the same manner as in the isotropic case. The times are then multiplied by precalculated group slowness factors. In performing VTI calculations there is a very little increase in computing time compared to isotropic calculations.

Acknowledgement

This work was partially supported by the GIF project I-524-18.8/1997. The authors wish to thank Paradigm Geophysical for the permission to publish this work.

Appendix: Calculation of the Slowness Vector from the Ray Direction for a VTI Medium

Let $\mathbf{l} = (l_1, l_2, l_3)$ denote the unit vector such that $\mathbf{p} = \frac{1}{v_p}\mathbf{l}$ with v_p the phase velocity. Given the components of the ray direction vector $\mathbf{n}$, the objective is to find the corresponding components of $\mathbf{l}$ and through them the slowness vector $\mathbf{p}$. Expressing (4) in terms of the components of $\mathbf{l}$ and solving for v_p gives,

$$v_p^2 = \frac{K' + \sqrt{K'^2 - 4L'}}{2} \quad , \tag{A.1}$$

(FARRA, 1989) where,

$$K' = (1+f)S_l^{-2} + 2\epsilon S_l^{-2} l_H^2 z \quad ,$$

and,

$$L' = S_l^{-4}\left(\left((1+2\epsilon)l_H^2 + fl_z^2\right)\left(fl_H^2 + l_z^2\right) - \left(2\delta(1-f) + (1-f)^2\right)l_H^2 l_z^2\right) \quad ,$$

where $l_H = (l_1^2 + l_2^2)^{1/2}$ and $l_z = l_3$.

According to equation (6), the ratio of the horizontal ray direction $n_H = (n_1^2 + n_2^2)^{1/2}$ to the vertical direction $n_z = n_3$ is given by,

$$r = \frac{n_H}{n_z} = \frac{(\nabla_{\mathbf{p}}G)_H}{(\nabla_{\mathbf{p}}G)_z} \quad , \tag{A.2}$$

where after differentiating (4),

$$(\nabla_{\mathbf{p}}G)_H = 2(1 + f + 2\epsilon)S_l^2 p_H - 4p_H\left((1+2\epsilon)fp_H^2 - p_z^2(\epsilon+f)\right) + 4\delta(1-f)p_H p_z^2 \quad , \tag{A.3}$$

$$(\nabla_{\mathbf{p}}G)_z = 2(1+f)S_l^2 p_z - 4p_z(fp_z^2 + p_H^2(\epsilon+f)) + 4\delta(1-f)p_H^2 p_z \quad . \tag{A.4}$$

For a given ratio r, equation (A.2) with (A.3) and (A.4) and the relation $l_z = (1 - l_H^2)^{1/2}$ defines a nonlinear equation for l_H. Using an initial estimate $l_H = n_H$, this equation can be readily solved by the Newton-Raphson method.

REFERENCES

ČERVENÝ, V. (1972), *Seismic Rays and Ray Intensities in Inhomogeneous Anisotropic Media*, Geophys. J.R. astr. Soc. 29, 1–13.

ČERVENÝ, V. (1989), *Ray Tracing in Factorized Anisotropic Media*, Geophys. J. Int. 99, 91–100.

FARRA, V. (1989), *Ray Perturbation Theory for Heterogeneous Hexagonal Anisotropic Media*, Geophys. J. Internat. 99, 723–737.

LEIDENFROST, A., ETTRICH, N., GAJEWSKI, D., and KOSLOFF, D. (1999), *Comparison of Six Different Methods for Calculating Travel Times*, Geophys. Prospect. 47, 269–298.

NICHOLS, D., FARMER, P., and PALACHARIA, G. (1998), *Improving Prestack Imaging by Using a New Ray Selection Method*, SEG Annual Meeting, expanded abstracts.

PAYTON, R.G. (1983), *Elastic Wave Propagation in Transversely Anisotropic Media*, Martinus Nijhoff Publishers.

THOMSEN, L. (1986), *Weak Elastic Anisotropy*, Geophysics *51*, 1954–1966.

(Received September 28, 2000, revised December 12, 2000, accepted February 8, 2001)

To access this journal online:
http://www.birkhauser.ch

Pure appl. geophys. 159 (2002) 1583–1599
0033–4553/02/081583–17 $ 1.50 + 0.20/0

❙Pure and Applied Geophysics

True Amplitude Migration Weights from Travel Times

CLAUDIA VANELLE[1] and DIRK GAJEWSKI[1]

Abstract — 3-D amplitude preserving prestack migration of the Kirchhoff type is a task of high computational effort. A substantial part of this effort is spent on the calculation of proper weight functions for the diffraction stack. We propose a new strategy to compute the migration weights directly from coarse gridded travel-time data which are in any event needed for the summation along diffraction time surfaces. The technique employs second-order travel-time derivatives that contain all necessary information on the weight functions. Their determination alone from travel times significantly reduces the requirements in computational time and particularly storage, since it is done on the fly. Application of the method shows good accordance between numerical and analytical results for the simple types of models considered in this study.

Key words: Kirchhoff migration, true amplitude, travel times, hyperbolic interpolation.

1. Introduction

Kirchhoff migration is a standard technique in seismic imaging. Whereas the conventional diffraction stack migration produces "only" an image of reflectors in the subsurface, a modified diffraction stack can be used to perform AVO analysis, lithological interpretation and reservoir characterization. In the modified diffraction stack, specific weight functions are applied which compensate the effect of geometrical spreading. Thus, amplitudes in the resulting migrated image are proportional to the reflector strength if the weight functions are chosen correctly. Three different theoretical approaches (BLEISTEIN, 1987; KEHO and BEYDOUN, 1988; SCHLEICHER *et al.*, 1993) have led to formulations of the weight functions. As DOCHERTY (1991) and HANITZSCH (1997) have shown, these results are closely related.

Since weight functions can be expressed in terms of second-order spatial derivatives of travel times, they were until now computed using dynamic ray tracing (ČERVENÝ and DECASTRO, 1993; HANITZSCH *et al.*, 1994). Although SCHLEICHER *et al.* (1993) state that the modulus of the weight function can also be determined from travel times, HANITZSCH (1997) points out that computing travel-time derivatives along dipping surfaces is expensive and numerically unstable. In this

---

[1] Institute of Geophysics, University of Hamburg, Bundesstr. 55, 20146 Hamburg, Germany.

paper we propose an alternative strategy for determining travel-time derivatives that does not suffer from the instability of numerical differentiation. It is based on a local spherical approximation of the wavefront, leading to a hyperbolic expansion of the travel times. We determine the complete weight functions on the fly from travel times sampled on a coarse grid, that are at the same time used for the computation of the diffraction time surface needed for the stack. This makes the algorithm computationally efficient in time and particularly in storage. Dynamic ray tracing is not required. For the special case of 2.5-D symmetry we use a simple expression for the out-of-plane spreading that can also be determined from travel times, and not, as it is usually done, from an integral along the raypath.

Following an outline of true amplitude migration using a weighted diffraction stack, we will give an expression for the actual form of migration weight functions as they were employed in this work. We will then demonstrate our method by applying it to two examples and comparing our results to analytical values. Our conclusions summarize the work.

2. Method

SCHLEICHER et al. (1993) show that a diffraction stack of the form

$$V(M) = -\frac{1}{2\pi} \int_A \int d\xi_1 d\xi_2 W_{3D}(\xi_1, \xi_2, M) \frac{\partial U(\xi_1, \xi_2, t)}{\partial t}\bigg|_{\tau_D(\xi_1, \xi_2, M)} \tag{1}$$

yields a true amplitude migrated trace if proper weight functions $W_{3D}(\xi_1, \xi_2, M)$ are applied. In equation (1) A is the aperture of the experiment (assumed to provide sufficient illumination), $\partial U(\xi_1, \xi_2, t)/\partial t|_{\tau_D(\xi_1, \xi_2, M)}$ is the time derivative of the input seismic trace in terms of the trace coordinates (ξ_1, ξ_2) (see Appendix A) at the diffraction travel time τ_D for a diffractor at a subsurface point M. The seismic data $U(\xi_1, \xi_2, t)$ is assumed to have the form

$$U(\xi_1, \xi_2, t) = \frac{\mathcal{R}\mathcal{A}}{\mathcal{L}} F(t - \tau_R(\xi_1, \xi_2)) \ . \tag{2}$$

In equation (2), $F(t)$ is the shape of the analytic source pulse, τ_R the reflection travel time and $\mathcal{L}$ the geometrical spreading. $\mathcal{R}$ is the plane wave reflection coefficient and $\mathcal{A}$ expresses transmission losses. The integral (1) cannot generally be analytically solved. It can, however, be transformed to the frequency domain and for high frequencies be approximately evaluated by the stationary phase method. This solution is then transformed back to the time domain and compared to the analytic true amplitude signal

$$U_{TA}(t) = \mathcal{L}U(\xi_1, \xi_2, t + \tau_R(\xi_1, \xi_2)) = \mathcal{R}\mathcal{A}F(t) \ . \tag{3}$$

The comparison of $U_{TA}(t)$ and the result of (1) shows that equation (1) yields indeed a true amplitude trace if the weight functions are chosen to be (SCHLEICHER et al., 1993)

$$W_{3D}(\xi_1, \xi_2, M) = \mathscr{L}\sqrt{|\det \underline{\mathbf{H}}_F|}\, e^{i\frac{\pi}{2}\left(1 - \frac{\operatorname{sgn}\underline{\mathbf{H}}_F}{2}\right)} \ . \qquad (4)$$

The matrix $\underline{\mathbf{H}}_F$ is the Hessian matrix of the difference $\tau_F = \tau_D - \tau_R$ between diffraction and reflection travel times at the stationary point (ξ_1^*, ξ_2^*), where $\nabla_{\tau_F} = 0$, meaning that in this point the diffraction and reflection travel-time curves are tangent to each other. $\underline{\mathbf{H}}_F$ can be expressed in terms of second-order spatial derivative matrices of travel times, which until now had to be computed by dynamic ray tracing. This is, however, not necessary since these derivatives can be extracted from travel-time data, that is required for the construction of the diffraction travel-time surface for the stack in any event. Using these derivatives also leads to an effective and highly accurate algorithm for interpolating travel times from the coarse input grid onto the fine migration grid (VANELLE and GAJEWSKI, 2002). The geometrical spreading can also be written in terms of second-order derivatives of travel times (see Appendix B). The resulting expression for the weight function is

$$W_{3D}(\xi_1, \xi_2, M) = \frac{\sqrt{\cos \alpha_s \cos \alpha_g}}{v_s}\, \frac{\left|\det[\underline{\mathbf{N}}_1^\top \underline{\boldsymbol{\Sigma}} + \underline{\mathbf{N}}_2^\top \underline{\boldsymbol{\Gamma}}]\right|}{\sqrt{|\det \underline{\mathbf{N}}_1 \det \underline{\mathbf{N}}_2|}}\, e^{-i\frac{\pi}{2}(\kappa_1 + \kappa_2)} \ . \qquad (5)$$

The angles α_s and α_g are the emergence and incidence angles at the source and receiver, v_s is the velocity at the source and $\underline{\mathbf{N}}_1$ and $\underline{\mathbf{N}}_2$ are second-order derivative matrices of the travel times. These quantities and their determination from travel times are explained in detail in Appendix C. κ_1 and κ_2 are KMAH indices. The matrices $\underline{\boldsymbol{\Sigma}}$ and $\underline{\boldsymbol{\Gamma}}$ (see Appendix A) describe the measurement configuration (e.g., common shot).

3. Applications

In this section we will apply our method to two velocity models. Simple examples were chosen in order to allow for comparison of numerically and analytically computed amplitudes. The method is, however, not limited to homogeneous velocity layer models. For convenience reasons we have restricted our examples to what is commonly referred to as a 2.5-D geometry (BLEISTEIN, 1986). The need to introduce this concept arises when seismic data is only available for sources and receivers constrained to a single straight acquisition line. Processing of these data with techniques based on 2-D wave propagation does not yield satisfactory results because the (spherical) geometrical spreading in the data caused by the 3-D earth does not agree with the cylindrical (i.e., line source) spreading implied by the 2-D wave equation. The problem can be dealt with by assuming the subsurface to be invariant in the off-line direction. This symmetry is denoted 2.5-dimensional. Apart from the geometrical spreading, the properties involved do not depend on the out-of-plane variable and can be computed with 2-D techniques. The geometrical spreading is split into an in-plane part that is equal to the 2-D spreading and an out-of-plane contribution. For the described symmetry, the product of both equals the spreading in a true 3-D medium. We determine the out-of-plane spreading along with the migration weight functions from travel times only. The actual expression and its derivation are given in Appendix B.

For both examples the velocity model was only used to compute the travel times with a finite-difference eikonal solver (FDES, VIDALE, 1990) using an implementation of LEIDENFROST (1998). These travel times were resampled from the original 10 m fine grid required by FDES for sufficient accuracy and stored on a coarse grid of 50 m in either direction. Diffraction travel times were interpolated from this coarse grid onto a fine migration grid of 5 m in z direction using the hyperbolic approximation as described in Appendix C (for a more detailed description, see VANELLE and GAJEWSKI, 2002). The migration weights were also computed from the coarse grid travel times using the coefficients determined from the hyperbolic approximation.

The first model has a planar horizontal reflector at a depth of 2500 m. The P velocity is $v_P = 5$ km/s in the upper part of the model and $v_P = 6$ km/s below the reflector. The S velocities $v_S = v_P/\sqrt{3}$ were used for reflection coefficients; only PP reflections were considered. The density is $\rho[\text{g/cm}^3] = 1.7 + 0.2v_P[\text{km/s}]$. Ray synthetic seismograms were computed for a receiver line consisting of 100 equidistantly positioned receivers with a spacing of 50 m and the first receiver 50 m away from the point source. The resulting common shot section is shown in Figure 1.

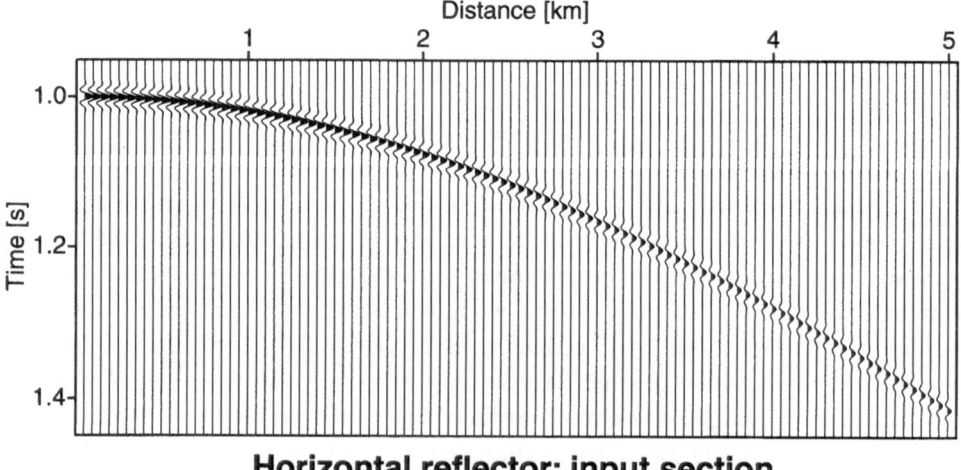

Horizontal reflector: input section

Figure 1
Synthetic common shot section for the first example, a two-layer model with a horizontal reflector.

Figure 2 shows the relevant part of the resulting migrated depth section together with the model. The reflector was migrated to the correct position and the source pulse, a Gabor wavelet, was reconstructed. Since there are no transmission losses caused by the overburden, the amplitudes of the migrated section coincide with the reflection coefficients. Figure 3 (top) shows the accordance between amplitudes picked from the migrated section in Figure 2 and theoretical values. Apart from the peaks at 300 m and 2000 m distance, the two curves coincide. These peaks are aperture effects caused by the limited extent of the receiver line. Figure 3 (middle) displays relative errors of the reflection coefficients. As a comparison, errors of reflection coefficients that were computed with our migration routine, but using analytical travel times as input data instead of FD travel times, are also given in Figure 3 (middle). Whereas the average error in the reflection coefficients from FD travel times is 2%, the error from analytical travel times is a magnitude smaller (0.22%; for both cases only values that are not affected by aperture effects were taken into account). The higher error for FD travel times as input data is due to the systematical errors that are inherent to the Vidale algorithm, as Figure 3 (bottom) illustrates: the maximum errors in the reflection coefficients coincide with the region of highest errors in the travel times.

The second model has the same velocities and densities as the first model, however, here a dipping plane reflector with an inclination angle of 14° separates the two velocity layers. The reflector depth under the source is 2500 m. Ray synthetic seismograms were computed for 80 receivers with 50 m distance, starting at 50 m from the point source. In order to ensure causality in the Vidale algorithm, this

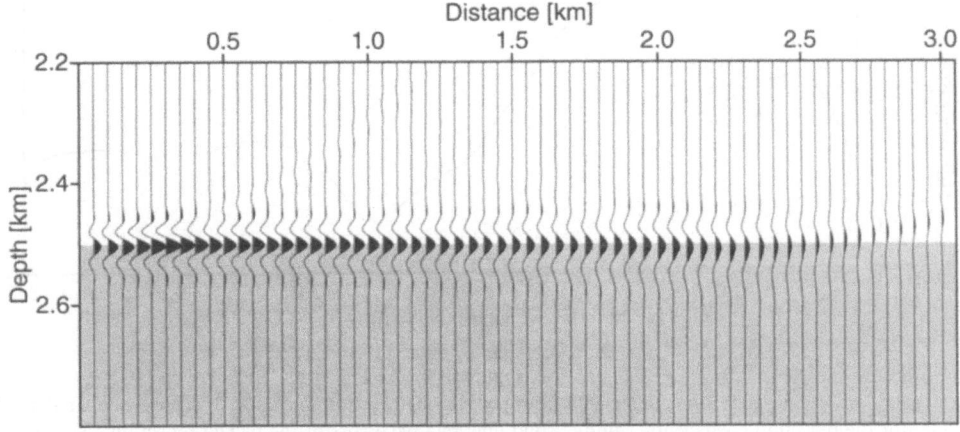

Horizontal reflector: migrated depth section

Figure 2
Migrated depth section for the first example. The gray area indicates the lower layer. The reflector was migrated to the correct position and the source pulse, a Gabor wavelet, was reconstructed.

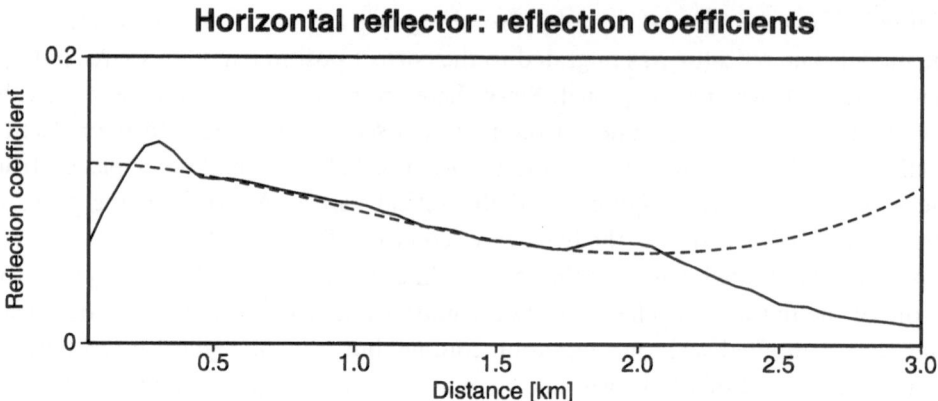

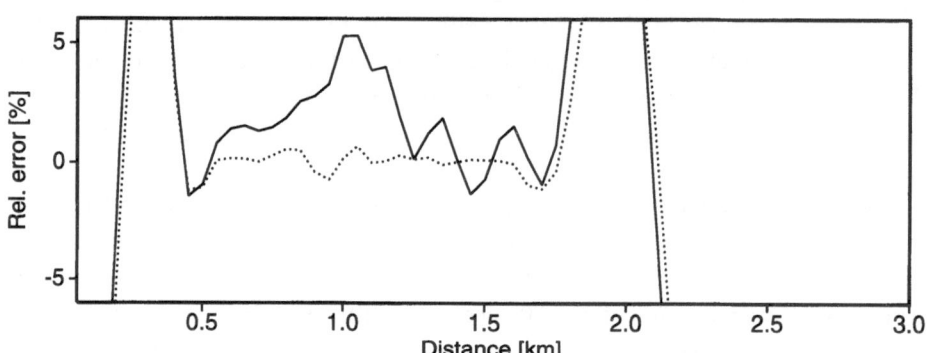

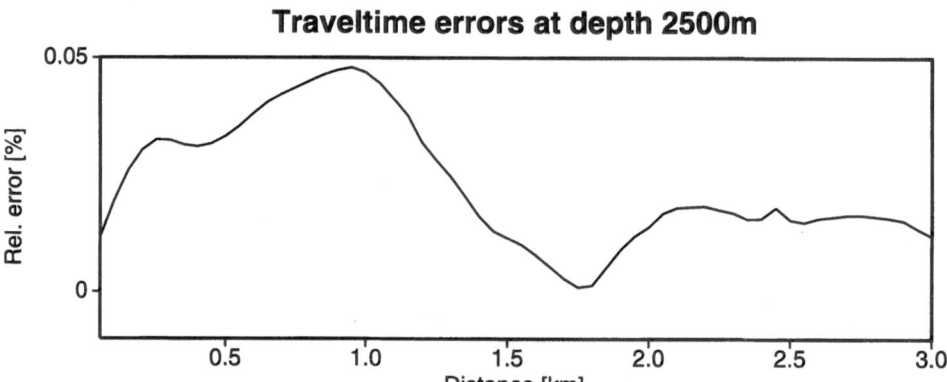

Figure 3

Top: solid line: picked reflection coefficients from the migrated section in Figure 2; dashed line: analytical values for the reflection coefficients. Middle: solid line: relative errors of the picked reflection coefficients; dotted line: relative errors of reflection coefficients if analytical travel times are used as input. Bottom: relative errors of the input travel times computed with an FD eikonal solver taken at the reflector at a depth of 2500 m. The maximum travel-time error coincides with the maximum error in the reflection coefficients.

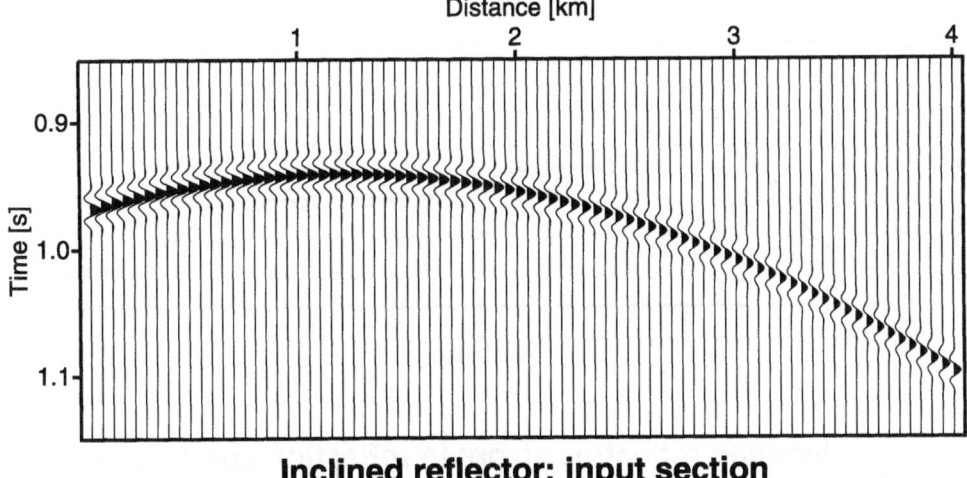

Inclined reflector: input section

Figure 4
Synthetic common shot section for the second example, a two-layer model with an inclined reflector.

velocity model had to be 10-fold smoothed. Figure 4 shows the common shot section and Figure 5 the migrated depth section together with the original reflector position. Again we find the reflector migrated to the correct depth and inclination. As for the first example, we picked the reflection coefficients from the migrated section and compared them to analytical values. Figure 6 (top) presents the result. The peaks at 900 m and 2500 m are again due to boundary effects. Figure 6 (bottom) displays the relative errors of the picked reflection coefficients. The average error for reflection coefficients from FD travel times is 2.9% and 0.27% for analytical travel times as input. In addition to the already discussed errors caused by the FD routine, the higher

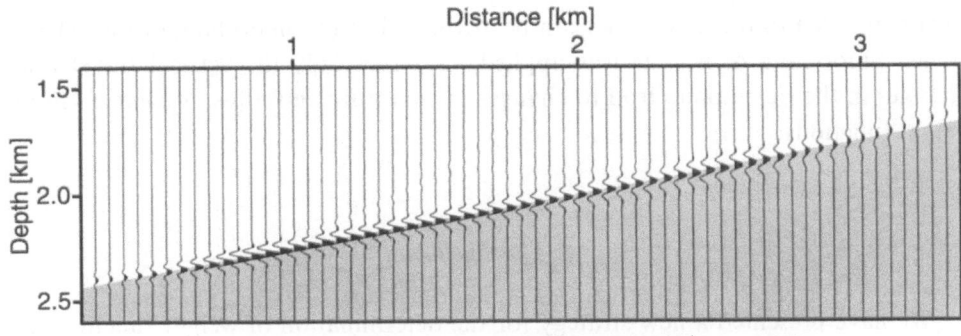

Inclined reflector: migrated depth section

Figure 5
Migrated depth section for the second example. The gray area indicates the lower layer. The reflector was migrated to the correct depth and inclination. The source pulse was correctly reconstructed.

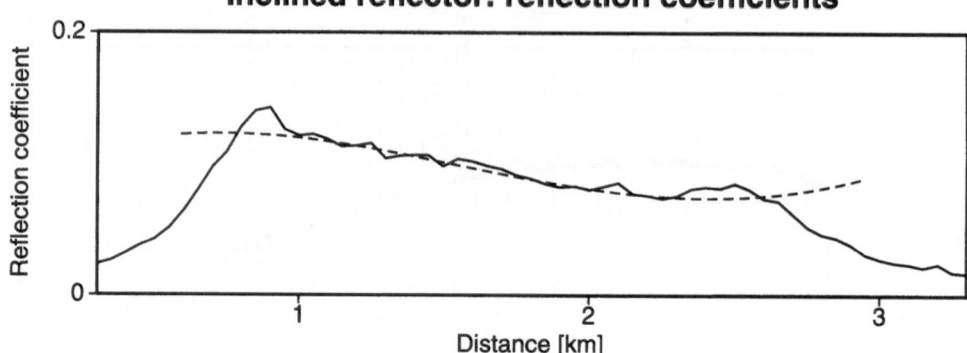

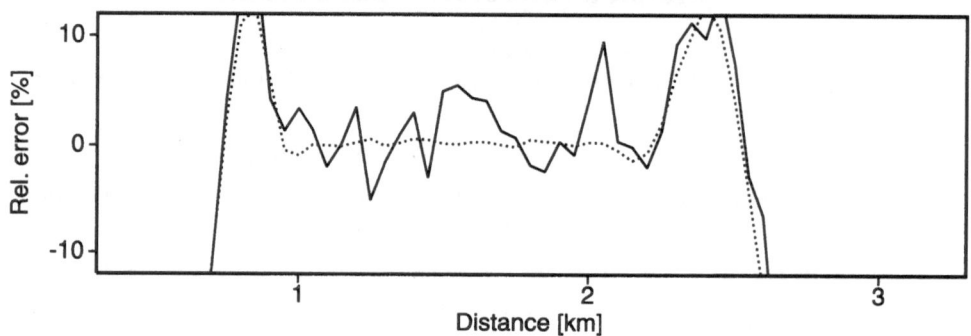

Figure 6

Top: solid line: picked reflection coefficients from the migrated section in Figure 5; dashed line: analytical values for the reflection coefficients. Bottom: solid line: relative errors of the picked reflection coefficients; dotted line: relative errors of reflection coefficients if analytical travel times are used.

error compared to the first model can be attributed to the smoothing of the velocity model. Smoothing of v^{-1}, as was applied, preserves only the vertical travel time. Changes in the travel times lead to changes in curvature and, thus, to changes in the weight functions.

4. Conclusions

We have presented a new strategy for the determination of weight functions for an amplitude preserving migration. Travel times on coarse grids are the only necessary input data. Since every required quantity can be computed on the fly from this coarse grid data alone, the technique is very efficient in computational time and storage. Dynamic ray tracing is not required. The method is particularly suited to be used in connection with techniques for travel-time computation that can directly provide coarse gridded data, such as, e.g. the wavefront construction method, which does not require a fine grid for sufficient accuracy of travel times as, e.g., FD eikonal solvers do. The examples show good accordance between the reconstructed reflectors and theoretical values in terms of position as well as in amplitudes.

A further significant reduction in computational time can be achieved by limiting the required migration aperture to a minimum. This optimum aperture can also be determined from coarse gridded travel times. A detailed discussion is given in a follow-up paper (VANELLE and GAJEWSKI, 2001a).

Future tasks involve attempts to more complex models as well as to real data. With a slight modification to the weight function, the method can also be applied to *PS* converted waves (VANELLE and GAJEWSKI, 2001b). This will be followed by a 3-D implementation and a comparison of our method to true-amplitude migration using dynamic ray tracing.

Acknowledgements

We thank the members of the Applied Geophysics Group in Hamburg for continuous and helpful discussions. Special thanks go to Andrée Leidenfrost for providing an FD eikonal solver and thus the necessary input travel times. Critical reading and suggestions from Christian Hanitzsch, Jörg Schleicher and Ivan Pšenčík improved this paper. This work was supported by the German Research Foundation (DFG, Ga 350-10) and the sponsors of the Wave Inversion Technology (WIT) consortium.

Appendix A

Measurement Configuration Matrices $\underline{\Sigma}$ and $\underline{\Gamma}$

Throughout this work we have expressed locations in trace coordinates (ξ_1, ξ_2). The relationship between them and the vectors of the source and receiver coordinates **s** and **g** are given by

$$\mathbf{s} = \mathbf{s}_0 + \underline{\Sigma}(\xi - \xi^*)$$
$$\mathbf{g} = \mathbf{g}_0 + \underline{\Gamma}(\xi - \xi^*) \, , \tag{A.1}$$

where

$$\Sigma_{ij} = \frac{\partial s_i}{\partial \xi_j}, \qquad \Gamma_{ij} = \frac{\partial g_i}{\partial \xi_j} \ . \tag{A.2}$$

ξ^* is the coordinate of the stationary point and the constants s_0, g_0 are given by

$$s_0 = s(\xi^*) \quad \text{and} \quad g_0 = g(\xi^*) \ . \tag{A.3}$$

For the special case of 2.5-D we assume ξ_1 to be in-plane and ξ_2 out-of-plane direction with the sources and receivers both being positioned at ξ_2^*. This corresponds to a zero-offset configuration in ξ_2 direction and therefore in this case the configuration matrices $\underline{\Sigma}$ and $\underline{\Gamma}$ reduce to

$$\underline{\Sigma} = \begin{pmatrix} \Sigma_{11} & 0 \\ 0 & 1 \end{pmatrix} \qquad \underline{\Gamma} = \begin{pmatrix} \Gamma_{11} & 0 \\ 0 & 1 \end{pmatrix}, \tag{A.4}$$

where only the first elements are configuration dependent. In a common shot experiment as used for the examples we find

$$\Sigma_{11} = 0 \qquad \Gamma_{11} = 1 \ . \tag{A.5}$$

Matrix elements for other frequently used configurations can be found in SCHLEICHER et al. (1993) and VERMEER (1995).

Appendix B

2.5-D Weight Functions and Geometrical Spreading

Equations (4) and (5) are expressions for weight functions if the diffraction stack is carried out over the aperture in ξ_1 and ξ_2. Since in the 2.5-D case we have data from a single acquisition line (assumed to coincide with the ξ_1 coordinate), we only integrate over ξ_1. In this case we have $U(\xi_1, \xi_2, t) = U(\xi_1, \xi_2^*, t)$ where the asterisk denotes the stationary point. Inserting expression (2) for the input traces into the stack integral (1) leads to

$$V(M) = -\frac{1}{2\pi} \int_A d\xi_1 \int_{-\infty}^{\infty} d\xi_2 W_{3D}(\xi_1, \xi_2, M) \frac{\mathscr{R}\mathscr{A}}{\mathscr{L}} \frac{\partial F(t)}{\partial t} \bigg|_{\tau_F(\xi_1,\xi_2,M)} . \tag{B.1}$$

After carrying out the integration over ξ_2 in the frequency domain by applying a stationary phase method, following MARTINS et al. (1997), we get

$$V(M) = \frac{1}{\sqrt{2\pi}} \int_A d\xi_1 W_{3D}(\xi_1, \xi_2^*, M) \left(\frac{\partial^2 \tau_D}{\partial \xi_2^2} \bigg|_{\xi_2^*} \right)^{-\frac{1}{2}} e^{-i\frac{\pi}{4}} \frac{\mathscr{R}\mathscr{A}}{\mathscr{L}} f[F(t)] \bigg|_{\tau_F(\xi_1, \xi_2^*, M)}$$

$$= \frac{1}{\sqrt{2\pi}} \int_A d\xi_1 W_{3D}(\xi_1, \xi_2^*, M) \left(\frac{\partial^2 \tau_D}{\partial \xi_2^2} \bigg|_{\xi_2^*} \right)^{-\frac{1}{2}} e^{-i\frac{\pi}{4}} f[U(\xi_1, \xi_2^*, t + \tau_D(\xi_1, \xi_2^*, M))]$$

$$= \frac{1}{\sqrt{2\pi}} \int_A d\xi_1 W_{2.5D}(\xi_1, \xi_2^*, M) f[U(\xi_1, \xi_2^*, t + \tau_D(\xi_1, \xi_2^*, M))] \tag{B.2}$$

with the function $f[U(t)]$ corresponding to application of a $\sqrt{i\omega}$ filter operation in the frequency domain (commonly called *half derivative*). The 2.5-D weight function is

$$W_{2.5D}(\xi_1, \xi_2^*, M) = W_{3D}(\xi_1, \xi_2^*, M) \left(\frac{\partial^2 \tau_D}{\partial \xi_2^2} \bigg|_{\xi_2^*} \right)^{-\frac{1}{2}} e^{-i\frac{\pi}{4}} . \tag{B.3}$$

We will now express the involved quantities in terms of travel-time derivatives as they are introduced in Appendix C, and apply the simplifications from the 2.5-D symmetry, also derived in Appendix C. We find

$$\det \underline{\mathbf{H}}_{F_{3D}} = \frac{(\det[\mathbf{N}_1^\top \underline{\mathbf{\Sigma}} + \mathbf{N}_2^\top \underline{\mathbf{\Gamma}}])^2}{\det[\underline{\mathbf{G}}_1 + \underline{\mathbf{G}}_2]} \tag{B.4}$$

$$\det \underline{\mathbf{H}}_{F_{2.5D}} = \frac{(N_{1_{11}} \Sigma_{11} + N_{2_{11}} \Gamma_{11})^2}{G_{1_{11}} + G_{2_{11}}} (N_{1_{22}} + N_{2_{22}}) , \tag{B.5}$$

where the first index denotes the downgoing ("1") or upgoing ("2") branch of the reflected ray and the second (double) index labels the corresponding matrix element. The matrices $\underline{\mathbf{\Sigma}}$ and $\underline{\mathbf{\Gamma}}$ depend on the measurement configuration and are explained in detail in Appendix A.

HUBRAL *et al.* (1992) give the normalized geometrical spreading in terms of the matrices of second-order derivatives of travel times. This is

$$\mathscr{L}_{3D} = \frac{1}{v_s} \sqrt{\frac{\cos \alpha_s \cos \alpha_g}{|\det \underline{\mathbf{N}}|}} e^{-i\frac{\pi}{2}\kappa} , \tag{B.6}$$

where the angles α_s, α_g are the emergence angle at $\mathbf{s}_0$ and the incidence angle at $\mathbf{g}_0$. They can be determined from the slownesses at these positions. κ is the KMAH-index of the ray connecting $\mathbf{s}$ and $\mathbf{g}$.

For a 2.5-D situation we find that $\det \underline{\mathbf{N}} = N_{11} \cdot N_{22}$ (see Appendix C) and therefore expression (B.6) can be reduced to the simple form

$$\mathscr{L}_{2.5D} = \frac{1}{v_s} \sqrt{\frac{\cos \alpha_s \cos \alpha_g}{|N_{11}|}} \frac{1}{\sqrt{N_{22}}} e^{-i\frac{\pi}{2}\kappa} . \tag{B.7}$$

This result is not surprising since BLEISTEIN (1986) found the relationship between the out-of-plane spreading (commonly denoted by σ) and the second-order travel-time derivative in out-of-plane direction. Thus we have

$$\sigma = \frac{1}{N_{22}} \ . \tag{B.8}$$

To date it was common practice to determine the out-of-plane spreading from the integral along the ray from $\mathbf{x}$ to $\mathbf{x}'$

$$\sigma = \int_{\mathbf{x}}^{\mathbf{x}'} ds v \tag{B.9}$$

with s being the arclength and v the velocity. We compute this quantity from travel times using equation (B.8) and do not have to trace rays to determine σ. For the out-of-plane spreading of the reflected ray we find

$$\left. \frac{\partial^2 \tau_D}{\partial \xi_2^2} \right|_{\xi_2^*} = -S_{1_{22}} - S_{2_{22}} = N_{1_{22}} + N_{2_{22}} = \frac{1}{\sigma} \ . \tag{B.10}$$

With this we get the spreading of the reflected ray

$$\mathscr{L}_{3D} = \frac{\sqrt{\cos \alpha_s \cos \alpha_g}}{v_s} \sqrt{\left| \frac{\det[\mathbf{G}_1 + \mathbf{G}_2]}{\det \underline{\mathbf{N}}_1 \det \underline{\mathbf{N}}_2} \right|} \, e^{-i\frac{\pi}{2}\kappa} \tag{B.11}$$

$$\mathscr{L}_{2.5D} = \frac{\sqrt{\cos \alpha_s \cos \alpha_g}}{v_s} \sqrt{\left| \frac{G_{1_{11}} + G_{2_{11}}}{N_{1_{11}} N_{2_{11}}} \right|} \sqrt{\frac{N_{1_{22}} + N_{2_{22}}}{N_{1_{22}} N_{2_{22}}}} \, e^{-i\frac{\pi}{2}\kappa} \ . \tag{B.12}$$

The resulting expression for the 2.5-D weight function is then

$$W_{2.5D}(\xi_1, \xi_2^*, M) = \frac{\sqrt{\cos \alpha_s \cos \alpha_g}}{v_s} \frac{|N_{1_{11}} \Sigma_{11} + N_{2_{11}} \Gamma_{11}|}{\sqrt{|N_{1_{11}} N_{2_{11}}|}} \sqrt{\frac{N_{1_{22}} + N_{2_{22}}}{N_{1_{22}} N_{2_{22}}}} \, e^{-i\frac{\pi}{4}(2\kappa + \text{sgn}\,\underline{\mathbf{H}}_F - 1)} \ . \tag{B.13}$$

We can further insert the KMAH indices κ_1 and κ_2 of the two ray branches with (SCHLEICHER et al., 1993)

$$\kappa_1 + \kappa_2 = \kappa - \left(1 - \frac{\text{sgn}\,\underline{\mathbf{H}}_F}{2} \right) \ . \tag{B.14}$$

The κ_i are not determined from the travel times themselves but we can use a suitable algorithm for computing travel times, as for example the wavefront construction method in the implementation introduced by COMAN and GAJEWSKI (2000), that outputs multi-valued travel times sorted for the KMAH index.

Please note that the diffraction stack (B.2) with a 2-D weight function $W_{2D}(\xi, M)$ instead of $W_{2.5D}(\xi_1, \xi_2^*, M)$ yields a true amplitude trace for a 2-D medium with 2-D geometrical spreading (line source) and

$$W_{2D}(\xi, M) = \frac{\sqrt{\cos \alpha_s \cos \alpha_g}}{v_s} \frac{|\underline{N}_1 \underline{\Sigma} + \underline{N}_2 \underline{\Gamma}|}{\sqrt{|\underline{N}_1 \underline{N}_2|}} \, e^{-i\frac{\pi}{4}(2\kappa + \mathrm{sgn}\underline{H}_F - 1)} \; . \qquad (B.15)$$

Here, the matrices $\underline{N}$, $\underline{\Sigma}$, $\underline{\Gamma}$ and $\underline{H}_F$ reduce to scalars.

Appendix C

Travel Times and Second-order Derivative Matrices

To compute the diffraction travel times for the stack (1) we use a second-order interpolation scheme from a coarse grid, on which input travel times are assumed to be given, onto a fine migration grid. If continuous first- and second-order derivatives of the travel-time field exist, the travel times can be expanded into a Taylor series until second degree. Provided that the distance to the expansion point is small, the Taylor series up to second order yields a good approximation for the original travel-time function. The definition of *small* depends on the velocity model under consideration. For a simple model, the sensitivity of the interpolation with regard to the coarse grid to fine grid ratio was investigated in VANELLE and GAJEWSKI (2002). For a multi-valued travel-time field the Taylor expansion is valid if the different branches of the travel-time curve are treated separately. Introducing a source position vector $\hat{\mathbf{x}} = (x_1, x_2, x_3)^\top = (x, y, z)^\top$ and a receiver position vector $\hat{\mathbf{x}}' = (x_1', x_2', x_3')^\top = (x', y', z')^\top$, the expansion of the travel time $\tau(\hat{\mathbf{x}}, \hat{\mathbf{x}}')$ from $\hat{\mathbf{x}}$ to $\hat{\mathbf{x}}'$ in the vicinity of $\hat{\mathbf{x}}_0$ and $\hat{\mathbf{x}}_0'$ is

$$\tau(\hat{\mathbf{x}}, \hat{\mathbf{x}}') = \tau_0 - \hat{\mathbf{p}}_0 \Delta \hat{\mathbf{x}} + \hat{\mathbf{q}}_0 \Delta \hat{\mathbf{x}}' - \Delta \hat{\mathbf{x}}^\top \underline{\hat{N}} \Delta \hat{\mathbf{x}}' - \tfrac{1}{2} \Delta \hat{\mathbf{x}}^\top \underline{\hat{S}} \Delta \hat{\mathbf{x}} + \tfrac{1}{2} \Delta \hat{\mathbf{x}}'^\top \underline{\hat{G}} \Delta \hat{\mathbf{x}}' \; , \qquad (C.1)$$

with the slownesses

$$\hat{p}_{0i} = -\frac{\partial \tau}{\partial x_i}\Big|_{\hat{\mathbf{x}}_0, \hat{\mathbf{x}}_0'} \quad \text{and} \quad \hat{q}_{0i} = \frac{\partial \tau}{\partial x_i'}\Big|_{\hat{\mathbf{x}}_0, \hat{\mathbf{x}}_0'} \qquad (C.2)$$

and the second-order derivative matrices

$$\hat{S}_{ij} = -\frac{\partial^2 \tau}{\partial x_i \partial x_j}\Big|_{\hat{\mathbf{x}}_0, \hat{\mathbf{x}}_0'}, \quad \hat{G}_{ij} = \frac{\partial^2 \tau}{\partial x_i' \partial x_j'}\Big|_{\hat{\mathbf{x}}_0, \hat{\mathbf{x}}_0'} \quad \text{and} \quad \hat{N}_{ij} = -\frac{\partial^2 \tau}{\partial x_i \partial x_j'}\Big|_{\hat{\mathbf{x}}_0, \hat{\mathbf{x}}_0'} \qquad (C.3)$$

where indices i, j take the values 1, 2, 3 and τ_0 is the travel-time from $\hat{\mathbf{x}}_0$ to $\hat{\mathbf{x}}_0'$. It can, however, be shown (VANELLE and GAJEWSKI, 2002), that a hyperbolic approximation of travel times is superior to the parabolic expansion (C.1). Expansion of $\tau^2(\hat{\mathbf{x}}, \hat{\mathbf{x}}')$ yields the hyperbolic approximation

$$\tau^2(\hat{\mathbf{x}}, \hat{\mathbf{x}}') = (\tau_0 - \hat{\mathbf{p}}_0 \Delta \hat{\mathbf{x}} + \hat{\mathbf{q}}_0 \Delta \hat{\mathbf{x}}')^2 + \tau_0 \left(-2 \Delta \hat{\mathbf{x}}^\top \hat{\underline{\mathbf{N}}} \Delta \hat{\mathbf{x}}' - \Delta \hat{\mathbf{x}}^\top \hat{\underline{\mathbf{S}}} \Delta \hat{\mathbf{x}} + \Delta \hat{\mathbf{x}}'^\top \hat{\underline{\mathbf{G}}} \Delta \hat{\mathbf{x}}' \right) .$$

$$(C.4)$$

Since multifold travel-time data is required for the migration, we assume it to be sampled on coarse grids. We employ (C.4) first for the determination of its coefficients from this coarse gridded data, and then use the coefficients for the interpolation onto the fine migration grid. To give an example, we explain how to determine the 1-components of $\hat{\mathbf{q}}_0$ and $\hat{\underline{\mathbf{G}}}$. We need only the travel times $\tau_0 = \tau(\hat{\mathbf{x}}_0, \hat{\mathbf{x}}'_0)$, $\tau_1 = \tau(\hat{\mathbf{x}}_0, \hat{\mathbf{x}}'_0 - \Delta x'_1)$ and $\tau_2 = \tau(\hat{\mathbf{x}}_0, \hat{\mathbf{x}}'_0 + \Delta x'_1)$. We insert τ_1 and τ_2 into the hyperbolic expansion (C.4), respectively. Building the sum and the difference of the resulting expressions and solving for $\hat{q}_{0_1}$ and $\hat{G}_{11}$ yields

$$\hat{q}_{0_1} = \frac{\tau_2^2 - \tau_1^2}{4\tau_0 \Delta x'_1} \quad \text{and} \quad \hat{G}_{11} = \frac{\tau_2^2 + \tau_1^2 - 2\tau_0^2}{2\tau_0 \Delta x_1'^2} - \frac{\hat{q}_{0_1}^2}{\tau_0} . \quad (C.5)$$

The remaining coefficients can be found in a similar way. For further information and examples for hyperbolic and parabolic travel-time interpolation including a 3-D version of the Marmousi model please refer to VANELLE and GAJEWSKI (2002). The weight function in equation (4) contains the Hessian matrix of the difference between diffraction (τ_D) and reflection (τ_R) travel times. To write $\underline{\mathbf{H}}_F$ in terms of second derivatives of travel times we will now derive expressions for τ_D and τ_R containing first and second derivatives. Let a source be positioned at $\mathbf{x}_0$ in a surface, which we denote the source surface. We will describe the variation in source position by the vector $\Delta \mathbf{x} = \mathbf{x} - \mathbf{x}_0$, with $\mathbf{x}$ also lying in the source surface, or, more precisely, in its tangent plane at $\mathbf{x}_0$, if the surface is curved. We also assume $\mathbf{x}'$ and $\mathbf{x}'_0$ to lie in a surface, named the reflector surface, or the reflector's tangent plane at $\mathbf{x}'_0$, if $\mathbf{x}'_0$ is on a curved reflector. A Taylor expansion of $\tau(\mathbf{x}, \mathbf{x}')$ similar to equation (C.1) will be carried out, however into the surfaces instead of in three dimensions. To distinguish between expansion into three spatial coordinates as in (C.1) and expansion into surfaces, coefficients and vectors are denoted by a hat (ˆ) in three dimensions. The travel-time expansion into the two surfaces looks as follows:

$$\tau(\mathbf{x}, \mathbf{x}') = \tau_0 - \mathbf{p}_0 \Delta \mathbf{x} + \mathbf{q}_0 \Delta \mathbf{x}' - \tfrac{1}{2} \Delta \mathbf{x}^\top \underline{\mathbf{S}} \Delta \mathbf{x} + \tfrac{1}{2} \Delta \mathbf{x}'^\top \underline{\mathbf{G}} \Delta \mathbf{x}' - \Delta \mathbf{x}^\top \underline{\mathbf{N}} \Delta \mathbf{x}' , \quad (C.6)$$

where the first-order travel-time derivatives

$$p_{0l} = -\left. \frac{\partial \tau}{\partial x_l} \right|_{\mathbf{x}_0, \mathbf{x}'_0} \qquad q_{0l} = \left. \frac{\partial \tau}{\partial x'_l} \right|_{\mathbf{x}_0, \mathbf{x}'_0} \qquad (C.7)$$

are the slowness vectors at $\mathbf{x}$ and $\mathbf{x}'$, respectively. The second-order derivatives are given by the matrices $\underline{\mathbf{S}}$, $\underline{\mathbf{G}}$ and $\underline{\mathbf{N}}$ with

$$S_{IJ} = -\left.\frac{\partial^2 \tau}{\partial x_I \partial x_J}\right|_{\mathbf{x}_0, \mathbf{x}'_0} , \quad G_{IJ} = \left.\frac{\partial^2 \tau}{\partial x'_I \partial x'_J}\right|_{\mathbf{x}_0, \mathbf{x}'_0} \quad \text{and} \quad N_{IJ} = -\left.\frac{\partial^2 \tau}{\partial x_I \partial x'_J}\right|_{\mathbf{x}_0, \mathbf{x}'_0} , \quad \text{(C.8)}$$

where indices I, J take the values 1, 2 and τ_0 is the travel time from $\mathbf{x}_0$ to $\mathbf{x}'_0$.

Once $\hat{\mathbf{S}}, \hat{\mathbf{G}}$ and $\hat{\mathbf{N}}$ are determined, the desired 2×2 matrices $\mathbf{S}, \mathbf{G}$ and $\mathbf{N}$ can be computed by projecting $\hat{\mathbf{S}}, \hat{\mathbf{G}}$ and $\hat{\mathbf{N}}$ from the Cartesian coordinates into the tangent planes of the corresponding surfaces. Since we assume a velocity model to compute travel times, we can also make use of it to extract the reflector position and geometry from it. An example for a 2.5-D symmetry: if the source-receiver line is equal to the x direction of the Cartesian system associated with the input travel times, the 11-components of $\mathbf{S}, \mathbf{G}$ and $\mathbf{N}$ are computed as follows:

$$N_{11} = \hat{N}_{xx} \cos \varphi - \hat{N}_{xz} \sin \varphi$$
$$G_{11} = \hat{G}_{xx} \cos^2 \varphi + \hat{G}_{zz} \sin^2 \varphi - 2\hat{G}_{xz} \sin \varphi \cos \varphi \qquad \text{(C.9)}$$
$$S_{11} = \hat{S}_{xx} ,$$

where φ is the inclination angle of the reflector's tangent plane against the source-receiver line.

We will now derive expressions in terms of (C.6) for diffraction and reflection travel times as they are needed for the weight functions. Since $\tau(\mathbf{x}_1, \mathbf{x}_2) = \tau(\mathbf{x}_2, \mathbf{x}_1)$ we can use (C.6) twice for the down- and upgoing ray branches of the reflected arrival with the travel time τ_R and the diffraction travel time τ_D. Similar as for the sources, we assume the receivers to be placed in a receiver surface. Source coordinates will be denoted by $\mathbf{s}$, receivers by $\mathbf{g}$ and subsurface points by $\mathbf{r}$. The travel time for the ray from $\mathbf{s}$ to $\mathbf{r}$ is

$$\tau_1(\mathbf{s}, \mathbf{r}) = \tau_{01} - \mathbf{p}_{01}\Delta\mathbf{s} + \mathbf{q}_{01}\Delta\mathbf{r} - \tfrac{1}{2}\Delta\mathbf{s}^\top \underline{\mathbf{S}}_1 \Delta\mathbf{s} + \tfrac{1}{2}\Delta\mathbf{r}^\top \underline{\mathbf{G}}_1 \Delta\mathbf{r} - \Delta\mathbf{s}^\top \underline{\mathbf{N}}_1 \Delta\mathbf{r} \qquad \text{(C.10)}$$

and from $\mathbf{g}$ to $\mathbf{r}$:

$$\tau_2(\mathbf{g}, \mathbf{r}) = \tau_{02} - \mathbf{p}_{02}\Delta\mathbf{g} + \mathbf{q}_{02}\Delta\mathbf{r} - \tfrac{1}{2}\Delta\mathbf{g}^\top \underline{\mathbf{S}}_2 \Delta\mathbf{g} + \tfrac{1}{2}\Delta\mathbf{r}^\top \underline{\mathbf{G}}_2 \Delta\mathbf{r} - \Delta\mathbf{g}^\top \underline{\mathbf{N}}_2 \Delta\mathbf{r} . \qquad \text{(C.11)}$$

The sums of equations (C.10) and (C.11) give us τ_D and τ_R. For the diffraction travel time the diffractor position is fixed at $\mathbf{r}_0$ and thus with $\Delta\mathbf{r} = 0$ we get

$$\tau_D = \tau_0 - \mathbf{p}_{01}\Delta\mathbf{s} - \mathbf{p}_{02}\Delta\mathbf{g} - \tfrac{1}{2}\Delta\mathbf{s}^\top \underline{\mathbf{S}}_1 \Delta\mathbf{s} - \tfrac{1}{2}\Delta\mathbf{g}^\top \underline{\mathbf{S}}_2 \Delta\mathbf{g} \qquad \text{(C.12)}$$

$(\tau_0 = \tau_{01} + \tau_{02})$. For the reflection travel time we must take into account that variation of source and/or receiver positions will result in a different reflection point $\mathbf{r}$. Aiming for an expression containing $\Delta\mathbf{s}$ and $\Delta\mathbf{g}$ only, we make use of Snell's law stating that $\mathbf{q}_{01} + \mathbf{q}_{02} = \nabla_r \tau_R = \mathbf{0}$. This we can solve for $\mathbf{r}$ and eliminate $\Delta\mathbf{r}$ from the sum of (C.10) and (C.11) resulting in

$$\tau_R = \tau_0 - \mathbf{p}_{01}\Delta\mathbf{s} - \mathbf{p}_{02}\Delta\mathbf{g} - \tfrac{1}{2}\Delta\mathbf{s}^\top \underline{\mathbf{S}}\Delta\mathbf{s} + \tfrac{1}{2}\Delta\mathbf{g}^\top \underline{\mathbf{G}}\Delta\mathbf{g} - \Delta\mathbf{s}^\top \underline{\mathbf{N}}\Delta\mathbf{g} , \qquad \text{(C.13)}$$

where we introduced the following matrices to bring (C.13) into the same form as (C.6):

$$\underline{S} = \underline{S}_1 + \underline{N}_1(\underline{G}_1 + \underline{G}_2)^{-1}\underline{N}_1^{\mathsf{T}}$$
$$\underline{G} = -\underline{S}_2 - \underline{N}_2(\underline{G}_1 + \underline{G}_2)^{-1}\underline{N}_2^{\mathsf{T}} \qquad (C.14)$$
$$\underline{N} = \underline{N}_1(\underline{G}_1 + \underline{G}_2)^{-1}\underline{N}_2^{\mathsf{T}} \ .$$

In a situation with a 2.5-D symmetry as considered in the numerical examples, the components of the slowness vectors $\mathbf{p}_0, \mathbf{q}_0$, and second-order derivative matrices $\underline{S}, \underline{G}$ and $\underline{N}$ simplify. Let the out-of-plane direction have index 2 – coinciding with the y axis of the Cartesian system used for the determination of $\hat{\underline{S}}, \hat{\underline{G}}$ and $\hat{\underline{N}}$ – with $y_0 = s_{20} = g_{20} = r_{20}$ the y position of the sources and receiver line. Then we have

$$N_{22}|_{y_0} = \hat{N}_{yy}|_{y_0}, \quad G_{22}|_{y_0} = \hat{G}_{yy}|_{y_0} \quad \text{and} \quad S_{22}|_{y_0} = \hat{S}_{yy}|_{y_0} \ . \qquad (C.15)$$

From the symmetry we can easily see that the y components of the slownesses vanish at y_0:

$$\left.\frac{\partial \tau}{\partial y_s}\right|_{y_0} = \left.\frac{\partial \tau}{\partial y_g}\right|_{y_0} = \left.\frac{\partial \tau}{\partial y_r}\right|_{y_0} = 0 \ . \qquad (C.16)$$

From this follows that the matrices $\underline{S}, \underline{G}$ and $\underline{N}$ consist only of diagonal elements. Furthermore, for the yy- or 22-components we get

$$N_{22}|_{y_0} = G_{22}|_{y_0} = -S_{22}|_{y_0} \qquad (C.17)$$

and the sign of $N_{22}|_{y_0}$ is positive; i.e., $\mathrm{sgn}(N_{22}|_{y_0}) = +1$.

REFERENCES

BLEISTEIN, N. (1986), *Two-and-one-half Dimensional In-plane Wave Propagation*, Geophys. Prospect. *34*, 686–703.

BLEISTEIN, N. (1987), *On the Imaging of Reflectors in the Earth*, Geophys. *52*, 931–942.

ČERVENÝ, V. and DECASTRO, M. A. (1993), *Application of Dynamic Ray Tracing in the 3-D Inversion of Seismic Reflection Data*, Geophys. J. Internat. *113*, 776–779.

COMAN, R. and GAJEWSKI, D. (2000), *3-D Multi-valued Travel-time Computation Using a Hybrid Method*, 70th Ann. Internat. Mtg., Soc. Expl. Geophys., Expanded Abstracts, 2309–2312.

DOCHERTY, P. (1991), *A Brief Comparison of Some Kirchhoff Integral Formulas for Migration and Inversion*, Geophys. *56*, 1164–1169.

HANITZSCH, C. (1997), *Comparison of Weights in Prestack Amplitude-preserving Depth Migration*, Geophys. *62*, 1812–1816.

HANITZSCH, C., SCHLEICHER, J., and HUBRAL, P. (1994), *True-amplitude Migration of 2-D Synthetic Data*, Geophys. Prospect. *42*, 445–462.

HUBRAL, P., SCHLEICHER, J., and TYGEL, M. (1992), *Three-dimensional Paraxial Ray Properties, Part 1: Basic Relations*, J. Seismic Expl. *1*, 265–279.

KEHO, T. H. and BEYDOUN, W. B. (1988), *Paraxial Kirchhoff Migration*, Geophys. *53*, 1540–1546.

LEIDENFROST, A. (1998), *Fast Computation of Travel Times in Two and Three Dimensions*, Ph.D. Thesis, University of Hamburg.

MARTINS, J. L., SCHLEICHER, J., TYGEL, M., and SANTOS, L. (1997), *2.5-D True-amplitude Migration and Demigration*, J. Seismic Expl. *6*, 159–180.

SCHLEICHER, J., TYGEL, M., and HUBRAL, P. (1993), *3-D True amplitude Finite-offset Migration*, Geophys. *58*, 1112–1126.

VANELLE, C. and GAJEWSKI, D. (2001a), *Determining the Optimum Migration Aperture from Travel Times*, J. Seismic Expl. *10*, 205–224.

VANELLE, C. and GAJEWSKI, D. (2001b), *True–amplitude Migration of PS Converted Waves*, 71st Ann. Internat. Mtg., Soc. Expl. Geophys., Expanded Abstracts, 288–291.

VANELLE, C. and GAJEWSKI, D. (2002), *Second-order Interpolation of Travel times*, Geophys. Prospect. *50*, 73–83.

VERMEER, G. J. O. (1995), *Discussion on "3-D True-amplitude Finite-offset Migration"* by J. SCHLEICHER, M. TYGEL *and* P. HUBRAL (1993), Geophys. *58*, 1112–1126, *with reply by the authors*, Geophys. *60*, 921–923.

VIDALE, J. (1990), *Finite-difference Calculation of Travel Times in Three Dimensions*, Geophys. *55*, 521–526.

(Received October 7, 2000, revised December 12, 2000, March 3, 2001)

 To access this journal online:
http://www.birkhauser.ch

Pure appl. geophys. 159 (2002) 1601–1616
0033–4553/02/081601–16 $ 1.50 + 0.20/0

© Birkhäuser Verlag, Basel, 2002

❙ Pure and Applied Geophysics

Model-independent Travel-time Attributes for 2-D, Finite-offset Multicoverage Reflections

Y. ZHANG,[1,2] S. BERGLER,[1] M. TYGEL,[3] and P. HUBRAL[1]

Abstract — In this paper, we provide a 5-parameter stacking formula to transform 2-D prestack data into a particular common-offset section. This requires the knowledge of the near-surface velocity only and it is expected that ray theory holds to describe primary reflections. The earth model can be arbitrarily inhomogeneous. The new stacking approach can be viewed as a generalization of the 3-parameter common-reflection-surface (CRS) stack, by which 2-D multicoverage data are stacked into a simulated zero-offset section. The new 5-parameter formula can handle *P-P*, *P-S* and *S-S* reflections.

Key words: Travel-time attributes, stacking formula, common-reflection-surface (CRS) stack.

1. Introduction

Recently, novel concepts of simulating zero-offset (ZO) sections from multi-coverage prestack data were introduced. The POLYSTACK method (see e.g., DE BAZELAIRE, 1988; THORE *et al.*, 1994), as well as the Multifocus method (see e.g., GELCHINSKY *et al.*, 1999) require only the near-surface velocity. The recently established common-reflection-surface (CRS) stack method (see e.g., MANN *et al.*, 1999) also provides a simulated ZO stack section using only the near-surface velocity. Consequently, all these methods belong to the class of macro-model-independent stacking techniques. Various aspects of macro-model independent ZO simulation methods and kinematic wavefield attribute extractions are discussed in HUBRAL (1999).

In a two-dimensional medium, the travel-time moveout expressions that are used in the POLYSTACK, Multifocus and CRS method depend on three parameters that are extracted from the multicoverage data. As opposed to the one-parametric conventional stacking methods, such as CMP or NMO/DMO stacking methods, these multiparametric moveout travel times provide a much better fit to primary reflection events.

[1] Geophysikalisches Institut, Universität Karlsruhe, Hertzstr. 16, 76187 Karlsruhe, Germany.
[2] Faculty for Physics, Lanzhou University, Lanzhou 730000, Gansu, P. R. China.
[3] Department of Applied Mathematics, State University of Campinas, 13081–970, SP, Brazil.

The kinematic stacking attributes of the above macro-velocity independent ZO simulation methods take not only the orientation and dip of the illuminated reflector segment into consideration, but also the curvature. As a consequence, their travel-time moveout curves will have a much better fit to the actual reflection events than the conventional operators. For instance, the NMO/DMO stack expects reflection events resulting from interfaces represented as the envelope of ZO isochrones (TYGEL *et al.*, 1996), whereas the prestack migration assumes the subsurface to be described in terms of diffraction points. Moreover conventional imaging procedures such as the NMO/DMO require a sufficiently accurate macro-velocity model.

The three parameters that define the travel-time moveout expression in the CRS method are the angle of emergence of the ZO ray and the radii of curvature of the normal (*N*) wave and the normal-incidence-point (*NIP*) wave. These are the two hypothetical eigenwaves introduced by HUBRAL (1983). The CRS travel-time moveout expression is another representation of the hyperbolic travel-time formula (see, e.g., SCHLEICHER *et al.*, 1993), that is applied to source-receiver pairs in the vicinity of a fixed, central ZO ray.

The three CRS stacking attributes are fully automatically obtained by means of a coherency analysis where the multicoverage seismic reflection data itself are used to determine the stacking operator. In other words, there is no need for computing an *a priori* macro-velocity model. Since the three stacking attributes are data-derived wavefield attributes which characterize the subsurface, they can, apart from other applications, be used for a subsequent inversion to determine an approximation of the macro-velocity model.

In this paper we introduce a model-independent 5-parameter travel-time formula that can be used to simulate any specified finite-offset (FO) section (e.g., a common-offset (CO) stack section) from prestack multicoverage reflection data. This travel-time expression is a hyperbolic formula which corresponds to paraxial rays around a fixed central ray with finite offset. As in the previous case when the central ray is a ZO ray, only the near-surface velocity in the vicinity of source and receiver has to be known. The CRS stacking operator for the simulation of a ZO stack section is a special case of this FO stacking operator. Thus, we call this method of simulating CO stack section from multicoverage reflection data, the generalized CRS stack, or FO CRS stack.

Furthermore, we will show that the five data-derived wavefield attributes can be used in a number of seismic applications. These include, for instance, (a) the computation of the geometrical spreading factor of the central finite-offset ray, (b) separation of diffractions from reflections, and (c) pseudo model-independent time migration.

2. *Travel-time Approximations*

As shown in BORTFELD (1989) for a specific seismic system (i.e., a system of constant-velocity layers with curved interfaces) and generalized in HUBRAL *et al.*

(1992) to laterally inhomogeneous layered media, the elements of the surface-to-surface propagator matrix of a central ray can be used to obtain useful second-order approximations for the two point travel time of a paraxial ray in the vicinity of a central ray (see also, ČERVENÝ, 2001). These are the *parabolic two point travel-time formula*

$$t = t_0 + p_G \Delta x_G - p_S \Delta x_S - \Delta x_S B^{-1} \Delta x_G + \tfrac{1}{2} \Delta x_S B^{-1} A \Delta x_S + \tfrac{1}{2} \Delta x_G D B^{-1} \Delta x_G , \quad (1a)$$

and the *hyperbolic two-point travel-time formula*

$$T^2 = (t_0 + p_G \Delta x_G - p_S \Delta x_S)^2$$
$$+ 2t_0 \left(-\Delta x_S B^{-1} \Delta x_G + \tfrac{1}{2} \Delta x_S B^{-1} A \Delta x_S + \tfrac{1}{2} \Delta x_G D B^{-1} \Delta x_G \right) , \quad (1b)$$

the hyperbolic form being obtained by squaring both sides of equation (1a) and neglecting the higher-order terms than the second. In the 2-D situation considered in this work the quantities A, B, and D in equation (1) are the constant elements of the ray propagator matrix given in BORTFELD (1989),

$$\underline{\mathbf{T}} = \begin{pmatrix} A & B \\ C & D \end{pmatrix} , \quad (2)$$

of a fixed central ray.

The matrix $\underline{\mathbf{T}}$ sets a linear relationship (see Fig. 1)

$$\begin{pmatrix} \Delta x_G \\ \Delta p_G \end{pmatrix} = \underline{\mathbf{T}} \begin{pmatrix} \Delta x_S \\ \Delta p_S \end{pmatrix} , \quad (3)$$

where

$$\Delta x_S = \bar{x}_S - x_S \quad \text{and} \quad \Delta x_G = \bar{x}_G - x_G , \quad (4)$$

$$\Delta p_S = \bar{p}_S - p_S \quad \text{and} \quad \Delta p_G = \bar{p}_G - p_G . \quad (5)$$

and possesses the symplectic property

$$AD - BC = 1 . \quad (6)$$

The latter equation has been used in the construction of equation (1) to eliminate the element C from the travel-time formulas.

In equations (1a) and (1b), t_0 is the travel time along the central ray, Δx_S and Δx_G are the lateral distance between the paraxial ray and the central ray at the source and receiver, respectively, and Δp_S and Δp_G are the differences of the horizontal slownesses of the paraxial ray from those of the central ray at the source and the receiver, respectively. Referring to Fig. 1, we have that

$$p_S = \frac{\sin(\beta_S)}{v_S} \quad \text{and} \quad p_G = \frac{\sin(\beta_G)}{v_G} , \quad (7)$$

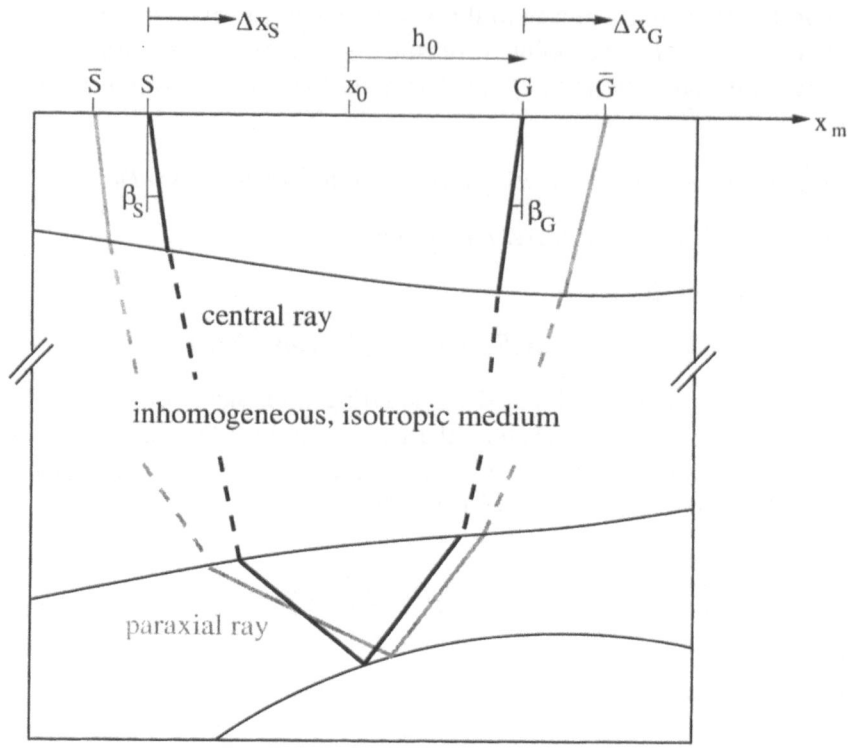

Figure 1

The central ray with a fixed offset in an inhomogeneous layered medium with half offset $h = h_0$ and midpoint $x_m = x_0$ is depicted in black, a paraxial ray in its close vicinity is shown in light gray.

where v_S and β_S are the medium velocity and incidence angle of the central ray with respect to the normal of the measurement surface in the depth direction at the source S, respectively, with analogous definitions for the quantities v_G and β_G at the receiver G.

It is also instructive to express the above travel-time formulas in terms of the midpoint and half-offset coordinates

$$\begin{cases} x_m = \frac{1}{2}(x_G + x_S) \\ h = \frac{1}{2}(x_G - x_S) \end{cases} \quad \text{and} \quad \begin{cases} \bar{x}_m = \frac{1}{2}(\bar{x}_G + \bar{x}_S) \\ \bar{h} = \frac{1}{2}(\bar{x}_G - \bar{x}_S) \end{cases}. \tag{8}$$

These give rise to the midpoint and half-offset perturbations

$$\Delta x_m = \bar{x}_m - x_m \quad \text{and} \quad \Delta h = \bar{h} - h , \tag{9}$$

so that

$$\Delta x_m = \frac{1}{2}(\Delta x_G + \Delta x_S) \quad \text{and} \quad \Delta h = \frac{1}{2}(\Delta x_G - \Delta x_S) , \tag{10}$$

as well as

$$\Delta x_S = \Delta x_m - \Delta h \quad \text{and} \quad \Delta x_G = \Delta x_m + \Delta h \ . \tag{11}$$

Substituting equations (7) and (11) into equations (1) yields

$$t = t_0 + \left(\frac{\sin(\beta_G)}{v_G} - \frac{\sin(\beta_S)}{v_S}\right)\Delta x_m + \left(\frac{\sin(\beta_G)}{v_G} + \frac{\sin(\beta_S)}{v_S}\right)\Delta h$$

$$+ \Delta x_m(DB^{-1} - B^{-1}A)\Delta h + \frac{1}{2}\Delta x_m(B^{-1}A + DB^{-1} - 2B^{-1})\Delta x_m$$

$$+ \frac{1}{2}\Delta h(B^{-1}A + DB^{-1} + 2B^{-1})\Delta h \ , \tag{12a}$$

and

$$T^2 = \left[t_0 + \left(\frac{\sin(\beta_G)}{v_G} - \frac{\sin(\beta_S)}{v_S}\right)\Delta x_m + \left(\frac{\sin(\beta_G)}{v_G} + \frac{\sin(\beta_S)}{v_S}\right)\Delta h\right]^2$$

$$+ 2t_0\left[\Delta x_m(DB^{-1} - B^{-1}A)\Delta h + \frac{1}{2}\Delta x_m(B^{-1}A + DB^{-1} - 2B^{-1})\Delta x_m\right.$$

$$+ \frac{1}{2}\Delta h(B^{-1}A + DB^{-1} + 2B^{-1})\Delta h\right] \ , \tag{12b}$$

respectively.

The interesting feature of the above-described parabolic and hyperbolic travel-time expressions is the direct relationship of their coefficients to the propagator matrix of the central ray. That propagator matrix, which is taken as a black box that encompasses the propagation characteristics of the unknown medium, is the key quantity to be recovered from the multicoverage input data. The travel-time expressions (12a) and (12b) are, thus, primarily designed for travel-time inversion purposes. Note that, in the famework of a forward-modeling approach, a number of basically similar second-order travel-time expressions are well known in the literature (see, e.g., ČERVENÝ et al., 1977; URSIN, 1982; GOLDIN, 1986). Those travel-time formulas, however, have coefficients that are expressed in terms of the subsurface velocity model, assumed to be given.

3. Simplified Travel-time Expressions

The travel-time expressions above will be used to determine the propagator matrix components A, B, C and D. In view of the symplectic property given by equation (6), only three components are needed. These will be expressed by means of suitably specified travel-time formulas in specific measurement configurations. Without loss of generality, we restrict ourselves to the parabolic travel-time expression (12a).

Travel Time for Four Data Configurations

We consider the following four data configurations:

(1) The *common midpoint (CMP) configuration*, in which the paraxial source $\bar{S}$ and receiver $\bar{G}$ are located symmetrically with respect to the corresponding points S and G on the central ray. In terms of the perturbations in equation (4), the CMP condition reads

$$\Delta x_S = -\Delta x_G \quad \text{or} \quad \Delta x_m = 0 . \tag{13}$$

The substitution of the above condition into equation (12a) gives rise to the parabolic CMP travel time

$$t_{CMP} = t_0 + \left(\frac{\sin(\beta_G)}{v_G} + \frac{\sin(\beta_S)}{v_S} \right) \Delta h + \frac{1}{2} \Delta h (B^{-1}A + DB^{-1} + 2B^{-1}) \Delta h . \tag{14}$$

(2) The *common offset (CO) configuration*, in which the paraxial source $\bar{S}$ and receiver $\bar{G}$ are shifted by the same amount and the same direction with respect to the corresponding points S and G on the central ray. The CO condition reads

$$\Delta x_S = \Delta x_G \quad \text{or} \quad \Delta h = 0 . \tag{15}$$

The above condition gives rise to the parabolic CO travel time

$$t_{CO} = t_0 + \left(\frac{\sin(\beta_G)}{v_G} - \frac{\sin(\beta_S)}{v_S} \right) \Delta x_m + \frac{1}{2} \Delta x_m (B^{-1}A + DB^{-1} - 2B^{-1}) \Delta x_m . \tag{16}$$

(3) The *common shot (CS) configuration*, in which the paraxial source always coincides with the source of the central ray. The CS condition reads

$$\Delta x_S = 0 \quad \text{or} \quad \Delta x_m = \Delta h = \Delta x_G/2 . \tag{17}$$

The substitution of this condition into equation (1a) gives rise to the parabolic CS travel time

$$t_{CS} = t_0 + \frac{\sin(\beta_G)}{v_G} \Delta x_G + \frac{1}{2} \Delta x_G DB^{-1} \Delta x_G . \tag{18}$$

(4) The *common receiver (CR) configuration*, in which the paraxial receiver always coincides with the receiver of the central ray. The CR condition reads

$$\Delta x_G = 0 \quad \text{or} \quad \Delta x_m = \Delta x_S/2 \quad \text{and} \quad \Delta h = - \Delta x_S/2 . \tag{19}$$

The above condition gives rise to the parabolic CR travel time

$$t_{CR} = t_0 - \frac{\sin(\beta_S)}{v_S} \Delta x_S + \frac{1}{2} \Delta x_S B^{-1}A \Delta x_S . \tag{20}$$

Structure of the Travel-time Expressions

The travel times of the above four configurations have the same structure of a one-variable quadratic polynomial, namely

$$t(x) = t_0 + a\Delta x + \tfrac{1}{2}\Delta x b\Delta x \ , \tag{21}$$

with the appropriate meaning of the coefficients a and b, as well as the polynomial variable Δx. For the various configurations, the linear coefficients, a, are given as linear combinations of the horizontal slownesses of the central ray at S and G. Corresponding to the data configurations, we have

$$a_{CMP} = \left.\frac{\partial t_{CMP}}{\partial(\Delta h)}\right|_{\Delta h=0} = \left(\frac{\sin(\beta_G)}{v_G} + \frac{\sin(\beta_S)}{v_S}\right) \ , \tag{22a}$$

$$a_{CO} = \left.\frac{\partial t_{CO}}{\partial(\Delta x_m)}\right|_{\Delta x_m=0} = \left(\frac{\sin(\beta_G)}{v_G} - \frac{\sin(\beta_S)}{v_S}\right) \ , \tag{22b}$$

$$a_{CS} = \left.\frac{\partial t_{CS}}{\partial(\Delta x_G)}\right|_{\Delta x_G=0} = \frac{\sin(\beta_G)}{v_G} \ , \tag{22c}$$

$$a_{CR} = \left.\frac{\partial t_{CR}}{\partial(\Delta x_S)}\right|_{\Delta x_S=0} = -\frac{\sin(\beta_S)}{v_S} \ . \tag{22d}$$

Correspondingly, the quadratic coefficients, b, are linear combinations of the second-order time derivatives $B^{-1}A$, DB^{-1} and B^{-1}. We have

$$b_{CMP} = \left.\frac{\partial^2 t_{CMP}}{\partial(\Delta h)^2}\right|_{\Delta h=0} = B^{-1}A + DB^{-1} + 2B^{-1} \ , \tag{23a}$$

$$b_{CO} = \left.\frac{\partial^2 t_{CO}}{\partial(\Delta x_m)^2}\right|_{\Delta x_m=0} = B^{-1}A + DB^{-1} - 2B^{-1} \ , \tag{23b}$$

$$b_{CS} = \left.\frac{\partial^2 t_{CS}}{\partial(\Delta x_G)^2}\right|_{\Delta x_G=0} = DB^{-1} \ , \tag{23c}$$

$$b_{CR} = \left.\frac{\partial^2 t_{CR}}{\partial(\Delta x_S)^2}\right|_{\Delta x_S=0} = B^{-1}A \ . \tag{23d}$$

As indicated earlier, the travel-time expressions significantly simplify in the case the central ray is a normal reflection ray. This situation is referred to as the ZO CRS case. As shown in the Appendix, the obtained FO CRS travel times naturally reduce to their counterpart ZO CRS formulas as given in the literature. Also described in the Appendix is the simplification that occurs in the FO CRS travel times in the case of a horizontally stratified medium.

Interpretation of the Quadratic Coefficients

Following the classical literature (see, e.g., ČERVENÝ, 2001) it is convenient to introduce the ray-centered coordinate system $(q, p^{(q)})$ along the fixed central ray. As shown in Figure 2, for a paraxial ray in the vicinity of the source S, the displacement Δx_S and the slowness difference Δp_S with respect to the central ray are related to their

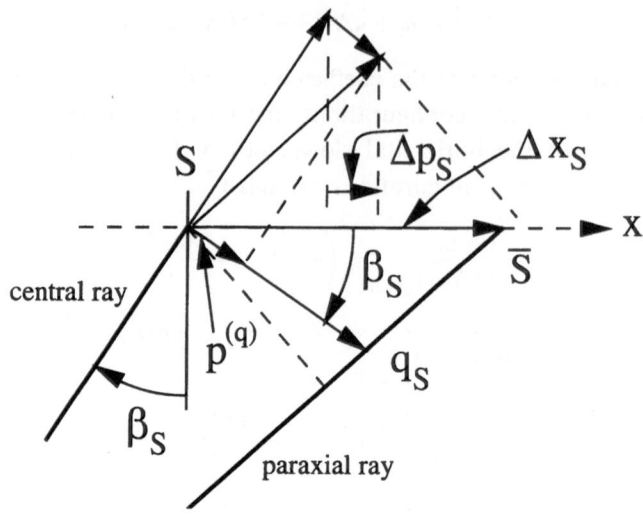

Figure 2
The ray-centered coordinate system and the local Cartesian coordinate system.

corresponding quantities q_S and $p_S^{(q)}$ in the ray-centered coordinate system by the formula

$$\begin{pmatrix} \Delta x_S \\ \Delta p_S \end{pmatrix} = \begin{pmatrix} \phi^{-1} & 0 \\ 0 & \phi \end{pmatrix} \begin{pmatrix} q_S \\ p_S^{(q)} \end{pmatrix} , \tag{24}$$

where

$$\phi = \cos(\beta_S) . \tag{25}$$

Moreover in the ray-centered coordinate system, we have (see e.g., HUBRAL, 1983)

$$p_S^{(q)} = \frac{q_S K_S}{v_S} \equiv q_S M_S , \tag{26}$$

where K_S is the curvature of the incidence wave at the source and

$$M_S = \frac{\partial^2 t}{\partial (q_S)^2}\bigg|_{q_S=0} . \tag{27}$$

From equations (24), (25) and (26) we obtain

$$\Delta p_S = M_S \Delta x_S \cos^2(\beta_S) . \tag{28}$$

Correspondingly, we obtain at the receiver G

$$\Delta p_G = M_G \Delta x_G \cos^2(\beta_G) . \tag{29}$$

Inserting equations (28) and (29) into equation (3) we get

$$\begin{pmatrix} \Delta x_G \\ M_G \Delta x_G \cos^2(\beta_G) \end{pmatrix} = \mathbf{T} \begin{pmatrix} \Delta x_S \\ M_S \Delta x_S \cos^2(\beta_S) \end{pmatrix} . \tag{30}$$

Let us now note that all configurations introduced above can be recast into the form

$$\Delta x_G = l \Delta x_S \quad \text{or} \quad \Delta x_S = \frac{1}{l} \Delta x_G \quad (l \neq 0) . \tag{31}$$

In fact, from equations (13), (15) and (19) we see that $l = -1$, $l = 1$, and $l = 0$ give rise to the CMP, CO, and CR configuration, respectively. Moreover, we interpret that the CS configuration is provided also by equation (31) upon the consideration of $l = \infty$.

For a general configuration defined by equation (31) where l is the fixed real number (including $l = \infty$ that provides the CS configuration) let

$$K_l^S = v_S M_l^S \quad \text{and} \quad K_l^G = v_G M_l^G \tag{32}$$

denote the corresponding wavefront curvatures at S and G, respectively.

Alternatively we can set

$$\Delta h = \Delta x_m \tan(\alpha) , \tag{33}$$

to describe the same measurement configurations (see Fig. 3). The relation between $\tan(\alpha)$ and l is easily seen to be

$$\tan(\alpha) = \frac{l - 1}{l + 1} . \tag{34}$$

Substituting equation (31) and equation (32) into equation (30) we get

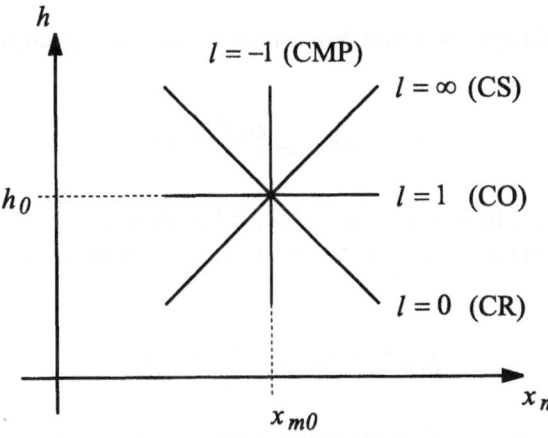

Figure 3
Specific measurement configurations specified by l are illustrated in the $(x_m - h)$–plane.

$$K_l^S \frac{\cos^2 \beta_S}{v_S} = B^{-1}(l - A) \tag{35a}$$

$$K_l^G \frac{\cos^2 \beta_G}{v_G} = \left(D - \frac{1}{l}\right)B^{-1} . \tag{35b}$$

Setting $l = -1$ in equation (35) (that provides the CMP configuration) yields

$$K_{CMP}^S \frac{\cos^2 \beta_S}{v_S} = -B^{-1}(A + 1) \quad \text{and} \quad K_{CMP}^G \frac{\cos^2 \beta_G}{v_G} = (D + 1)B^{-1} . \tag{36}$$

Here K_{CMP}^S and K_{CMP}^G can be interpreted as the wavefront curvatures observed at S and G, respectively, of an (fictitious elementary) *eigenwave* for which each paraxial ray that starts at the horizontal x axis at a location Δx_S with respect to the origin on its way down, hits, after reflection, the same x axis at the location $\Delta x_G = -\Delta x_S$ with respect to G. Also

$$b_{CMP} = K_{CMP}^G \frac{\cos^2 \beta_G}{v_G} - K_{CMP}^S \frac{\cos^2 \beta_S}{v_S} . \tag{37}$$

In the same way, if we set $l = 1$ (that provides the CO configuration), we have

$$K_{CO}^S \frac{\cos^2 \beta_S}{v_S} = B^{-1}(1 - A) \quad \text{and} \quad K_{CO}^G \frac{\cos^2 \beta_G}{v_G} = (D - 1)B^{-1} . \tag{38}$$

These are the wavefront curvatures observed at S and G, respectively, of an (fictitious elementary) *eigenwave* for which each paraxial ray that starts at the horizontal x axis at a location Δx_S with respect to the origin on its way down, hits, after reflection, the same x axis at the location $\Delta x_G = \Delta x_S$ with respect to G. Also

$$b_{CO} = K_{CO}^G \frac{\cos^2 \beta_G}{v_G} - K_{CO}^S \frac{\cos^2 \beta_S}{v_S} . \tag{39}$$

The quadratic coefficients b_{CS} and b_{CR} admit simpler interpretations as elementary wavefront curvatures.

$$b_{CS} = DB^{-1} = \frac{\cos^2 \beta_G}{v_S} K_{CS}^G , \tag{40}$$

where K_{CS}^G is the wavefront curvature observed at G of the wave that starts from a point source at S and is obtained by setting $l = \infty$ in equation (35). Finally by setting $l = 0$, we get

$$b_{CR} = B^{-1}A = \frac{\cos^2 \beta_S}{v_S} K_{CR}^S , \tag{41}$$

where K_{CR}^S is the wavefront curvature observed at S of the wave that starts from a point source at G ($\Delta x_S = 0$).

4. Determination of the Propagator Matrix

From the previous section, we see that all travel-time attributes, namely the two linear attributes (slowness), $\sin(\beta_S)/v_S$ and $\sin(\beta_G)/v_G$, and the three quadratic attributes (curvatures) $B^{-1}A$, DB^{-1} and B^{-1}, of the central ray can be easily derived, once the coefficients of three of the described travel-time experiments have been obtained. Using, as one of the possibilities, the CMP, CO and CS experiments, we readily find

$$\frac{\sin(\beta_S)}{v_S} = \frac{1}{2}(a_{CMP} - a_{CO}) = -a_{CR} \quad \text{and} \quad \frac{\sin(\beta_G)}{v_G} = \frac{1}{2}(a_{CMP} + a_{CO}) = a_{CS} \ , \quad (42)$$

as well as,

$$B^{-1}A = \tfrac{1}{2}(b_{CMP} + b_{CO}) - b_{CS} = b_{CR}, \quad DB^{-1} = b_{CS}, \text{ and}$$
$$B^{-1} = \tfrac{1}{4}(b_{CMP} - b_{CO}) \ . \quad (43)$$

The obtained coefficients together with equation (6) determine, in turn, the elements of the propagator matrix, namely

$$A = \frac{2(b_{CMP} + b_{CO} - 2b_{CS})}{b_{CMP} - b_{CO}} \quad B = \frac{4}{b_{CMP} - b_{CO}}$$
$$C = -\frac{(4b_{CS} - b_{CMP} + b_{CO})^2 - 16b_{CS}b_{CO}}{4(b_{CMP} - b_{CO})} \quad D = \frac{4b_{CS}}{b_{CMP} - b_{CO}} \ . \quad (44)$$

The coefficients a and b of the various travel-time experiments can be estimated upon the application of two-attribute stack-based coherency analysis (e.g., semblance) algorithms directly on multicoverage seismic data sets. The various practical aspects of feasibility, stability and computer costs involved in the procedures are beyond the scope of this work.

5. Applications

We now consider some applications of the kinematic attributes that can be realized upon the knowledge of the elements of the propagator matrix.

FO CRS Stack

Once we determined the elements of the ray propagator matrix, all the attributes in the travel-time formulas (1) are available. Equations (1) can then be used as 5-parameter FO CRS stacking operator for the simulation of a more generalized stack section than the ZO section. The FO CRS operator approximates the reflection travel-time surface in the vicinity of the central point $P_0(t_0, x_{m0}, h_0)$ which corresponds to the central ray in the 3-D travel-time midpoint-offset $(t - x_m - h)$

space. To perform the FO CRS stack, we stack the data along this surface and assign the stacked value to the point P_0 (see Fig. 4). If we repeat the procedure for all the points on the CO plane, $h = h_0$, then at the end of the operation we derive a simulated CO section. Since the travel-time formula is valid not only for the *P-P* wave, but also for *S-S* wave, *P-S* or *S-P* converted waves, the FO CRS can be applied to any type of reflections.

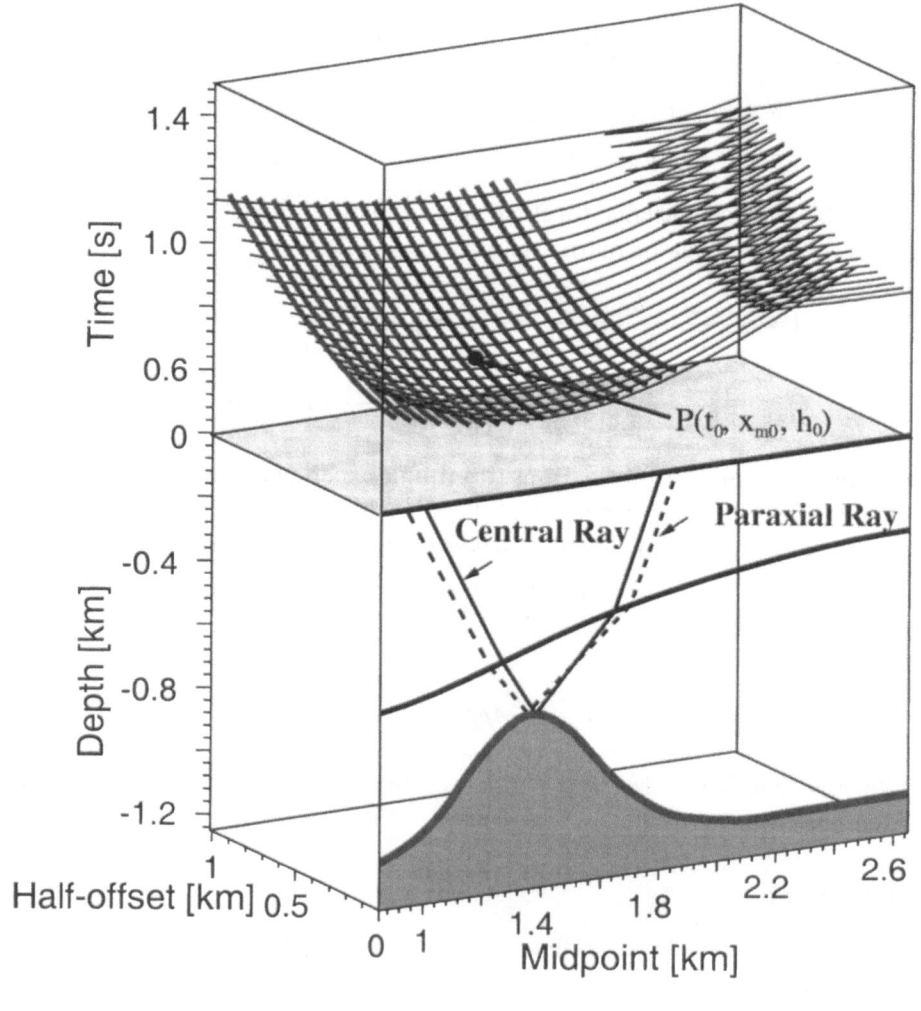

Figure 4

3-D volume for a 2-D multicoverage data set: The travel time of the rays reflected by the dome-like interface are represented by the CO travel-time curves in black. These are approximated around $P_0(t_0, x_{m0}, h_0)$ by a CRS surface displayed by means of CMP travel-time curves in thick grey.

Geometrical Spreading Factor

Recalling the well-known expression (HUBRAL *et al.*, 1993)

$$\mathscr{L}(G,S) = \sqrt{\frac{\cos \beta_S \cos \beta_G}{v_s v_g}} \, B^{\frac{1}{2}} \, e^{\frac{-i\pi\sigma}{2}} \, , \tag{45}$$

where $\mathscr{L}(G,S)$ is the normalized in-line geometrical-spreading factor as observed at G due to a point source at S. The parameter σ is the so-called KMAH index, which can only be determined by ray tracing. We see that $\mathscr{L}(G,S)$ can be determined from the knowledge of the quadratic coefficients from different travel times except for a phase coefficient. More explicitly, we have

$$\mathscr{L}(G,S) = \sqrt{\frac{\cos \beta_S \cos \beta_G}{v_s v_g}} \, \sqrt{\frac{|b_{CMP} - b_{CO}|}{4}} \, e^{\frac{-i\pi\sigma}{2}} \, . \tag{46}$$

Separation of Diffractions from Reflections

We first examine the ZO case and consider the two hypothetical waves, namely the N wave and the *NIP* wave, once again. We can observe that if a point in the depth is a diffraction point, then the two wavefront curvatures of the emergence waves, namely K_N and K_{NIP} should be identical, and therefore the radii of these two wavefront curvatures are also identical. This leads to the idea of using these two radii to classify the seismic events as reflections and (quasi) diffractions (MANN *et al.*, 2000; GELCHINSKY, 1997). To accomplish this MANN *et al.* (2000) used the factor $|K_N/K_{NIP}|$. They point out that for a "pure" diffraction event this factor should be unity and for a reflection event it can be far from 1. One can imagine that the smoother the reflector segment is, the larger the departure of that factor from unity.

For the CO case, we can also find a similar criterion. Recall the two hypothetical waves assumed in the CO and the CMP configurations. Again, we find that if the point on the reflector is a diffraction point, then K_{CO}^G should be equal to K_{CMP}^G and the ratio of the radii of these two curvatures serves as the criterion to classify the diffraction and the reflection events just as in the ZO case.

6. Conclusions

For a central finite-offset ray on a given seismic line above a 2-D isotropic medium, the travel times of paraxial rays between arbitrary locations on the seismic line depend on five attributes that refer only to the source and receiver locations of the central ray. The number of model-independent attributes can be reduced to two, upon the restriction to paraxial rays that refers to sources and receivers connected by means of certain conventional seismic configurations, in particular, the CMP, CO, CS and CR configurations.

The travel times for each of these configurations exhibit the same structure of a one-dimensional quadratic polynomial. This common structure is very attractive for the design of a unifying strategy to extract the five attributes from reflections in a given multi-coverage data set.

Extraction of the kinematic attributes, using individually the travel times in any three of the considered configurations, leads to the determination of the whole set of five attributes required by the travel-time formula for arbitrary source and receiver pairs in the vicinity of the central ray. This leads to a powerful generalization of the CRS stacking method to simulate seismic sections other than the ZO section.

Moreover, the complete propagator matrix of the central ray is also determined by the travel-time attributes. This means that, in particular, true-amplitude sections, other than ZO sections, can also be produced.

Acknowledgments

The author Y. Zhang would like to thank Jörg Zaske, Jürgen Mann, Kai-Uwe Vieth, Andreas Kirchner and Pedro Chira for instructive discussions and comments. This research has been partially supported by the Research Foundation of the State of São Paulo (FAPESP/Brazil).

Appendix: Particular Cases

In this appendix, we consider two situations where the travel-time formulas provided in the text significantly simplify. These are the cases of a horizontally stratified medium and the zero-offset situation.

Horizontally Stratified Medium

In this case, one can readily verify that, due to the particular geometrical symmetry involved, we have for any primary reflection ray

$$\beta_S = \beta_G \ . \tag{A.1}$$

Also, for the same reflector, the travel time remains unaltered when Δh is fixed and Δx_m varies. Owing to equation (12a), this implies the conditions

$$DB^{-1} - B^{-1}A = 0 , \quad \text{and} \quad B^{-1}A + DB^{-1} - 2B^{-1} = 0 , \tag{A.2}$$

so that

$$A = D = 1 \ . \tag{A.3}$$

Substituting into the travel-time formula (12a) we find

$$t = t_0 + 2 \frac{\sin \beta_S}{v_S} \Delta h + 2B^{-1}\Delta h^2 \ . \tag{A.4}$$

We see that for the particular model of homogeneous layers bounded by horizontal interfaces, the travel time is dependent on just two parameters, especially the second-order term is related to the geometrical spreading only.

Zero Offset

We now consider the case in which the central reflection ray is the zero-offset ray. This means that the source and receiver of the fixed central ray coincide. In this case we have the following simplifications: The coincidence of source and receiver in the central ray leads to the requirement of only one horizontal slowness $\sin \beta_0 / v_0$. Let this coincide with the horizontal slowness at S. Then we can write

$$\frac{\sin \beta_0}{v_0} = \frac{\sin(\beta_S)}{v_S} = -\frac{\sin(\beta_G)}{v_G} \ . \tag{A.5}$$

The ZO condition translates itself into the additional condition for the components of the propagator matrix (HUBRAL, 1983)

$$A = D \ . \tag{A.6}$$

As a consequence, only two instead of three elements of the propagator matrix need to be determined.

The elements of the propagator matrix can be conveniently expressed in terms of the elements of the propagator matrix that corresponds to the one-way ray that connects the reflection point to the coincident source-receiver location. Note that the incident and reflection segments of the ZO ray are the same ray paths in opposite direction. The reflection point of the ZO ray is referred to in the literature as the *normal incidence point* *(NIP)*. Denoting the propagator matrix of the downgoing ray segment by

$$\mathbf{T}_0 = \begin{pmatrix} A_0 & B_0 \\ C_0 & D_0 \end{pmatrix} , \tag{A.7}$$

the propagator matrix of the two-way ZO (i.e., the normal ray) ray can be written (see, e.g., HUBRAL *et al.*, 1992)

$$\mathbf{T} = \begin{pmatrix} A & B \\ C & D \end{pmatrix} = \begin{pmatrix} A_0 D_0 + B_0 C_0 & 2B_0 D_0 \\ 2A_0 C_0 & A_0 D_0 + B_0 C_0 \end{pmatrix} \ . \tag{A.8}$$

The parabolic and hyperbolic travel-time expressions become

$$t = t_0 - 2\frac{\sin(\beta_0)}{v_0} \Delta x_m + \Delta x_m D_0^{-1} C_0 \Delta x_m + \Delta h B_0^{-1} A_0 \Delta h \ , \tag{A.9a}$$

and

$$T^2 = \left[t_0 - 2\frac{\sin(\beta_0)}{v_0} \Delta x_m \right]^2 + 2t_0 \left[\Delta x_m D_0^{-1} C_0 \Delta x_m + \Delta h B_0^{-1} A_0 \Delta h \right] \ , \tag{A.9b}$$

respectively. The CMP and CO eigenwave curvatures get the well-known simple and attractive forms. We have

$$b_{CMP}^{ZO} = D_0^{-1} C_0 = \frac{\cos^2(\beta_0)}{v_0} K_N \ , \tag{A.10a}$$

and

$$b_{CO}^{ZO} = B_0^{-1} A_0 = \frac{\cos^2(\beta_0)}{v_0} K_{NIP} \ . \tag{A.10b}$$

REFERENCES

BORTFELD, R. (1989), *Geometrical Ray Theory: Rays and Travel Times in Seismic Systems (Second-order Approximations of the Travel Time)*, Geophysics *48*, 342–349.

ČERVENÝ, V., *Seismic Ray Theory* (Cambridge University Press 2001).

DE BAZELAIRE, E. (1988), *Normal Moveout Revisited – Inhomogeneous Media and Curved Interfaces*, Geophysics *53*, 143–157.

GELCHINSKY, B. (1997), *Special Course on Homeomorphic Imaging*, Wave Inversion Technology Consortium, Geophysical Institute, Karlsruhe University.

GELCHINSKY, B., BERKOVITCH, A., and KEYDAR, S., editors (1999), *Special Issue on Macro-model Independent Seismic Reflection Imaging*, vol. 42, J. Appl. Geophys., Amsterdam. Elsevier.

HUBRAL, P. (1983), *Computing True Amplitude Reflections in a Laterally Inhomogeneous Earth*, Geophysics *48*, 1051–1062.

HUBRAL, P., editor (1999), *Special Issue on Macro-model Independent Seismic Reflection Imaging*, vol. 42, J. Appl. Geophys, Amsterdam. Elsevier.

HUBRAL, P., SCHLEICHER, J., and TYGEL, M. (1992), *Three–dimensional Paraxial Ray Properties, Part I: Basic Relations*, J. Seismic Expl. *1*, 265–279.

HUBRAL, P., SCHLEICHER, J., and TYGEL, M. (1993), *Three-dimensional Primary Zero-offset Reflections*, Geophysics *58*, 692–702.

MANN, J., HUBRAL, P., TRAUB, B., GERST, A., and MEYER, H. (2000), *Macro-model independent approximative prestack time migration*. In *Expanded Abstracts of the 62th EAGE Conf. and Tech. Exhibition*, Glasgow, pages B–52.

MANN, J., JÄGER, R., MÜLLER, T., HÖCHT, G., and HUBRAL, P. (1999), *Common-reflection-surface Stack – A Real Data Example*, J. Appl. Geophys. *42*, 301–318.

SCHLEICHER, J., TYGEL, M., and HUBRAL, P. (1993), *Parabolic and Hyperbolic Paraxial Two-point Travel Times in 3-D Media*, Geophys. Prosp. *41*, 495–514.

THORE, P. D., DE BAZELAIRE, E., RAY, M. P. (1994), *Three-parameter Equation: An Efficient Tool to Enhance the Stack*, Geophysics *59*, 297–308.

TYGEL, M., SCHLEICHER, J., and HUBRAL, P. (1996), *2.5-D Kirchhoff MZO in laterally inhomogeneous media*. In *Annual Meeting Abstracts*, 483–486, Society of Exploration Geophysicists.

(Received October 27, 2000, accepted November 2, 2001, accepted August 3, 2001)

 To access this journal online:
http://www.birkhauser.ch

Pure appl. geophys. 159 (2002) 1617–1635
0033–4553/02/081617–19 $ 1.50 + 0.20/0

❙ Pure and Applied Geophysics

Statistical Travel-time Tomography in Terms of Stacking Velocity

DAVID GERAETS[1] and ALAIN GALLI[1]

Abstract — Velocity evaluation is a key step in seismic analysis. The covariance of the true velocity field must be known when interpolating or simulating velocities from well measurements using geostatistical methods. In addition, inversion procedures often require information pertaining to this covariance. Traditionally it has been taken to be the covariance of stacking velocities. We present a simple example to show that this approximation can lead to significant errors. Better methods, such as those of TOUATI (1996) and IOOSS (1998), use the variance of prestack picked travel times as a function of offset to infer that of the velocities. In this paper we extend their results on the estimation of the covariance of the reflected traveltimes, and obtain an explicit expression for the covariance of the square of the stacking slowness as a function of the covariance of the velocities. Although we are not able to invert the formula analytically to yield an explicit estimator for these parameters, the results obtained using it furnish a good and quick estimation of the velocity's covariance. This is illustrated with synthetic examples.

Key words: Velocity, correlation, geostatistics.

1. Introduction

A critical step when processing seismic data is the estimation of an *instantaneous velocity field*, $V(x,y,z)$, defined in every point (x,y,z) of the space (AL-CHALABI, 1997). Generally, a velocity field is constructed by interpolating well-log velocities between wells, using the interval velocities deduced from the stacking velocities to guide the process. Given our limited information, it is impossible to find the true earth model exactly. One common solution is to ignore the fine detail available at the wells and construct a very simple blocky or layered model of interval velocities, even though we know this is certainly an oversimplification of the real earth.

Geostatistical methods approach the uncertainty problem differently. They treat the true instantaneous velocity field as one member of a universe of possible velocity fields, each of which is geologically plausible and consistent with our data. Geostatistical methods allow us to find an instantaneous velocity field *simulation*: a representative heterogeneous velocity model that honors the well-log

[1] Centre de Géostatistique, Ecole des Mines de Paris, Fontainebleau (France)
E-mail: geraets@cg.ensmp.fr; galli@cg.ensmp.fr

and interval-velocity information exactly, and "looks like" the real earth between wells, in a statistical sense (even though it cannot be accurate in the details). Geostatistical methods can also provide a measure of the uncertainty: how big is the space of all possible velocity models, and how much should we expect the real earth to deviate from the instantaneous velocity field calculated by our interpolation between wells?

To produce these results, geostatistical tools require the knowledge of the covariance or variogram of the true velocity field. This covariance is also required in order to estimate the velocity shift observed for rays travelling in a random media (SNIEDER and SPENCER, 1993; WITTE et al., 1996). In addition to this, it is often useful to have a reasonable estimation of this covariance to control inversion and or to restrict the dimensionality of the search space. For all these reasons it is important to know the covariance of the velocity field.

The question is how to estimate it from experimental data. Many authors assume that the covariance of stacking velocities behaves in the same way as the covariance of the instantaneous velocities does. See for example CHU et al. (1994). Unfortunately this is wrong; the difference is generally significant. However, more accurate methods have been developed. Following work by RYTOV et al. (1989) and using the variance of the prestack travel times TOUATI et al. (1997), and IOOSS (1998) were able to obtain good estimates of this covariance of instantaneous velocities. The drawback of their methods is the need for a prestack picking. To avoid this step (or at least to have an alternative approach), we need to find the relationship between it and the covariance of stacking velocities within the same framework (geometrical optics and first-order perturbations). In this article, we present generalizations of some of their results relating reflected time covariance and velocity covariance, which lead to a closed form relationship between the covariance of the inverse of the square of the stack velocities and that of the instantaneous velocities.

The paper starts by giving numerical examples to illustrate how the correlation length measured from the stacking velocities depends on the acquisition geometry. It turns out to be a very hazardous estimator of the covariance of the velocity field. Then after reviewing the existing results, we show how to compute the covariance of the reflected travel times for any source/receiver geometry. We obtain an explicit expression for the structure of the slowness field (inverse of the velocities) in terms of the target covariance. Finally, using synthetic experiments with known results, we explore the validity and limitations of our new method.

2. Introductory Experimental Results

We consider two types of velocities: stacking and instantaneous (which are merely the complete velocity field we integrate on layers when considering interval velocities). To simplify the terminology we will use "velocities" for the instantaneous velocities. Consider a 2-D-onshore seismic experiment. Using a flat horizon at a depth of 1500 m, we propagate waves through a known velocity field, using Virieux's program (VIRIEUX, 1986) for the elasto-dynamic waves. The velocity fields are defined at each point by

$$V(x) = V_0 - \varepsilon(x) \qquad (1)$$

where V_0 is assumed to be a constant equal to 3000 m/s, and x stands for the 2-D coordinates.

The perturbations $\varepsilon(x)$ of the velocity field are assumed to be stationary and Gaussian, with Gaussian variogram and with a standard deviation $\sigma = 20\,\text{m/s}$ (see Appendix A for variograms). In order to reproduce the approximate layering of the natural media, we use an anisotropic model that is more continuous horizontally $a = 250\,\text{m}$ than vertically $b = 50\,\text{m}$. $(a/b = 5)$. We generate three independent simulations of this velocity field using the turning bands program written by LANTUÉJOUL (1994). The shots are performed every 10 meters, and the receivers are displayed every 20 meters, extending to a maximum distance of 1000 meters from the source. We picked the first arrival time, and estimate the stacking velocity for each CMP collection, using the two-term hyperbolic least-squares approximation. The three-term approximation suggested by AL-CHALABI (1973) was tested but not retained.

MUKERJI (1995) illustrated how the processing can vitally impact the values of correlation length inferred from the seismic image. We will show here, in a simpler case where no processing is needed, that the values inferred also depend on the acquisition geometry (in our case, the maximal offset used for the velocity analysis). This allows us to conclude on the inadequacy of the correlation length of the stacking velocities for the estimation of the covariance of the instantaneous velocity field.

Figure 1a shows the estimated stacking velocities obtained for the three simulations, as a function of the maximum offset used for the fitting.

As expected, the results show that the estimated velocities become more continuous and less scattered as the maximum offset used for the velocity analysis increases (the variance of velocities is three to five times greater for the 500m-maximal offset stacking velocities collections than for the 1000m-maximal offset stacking velocities collections). In Figure 1b we present the corresponding experimental variograms, normalized by their maximum (in order to compare the structures). The range of the stacking velocities clearly decreases as we reduce the maximum offset used for the stacking analysis. This confirms the impact of the acquisition geometry on the estimation of stacking velocity obtained from the data, as well as on the spatial aspects of its variability. We also show the variogram model of the instantaneous velocity field, in order to demonstrate that none of the three experimental variograms reproduce it correctly. Furthermore, experiments performed using different models of perturbations of the velocity field (varying a and b) indicated no obvious links between the structure of the perturbations of the velocity model and the structure of the stacking velocities.

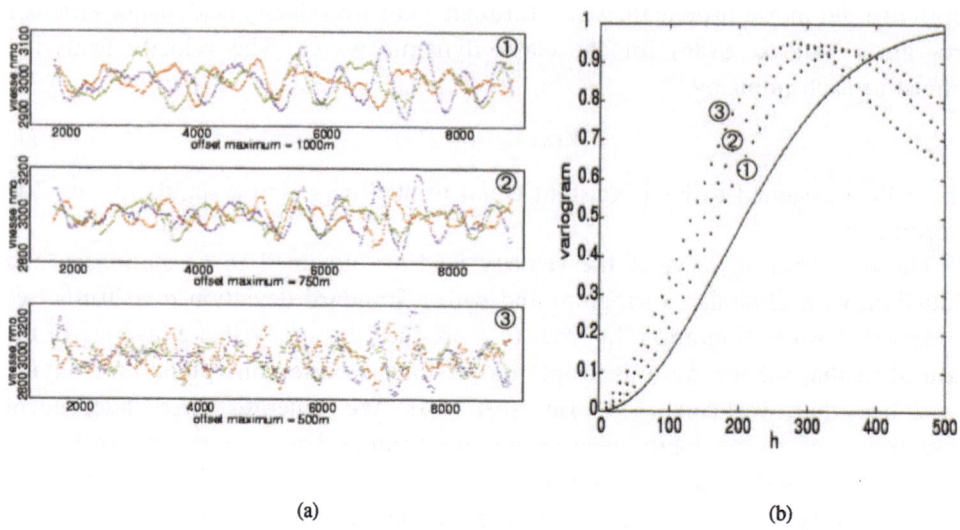

(a)   (b)

Figure 1

Spatial behavior of the stack velocity for each of the three simulations and for three different maximum offsets, and the corresponding variograms (in dashed lines, the variograms corresponding to the stack velocities and in solid line the variogram corresponding to the instantaneous velocity model).

Having demonstrated the influence of the offset on the results and shown that the experimental variograms of stacking velocities are far from the correct one, we now introduce the general framework used for the theoretical developments and present the results from TOUATI (1996) and IOOSS (1998). We then proceed to generalize these in order to obtain an explicit expression of the covariance of the stacking slowness in terms of the covariance of the instantaneous velocity field.

3. Geometrical Optics Hypothesis and Touati-Iooss Main Results

Our work was carried out within the scope of geometrical optics, which is a valid approximation for wave propagation if the wavelength of the propagated signal λ_0 is small compared to the minimum correlation length of the instantaneous velocities. As the horizontal correlation distance is generally far greater than the vertical one, this involves only the scale parameter a of the covariance or variogram. Formally, this requires

$$a \gg \lambda_0 \text{ and } b \gg \lambda_0 \text{ and } a \gg \sqrt{\lambda_0 L} \tag{2}$$

where L is the depth of the reflector we consider.

If we write the square of the slowness

$$\frac{1}{V^2(x)} = \frac{1}{V_0^2}[1 + \varepsilon(x)] . \tag{3}$$

Using the ε expansion of the travel time $T(x) = T_0(x) + T_1(x) + T_2(x) + \cdots$ (with T_1 the first-order term in ε), this leads to the well-known first-order expression in ε for the first arrival times in the geometrical optics approximation:

$$T(x) = T_0(x) + T_1(x) = \int_x \frac{dz}{V_0} + \int_x \frac{\varepsilon}{2V_0} dz \ . \tag{4}$$

Starting from these hypotheses, TOUATI (1996), IOOSS (1998) and IOOSS and GALLI (2000) expressed the variance of the first reflected arrival times in terms of the correlation of the 2-D velocity model. They expressed the normalized variance of the reflected travel times measured along a seismic line, as a function of the offset x:

$$\frac{\text{Var}[T(x)]}{\text{Var}[T(0)]} = \frac{1}{2} \left[1 + \frac{1}{x} \frac{\displaystyle\int_0^x \int_0^\infty C_\varepsilon(u,z) \, dz \, du}{\displaystyle\int_0^\infty C_\varepsilon(0,z) \, dz} \right]. \tag{5}$$

Using the Abel transform, Iooss found that:

$$C_\varepsilon\left(\frac{x}{a}\right) = \left[\int_x^\infty \frac{I_R(x')}{\sqrt{x'^2 - x^2}} \, dx' \right] \Big/ \left[\int_0^\infty \frac{I_R(x')}{x'} \, dx' \right] \tag{6}$$

where $I_R(x) = \frac{\partial^2}{\partial x^2} \{x \, \text{Var}[T(x)]\}$, $\text{Var}[T(x)]$ is the variance of the first reflected arrival times at offset x, and a is the horizontal scale parameter of the velocity perturbation. (See Appendix A for information relative to variograms and covariances and general references concerning geostatistics).

This leads to a good estimate of the horizontal correlation length. Tests carried out by Iooss (1998) give at worst a 20% inaccuracy in the estimation of a. Although this method is very efficient and powerful, it requires the value of the prestack travel-times for the reflectors considered. It is also limited to sub-horizontal reflectors and to 2-D geometry.

Our objective here is to extend this method in order to avoid the prestack picking requirement. We first extend the previous results to the more general case of the covariance of the travel times from different sources and with varying offsets. Then we use these travel-time covariances to express the covariance of the square of the "stacking slowness" (inverse of the stacking velocity) in terms of the covariance of the instantaneous velocity model.

4. Covariance of First Arrival Times with Different Sources and Offsets

In the rest of the paper, we consider a stationary velocity field (with horizontal and vertical scaling factors, a and b), and a horizontal reflector at depth L. Let us consider two different shots (Fig. 8, in Appendix B), with sources S_1 and S_2, and offsets h_1 and h_2. Let r_1 be the ray emitted by S_1 with components r_{11} and r_{12}, and r_2 be the one emitted by S_2. We are going to express the covariance between the travel times of these two shots, written as $\text{cov}(T(r_1), T(r_2))$ for short (without showing the other parameters), in terms of the covariance of the instantaneous velocities. Here C_V refers to the covariance of instantaneous velocities and C_ε to that of the first-order perturbations. As is shown in Appendix A we have

$$C_V(x_1 - x_2, z_1 - z_2) = \tfrac{1}{4}C_\varepsilon(x_1 - x_2, z_1 - z_2) . \tag{7}$$

We show in Appendix B that $\text{cov}(T(r_1), T(r_2))$ can be rewritten as the sum of four components of the form

$$\frac{ab}{2a_{vi}} \int_{-\infty}^{\infty} \varphi(\min_i, \max_i, r)C_\varepsilon(r)r\,dr , \tag{8}$$

where $\varphi(\min_i, \max_i, r)$ depends on the acquisition parameters L, h_1, and h_2, and on the variable r.

This allows us to obtain a closed form for $\text{cov}(T(r_1), T(r_2))$. We will now use it to estimate the correlation of the square of the stacking slowness.

Before starting we note that the particular case of $h_1 = h_2 = h \neq 0$ and $\Delta S = 0$ brings us back to a result

$$\begin{aligned}
\text{var}(T(x)) &= \frac{1}{2V_0^2} \iint_{\text{Dom}} 2C_\varepsilon\left(u\frac{h}{L}, 2u\right) + 2C_\varepsilon\left(h - v\frac{h}{L}, 2u\right)du\,dv \\
&= \frac{L}{V_0^2}\left[\int_{-\infty}^{\infty} C_\varepsilon\left(u\frac{h}{L}, 2u\right)du + \frac{ab}{h}\int_0^h C_\varepsilon(r)r\,dr + \frac{ab}{h}\int_h^{\infty} \sin^{-1}\left(\frac{h}{r}\right)C_\varepsilon(r)r\,dr \right]
\end{aligned} \tag{9}$$

equivalent to the one obtained by Iooss, using the spectral measure associated with the covariance.

5. Covariance of the Inverse of Stacking Velocities (Slowness)

In practice the data usually produced by the processing are the stacking velocity. Therefore we want the relationship between their covariance and that of the instantaneous velocities. It becomes much easier to find the relationship between the inverse of the square of the stacking velocities and the instantaneous velocities.

In appendix C it is shown that

$$\text{cov}\left(\frac{1}{V_{S1}^2}, \frac{1}{V_{S2}^2}\right) \approx \sum_{i=1}^{n}\sum_{j=1}^{n} k(i, j, n)[E^2(T_{1i})\text{var}(T_{2j}) + 4\,E(T_{1i})E(T_{2j})\text{cov}(T_{1i}, T_{2j})$$
$$+ E^2(T_{2j})\text{var}(T_{1i})] , \tag{10}$$

where $k(i, j, n)$ is a function of the acquisition geometry (C-2).

Note that to obtain the theoretical development we worked in terms of covariance, whereas the illustrations are expressed in terms of variograms. It is important to remember that since we assumed the velocity field to be stationary, these two representations are equivalent (eq. A-2). By working in terms of a variogram we avoid the difficulty of estimating the mean arrival time, and yet obtain results that are valid under less restrictive assumptions. (In this case, we do not have to estimate the mean value of the times).

We are going to use a Gaussian model for the covariance of the instantaneous velocities because it has the advantages of being smooth and of mimicking a type of layered system often encountered in the subsurface. The results are strongly influenced by the correlation distances (ranges) of the instantaneous velocity field and therefore by the scaling factors a and b (Fig. 2), and by the survey geometry of the acquisition (as has already been shown in Fig. 1).

We conclude that, once the geometry of acquisition is known, the correlation of the square of stacking slowness provides us with a tool to infer the parameters of the covariance of the instantaneous velocity field, nonetheless experience has shown that it is more difficult to infer the type of covariance to use.

6. Experimental Results

Synthetic tests have been carried out to validate these theoretical results. In all cases we fix the covariance of instantaneous velocities to be a Gaussian model and vary its parameters. Initially, we compare our theoretical results for the time arrivals.

In our experiments we use two types of sets of points: firstly, the propagation values (three series of 800 shots) obtained using the Virieux's elasto-dynamic program (already used in section 2), and secondly, values obtained considering propagation along nonperturbed rays. This latter method allows us to get considerably larger collections of shots (here we consider series of 10,000 shots), in a more reasonable computational time. Due to the small amplitude of the velocity perturbations, this simplification (which is correct to the first order of the times) has no significant impact on the times recovered compared to the definition of the gridding of the numerical propagation model (we used a time increment ≈ 1.6 ms). The dominant wave length of the signal is 50 m.

In Figure 3, we compare theoretical and experimental results for the covariance of the square of the stacking slowness. In this case, 800 shots are insufficient to obtain a correlation curve that matches the theoretical one, but using 10,000 shots (obtained by propagation along nonperturbed rays) gives better results, independent of the geometry considered. However, if we look more carefully at the first curve, computed with 800 shots, we can see that the general shape is correct, particularly for

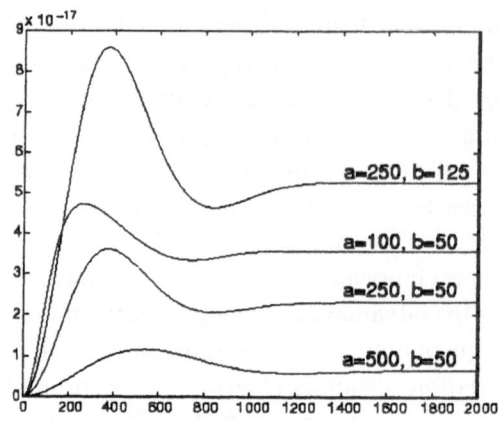

Figure 2
Covariance models of square of slowness with varying a and b, for a Gaussian covariance model.

the most important part corresponding to the maximum. So the mismatch only concerns the variance.

It is worthwhile discussing more precisely the data requirement to obtain a very good fit between the model and the experimental variogram. Here the experiments were made along a line, although in real cases we would be working mainly with 3-D seismics for which we would have hundreds of shots per line and more than 50 or 100 lines. When computing variograms along one direction we generally consider that the mean variogram along all the lines is a meaningful statistic. This is valid under the hypothesis of stationarity or if the trend in the velocity is almost perpendicular to the direction of computation. Doing so we would in most cases use several thousand shot points. This is precisely the point we illustrate in the next figure.

In Figure 4, we show the same results for a borderline case $(a = 250 \approx \sqrt{\lambda_0 L}$ and $b = 50\,\text{m})$, for two sets of samples: one line with 10,000 points and the average on three lines each with 800 shot points. We see that in both cases, the match between the theoretical and experimental values is still very good, and that using the mean results from the three series of 800 shot times gives us satisfying results. This is an efficient way of obtaining the covariance of the stacking velocities, while having more reasonable requirements.

In order to illustrate how the experimental curve moves away from the theoretical one when we depart from the geometrical optics hypothesis (2), we show the general behavior of the results for five different values of the horizontal correlation length a (resp. $a = 100, 200, 300, 400$ and 500). With smaller values of a, the variance of the stacking velocities is larger, thus the variogram curve appears to stabilize at a higher sill, but with a shorter range.

We see in Figures 5 and 6 that, for shorter values of a, the quality of the match deteriorates. However, the values found by our theoretical expression are still in a

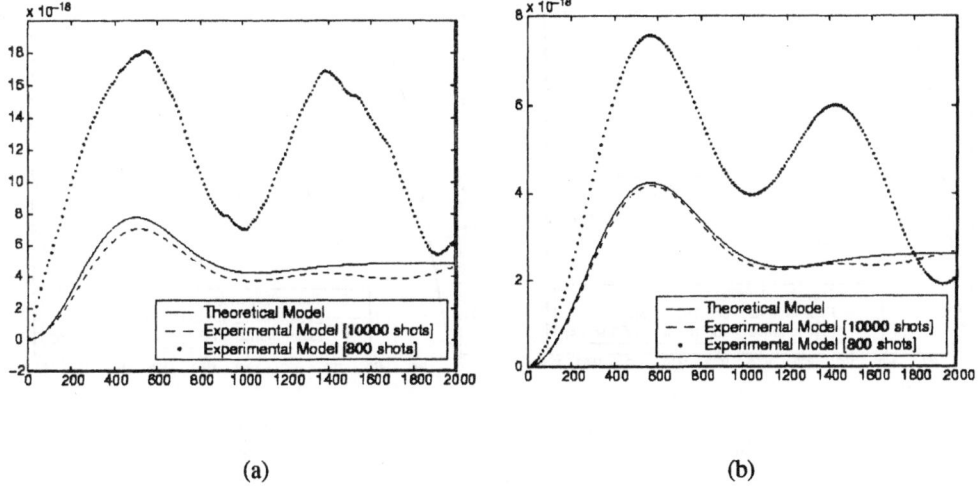

(a)   (b)

Figure 3

Results in the Gaussian case, for (a) 51 and (b) 101 receivers, with the correlation lengths $a = 500$ m and $b = 50$ m.

correct order of magnitude, and give a good estimate of the correlation parameter, even when the initial hypothesis is no longer satisfied. It seems that the major impact comes from the anisotropy in the velocity structure rather than the fact that the assumptions of the geometrical optics approximation are satisfied or not. This was also observed by MATHERON (1991) for the estimation of the uncertainty on the position of a reflecting point.

Although we are not able, at this point, to invert analytically the formula found for the covariance of the stacking slowness, we believe that the use of an (extremely simple) iterative matching process, provides a satisfactory estimate of the horizontal correlation length of the instantaneous velocity field.

7. Conclusion

Starting from the results by TOUATI et al. (1997) and IOOSS (1998), under the geometrical optic framework and using them as a first-order perturbation method, we have been able to calculate the covariance of reflected travel times, and then the covariance of the square of stacking slowness as a function of the velocity's covariance and the acquisition geometry. This allows us to infer the correlation parameters from the velocities without any need for a prestack picking.

The results obtained using the covariance of the stacking slowness are probably less accurate than those obtained using the covariance of the travel times. This is a logical result of the smoothing effect of the velocity analysis. But this method might be more robust against the hypotheses than that of Iooss and Touati, as is the

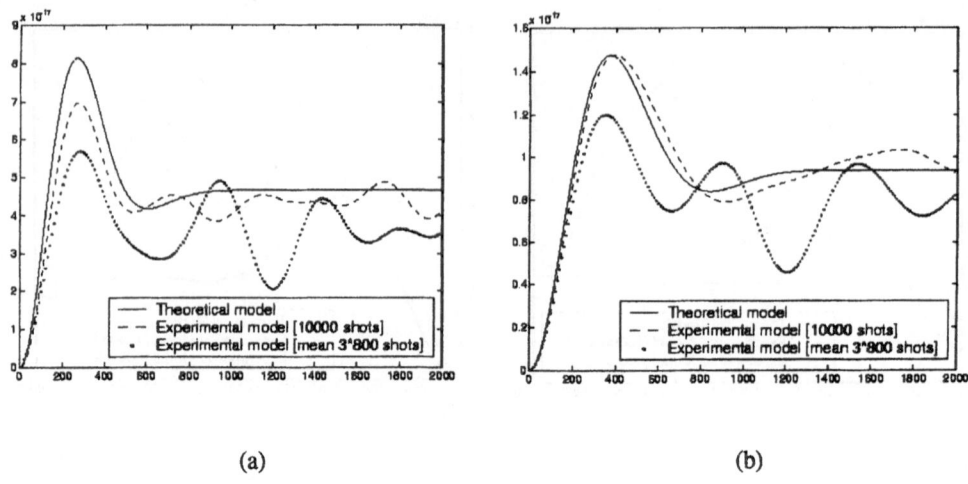

(a)   (b)

Figure 4
Results in the Gaussian case, for (a) 51 and (b) 101 receivers, with the correlation lengths $a = 250$ m and $b = 50$ m.

stacking process. Also, it is not extremely sensitive to the validity of the geometrical optic approximation. The main factor appears to be the ratio between the vertical and horizontal correlation parameters. The more anisotropic the medium is, the better it seems to work. We believe that the methodology developed here provides a fast and easy way to gain sense of the covariance, and can therefore constitute a very useful tool for the geophysicist working on velocity.

Acknowledgements

The authors wish to thank B. Iooss for his discussions. We are also very grateful to the reviewers, especially Joe Dellinger for his many constructive suggestions, and finally to I. Pšenčík, for his support and assistance on the document.

Appendices

Appendix A: Introduction to Geostatistical Tools

In this Appendix we introduce the general hypotheses of the geostatistic framework, and define the covariance and variogram functions, with some of their important properties.

Let us start with a random function $Z(x)$ where x stands for the coordinates in a 2-D or 3-D space. A random function is just a collection of random variables indexed

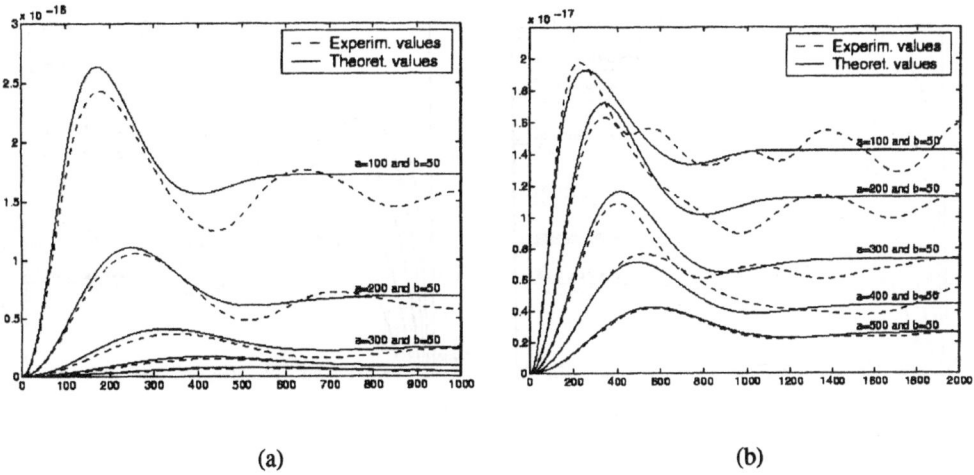

(a)   (b)

Figure 5

Results in the Gaussian case, for (a) 51 and (b) 101 receivers, with various correlation lengths a, and
$b = 50$ m.

by location. The random functions considered here will be the instantaneous velocities $V(x)$, the stacking velocities $V_S(x)$ and the perturbation of the instantaneous velocities $\varepsilon(x)$. Classically we can define the expectation (mean value) of $Z(x)$ as: $m(x) = E(Z(x))$, and the variance as: $\mathrm{var}[Z(x)] = E[(Z(x) - m(x))^2]$. Its covariance is by definition:

$$\mathrm{cov}(Z(x), Z(y)) = E[(Z(x) - m(x))(Z(y) - m(y))]. \tag{A-1}$$

One important case is when the mean does not depend on the location and when the covariance depends only on the separation vector $x - y$, that is $C(x,y) = \mathrm{cov}(Z(x), Z(y)) = C(x - y)$. In that case we say that the random function $Z(x)$ is *second-order stationary*. A slightly more general case is when the expectation does not depend on location and the variogram, $\gamma(x,y) = \frac{1}{2}\mathrm{var}[(Z(x) - Z(y))^2]$ depends only on the separation distance, that is $\gamma(x,y) = \gamma(x - y)$. In the later case we say that the random function is *intrinsic*. The second case is more general than the first, the second-order stationarity random functions are intrinsic but the converse is not true. The variogram itself has the advantage that we do not need to estimate the mean. When the random function is second-order stationary, there is a well-known relationship between variograms and covariances

$$\gamma(x - y) = C(0) - C(x - y) . \tag{A-2}$$

To give an example, let us compute the covariance of the instantaneous velocities under the first-order perturbation model used in this paper. Let C_V be the covariance of instantaneous velocities and C_ε that of the first-order perturbations.

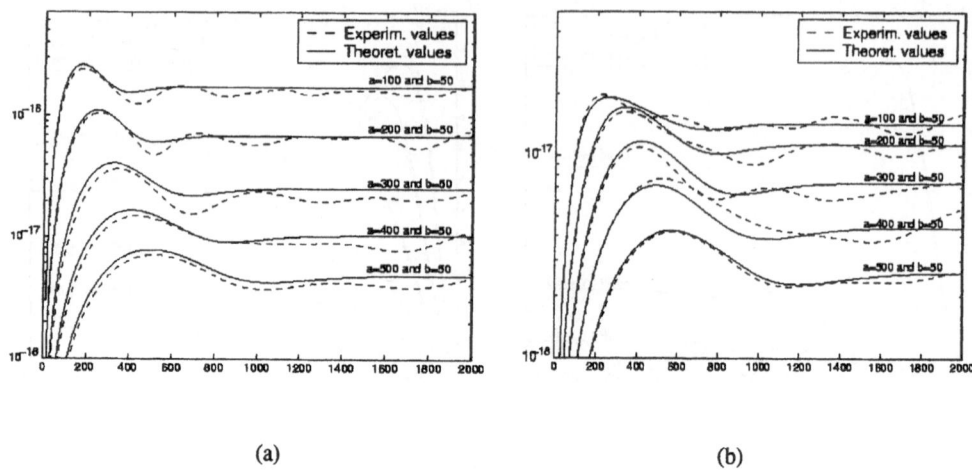

Figure 6
Results in the Gaussian case, for (a) 51 and (b) 101 receivers, with varying correlation lengths a, and $b = 50$ m: detail from Figure 5, in logarithmic scale.

First note that for velocities, the model we consider is equivalent to $V(x) = V_0 - \varepsilon(x)/2$, or if we want to stress the fact we are working in 2-D $V(x,z) = V_0 - \varepsilon(x,z)/2$

$$
\begin{aligned}
C_V(x_1, z_1; x_2, z_2) &= \text{cov}((V(x_1, z_1), \quad V(x_2, z_2)) \\
&= \text{cov}(V_0 - \varepsilon(x_1, z_1)/2, \quad V_0 - \varepsilon(x_2, z_2)/2) \\
&= 1/4 \ E(\varepsilon(x_1, z_1), \quad \varepsilon(x_2, z_2)) \\
&= 1/4 \ \text{cov}(\varepsilon(x_1, z_1), \quad \varepsilon(x_2, z_2)) \\
&= 1/4 \ C_\varepsilon(x_1, z_1; x_2, z_2)
\end{aligned}
$$

(because $E(\varepsilon(x,z))$ is 0). Finally we see that if we assume V to be stationary (as is the case in this model) both covariances depend only on the separation vector, that is

$$
C_V(x_1 - x_2, z_1 - z_2) = 1/4 \ C_\varepsilon(x_1 - x_2, z_1 - z_2) \tag{A-3}
$$

and we have the same relationship between the variograms.

Note that not all functions can be a covariance or a variogram: for example, the variogram must be negative definite

$$
-\sum_\alpha \sum_\beta \lambda^\alpha \lambda^\beta \gamma(x_\alpha - x_\chi) > 0 \quad \forall \lambda^\alpha \text{ such that } \sum_\alpha \lambda^\alpha = 0 . \tag{A-4}
$$

It is easy to show that any positive linear combination of variograms (or covariances) is a variogram (or a covariance).

The experimental variogram

$$\hat{\gamma}(h) = \frac{1}{2n(A_h)} \sum_{x \in A_h} [z_x - z_{x+h}]^2 \qquad \text{(A-5)}$$

is an approximation of this variogram (here A_h represents the set of values z_x for which the value z_{x+h} also belongs to the set of data, and $n(A_h)$ the number of pairs (z_x, z_{x+h}) in this set).

When the set A_h is restricted to pairs of points in certain classes of directions we obtain directional variograms. They allow us to check and, later on, to model anisotropies in data.

For the examples in this paper, we use only one kind of variogram called the gaussian variogram

$$\gamma(x_1 - x_2, z_1 - z_2) = 1 - \exp\left[-\left(\frac{x_1 - x_2}{a}\right)^2 + \left(\frac{z_1 - z_2}{b}\right)^2\right], \qquad \text{(A-6)}$$

where the x coordinates stand for the horizontal and the z for the vertical. The parameters a and b are scale factors for these two directions. They are equal to one third of the *practical range*, that is the distance at which the variogram stabilises.

Note that this model is well suited to represent velocities which are extensions of the classical layered 1-D velocities in which the layering appears only on average. In practice the anisotropy ratio between the horizontal and vertical correlation parameters a/b is quite large (5 or 10 or even more). The velocities described with this model also have the advantage of being twice differentiable.

Once we have a variogram or covariance model, estimates or simulations can be made. Figure 7 shows two sets of two realizations of 2-D instantaneous velocities with different anisotropy factors.

More information on geostatistics is given in MATHERON (1970), ARMSTRONG (1998), CHILÈS and DELFINER (1999) and GOOVAERTS (1997).

Appendix B: Correlation of the First Arrival Times

In this Appendix we show how the covariance of the travel times of two rays r_1 and r_2 can be expressed in terms of the covariance C_ε of the velocity field's first-order perturbations and of indicator functions.

The acquisition geometry we consider is illustrated in Figure 8.

If we split the covariance of the travel times into four terms, in terms of the covariances along the different parts of the rays, we can rewrite the expression for the covariance as the sum of four integrals

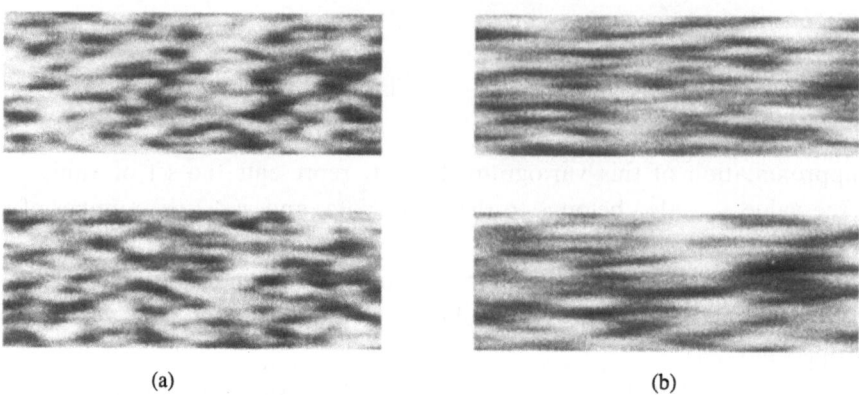

(a)   (b)

Figure 7

Simulations of instantaneous velocity fields, with varying anisotropy ratios: (a) $a/b = 5$ and (b) $a/b = 10$.

$$\mathrm{cov}(T(r_1), T(r_2)) = E\left[\int_{r_{11}, r_{21}} \frac{\varepsilon(x_1(z_1), z_1)}{2V_0} \frac{\varepsilon(x_2(z_2), z_2)}{2V_0} dz_1 dz_2\right.$$

$$+ \int_{r_{12}, r_{21}} \frac{\varepsilon(x_1(z_1), z_1)}{2V_0} \frac{\varepsilon(x_2(z_2), z_2)}{2V_0} dz_1 dz_2$$

$$+ \int_{r_{11}, r_{22}} \frac{\varepsilon(x_1(z_1), z_1)}{2V_0} \frac{\varepsilon(x_2(z_2), z_2)}{2V_0} dz_1 dz_2$$

$$\left. + \int_{r_{12}, r_{22}} \frac{\varepsilon(x_1(z_1), z_1)}{2V_0} \frac{\varepsilon(x_2(z_2), z_2)}{2V_0} dz_1 dz_2\right] .$$

$$(\text{B-1})$$

Interchanging the expectation sign and the integral signs along the four parts of rays gives

$$\mathrm{cov}(T(r_1), T(r_2)) = \frac{1}{4V_0^2} \int_0^L \int_0^L C_\varepsilon\left(\Delta S + \frac{h_2 z_2}{2L} - \frac{h_1 z_1}{2L}, z_1 - z_2\right) dz_1 dz_2$$

$$+ \frac{1}{4V_0^2} \int_0^L \int_0^L C_\varepsilon\left(\Delta S - h_1 + \frac{h_2 z_2}{2L} + \frac{h_1 z_1}{2L}, z_1 - z_2\right) dz_1 dz_2$$

$$+ \frac{1}{4V_0^2} \int_0^L \int_0^L C_\varepsilon\left(\Delta S + h_2 - \frac{h_2 z_2}{2L} - \frac{h_1 z_1}{2L}, z_1 - z_2\right) dz_1 dz_2$$

$$+ \frac{1}{4V_0^2} \int_0^L \int_0^L C_\varepsilon\left(\Delta S - h_1 + h_2 - \frac{h_2 z_2}{2L} + \frac{h_1 z_1}{2L}, z_1 - z_2\right) dz_1 dz_2 .$$

Writing $u = \frac{z_1 - z_2}{2}$, $v = \frac{z_1 + z_2}{2}$, and $\iint_{\text{Dom}} dv\, du = \int_0^{\frac{L}{2}} dv \int_{-v}^{v} du + \int_{\frac{L}{2}}^{L} dv \int_{v-L}^{L-v} du$, we obtain

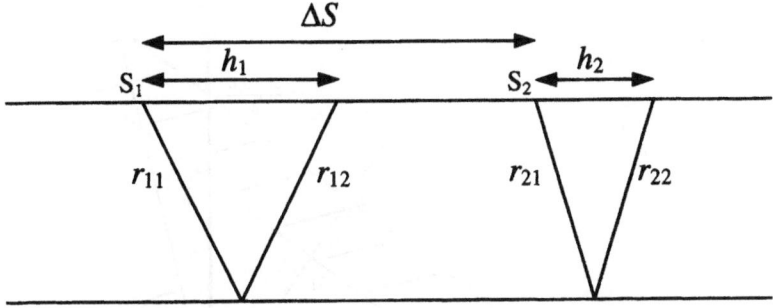

Figure 8
Acquisition geometry.

$$\mathrm{cov}(T(r_1), T(r_2)) = \frac{1}{V_0^2} \iint_{\mathrm{Dom}} \left[C_V \left(\Delta S + u \frac{-h_1 - h_2}{2L} + v \frac{-h_1 + h_2}{2L}, 2u \right) \right.$$
$$+ C_V \left(\Delta S - h_1 + u \frac{h_1 - h_2}{2L} + v \frac{h_1 + h_2}{2L}, 2u \right)$$
$$+ C_V \left(\Delta S + h_2 + u \frac{-h_1 + h_2}{2L} + v \frac{-h_1 - h_2}{2L}, 2u \right)$$
$$\left. + C_V \left(\Delta S - h_1 + h_2 + u \frac{h_1 + h_2}{2L} + v \frac{h_1 - h_2}{2L}, 2u \right) \right] 2 \, dv \, du \ . \tag{B-2}$$

Finally, using Rytov's approximation (RYTOV *et al.*, 1989, pp. 6–7), we can approximate these four integrals with four terms of the form

$$\int\limits_0^L \int\limits_{-\infty}^\infty C_V(a_{0i} + a_{ui}u + a_{vi}v, 2u) du \, dv \tag{B-3}$$

where a_{0i}, a_{ui} and a_{vi} depend on h_1, h_2, ΔS and L.

These terms can be rewritten in a more suitable form, using a change of variable

$$\begin{cases} t = \frac{2u}{b} \\ s = \frac{a_0 + a_v v + a_u u}{a} \end{cases} .$$

The four terms then become

$$\int\limits_0^L \int\limits_{-\infty}^\infty C_V(a_{0i} + a_{ui}u + a_{vi}v, 2u) du \, dv = \frac{2ab}{a_{vi}} \int_{-\infty}^\infty \int\limits_{\frac{a_0 + a_u \frac{t}{2} b}{a}}^{\frac{a_0 + a_v L + a_u \frac{t}{2} b}{a}} C_V(\sqrt{s^2 + t^2}) ds \, dt. \tag{B-4}$$

This suggests using polar coordinates.

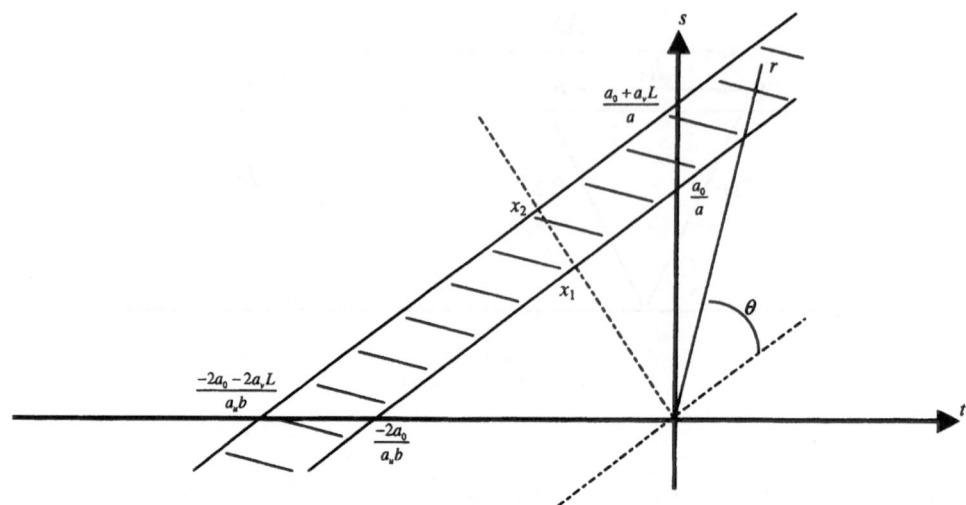

Figure 9

Cartesian (s, t) to polar (θ, r) coordinates transformation, the dashed area corresponds to the domain of integration. Remark that if $x_1.x_2 < 0$, the integration domain comprises the origin $(0, 0)$.

Setting $\begin{cases} t = r\cos\theta \\ s = r\sin\theta \end{cases}$, we obtain an expression for the four terms of the covariance that we can integrate in θ, with varying boundaries (see Fig. 9).

$$\int_0^L \int_{-\infty}^{\infty} C_V(a_{0i} + a_{ui}u + a_{vi}v, \; 2u)du \; dv = \frac{ab}{2a_{vi}} \int_{-\infty}^{\infty} \varphi(\min_i, \max_i, r)C_\varepsilon(r)r \; dr, \quad \text{(B-5)}$$

where the term $\varphi(\min_i, \max_i, r)$ is equal to the sum of indicator functions ($1_{[a,b]}(x) = 1$ if $x \in [a, b]$ and 0 otherwise)

$\varphi(\min, \max, r) =$

$$\begin{cases} 1_{[\min,\max]}\left(\frac{\pi}{2} - \sin^{-1}\left(\frac{\min}{r}\right)\right) + 1_{[\max,\infty]}\left(\sin^{-1}\left(\frac{\max}{r}\right) - \sin^{-1}\left(\frac{\min}{r}\right)\right) & \text{if } x_1.x_2 > 0, \\ 1_{[0,\min]}(\pi) + 1_{[\min,\max]}\left(\frac{\pi}{2} + \sin^{-1}\left(\frac{\min}{r}\right)\right) + 1_{[\max,\infty[}\left(\sin^{-1}\left(\frac{\max}{r}\right) + \sin^{-1}\left(\frac{\min}{r}\right)\right) & \text{otherwise.} \end{cases}$$

$$\text{(B-6)}$$

Here $\min_i$ and $\max_i$ are respectively the minimum and maximum of $(|x_{1i}|, |x_{2i}|)$ defined as

$$\begin{cases} x_{1i} = \frac{2a_{0i}}{\sqrt{4a^2 + a_{ui}^2 b^2}} \\ x_{2i} = \frac{2a_{0i} + 2a_{vi}L}{\sqrt{4a^2 + a_{ui}^2 b^2}} \end{cases} \quad . \quad \text{(B-7)}$$

As a remark, observe that the sum of the four terms can be expressed in terms of the spectral form of the covariance

$$\mathrm{cov}(T(r_1), T(r_2)) = \frac{1}{V_0^2} \int_0^L \int_{-\infty}^{\infty} \int_{-\infty}^{\infty} \int_{-\infty}^{\infty} \Phi(k_x, k_z) \cos\left(\left(\Delta S + \frac{h_2 - h_1}{2}\right)k_x + 2uk_z\right)$$

$$\cos\left(\left(\frac{-h_2}{2} + u\frac{-h_2}{2L} + v\frac{h_2}{2L}\right)k_x\right)$$ \hfill (B-8)

$$\cos\left(\left(\frac{h_1}{2} + u\frac{-h_1}{2L} + v\frac{-h_1}{2L}\right)k_x\right) dk_x dk_z dv du$$

with $\Phi(k_z, k_x)$ the Fourier transform of $C_v(x, z)$.

Appendix C: Covariance of the Stacking Slowness

In this Appendix we give the first-order expression of the covariance of the inverse of the square of the stacking velocities, in terms of the expectations, variances and covariances of the travel times, weighted by a function k depending on the acquisition geometry.

Starting from the expression for the square of the stacking slowness at point p_1 and assuming that the arrival times have a hyperbolic behavior versus offset,

$$\frac{1}{V_{S1}^2} = \frac{n\sum_{i=1}^{n} T_{1i}^2 x_i^2 - \sum_{i=1}^{n} T_{1i}^2 \sum_{i=1}^{n} x_i^2}{n\sum_{i=1}^{n} x_i^4 - \sum_{i=1}^{n} x_i^2 \sum_{i=1}^{n} x_i^2}$$

where n is the number of offsets used for the hyperbolic fitting, x_i is the offset for the shot i, we can express the covariance of the square of the inverse of the stacking velocities in terms of the covariance of the velocity field

$$\mathrm{cov}\left(\frac{1}{V_{S1}^2}, \frac{1}{V_{S2}^2}\right) = E\left(\frac{1}{V_{S1}^2} \frac{1}{V_{S2}^2}\right) - E\left(\frac{1}{V_{S1}^2}\right) E\left(\frac{1}{V_{S2}^2}\right) .$$

This leads to

$$\mathrm{cov}\left(\frac{1}{V_{S1}^2}, \frac{1}{V_{S2}^2}\right) = \sum_{i=1}^{n} \sum_{j=1}^{n} \mathrm{cov}(T_{1i}^2, T_{2j}^2) \quad k(i, j, n)$$

$$\mathrm{var}\left(\frac{1}{V_{S1}^2}\right) = \sum_{i=1}^{n} \sum_{j=1}^{n} \mathrm{cov}(T_{1i}^2, T_{1j}^2) \quad k(i, j, n)$$
\hfill (C-1)

where $k(i, j, n)$ is a function of the acquisition geometry

$$k(i,j,n) = \frac{n^2 x_i^2 x_j^2 - n x_i^2 \left(\sum_{k=1}^{n} x_k^2\right) - n x_j^2 \left(\sum_{k=1}^{n} x_k^2\right) + \left(\sum_{k=1}^{n} x_k^2\right)\left(\sum_{k=1}^{n} x_k^2\right)}{\left[n \sum_{k=1}^{n} x_k^4 - \left(\sum_{k=1}^{n} x_k^2\right)\left(\sum_{k=1}^{n} x_k^2\right)\right]^2} \quad \text{(C-2)}$$

and $\mathrm{cov}(T_{1i}^2, T_{2j}^2)$ approximated by

$$\begin{aligned}
\mathrm{cov}(T_{1i}^2, T_{2j}^2) &= E(T_{1i}^2 \, T_{2j}^2) - E(T_{1i}^2)\, E(T_{2j}^2) \\
&= E\left[\left[\int_{r_1,r_2} \left(\frac{1}{V_0} + \frac{\varepsilon(x_1,z_1)}{2V_0}\right)\left(\frac{1}{V_0} + \frac{\varepsilon(x_2,z_2)}{2V_0}\right) dz_1 dz_2\right]^2\right] - E(T_{1i}^2)E(T_{2j}^2) \\
&\approx E(T_{1i}^2)\, E(T_{2j}^2) + E^2(T_{1i})\, \mathrm{var}(T_{2j}) + 4\, E(T_{1i})E(T_{2j})\, \mathrm{cov}(T_{1i}, T_{2j}) \\
&\quad + E^2(T_{2j})\, \mathrm{var}(T_{1i}) - E(T_{1i}^2)E(T_{2j}^2) \\
&= E^2(T_{1i})\, \mathrm{var}(T_{2j}) + 4\, E(T_{1i})E(T_{2j})\, \mathrm{cov}(T_{1i}, T_{2j}) + E^2(T_{2j})\, \mathrm{var}(T_{1i}) \ .
\end{aligned}$$

$$\text{(C-3)}$$

REFERENCES

AL-CHALABI, M. (1973), *Series Approximation in Velocity and Travel-time Computations*, Geophysical Prospecting *21*, 783–795.

AL-CHALABI, M. (1997), *Instantaneous Slowness versus Depth Functions*, Geophysics *62*, 270–273.

ARMSTRONG, M., *Basic Linear Geostatistics* (Springer, Berlin 1998).

CHILÈS, J.P. and DELFINER, P., *Geostatistics: Modelling Spatial Uncertainty* (Wiley, New York 1999).

CHU, J., XU, W., and JOURNEL, A.G., *3-D implementation of geostatistical analyses – the Amoco case study*. In *Stochastic Modelling and Geostatistics; Principles, Methods and Case Studies*, AAPG Computer Applications in Geology *3*, 201–217 (eds. Yarus J. and Chambers R.) (1994).

GOOVAERTS, P., *Geostatistics for Natural Resources Evaluation* (Oxford University Press, Oxford 1997).

IOOSS, B. (1998), *Seismic Reflection Travel Times in Two-dimensional Statistically Anisotropic Random Media*, Geophys. J. Int. *135*, 999–1010.

IOOSS, B., and GALLI, A. (2000), Statistical tomography for seismic reflection data. In *6th International Geostatistics Congress*, Cape Town, South Africa, 10 p. (eds. Kleingeld W. and Krige D.), in press.

LANTUÉJOUL, CH., *Nonconditional simulation of stationary isotropic multi-Gaussian random functions*. In *Geostatistical Simulations*, (eds. Armstrong M. and Dowd P.A.) (Kluwer, Dordrecht 1994) 147–177.

MATHERON, G., *The theory of regionalised variables and its applications*, Cahiers du Centre de Morphologie Mathématique (Ecole des Mines de Paris, Paris 1970).

MATHERON, G. (1991), *Géodésiques aléatoires: application à la prospection sismique*, Cahiers de Géostatistique *1*, 1–17.

MUKERJI, T., RIO, P., and MAVKO, G.M., *Impact of seismic resolution on geostatistical integration technique*, 65th SEG Annual Meeting Expanded Technical Program Abstracts with Biographies, 215–218 (SEG, Tulsa 1995).

RYTOV, S. M., KRAVTSOV, YU A., and TATARSKII, V. I., *Principles of Statistical Radiophysics, vol 4: Wave Propagation through Random Media* (Springer-Verlag, Berlin 1989).

SNIEDER, R. and SPENCER, C. (1993), *A Unified Approach to Ray Bending, Ray Perturbation and Paraxial Ray Theories*, Geophys. J. Int. *115*, 456–470.

TOUATI, M., GALLI, A., RUFFO, P., and DELLA ROSSA, E., *Migration uncertainties: a probabilistic approach*. In *Geostatistics Wollongong '96*, (eds. Baali, E. Y. and Schofield, N. A.) (Kluwer, Dordrecht 1997) *1*, 597–608.

TOUATI, M., IOOSS, B., and GALLI, A. (1999), *Quantitative Control of Migration: A Geostatistical Attempt*, Mathematical Geology *31*, 277–295.

VIRIEUX, J. (1986), *P-SV Wave Propagation in Heterogeneous Media: Velocity-stress Finite-difference Method*, Geophysics *51*, 889–901.

WITTE, O., ROTH, M., and MULLER, G. (1996), *Ray Tracing in Random Media*, Geophys. J. Int. *124*, 159–169.

(Received September 18, 2000, revised January 1, 2001, accepted April 19, 2001)

Pure appl. geophys. 159 (2002) 1637–1679
0033–4553/02/081637–43 $ 1.50 + 0.20/0

© Birkhäuser Verlag, Basel, 2002

❘ Pure and Applied Geophysics

Applications of Directional Wavefield Decomposition, Phase Space, and Path Integral Methods to Seismic Wave Propagation and Inversion

Louis Fishman[1]

Abstract — Recently, de Hoop and coworkers developed an asymptotic, seismic inversion formula for application in complex environments supporting multi-pathed and multi-mode wave propagation (DE HOOP *et al.*, 1999; DE HOOP and BRANDSBERG-DAHL, 2000; STOLK and DE HOOP, 2000). This inversion is based on the Born/Kirchhoff approximation, and employs the global, uniform asymptotic extension of the geometrical method of "tracing rays" to account for caustic phenomena. While this approach has successfully inverted the multicomponent, ocean-bottom data from the Valhall field in Norway, accounting for severe focusing effects (DE HOOP and BRANDSBERG-DAHL, 2000), it is not able to account properly for wave phenomena neglected in the "high-frequency" limit (i.e., diffraction effects) and strong scattering effects. To proceed further and incorporate wave effects in a nonlinear inversion scheme, the theory of directional wavefield decomposition and the construction of the generalized Bremmer coupling series are combined with the application of modern phase space and path (functional) integral methods to, ultimately, suggest an inversion algorithm which can be interpreted as a method of "tracing waves." This paper is intended to provide the seismic community with an introduction to these approaches to direct and inverse wave propagation and scattering, intertwining some of the most recent new results with the basic outline of the theory, and culminating in an outline of the extended, asymptotic, seismic inversion algorithm. Modeling at the level of the fixed-frequency (elliptic), scalar Helmholtz equation, exact and uniform asymptotic constructions of the well-known, and fundamentally important, square-root Helmholtz operator (symbol) provide the most important results.

Key words: Seismic inversion, directional wavefield decomposition, phase space analysis, path integral representations, square-root Helmholtz operator symbol.

1. Introduction

The multicomponent, ocean-bottom data from the Valhall field in Norway are particularly difficult for seismic inversion due to the presence of a series of gas pockets which act as "lenses" to create severe focusing effects (DE HOOP and BRANDSBERG-DAHL, 2000). To account for such severe focusing effects, de Hoop and coworkers have recently developed an asymptotic, seismic inversion formula for application in complex environments supporting multi-pathed and multi-mode wave

[1] Code 7181, Naval Research Laboratory, Stennis Space Center, MS 39529, USA. Also at: Department of Physics, University of New Orleans, LA 70148, USA. E-mail: Shidi53@aol.com

propagation (DE HOOP et al., 1999; DE HOOP and BRANDSBERG-DAHL, 2000; STOLK and DE HOOP, 2000). This formulation is based on a separation of the functions characterizing the medium (i.e., the sound speed profile) into (1) a smooth component that defines the background medium (embedding) and (2) a singular component that defines the medium contrast and is compactly supported (i.e., of finite extent). The scattering data are generated by sources and recorded by receivers which are distributed in open patches on a surface surrounding the support (boundary) of the medium contrast. This asymptotic, seismic inversion carries out a mapping from the singular support of the scattering data (wavefront) to the singular support of the medium contrast (interfaces). From a mathematical perspective this is a "microlocal" analysis, in that it tracks the propagation of singularities such as wavefronts. Given a known, smooth background medium (embedding), the singular medium component is reconstructed in a sequence of by now, well-known operations: propagation — modeling — acquisition — imaging — resolution — inversion. This "microlocal" approach essentially extends the classical Kirchhoff-style imaging-inversion process in conjunction with migration-velocity analysis. Underlying the inversion are the assumptions that the (incident) wavefield in the background medium can be evaluated with the Maslov canonical operator (the background is sufficiently smooth), and the scattered wavefield can be evaluated in the Born/Kirchhoff approximation (for example, the contrast is sufficiently localized and small). This inversion algorithm can be viewed as the global, uniform asymptotic extension of the geometrical method of "tracing rays" in conjunction with the Born/Kirchhoff approximation. As such, caustic phenomena are properly incorporated, accounting for the successful inversion of the Valhall field data.

While the microlocal, asymptotic, seismic inversion formula can provide for severe focusing effects, it is still an inherently "high-frequency" approximation, which neglects wave effects, in addition to being incapable of treating the more rapidly-varying medium properties (as measured against a typical wavelength). This algorithm can be extended through the application of two constructions which are well known (at least physically and from a formal mathematical viewpoint) in the seismic literature: (1) directional wavefield decomposition and (2) the generalized Bremmer coupling series (FISHMAN, 1993; DE HOOP, 1996). While seismic wave propagation modeling is often most appropriately formulated in the time domain, numerical calculations are most often carried out, by computational necessity, in the frequency domain. Thus, the extension of the asymptotic, seismic inversion formula of de Hoop and coworkers is addressed here at the modeling level of the fixed-frequency (elliptic), scalar Helmholtz equation.

The square-root Helmholtz operator (in this frequency-domain formulation) is the central object in both the directional wavefield decomposition and the generalized Bremmer coupling series. It provides for the construction of the right- and left-traveling wavefield components, and, in exponentiated form, represents the formal, fundamental, one-way wavefield solution, or propagator, in the tracking of the multiple scattering in the Bremmer series. It would not be an overstatement to say that the square-root Helmholtz operator (and its generalization, the Dirichlet-to-Neumann (DtN) operator, along with the corresponding scattering (reflection and transmission) operators) are some of the most central and important objects in wave propagation, scattering, and inverse scattering theory. The construction, approximation, and understanding of the square-root Helmholtz operator have been major problems in wave propagation and scattering for, at least, the past fifty years in areas as diverse as acoustics, electromagnetics, optics, seismics, and pure, applied, and computational mathematics. Beyond its role in wave propagation and scattering, this operator plays a crucial role in the construction of computational boundary conditions and the corresponding stochastic formulations of the above-mentioned problems. Many people, over the years, in each of these fields, have tried to approximately construct and characterize this operator. Well-known names involved in this pursuit, with a connection to seismics, would include Majda, Engquist, Clayton, Collins, Tappert, Brooke, Trefethen, Zhang, Joly, Halpern, and Bamberger. (For a more complete history and exposition of this material, see the extensive reference citations in FISHMAN et al., 1987, 1997, 2000; FISHMAN, 1992.)

For the most part, all previous work resulted in formal operator rational approximations and (generalized) Taylor series. This formal operator approximation analysis is a type of formal "asymptotics." Over the past half century, theoretical mathematicians have developed a mathematically rigorous foundation and constructive approach for this "asymptotics" in the form of the theory of pseudodifferential (ΨDO) and Fourier integral operators (FIO), and, in particular, the elliptic pseudodifferential operator calculus. (For a more complete discussion of this point, see, again, the extensive reference citations in FISHMAN et al., 1987, 1997, 2000; FISHMAN, 1992.) From the outset, only part of the solution is retained; the entire theory is about approximation. In some cases, this is good enough; in other cases, it is not. For the hyperbolic wave propagation problems, such as the time-domain, plasma, and Klein-Gordon wave equations, for the corresponding square-root operators, it is good enough (FISHMAN, 1992; FISHMAN et al., 1997, 2000). There is a rigorous, asymptotic machinery, which is what, ultimately, justifies much of the purely formal operator manipulations in the hyperbolic case. The neglected part can be shown to be appropriately small in a well-defined sense. In the case of the elliptic wave propagation problems, which are of interest here, however, this is no longer true. The neglected part of the solution plays a key role, which must be taken into account. Thus, these problems lie outside of the rigorous asymptotic machinery — a corresponding, rigorous mathematical theory does not exist at this time. While there may be a superficial analogy between the elliptic and hyperbolic square-root operators at the purely formal, operator level, there is an underlying, fundamental difference which is not readily apparent at this level, and it is this difference that drives the elliptic wave propagation problems.

This means that the corresponding, formal operator manipulations, in the elliptic case, can be fraught with danger. With a formal, operator Taylor series, the expansion is nonuniform, and, in its infinite realization, singular. This expansion can never account for the fundamental oscillatory character and the correct singularity structure which characterize (underlie) the square-root Helmholtz operator (FISH-MAN *et al.*, 1997). The part of the solution that is required for this has been inherently omitted in the formal expansion procedure, although this is not readily apparent from the formal analysis. (It turns out that the formal operator rational approximations, if properly constructed, contain information about the oscillatory and singularity structure (FISHMAN, 1992; FISHMAN *et al.*, 1997, 2000). However, this information is implicit, and it is never transparent in the subsequent numerical solution of the resulting partial differential equations.)

While the asymptotic construction of the square-root Helmholtz operator lay outside of the mathematically rigorous, elliptic pseudodifferential operator calculus, it was recognized by Fishman and coworkers that the theory of pseudodifferential and Fourier integral operators (the so-called phase space methods), in conjunction with global path (functional) integral techniques, provided the appropriate framework to extend the Fourier methods, which are appropriate for homogeneous (translation-invariant) environments, to extended (over many wavelengths) inhomogeneous environments. The path integral representations express the fundamental wavefield solution as an infinite-dimensional integral. Such constructions were previously employed by Wiener in his study of Brownian motion and Feynman in his reformulation of quantum mechanics. Phase space provides the natural setting for both the operator analysis and the path integral representations (FISHMAN *et al.*, 1987; FISHMAN and WALES, 1987). Initially, this synthesis of the ideas produced explicit representations and defining equations for the appropriate wave propagation operator, wave equation, and fundamental wavefield solution (propagator) for the one-way Helmholtz equation appropriate in the transversely-inhomogeneous (range-independent) limit. This naturally led to phase space, marching computational algorithms. Through phase space and path integral methods, a concise, explicit, unified statement of the problem and subsequent, in-principle solution was available (FISHMAN *et al.*, 1987; FISHMAN and WALES, 1987). The formal, square-root Helmholtz operator was replaced by a corresponding phase space function, the operator symbol. The square-root Helmholtz operator symbol obeyed a well-defined equation and corresponding radiation condition, which were, subsequently, amenable to exact and asymptotic mathematical analysis. The operator symbol was the pivotal component in the path integral representation and the computational algorithm for the fundamental wavefield solution. This development was, subsequently, combined with invariant imbedding and Dirichlet-to-Neumann operator methods to incorporate generally-inhomogeneous environments within the same formalism (FISHMAN, 1993; FISHMAN *et al.*, 1997). The centerpiece of this generalization was the Dirichlet-to-Neumann operator symbol, which is the appropriate extension of the square-root Helmholtz operator symbol.

However, even in the earliest papers, it was clearly recognized and stated that an effective and computational theory depended upon the construction of uniformly-valid, phase space approximations of the appropriate operator symbols. The emphasis on the uniform approximations was another manifestation of the inadequacy of the elliptic pseudodifferential operator calculus in this case. Simply applying the calculus to the construction of the square-root Helmholtz operator symbol results in a nonuniform, singular approximation over phase space. The emphasis had shifted to complete operator symbols. Essentially, the uniform characterization over phase space and subsequent exploitation of the (singularity and oscillatory) structure of the operator symbols provide the focus for the direct and inverse analysis. New, uniform, high-frequency (UHF), asymptotic symbol expansions have been developed for the square-root Helmholtz operator (FISHMAN et al., 1997). This analysis uniformly incorporates the usual algebraic terms associated with the elliptic pseudodifferential operator calculus and the terms of exponential order, corresponding to the contributions from the infinitely-smooth part of the kernel, lying outside of the standard theory. In addition to this uniform asymptotic, operator symbol analysis, several exact operator symbol constructions, incorporating canonical features, have been derived (FISHMAN, 1992; FISHMAN et al., 2000). It is important to emphasize that the uniform asymptotic analysis on the operator symbols in the phase space correctly incorporates much more than the contribution from the evanescent waves. Relative to the nonuniform asymptotic analysis, the uniform treatment correctly incorporates the structure associated with both the high-angle propagation and near-evanescent local phase space regimes, in the process, completely accounting for the oscillatory component characterizing these operator symbols. The figures comparing the exact, nonuniform, and uniform, operator symbol constructions clearly demonstrate the significant phase space region involved. In highly-variable environments, the correct representation of this phase space regime is crucial for the accurate computation of the wavefield and the proper treatment of integrated energy-flux conservation, and for application in inverse problems.

While the early work was essentially conceptual and formal (the ideas, their formal expression, and their broad implications), the more recent work has been far more mathematically oriented (the details that actually make the initial formalism relevant and applicable, revealing the true insights contained in the approach). Both the exact and uniform asymptotic symbol constructions of the square-root Helmholtz operator are the only known examples for operators of this type, which lie outside of the standard asymptotic theory. This is the reason that theoretical mathematicians have an interest in these results. They now have a glimpse of what a complete, mathematically rigorous theory, for this case, might look like.

In making the constructions of the square-root Helmholtz operator and the corresponding fundamental wavefield solution (propagator) explicit, the above-outlined developments provide the basis to extend the asymptotic, seismic inversion procedure. Essentially, the Born/Kirchhoff approximation is replaced by the generalized Bremmer coupling series and the Maslov canonical operator is replaced by directional wavefield decomposition, with the one-way wave propagator written as a path integral (Trotter product) in terms of the square-root Helmholtz operator symbol. Thus, an inversion, which is based, in part, on the global, uniform asymptotic extension of the geometrical method of "tracing rays," is, in principle, generalized, ultimately, to suggest an inversion algorithm which can be interpreted as a method of "tracing waves" (FISHMAN et al., 2000; DE HOOP, 1996; DE HOOP et al., 2002).

While the extension of the de Hoop asymptotic, seismic inversion algorithm addresses wave propagation in a range-dependent environment through the generalized Bremmer coupling series, there is an alternative approach to range-dependent wave propagation in the context of the directional wavefield decomposition, phase space, and path integral analysis (FISHMAN et al., 1997). This previously-mentioned generalization constitutes an exact, well-posed, one-way reformulation of elliptic wave equations in terms of appropriate Dirichlet-to-Neumann operators, which provide the extensions of the square-root Helmholtz operator, exact for the range-independent case, to the range-dependent case. The DtN operator provides both the exact, one-way wave propagation operator and the means to construct the exact, initial total wavefield at the source location. The latter point involves the solution of the complete multiple-scattering problem between the "blocks" (half-spaces) to the left and right of the source. First, the first-order operator Riccati equation for the DtN operator, which is well-posed for marching, is solved. The resulting DtN operator is then used to construct the initial total wavefield at the source location, which provides the initial condition for the exact, one-way wave equation. Finally, the one-way wave equation, in terms of the just-constructed DtN operator and the initial condition, is then marched, in a well-posed manner, in the range direction to compute the total wavefield. There currently exist numerical codes (the "Riccati Method") which compute all three steps of this, in principle, exact solution (LU and MCLAUGHLIN, 1996; LU, 1999; LU et al., 2001). In the phase space and path integral approach, consistent with the range-independent case in terms of the square-root Helmholtz operator symbol, uniform asymptotic approximations of the DtN operator symbol are constructed, in the range-dependent case, to eliminate the first two numerical steps. The remaining computational step is then the one-way marching of the total wavefield in a manner analogous to that in the range-independent case.

In the seismic community, where ray-based methods are so deeply rooted, it must be emphasized that this phase space and path integral approach is not a ray-based method, but, rather, is inherently a "full wave" approach based on integrating over the entire phase space. In this sense, this work is viewed more properly as an extension of the well-known parabolic equation (PE) approximations to the full, elliptic wave equation formulation. The phase space and path integral formulation then provides the general framework to unify the (generalized) operator Taylor series, operator rational approximation, and generalized phase screen representations of the square-root Helmholtz operator, for example. These well-known approximations provide the link between the uniform asymptotic operator symbol expansions and the more practical computational algorithms. (For a more complete discussion of this point, see the extensive reference citations in FISHMAN, 1992; FISHMAN et al., 1997, 2000; DE HOOP et al., 2001.)

Given that the results presented in this paper are at the level of the scalar Helmholtz equation, it is fair for the seismic community to ask why the corresponding explicit formulas, for other more appropriate levels of seismic modeling (i.e., isotropic and anisotropic elastic, and poro-elastic), are not also available. The response is historically based. From the beginning of this work in the early 80s, it was quite clear that a basic decision on the line of development would be necessary. There were two choices. The first choice was to concentrate on the simplier scalar case and work all the way through, from the basic formalism through the operator symbol analysis and asymptotics (for both the square-root Helmholtz and DtN operator symbols), all the way to the applications. In other words, the goal was to demonstrate to the relevant scientific communities, in constructive and computational detail, that everything works and is useful, at least for the scalar case. With this "blueprint" in hand, others could make the appropriate generalizations to the vector theories of electromagnetism and elasticity, for example. The second choice was to address the scalar and vector cases more or less simultaneously, and to break the process into steps. The first step would be to get all of the cases to the formalism level; then, to get all of the cases through the operator symbol analysis stage; and, finally, to concentrate on the applications in all of the areas. Appreciating the difficulty involved in the operator symbol analysis step (which would be the rate determining step in this process), the first choice was made. Further, there was the strong feeling that a "completed" case, after a certain period of time, was more satisfying than several incomplete cases. The seismic community is certainly invited to work on the extension of the ideas and results presented in this paper, to the more appropriate model equations.

Beyond the restriction to the scalar Helmholtz equation, this paper focuses on the operator symbols (and, primarily, the square-root Helmholtz operator symbol at that), and, further, the exact construction cases correspond to, by necessity, relatively simple profiles. So, what is in this work, at this stage, for the majority of the seismic community? Why focus on operator symbols, and relatively simple, exact ones at that? There are several reasons.

(1) The square-root Helmholtz operator symbol is the generalized vertical slowness in geophysics; it is about the operator symbols. Moreover, for all of the compact formulas in the directional wavefield decomposition/phase space/path integral formalism, it all comes down to being able to uniformly approximate the operator symbols. Without this, there are no significant applications.

(2) The square-root Helmholtz operator plays a prominent role in factorization and subsequent backscatter formulations in the seismic community. Since the operator, in this seismic work, is invariably treated in a purely formal manner, and the resulting calculations are carried out only at the level of the numerical solution of partial differential equations, it would be extremely useful to see just what these operators actually "look like," where the information actually "lives," and what the formal operator manipulations actually account for, and, more importantly, omit. For this reason alone, this paper should be of interest to the seismic community.

(3) While the exact, operator symbol construction cases presented are, not surprisingly, for relatively simple environments, the great value of exact results should never be underestimated. This is especially true when such matters are usually treated at a purely formal level, which precludes connecting the operator to the underlying physics in a detailed, transparent manner. It is the operator symbol, in playing a role formally analogous to that of the Hamiltonian in the quantization of classical dynamical systems, which provides the most convenient encapsulation of the physical information, and, ultimately, the "microscopic" view of the wave propagation problem. Moreover, these relatively simple cases are, in some sense, canonical, in that they provide a qualitatively correct picture for more complex profiles. Furthermore, while the exact constructions illustrated may only represent rather simple cases, the square-root Helmholtz operator symbol for an arbitrary, piecewise-constant environment can be computed in principle. Most importantly, the UHF asymptotic construction presented applies to a rather general class of profiles, and this is the key to the applications. The exact and uniform asymptotic, operator symbol constructions provide both a level of understanding and the key to future applications.

(4) The extension of the de Hoop asymptotic, seismic inversion method is accomplished through the generalized Bremmer coupling series, with the fundamental wavefield solutions (propagators) being computed from the UHF, square-root Helmholtz operator symbol construction in conjunction with the path integral representations (Trotter products). Thus the analysis outlined in this paper, at the level of the square-root Helmholtz operator symbol, is indeed relevant to seismic inversion. To incorporate wave effects and nonlinearity in a (non-optimization-based) asymptotic inversion, even in a scalar formulation, is a reasonably relevant result for the seismic community.

Given the pivotal role played by the synthesis of directional wavefield decomposition, phase space, and path integral ideas and constructions in the extension of seismic inversion algorithms, the remainder of this paper intertwines some of the most recent new results with the basic outline of the theory. To be more specific, and to attempt to provide a "road map" for the reader, the essential points of this paper are the following.

(1) This Section has presented the asymptotic, seismic inversion method of de Hoop and coworkers, and noted its initial success. The fundamentally important role of the square-root Helmholtz and related operators was discussed in the context of

the ideas of directional wavefield decomposition and the generalized Bremmer coupling series. Directional wavefield decomposition, phase space, and path (functional) integral methods, developed over the past two decades, were briefly introduced, and applied to indicate how to extend the de Hoop approach to suggest, essentially, an inversion algorithm which can be interpreted as a method of "tracing waves."

(2) For fixed-frequency, elliptic, two-way wave propagation, the basic formalism of the wavefield decomposition/phase space/path integral approach is briefly outlined for the scalar Helmholtz formulation. While the exact, well-posed, one-way reformulation of the Helmholtz equation is stressed, the remainder of this paper will focus on the transversely-inhomogeneous (range-independent), limit, at the level of the square-root Helmholtz operator (symbol). This is review material, and is only intended to provide a "bare bones" context for the remainder of the paper. A considerably more detailed exposition is provided in the references. It is virtually impossible to provide a satisfying explanation, for all issues and constructions contained in this material, to non-specialists in a relatively short space. This is contained in Section 2 (Mathematical Formulation). Most importantly, the primary role of the operator symbols, in this approach, is stated at the end of this section.

(3) The important role of the square-root Helmholtz operator symbol is discussed at the beginning of Section 3 (Exact Operator Symbol Constructions). The crucial problem of the operator symbol approximation, and the subsequent need to go beyond the well-established asymptotic machinery is discussed. Exact square-root Helmholtz operator symbol constructions are presented to illustrate these points, and the ability of the operator symbol to encapsulate the physical information in the problem in a most convenient and insightful manner. All of the mathematical detail is referenced, and further the section attempts to use the operator symbol surface figures as the primary device to present and briefly establish these points.

(4) The fundamental importance of the uniform approximation of the operator symbol, particularly for applications, is discussed in some detail, establishing the limitations of the standard analysis (elliptic pseudodifferential operator calculus), and the subsequent need for an extended theory. Such a uniform, high-frequency result is presented and contrasted with the inadequate ΨDO analysis (again, with all mathematical details contained in the references). This is the most important point in the paper, and it is, again, illustrated with a pivotal figure, and further supported by the figures from Section 3. This material is contained in Section 4 (Uniform Asymptotic Operator Symbol Constructions).

(5) The basic structure and information content of the square-root Helmholtz operator symbol is reviewed and discussed in the context of connecting the elliptic (frequency-domain) and hyperbolic (time-domain) formulations of the problem. The crucial point, that the resulting uniform, high-frequency wave theory (following from the result in Section 4) is distinct from geometrical approximations (Klauder, Maslov) and is inherently a full-wave theory, is established. The connection to

computational, nonreflecting boundary conditions is made. Finally, and most importantly, an outline is formed how to combine the uniform asymptotic operator symbol results (at the level of the square-root Helmholtz operator) with the generalized Bremmer coupling series work of de Hoop to suggest the desired extension of the asymptotic, seismic inversion algorithm introduced in Section 1. This brings the paper full circle, completing the connection to seismic inversion. This material is contained in Section 5 (Discussion).

Finally, it is extremely important to re-emphasize that, in many respects, the figures in this paper play a primary role. It is through the figures that many of the essential points are made to this, primarily applied, audience. In this case, the picture is really worth a handful of formulas and thousands of words. (For a look at the type of analysis behind the results in Sections 3 and 4, see FISHMAN et al., 1997 and FISHMAN et al., 2000). If the most important message in this work is distilled to its pictorial essence, then it is contained in Figures 6, 7, 9, 10, and 13. At some intuitive level, the basic message from these pictures is: (1) it is about the operator symbols, (2) the established asymptotic theory fails in a significant way, and (3) the UHF constructions get it right.

2. Mathematical Formulation

For fixed-frequency (elliptic), two-way, wave propagation modeling, the inclusion of backscatter effects results in a coupling of the one-way wavefield components. Within the directional wavefield decomposition/phase space/path integral framework discussed in Section 1, there are two complementary approaches to exactly account, in principle, for the "inclusion of range-dependent effects" in the modeling. The first is the generalized Bremmer coupling series, while the second is the exact, well-posed, one-way reformulation of elliptic wave equations in terms of appropriate operators (FISHMAN et al., 2000). Since the former is probably much more familiar to the seismic community, the latter will be taken as the first step in the mathematical formulation and analysis. Following this formally exact reformulation, the phase space and path integral analysis provides for the explicit representation of the relevant operators and the fundamental solution. Finally, the explicit nature of the phase space analysis allows for the uniform asymptotic approximations of the operators, resulting in a true, one-way computational algorithm and an asymptotic relation between the data and the unknown function in an inverse analysis. The exact reformulation, along with the phase space constructions and the path integral representations, are outlined in this section; the construction of the uniform asymptotic approximations is presented in Section 4.

The modeling is provided by the d-dimensional, scalar Helmholtz equation,

$$\{\partial_x^2 + \nabla_t^2 + \bar{k}^2 K^2(\underline{x})\} \phi(\underline{x}) = -\delta(\underline{x} - \underline{x}'_s) , \qquad (1)$$

where $\bar{k}$ is a reference wave number, $K(\cdot)$ is the refractive index field, the range coordinate x denotes the principal propagation direction, $\underline{x}_t$ denotes the remaining $d-1$ Cartesian coordinates perpendicular to x, and the vector $\underline{x} = \{x, \underline{x}_t\}$. Furthermore, as depicted in Figure 1, the Euclidean space $\mathbb{R}^d$ is partitioned into three contiguous regions with respect to x, a left $(x < a)$ and right $(x > b)$, transversely-inhomogeneous half-space and a central transition region of general variability of width $b - a$. For seismic modeling, the figure can be rotated through ninety degrees.

Directional wavefield decomposition essentially decomposes the total wavefield into its right- and left-traveling components, which represent the exact, decoupled, one-way wavefields in a transversely-inhomogeneous environment (the half-spaces in Fig. 1), while only providing for a coupled, mathematical representation of the total wavefield in a generally-varying environment (the transition region in Fig. 1). The right- and left-traveling wavefield components are defined in terms of the total wavefield and its normal derivative, and are related by an appropriate set of scattering (reflection and transmission) operators. This decomposition provides a natural "scattering" formulation of the wave propagation process, employing invariant imbedding techniques to derive the governing equations for the scattering operators. Alternatively, the wave propagation process can be re-expressed in terms of the total wavefield and its normal derivative and the resulting set of Dirichlet-to-Neumann and associated propagator operators. This complementary representation of the wave propagation process provides a natural "boundary-value" formulation.

The "boundary-value," exact, one-way reformulation of the scalar Helmholtz equation, expressed in terms of the total wavefield, is outlined here. For the internal wavefield in the transition region (i.e., the center layer), given the exact, initial total

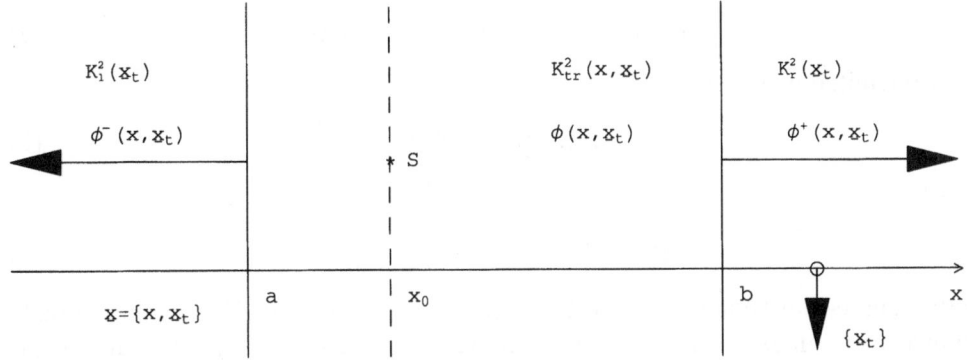

Figure 1
Model geometry. $\mathbb{R}^d$ is divided into three regions with respect to a principal, global propagation, or range, coordinate x. The remaining Cartesian coordinates perpendicular to x are denoted by $\{\underline{x}_t\}$ and S denotes the source. In the two half-spaces, K depends only on $\{\underline{x}_t\} (\Rightarrow$ a transversely-inhomogeneous environment), while in the transition region $a < x < b$, K is a function of both the range x and the transverse coordinates $\{\underline{x}_t\}$. For the scattering problem, sources are located in the half-spaces.

wavefield, $\phi(x_0, \underline{x}_t)$, (the determination of which is part of the problem), the propagation from x_0 in the $+x$ direction is governed by

$$((1/\bar{k})\partial_x + \Lambda^+(x,b))\phi(x,\underline{x}_t) = 0 , \tag{2}$$

while the propagation in the $-x$ direction from x_0 is governed by

$$((1/\bar{k})\partial_x - \Lambda^-(a,x))\phi(x,\underline{x}_t) = 0 , \tag{3}$$

where $\Lambda^\pm$ are the DtN operators acting on $\underline{x}_t$. Furthermore, Eqs. (2) and (3) provide exact reflecting boundary conditions, which follows directly from the definition of the DtN operators. Hence, for a subdomain $[x_1, x_2]$ of the transition region containing x_0, the Helmholtz equation (1) supplemented with the boundary conditions

$$\begin{cases} ((1/\bar{k})\partial_x + \Lambda^+(x_2,b))\phi(x_2,\underline{x}_t) = 0, \\ \\ \qquad\qquad\qquad\qquad\qquad\qquad\qquad x \in [x_1,x_2] \subset [a,b] , \\ \\ ((1/\bar{k})\partial_x - \Lambda^-(a,x_1))\phi(x_1,\underline{x}_t) = 0, \end{cases} \tag{4}$$

provide an exact statement of the original problem in the subdomain. The boundary conditions given in Eq. (4) account precisely for the reflected energy. Moreover, the DtN operators satisfy Riccati operator equations of the form

$$(1/\bar{k})\partial_x\Lambda^+(x,b) = (\Lambda^+(x,b))^2 + \mathbf{B}^2(x) \tag{5}$$

with the initial condition

$$\Lambda^+(b,b) = -i\mathbf{B}(b) , \tag{6}$$

and

$$-(1/\bar{k})\partial_x\Lambda^-(a,x) = (\Lambda^-(a,x))^2 + \mathbf{B}^2(x) \tag{7}$$

with the initial condition

$$\Lambda^-(a,a) = -i\mathbf{B}(a) , \tag{8}$$

where

$$\mathbf{B}(x) = [K^2(x,\underline{x}_t) + (1/\bar{k}^2)\nabla_t^2]^{1/2} \tag{9}$$

is the square-root Helmholtz operator. Equations (2), (3), and (5)–(9) are essentially equivalent to an exact factorization of the Helmholtz equation. The closure of this exact reformulation is provided by expressing the initial total wavefield, $\phi(x_0,\underline{x}_t)$, in terms of $\Lambda^\pm$ in the form

$$\phi(x_0,\underline{x}_t) = (1/\bar{k})\{\Lambda^+(x_0,b) + \Lambda^-(a,x_0)\}^{-1}\delta(\underline{x}_t - \underline{x}_t^S) , \tag{10}$$

which fully accounts for the multiple scattering.

This exact reformulation of the Helmholtz equation (1) is a particular mathematical realization of a double-sweep method. The original, two-way Helmholtz equation, which is ill-posed for marching in the range direction, has been transformed into one-way equations on the DtN operator and the total wavefield which are successively marched, in a well-posed manner, in opposite directions to solve the elliptic problem. This reformulation can also be viewed, equivalently, in terms of solving a succession of fictitious internal scattering problems in the complementary "scattering" representation. This viewpoint establishes the physical content of Eqs. (2)–(10). The construction of the DtN operators essentially can be reduced to the scattering problem of two, single blocks (representing the intervals $[a, x_0]$ and $[x_0, b]$), subsequently followed by the multiple-scattering problem between the two blocks when the individual solutions are "glued together" to obtain the correct initial total wavefield. The first sweep, Eqs. (5)–(8), effectively solves the series of fictitious internal scattering problems that the wave encounters on propagating into the block. The solution of the scattering problem exactly decouples the wavefield components, leading to the one-way wave equations. The solution of the multiple-scattering problem to establish the initial total wavefield is concisely given in Eq. (10), which exactly accounts for all of the reflections between the two blocks. The subsequent, one-way marching procedure in Eqs. (2) and (3) of the second sweep is well-posed because the solution of the internal scattering problems encountered by the propagating wave eliminates the ill-posed backpropagation required to simultaneously march the total wavefield and its normal derivative. Moreover, the initial data is consistent with the final one-way equation, unlike the case of the simultaneous marching of the total wavefield and its normal derivative, which requires non-independent initial data. This heuristic, physical picture is rigorously supported by the spectral analysis of the operators.

Figure 1 provides the basis for a thought experiment, which illustrates these points in the context of seismic wave propagation. Imagine wanting to compute the total wavefield, in either the transition or half-space regions, resulting from the point source in Figure 1. Equations (2) and (3) then provide the, in principle, exact, one-way wave equations for marching the general Helmholtz total wavefield to both the right and left of the source, respectively. Once the DtN operators have been constructed from Eqs. (5)–(8) and the initial total wavefield formed from Eq. (10), a single sweep (marching) in the range direction with Eqs. (2) and (3) completely provides for all of the internal reflections and transmissions normally associated with the multiple sweeps, in the tracking of the "multiples," in the generalized Bremmer coupling series formulation. The DtN operators contain all of the relevant information in the range-dependent problem, which accounts for their role in the exact reflecting boundary conditions in Eq. (4). In approximate formulations, the single sweep in the range direction, for the computation of the total wavefield, is always maintained; only the DtN operator is appropriately approximated. This is always a single sweep (marching) method at the level of the total wavefield.

The formal operator equations are explicitly constructed and analyzed in the Weyl pseudodifferential operator calculus. These explicit constructions enable the numerical computation of both the DtN, Riccati operator equations (5)–(8) and the initial total wavefield in Eq. (10) to, ultimately, be replaced by asymptotic methods, resulting in a true, one-way computational procedure for the inherently, two-way Helmholtz problem. The remainder of the paper considers only the transversely-inhomogeneous (range-independent) limiting case, with $K^2(x, \underline{x}_t) = K^2(\underline{x}_t)$ and $\Lambda^+ \to -i\mathbf{B}$. Equation (2) then reduces to

$$\left\{ (i/\bar{k})\partial_x + [K^2(\underline{x}_t) + (1/\bar{k}^2)\nabla_t^2]^{1/2} \right\} \phi^+(x, \underline{x}_t) = 0 \ , \tag{11}$$

where $\phi^+(x, \underline{x}_t)$ denotes the right-traveling wavefield, and represents the formally exact wave equation for propagation in a transversely-inhomogeneous half-space supplemented with appropriate right-traveling-wave radiation and initial-value conditions.

Equation (11) can be expressed as a Weyl pseudodifferential equation in the form

$$(i/\bar{k})\partial_x \phi^+(x, \underline{x}_t) + \left(\frac{\bar{k}}{2\pi}\right)^{d-1} \int_{\mathbb{R}^{2d-2}} d\underline{x}_t' \, d\underline{p}_t \, \Omega_\mathbf{B}(\underline{p}_t, (\underline{x}_t + \underline{x}_t')/2)$$
$$\times \exp[i\bar{k}\underline{p}_t \cdot (\underline{x}_t - \underline{x}_t')]\phi^+(x, \underline{x}_t') = 0 \ , \tag{12}$$

where $\Omega_\mathbf{B}(\underline{p}, \underline{q})$ is the symbol for the square-root Helmholtz operator $\mathbf{B} = [K^2(\underline{q}) + (1/\bar{k}^2)\nabla_q^2]^{1/2}$. The Helmholtz, Weyl composition equation in the pseudodifferential operator (ΨDO) calculus is given by

$$\Omega_{\mathbf{B}^2}(\underline{p}, \underline{q}) = K^2(\underline{q}) - \underline{p}^2 = \left(\frac{\bar{k}}{\pi}\right)^{2d-2} \int_{\mathbb{R}^{4d-4}} d\underline{t} \, d\underline{s} \, d\underline{v} \, d\underline{u}$$
$$\times \Omega_\mathbf{B}(\underline{t} + \underline{p}, \underline{s} + \underline{q}) \, \Omega_\mathbf{B}(\underline{v} + \underline{p}, \underline{u} + \underline{q}) \exp[2i\bar{k}(\underline{s} \cdot \underline{v} - \underline{t} \cdot \underline{u})] \ , \tag{13}$$

with $\Omega_{\mathbf{B}^2}(\underline{p}, \underline{q})$ the operator symbol associated with $\mathbf{B}^2$. Equation (13) and the corresponding right-traveling-wave radiation condition define the square-root Helmholtz operator symbol through the construction of the square-root of the operator from its square.

The exact pseudodifferential evolution equation (12) and the approximate equations derived from the analysis of the composition equation (13) are singular integrodifferential wave equations. Solution representations for such pseudodifferential equations can be directly expressed as infinite-dimensional functional, or path, integrals in terms of the operator symbols.

The path integral representation for the fundamental solution (propagator) to Eq. (12) is given by

$$G^+(x, \underline{x}_t; 0, \underline{x}_t') = \lim_{N \to \infty} \int_{\mathbb{R}^{(d-1)(2N-1)}} \prod_{j=1}^{N-1} \mathrm{d}\underline{x}_{jt} \prod_{j=1}^{N} (\bar{k}/(2\pi))^{d-1} \mathrm{d}\underline{p}_{jt}$$

$$\times \exp\left[i\bar{k} \sum_{j=1}^{N} (\underline{p}_{jt} \cdot (\underline{x}_{jt} - \underline{x}_{j-1t}) + (x/N)h_{\mathbf{B}}^s(\underline{p}_{jt}, \underline{x}_{jt})) \right] , \quad (14)$$

where

$$h_{\mathbf{B}}^s(\underline{p}, \underline{q}) = (\bar{k}/\pi)^{d-1} \int_{\mathbb{R}^{2d-2}} \mathrm{d}\underline{s}\, \mathrm{d}\underline{t}\, \Omega_{\mathbf{B}}(\underline{s}, \underline{t}) \exp\{-2i\bar{k}(\underline{q} - \underline{t}) \cdot (\underline{p} - \underline{s})\} \quad (15)$$

is the standard pseudodifferential operator symbol. While either the Weyl or the standard pseudodifferential operator calculus provides a complete description of the wave propagation problem, they are, in many respects, complementary, and can be used in conjunction to advantage. For example, the symmetry inherent in the Weyl calculus can often be exploited in operator symbol constructions and analysis (particularly that involving integrated energy-flux conservation calculations and that separating the effects due to anisotropy from those due to heterogeneity), while the standard calculus naturally results in more computationally efficient numerical algorithms.

The one-way, phase space, marching numerical algorithm is based upon (1) the marching range step (following from the path integral), (2) a detailed, uniform, operator symbol analysis (reflecting the study of Eq. (13)), and (3) Fourier component, or wave number, filtering in phase space (for increased efficiency, decreased computational time, and reduced error). Sufficiently accurate approximations of the pseudodifferential operator symbol over the relevant region of phase space result in very accurate numerical wavefield calculations. Computing Helmholtz equation wavefields as high (in principle, infinite)-dimensional integrals is in sharp contrast to the more traditional finite-difference and finite-element numerical algorithms.

The square-root Helmholtz operator symbol can be constructed in terms of spectral (modal) summations and contour-integral representations in the complex plane. For the particular case $d = 2$, the operator symbol $\Omega_{\mathbf{B}}$ is constructed from the composition relation

$$\Omega_{\mathbf{B}}(p, q) = \frac{\bar{k}^2}{\pi^2} \int_{\mathbb{R}^4} \mathrm{d}t\, \mathrm{d}s\, \mathrm{d}v\, \mathrm{d}u\, \Omega_{\mathbf{B}^{-1}}(t + p, s + q)\, \Omega_{\mathbf{B}^2}(v + p, u + q) \exp[2i\bar{k}(sv - tu)] ,$$

$$(16)$$

where

$$\Omega_{\mathbf{B}^2}(p, q) = K^2(q) - p^2 , \quad (17)$$

$$\Omega_{\mathbf{B}^{-1}}(p, q) = \int_{\mathbb{R}} \mathrm{d}u\, \exp[i\bar{k}pu]\, \mathscr{B}^{-1}\left(q - \frac{u}{2}, q + \frac{u}{2}\right) , \quad (18)$$

and

$$\mathscr{B}^{-1}(z, z') = -2i\bar{k}G(0, z; 0, z') \ . \tag{19}$$

Here, $\Omega_{\mathbf{B}^{-1}}(p, q)$ denotes the symbol for the inverse square-root Helmholtz operator with the corresponding kernel, $\mathscr{B}^{-1}(z, z')$, and the Helmholtz Green's function, $G(x, z; x', z')$, satisfies

$$\{\partial_x^2 + \partial_z^2 + \bar{k}^2 K^2(z)\} \ G(x, z; x', z') = -\delta(x - x')\delta(z - z') \ , \tag{20}$$

supplemented with an outgoing-wave radiation condition.

The primary aim of the pseudodifferential operator theory is to extend classical Fourier analysis of homogeneous media to inhomogeneous environments. (Indeed, this approach can be viewed as a natural extension of the theory of partial differential operators with variable coefficients.) The principal focus is on the operator symbol, which contains the complete spectral information in just the appropriate manner to lead immediately to the infinitesimal propagator. Essentially, the singular operator (kernel) calculus is replaced by a calculus for well-behaved functions (operator symbols). Moreover, operator symbols are a natural quantity to study. For example, they furnish the "generalized vertical slowness" in geophysics, the natural multidimensional extension of the scattering (reflection and transmission) coefficients in the one-dimensional formulation, and the suitable framework to quantize (semi-) classical theories in quantum physics. While the square-root Helmholtz operator symbol plays a role analogous to that of the Hamiltonian in the quantization of classical dynamics, its properties deviate in a significant way from most dynamical Hamiltonians. This will have profound ramifications in the subsequent asymptotic analysis. A more complete treatment of the material in this section can be found in the literature (FISHMAN et al., 1987, 1997, 2000; FISHMAN and WALES, 1987; FISHMAN, 1992, 1993; DE HOOP, 1996; LU and MCLAUGHLIN, 1996; LU, 1999; LU et al., 2001).

3. Exact Operator Symbol Constructions

In addition to its key role in the development of a "wave-tracing" seismic inversion algorithm, the square-root Helmholtz operator symbol plays a pivotal role in the parabolic equation method employed in seismic migration. The explicit construction of the one-way wave equation and initial total wavefield, path-integral solution representation and subsequent numerical algorithm, computational boundary conditions, and fundamental inverse relationship depend crucially upon the analysis and subsequent properties of the square-root Helmholtz operator symbol. The extension to the scattering and DtN operator symbols in the fully-coupled, two-way, elliptic formulation only reinforces this importance. From this perspective, the construction of exact, square-root Helmholtz operator symbols is of great value in illuminating the general mathematical propagation theory, in addition to providing benchmark solutions for both asymptotic and numerical operator symbol constructions. Furthermore, the explicit construction of these nontrivial symbols, corresponding to the fractional power of the indefinite, transverse Helmholtz operator, is of mathematical interest in its own right. Since the relevant (frequency-domain) operators lie outside of the well-developed theory of elliptic pseudodifferential operators, a new, asymptotic, operator symbol characterization is required. This construction is essentially accomplished by incorporating complex and spectral analyses into the calculus of pseudodifferential operators, following from the exact solution methods.

The construction of exact, square-root Helmholtz operator symbols, for specific profiles, follows from Eqs. (16)–(20) and the detailed methods illustrated in FISHMAN (1992), FISHMAN et al. (2000) for the two-dimensional case. The exact symbol constructions can be divided into smooth and nonsmooth profiles. Examples of the former include the generalized Epstein profile and the focusing and defocusing quadratic profiles, whereas those of the latter, which have significant application in physical modeling, are largely represented by the piecewise-constant cases. In this section, the exact, square-root Helmholtz operator symbols for the focusing and defocusing quadratic, delta function, discontinuity, and three-layer composite profiles are given and briefly analyzed through the display of their operator symbol surface plots. These profiles have wide application in acoustic, electromagnetic, seismic, optical, and quantum mechanical modeling, where layered environments and barrier and well potentials are common (FISHMAN et al., 1997, 2000).

3.1. Focusing and Defocusing Quadratic Profiles

The focusing and defocusing quadratic profiles are defined, respectively, by $K^2(q) = K_0^2 \mp \omega^2 q^2$, with $K_0, \omega > 0$. The operator symbols can be represented by both spectral (modal) summations and contour-integral representations.

For the focusing case, the spectral (modal) summation representation for the inverse square-root Helmholtz operator kernel is given by

$$\mathscr{B}^{-1}(z, z') = -\exp\left(\frac{3}{4}\pi i\right) \frac{\bar{k}}{\pi^{1/2}} \sum_{n=0}^{\infty} \frac{1}{n!} \frac{1}{[i(2n+1-Y)]^{1/2}} 2^{-n} \Theta_n(z, z') , \qquad (21)$$

with

$$\Theta_n(z, z') = \phi_n((\omega\bar{k})^{1/2}z)\phi_n((\omega\bar{k})^{1/2}z') , \quad \phi_n(\zeta) = \exp\left(-\tfrac{1}{2}\zeta^2\right)H_n(\zeta) , \qquad (22)$$

where $H_n(\cdot)$ is the Hermite polynomial of degree n, $Y = K_0^2/\varepsilon$, and $\varepsilon = \omega/\bar{k}$ is a dimensionless parameter which measures the medium variation on the wavelength scale. With the effective refractive index associated with mode n given by $[-\varepsilon(2n+1-Y)]^{1/2}$, the number of propagating modes, L, is obtained from the estimate $2L - 1 < Y < 2L + 1$. In Eq. (21), the principal value of the square-root function is taken, consistent with the right-traveling-wave condition, enforcing the radiation condition at infinity to be satisfied in the usual manner. The series representation is understood in the distributional sense.

The expression for the standard symbol $h_{\mathbf{B}}^s(p, q)$, of the square-root Helmholtz operator, can be written as an absolutely and uniformly convergent infinite series and a well-defined integral in the form

$$h_{\mathbf{B}}^{s;\text{foc}}(p, q) = (2\varepsilon)^{1/2} \sum_{n=0}^{\infty} \frac{1}{n!} \left(\frac{i}{2}\right)^n \Psi_n(p, q)\left\{[-(2n + 1 - Y)]^{1/2}\right.$$
$$+ 2i\left(\frac{\alpha}{\pi}\right)^{1/2}[-(2n + 1 - Y)]_1F_1(1/2; 3/2; -\alpha(2n + 1 - Y))\Big\}$$
$$- i\left(\frac{\varepsilon}{\pi}\right)^{1/2} \int_0^\alpha dt\, t^{-1/2} \exp(Yt)\, \mathrm{F}(t)\ , \tag{23}$$

where $\alpha > 0$ can be used to control the magnitude of terms in numerical evaluations,

$$_1F_1(1/2; 3/2; -\alpha(2n + 1 - Y)) = \frac{1}{2}\alpha^{-1/2} \int_0^\alpha dt\, t^{-1/2} \exp[-(2n + 1 - Y)t] \tag{24}$$

is the incomplete gamma function expressed as a confluent hypergeometric function,

$$\mathrm{F}(t) = \exp\left[-\tfrac{1}{2}\chi \tanh(2t) + iZ(\text{sech}(2t) - 1)\right] (\text{sech}(2t))^{1/2}$$
$$\times [Y - \chi (\text{sech}(2t))^2 - 2iZ \,\text{sech}(2t)\tanh(2t) - \tanh(2t)]\ , \tag{25}$$
$$\Psi_n(p, q) = \exp(-i\bar{k}\, pq)\phi_n((\omega\bar{k})^{1/2}q)\phi_n((\bar{k}/\omega)^{1/2}p)\ , \tag{26}$$

$\chi = \varepsilon^{-1}(\omega^2 q^2 + p^2)$, and $Z = \varepsilon^{-1}\omega\, qp$. The Weyl symbol, $\Omega_{\mathbf{B}}(p, q)$, of the square-root Helmholtz operator is found to be

$$\Omega_{\mathbf{B}}^{\text{foc}}(p, q) = 2\varepsilon^{1/2} \sum_{n=0}^{\infty} \Phi_n(\chi)\left\{[-(2n + 1 - Y)]^{1/2}\right.$$
$$+ 2i\left(\frac{\alpha}{\pi}\right)^{1/2}[-(2n + 1 - Y)]_1F_1(1/2; 3/2; -\alpha(2n + 1 - Y))\Big\}$$
$$- i\left(\frac{\varepsilon}{\pi}\right)^{1/2} \int_0^\alpha dt\, t^{-1/2} \exp(Yt)\, \tilde{\mathrm{F}}(t)\ , \tag{27}$$

where

$$\tilde{\mathrm{F}}(t) = \exp[-\chi \tanh(t)]\, \text{sech}(t)\, [Y - \chi (\text{sech}(t))^2 - \tanh(t)] \tag{28}$$

and

$$\Phi_n(\chi) = \exp(-\chi)(-)^n L_n^{(0)}(2\chi)\ , \tag{29}$$

with $L_n^{(0)}(\cdot)$ denoting the simple Laguerre polynomial of degree n. The singular part of the operator kernel is contained in the integral term in both Eqs. (23) and (27).

With $\zeta(\sigma, \Delta, \xi)$ denoting the Lerch transcendental function,

$$\zeta(\sigma, \Delta, \xi) = \sum_{n=0}^{\infty} (n+\Delta)^{-\sigma} \xi^n, \quad \Delta \neq 0, -1, -2, \ldots, \quad |\xi| < 1 \ , \tag{30}$$

and appropriate analytic continuations, contour-integral representations, equivalent to the spectral sums, can be constructed. The final results are

$$h_{\mathbf{B}}^{s;\mathrm{foc}}(p,q) = -\exp\left(\frac{1}{4}\pi i\right)\left(\frac{\varepsilon}{2}\right)^{1/2}\frac{1}{\pi}\int_{\mathscr{L}} d\tau\, \zeta(1/2, (-i/(2\pi))\tau, \exp(2\pi i Y))$$

$$\times \exp\left[Y\tau - \frac{1}{2}\chi\tanh(2\tau) + iZ(\mathrm{sech}(2\tau) - 1)\right](\mathrm{sech}(2\tau))^{1/2}$$

$$\times [Y - \chi(\mathrm{sech}(2\tau))^2 - 2iZ\,\mathrm{sech}(2\tau)\tanh(2\tau) - \tanh(2\tau)],$$

$$Y \neq 0, 1, 2, \ldots, \tag{31}$$

for the standard symbol and

$$\Omega_{\mathbf{B}}^{\mathrm{foc}}(p,q) = -\exp\left(\frac{1}{4}\pi i\right)\left(\frac{\varepsilon}{2}\right)^{1/2}\frac{1}{\pi}\int_{\mathscr{L}} d\tau\, \zeta(1/2, (-i/(2\pi))\tau, \exp(2\pi i Y))$$

$$\times \exp\left[Y\tau - \chi\tanh(\tau)\right]\mathrm{sech}(\tau)\,[Y - \chi(\mathrm{sech}(\tau))^2 - \tanh(\tau)],$$

$$Y \neq 0, 1, 2, \ldots, \tag{32}$$

for the Weyl symbol. In Eq. (32), the contour $\mathscr{L}$ starts at $\tau = 0$ and ends at $\tau = i2\pi$, keeping the integrand singularities, which include branch points (associated with ζ) at $\tau = -i2\pi n$ ($n = 0, 1, 2, \ldots$) with the associated branch lines chosen to lie on the negative imaginary τ axis, and isolated singularities (associated with $\mathrm{sech}(\cdot)$ and $\tanh(\cdot)$) at $\tau = i\pi(2n+1)/2$ ($n = \ldots, -2, -1, 0, 1, 2, \ldots$), "outside" the contour with respect to the half-plane $\mathrm{Re}[\tau] > 0$. The presence of the Lerch transcendental function and the integration endpoints of 0 and $i2\pi$ is a reflection of the periodicity of the underlying, quantum mechanical, harmonic oscillator problem.

For the defocusing case, the contour-integral representations follow immediately from the expressions in Eq. (31) and (32). The Weyl symbol, for example, is given by

$$\Omega_{\mathbf{B}}^{\mathrm{def}}(p,q) = -\left(\frac{\varepsilon}{2}\right)^{1/2}\frac{1}{\pi}\int_{\mathscr{L}} d\tau\, \zeta(1/2, (-i/(2\pi))\tau, \exp(-2\pi Y))$$

$$\times \exp\left[i(Y\tau + X\tanh(\tau))\right]\mathrm{sech}(\tau)\,[iY + iX(\mathrm{sech}(\tau))^2 - \tanh(\tau)],$$

$$Y \neq 0 \ , \tag{33}$$

and readily reduces to the simplified, real integral form,

$$\Omega_{\mathbf{B}}^{\text{def}}(p,q) = -\exp[i\pi/4](\varepsilon/\pi)^{1/2}$$

$$\times \int_0^\infty dt \, \exp[i(Yt + X\tanh(t))] \, t^{-1/2}\text{sech}(t)$$

$$\times [iY + iX(\text{sech}(t))^2 - \tanh(t)] \,, \tag{34}$$

where $X = \varepsilon^{-1}(\omega^2 q^2 - p^2)$. The contour-integral representations provide the analytic continuation between the focusing and defocusing cases.

The standard operator symbol surface for the focusing case is illustrated in Figure 2, where Eq. (31) is plotted for the choice of $K_0 = 1$, $\omega = 1$, and $\bar{k} = 10.5$. The surface in the entire phase space follows from the symmetry relation $h_{\mathbf{B}}^{s;\,\text{foc}}(p,q) = h_{\mathbf{B}}^{s;\,\text{foc}}(-p,-q)$, reflecting the asymmetry in the underlying operator-ordering inherent in the standard calculus. The real and imaginary parts of the surface essentially consist of an oscillatory structure superimposed on the locally-homogeneous medium surface. The oscillatory component is confined to the locally-propagating part of the spectrum in the phase space. The surface is regular; there is no square-root singularity. Asymptotic analysis establishes that the oscillatory period can be related to the number of propagating modes and the details of the underlying quadratic profile. Figure 3 depicts the analogous results for the Weyl operator symbol surface, following from Eq. (32), for the identical case. The surface is an even function of p, following from the inherent symmetry in the Weyl calculus, and an even function of q, following from the even symmetry in q of the quadratic profile. In this case, the surface is a radial function of the quadratic form. Similar comments to those characterizing the standard operator symbol surface apply in the Weyl case with regard to the singularity structure and the nature of the oscillatory and locally-homogeneous medium (algebraic) surface components. Following from Eq. (34), Figure 4 displays the Weyl operator symbol surface for the defocusing case, for the choice of $K_0 = 1$, $\omega = 1$, and $\bar{k} = 10.5$. The principal difference between the focusing and defocusing cases for the Weyl operator symbol surface is the confinement of the oscillatory component to the locally-evanescent part of the spectrum in the phase space in the defocusing case. This complementary structure is a reflection of the analytic continuation between the focusing and defocusing quadratic profiles, and is a general feature of the operator symbol surfaces connecting complementary focusing and defocusing models. A more detailed treatment of the focusing and defocusing quadratic profile cases is available in the literature (FISHMAN, 1992; FISHMAN et al., 2000; JORDAN, 2001).

3.2. Delta Function and Piecewise-constant Profiles

While the operator symbols for these cases can also be constructed through spectral (modal) summation and contour-integral representations in both the standard and Weyl calculi, only one realization of the integral representations for the Weyl calculus will be briefly discussed. The complete details of these constructions will be presented in the mathematical literature.

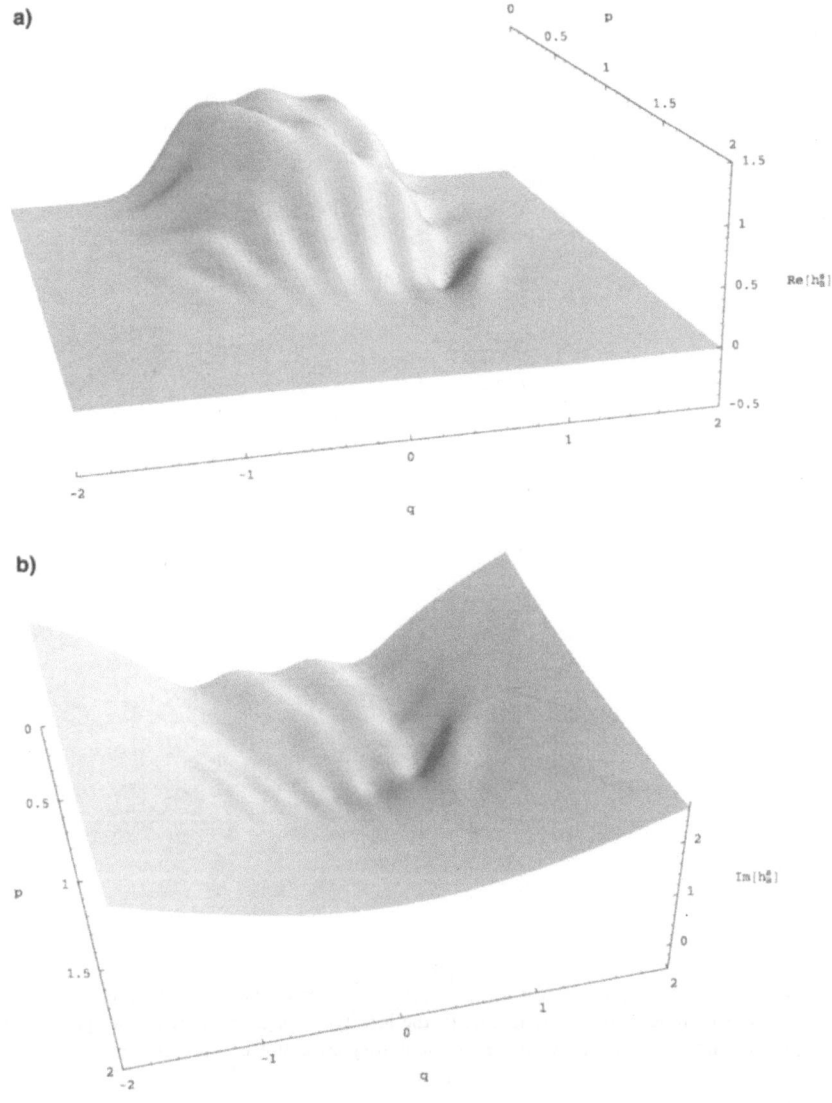

Figure 2
(a) $\mathrm{Re}[h_{\mathbf{B}}^{s;\,\mathrm{foc}}(p,q)]$ and (b) $\mathrm{Im}[h_{\mathbf{B}}^{s;\,\mathrm{foc}}(p,q)]$ for $K^2(q) = K_0^2 - \omega^2 q^2$, with $K_0 = \omega = 1$ and $\bar{k} = 10.5$. The oscillatory structure is confined, characteristically, to the locally-propagating part of the spectrum in the phase space for this smooth focusing profile. This oscillatory structure is representative of the trapped modes.

For the delta function profile, the square of the refractive index field is given by $K^2(q) = K_0^2 + 2\lambda\delta(q)$, where $\delta(\cdot)$ is the Dirac delta function and $K_0(>0)$ and λ are real constants. For this case, the exact operator symbol for the square-root Helmholtz operator is given by

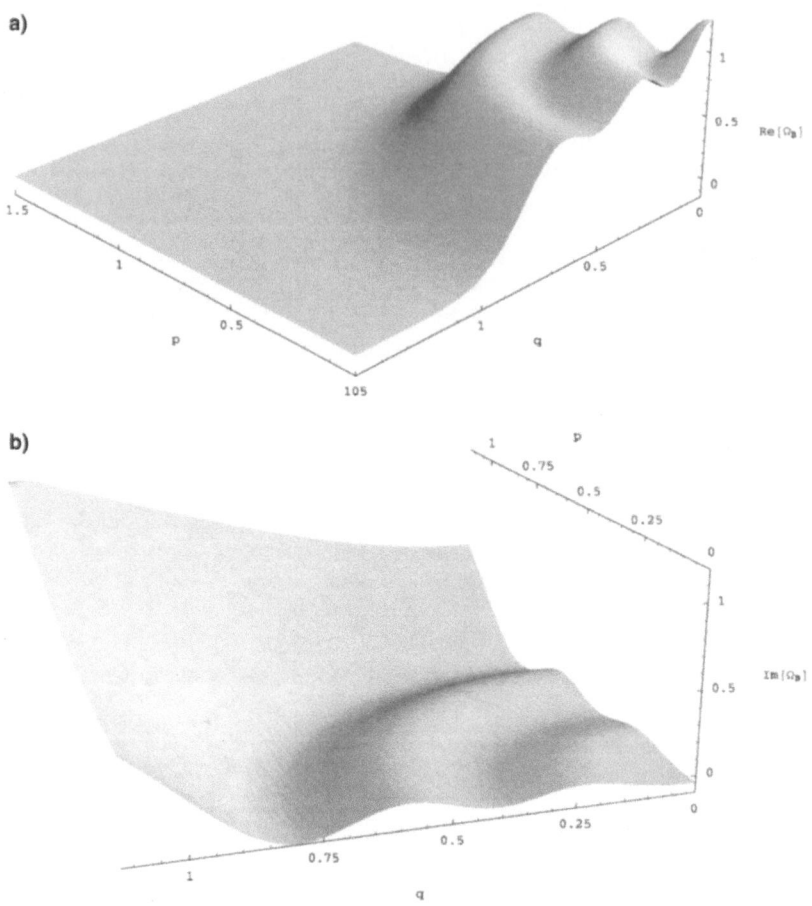

Figure 3

(a) $\mathrm{Re}[\Omega_{\mathbf{B}}^{\mathrm{foc}}(p,q)]$ and (b) $\mathrm{Im}[\Omega_{\mathbf{B}}^{\mathrm{foc}}(p,q)]$ for $K^2(q) = K_0^2 - \omega^2 q^2$, with $K_0 = \omega = 1$ and $\bar{k} = 10.5$. The oscillatory structure is confined, characteristically, to the locally-propagating part of the spectrum in the phase space for this smooth focusing profile. This oscillatory structure is representative of the trapped modes.

$$\Omega_{\mathbf{B}}(p,q) = [K_0^2 - p^2]^{1/2} - \frac{2i\lambda\bar{k}\,e^{2iK_0Q}}{\pi} \int_0^\infty dt \left\{ \frac{e^{-2Qt}t^{1/2}[t - 2iK_0]^{1/2}}{t - (\lambda\bar{k} + iK_0)} \right.$$

$$\left. \times \left[\frac{\sin[2pQ]}{p} + \frac{1}{2}\left(\frac{e^{2ipQ}}{t - i(K_0 + p)} + \frac{e^{-2ipQ}}{t - i(K_0 - p)} \right) \right] \right\} , \qquad (35)$$

where $Q = \bar{k}|q|$. The operator symbol surface is illustrated in Figure 5, where Eq. (35) is plotted for the choice of $K_0 = 1$, $\lambda = 10$, and $\bar{k} = 1$. The result in Eq. (35) is an even function of p, following from the inherent symmetry in the Weyl calculus, and an even function of q, following from the even symmetry in q of the delta function profile. The real and imaginary parts of the surface essentially consist of an intricate,

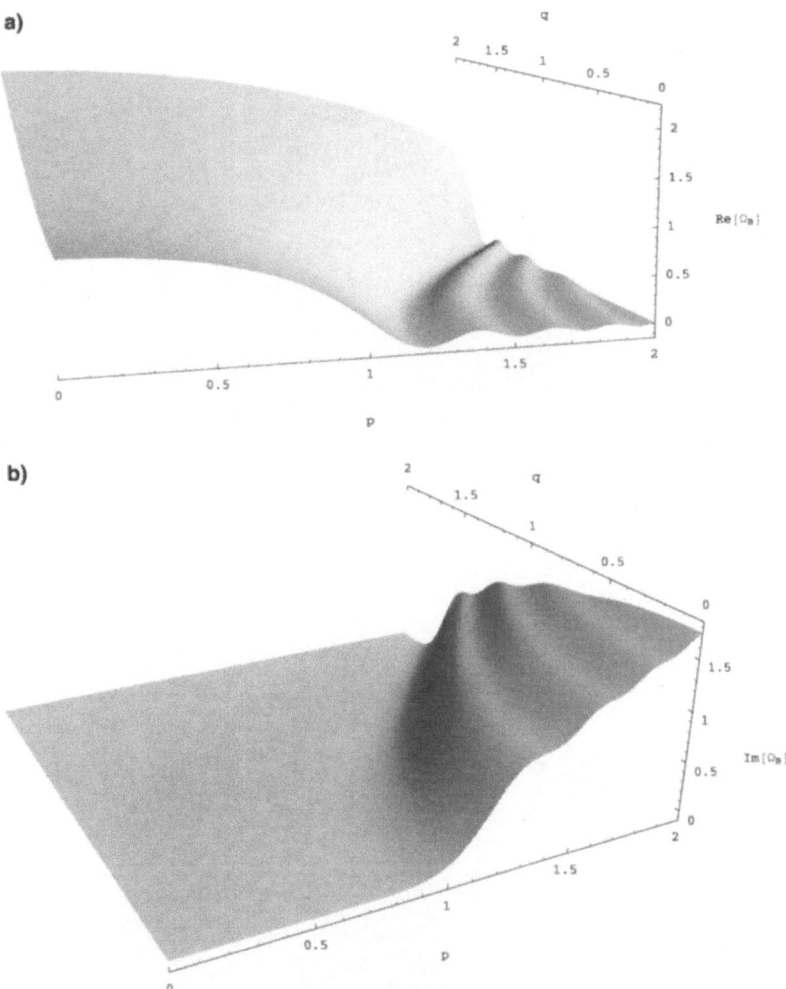

Figure 4

(a) $\mathrm{Re}[\Omega_{\mathbf{B}}^{\mathrm{def}}(p,q)]$ and (b) $\mathrm{Im}[\Omega_{\mathbf{B}}^{\mathrm{def}}(p,q)]$ for $K^2(q) = K_0^2 + \omega^2 q^2$, with $K_0 = \omega = 1$ and $\bar{k} = 10.5$. The oscillatory structure is confined, characteristically, to the locally-evanescent part of the spectrum in the phase space for this smooth defocusing profile. This oscillatory structure is connected to that in the focusing case by an analytic continuation, and is complementary in nature.

oscillatory structure and a weak, logarithmic singularity in q in the imaginary part, both, superimposed upon the homogeneous medium surface (the first term in Eq. (35)). In principle, the oscillatory structure exists throughout the phase space. The square-root singularity in p occurs at the value of the homogeneous background profile, K_0. The weak, logarithmic singularity at $q = 0$ in the operator symbol results from the singular nature of the delta function profile, and is not fully illustrated in Figure 5(b). The structure in the neighborhood of $q = 0$ is consistent with the single,

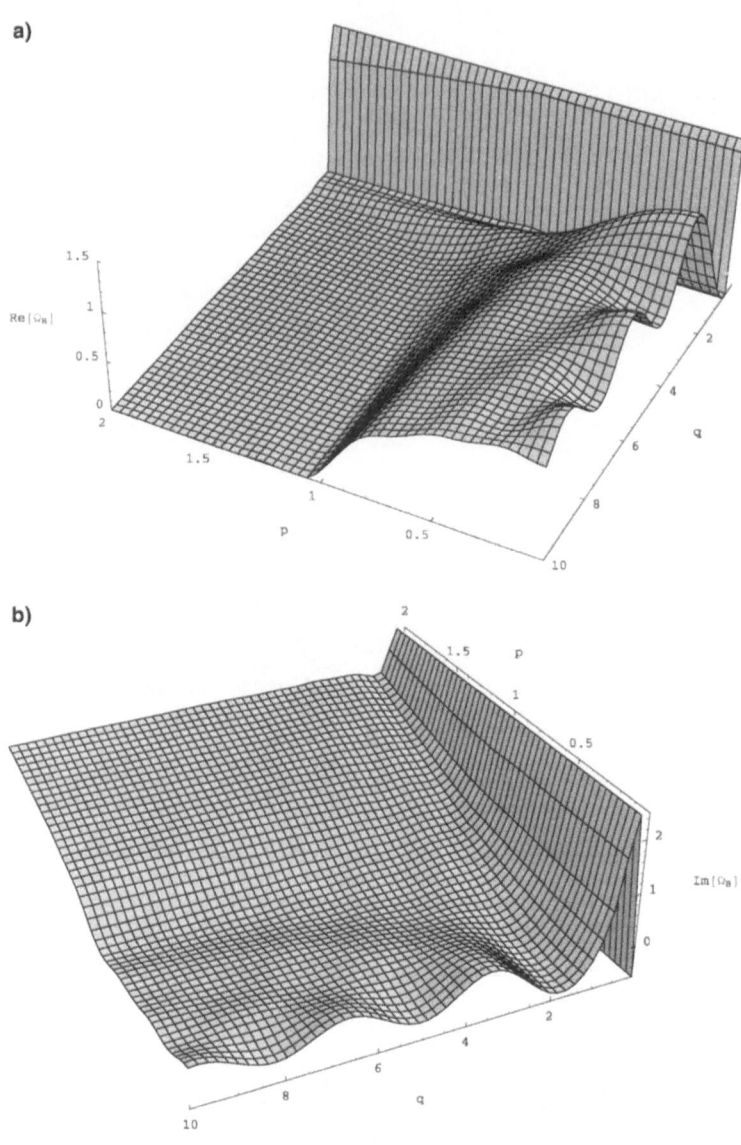

Figure 5

(a) $\mathrm{Re}[\Omega_{\mathbf{B}}(p,q)]$ and (b) $\mathrm{Im}[\Omega_{\mathbf{B}}(p,q)]$ for $K^2(q) = K_0^2 + 2\lambda\delta(q)$, with $K_0 = 1, \lambda = 10$, and $\bar{k} = 1$. The oscillatory structure exists throughout the phase space for this singular profile, while there is a weak, logarithmic singularity in the imaginary part at $q = 0$, and a square-root singularity in p occurring at the value of the homogeneous background refractive index.

trapped mode associated with the (attractive) delta function profile. Note, that as $|q|$ increases, the magnitude of the oscillations decreases, with the operator symbol surface approaching that of the background homogeneous medium. In effect, from a physical perspective, sufficiently far from the delta function, the effects of the disturbance become negligible, as expected. The traditional, elliptic pseudodifferential operator calculus, applied directly to this case, would only yield the homogeneous medium background surface and an undefined, singular quantity at $q = 0$.

The starting point for the piecewise-constant class of profiles is the discontinuity case. Fox this profile, $K^2(q) = K_1^2 + (K_2^2 - K_1^2)H(q)$, where K_1 and K_2 are positive constants and $H(\cdot)$ denotes the Heaviside unit step function. While the complete expression for the operator symbol will not be displayed, it is important to note that both the coordinate, q, and reference wave number, $\bar{k}$, dependencies enter solely through the dimensionless (scaled) variable Q. This reflects the fact that the operator symbol surface properties essentially depend upon the number of wavelengths from the discontinuity in each half-space, as expected. Figures 6(a) and 7(a) depict the exact operator symbol surface plotted as a function of p and Q for the specific choice of $K_1 = 3$, and $K_2 = 1$. Again, resulting from the inherent symmetry in the Weyl calculus, the symbol surface is an even function of p. Roughly speaking, the operator symbol surface can be viewed as the previous, intricate, oscillatory structure superimposed upon the respective homogeneous medium symbol surfaces in each half-space. However, it is important to note that the surface is continuous at $Q = 0$, unlike the original profile. While this result follows rigorously from the mathematical construction, it can be heuristically motivated. For a smooth (infinitely-differentiable) profile, no matter how rapid the variation, there always exists a (wavelength) scale, such that on this and all smaller scales, the profile is slowly changing. This situation holds continuously to the limit, and allows for a high-frequency expansion about the locally-homogeneous limiting form. For a discontinuity, which changes infinitely fast on all finite scales, no matter how small, however, no such scales exist. The result is that a high-frequency expansion about the locally-homogeneous limiting form is no longer appropriate. Moreover, the only surface singularity is a square-root singularity occurring when p is equal to the average value of the two, half-space refractive indices. The surface is nonsingular when p takes on the individual values of the half-space refractive indices. The $p = 0$ slice essentially exhibits an oscillatory and smoothed version of $K(q)$. The oscillatory structure, again, exists, in principle, throughout the phase space, with the symbol surface approaching the homogeneous medium limiting form appropriate in each half-space with increasing Q. The results in Figures 6(a) and 7(a) are compared in Figures 6(b) and 7(b) to the real and imaginary parts, respectively, of the locally-homogeneous medium, operator symbol surfaces in each half-space, $H(-q)[K_1^2 - p^2]^{1/2} + H(q)[K_2^2 - p^2]^{1/2}$, which, in some sense, can be viewed as the usable, nonsingular part resulting from the standard asymptotic analysis (elliptic pseudodifferential operator calculus). It is seen that many of the interesting and, ultimately, essential surface features (for accurate wavefield calculations and conservation of the integrated energy flux) lie outside of the elliptic calculus.

The $Q = 0$ slice of the exact, discontinuity symbol surface deserves special attention. The Weyl operator symbol for this case reduces to

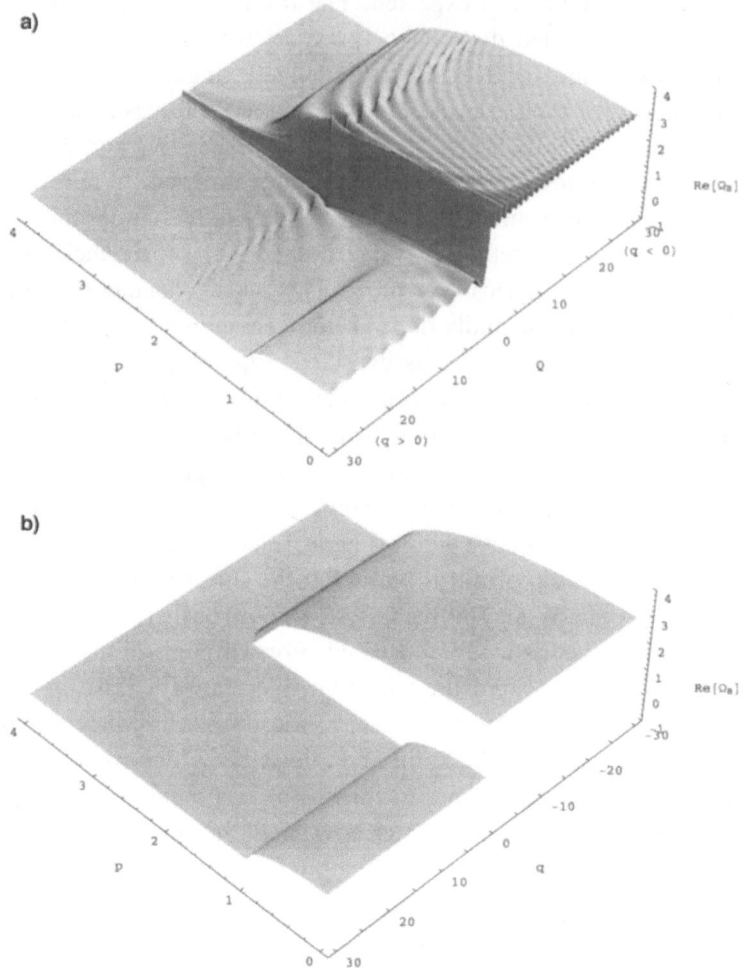

Figure 6

(a) $\text{Re}[\Omega_B(p,q)]$ for $K^2(q) = K_1^2 + (K_2^2 - K_1^2)H(q)$,   with $Q = \bar{k}|q|, K_1 = 3$,   and $K_2 = 1$.   (b) $\text{Re}[H(-q)[K_1^2 - p^2]^{1/2} + H(q)[K_2^2 - p^2]^{1/2}]$. The oscillatory structure exists throughout the phase space for this nonsmooth profile, while the surfaces are continuous at the point of discontinuity in the profile, and the square-root singularity in p occurs at the average value of the two, half-space refractive indices. The standard asymptotic results in (b) are clearly unable to capture the detailed oscillatory and correct singularity structure.

$$\Omega_B(p,0) = H(p^2 - K_A K_D)[K_A^2 - p^2]^{1/2}[1 - (K_D/p)^2]^{1/2}[1 - (K_A K_D/p^2)^2]$$
$$+ \frac{8(K_A + K_D)(K_A K_D)^2}{\pi[p^2 + K_A K_D]^2} \int_0^1 dt \left\{ \frac{t^{1/2}[1 - t]^{1/2}[1 - \alpha t]^{1/2}}{1 - \beta(p)t} \right\}, \tag{36}$$

where $K_{A,D} = [K_> \pm K_<]/2$, $K_> = \max[K_1, K_2], K_< = \min[K_1, K_2]$, and

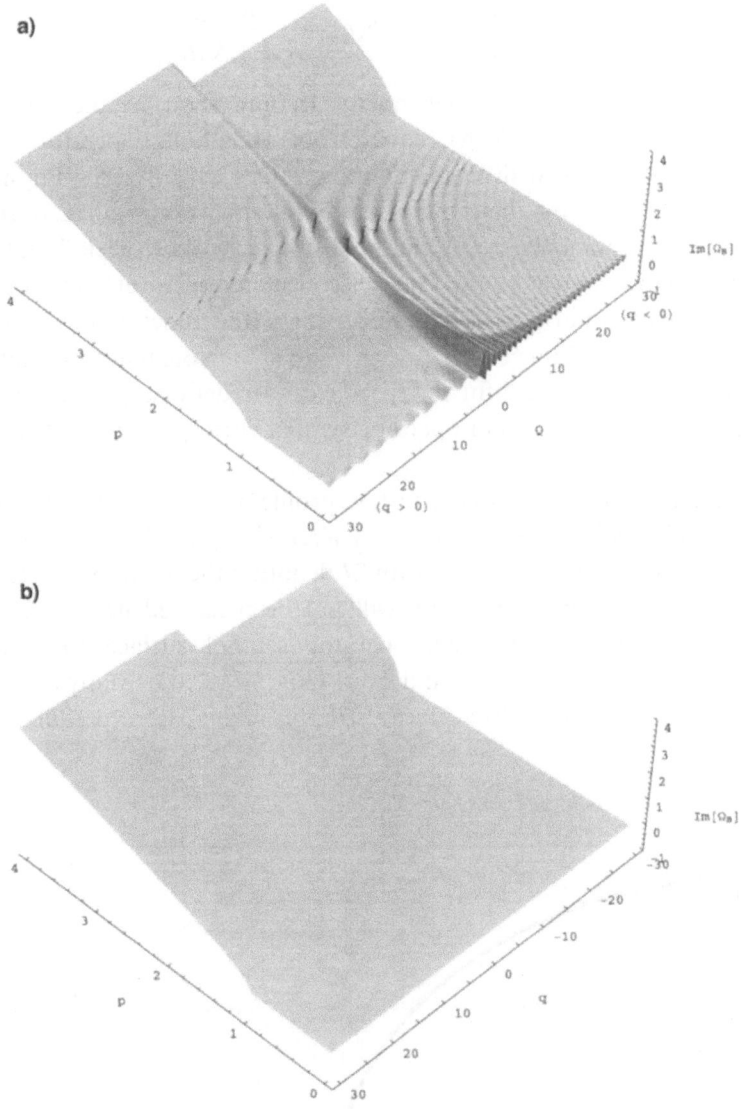

Figure 7

(a) $\text{Im}[\Omega_{\mathbf{B}}(p,q)]$ for $K^2(q) = K_1^2 + (K_2^2 - K_1^2)H(q)$, with $Q = \bar{k}|q|$, $K_1 = 3$, and $K_2 = 1$. (b) $\text{Im}[H(-q)[K_1^2 - p^2]^{1/2} + H(q)[K_2^2 - p^2]^{1/2}]$. The oscillatory structure exists throughout the phase space for this nonsmooth profile, while the surfaces are continuous at the point of discontinuity in the profile, and the square-root singularity in p occurs at the average value of the two, half-space refractive indicies. The standard asymptotic results in (b) are clearly unable to capture the detailed oscillatory and correct singularity structure.

$$\alpha = \frac{4K_A K_D}{(K_A + K_D)^2}, \qquad \beta(p) = \frac{4(K_A K_D)p^2}{(p^2 + K_A K_D)^2}. \tag{37}$$

Equation (36) provides the q-independent, leading term in the uniform, low-frequency expansion of the Weyl, square-root Helmholtz operator symbol for profiles which asymptote to different limiting values as $q \to \pm\infty$. Physically, this follows from the observation that, in the limit of infinite wavelength (zero frequency), only the two asymptotic half-spaces can be distinguished, or "seen," and, thus, the leading expansion term must be the exact discontinuity operator symbol in the limit of $\bar{k} \to 0$, i.e., evaluated at $Q = 0$. Not surprisingly, this function of p only depends upon the average and difference refractive indices, $K_{A,D}$. Note that this result, given in Eq. (36) and shown in Figure 8 for the specific case under discussion, is not equal to that of the homogeneous medium operator symbol corresponding to the average of the two refractive indices.

The last example of the piecewise-constant profiles presented here is the three-layer composite profile, $K^2(q) = K_{+\infty}^2 H(q - l) + K_1^2 H(l^2 - q^2) + K_{-\infty}^2 H(-q - l)$, where $K_1, K_{\pm\infty}$, and l are positive constants with $2l$ denoting the width of the center layer. The exact operator symbol expression will not be presented here; rather, Figures 9(a) and 10(a) illustrate the Weyl operator symbol surface for the case of $K_{+\infty} = 1$, $K_{-\infty} = 2$, $K_1 = 3$, $l = 5$, and $\bar{k} = 1$ (JORDAN, 2000). Many of the previous comments about the discontinuity surface also apply in this case, especially with regard

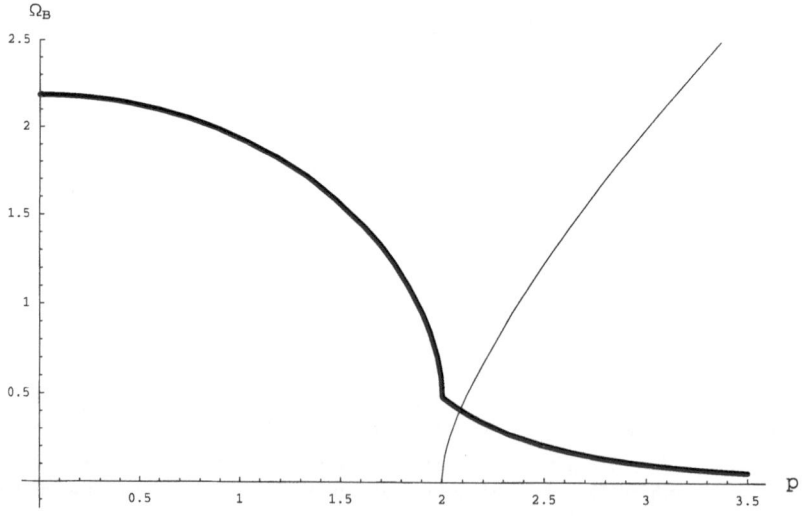

Figure 8

Bold: $\mathrm{Re}[\Omega_B(p,q)]$ vs. p and thin: $\mathrm{Im}[\Omega_B(p,q)]$ vs. p for $K^2(q) = K_1^2 + (K_2^2 - K_1^2)H(q)$, with $Q = \bar{k}|q|, K_1 = 3, K_2 = 1$, and $Q = 0$. These curves provide the leading term in the uniform, low-frequency expansion of the operator symbol for profiles which asymptote to different limiting values. Observe that this result is not equal to that of the homogeneous medium operator symbol corresponding to the average value of the two refractive indices.

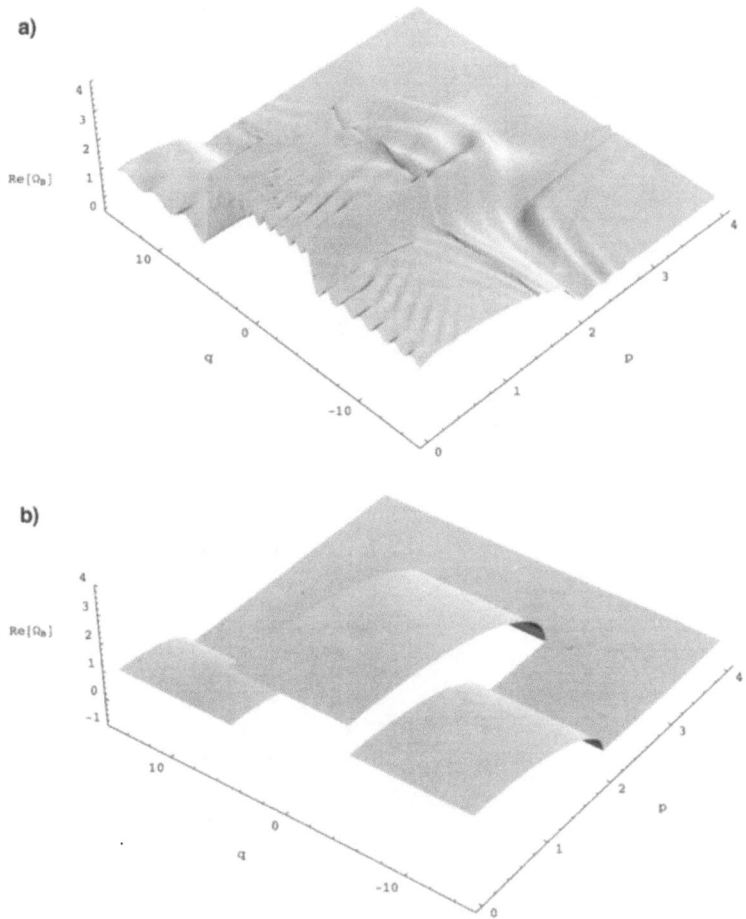

Figure 9

(a) $\mathrm{Re}[\Omega_{\mathbf{B}}(p,q)]$ for $K^2(q) = K^2_{+\infty}H(q-l) + K^2_1 H(l^2 - q^2) + K^2_{-\infty}H(-q-l)$, with $K_{-\infty} = 2, K_1 = 3, K_{+\infty} = 1$, $l = 5$, and $\bar{k} = 1$. (b) $\mathrm{Re}[H(q-l)[K^2_{+\infty} - p^2]^{1/2} + H(l^2 - q^2)[K^2_1 - p^2]^{1/2} + H(-q-l) [K^2_{-\infty} - p^2]^{1/2}]$. While many of the characteristics of the discontinuity surface apply in this case, the new feature is the oscillatory structure in the center layer, representative of the trapped modes for this focusing profile. The standard asymptotic results in (b) are clearly unable to capture the detailed oscillatory and correct singularity structure.

to the structure and asymptotic behavior corresponding to the two homogeneous half-spaces. For example, the only surface singularity is the square-root singularity occurring when p is equal to the average value of the two, half-space refractive indices. Further, the symbol surface is continuous at the points where the profile is discontinuous, and the $p = 0$ slice essentially exhibits an oscillatory and smoothed version of $K(q)$. The new feature is the oscillatory structure characteristic of the center layer. For this model profile, which supports trapped modes (i.e., bound states), the details of the oscillatory periods provide the trapped-mode information (corresponding

a)

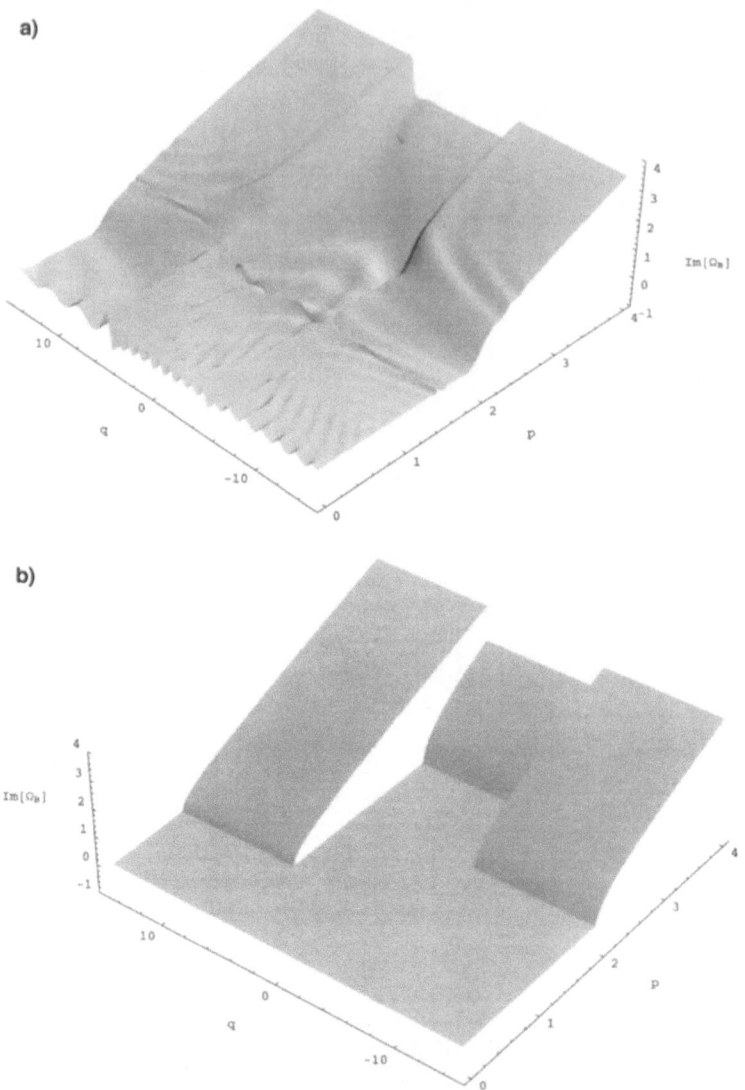

b)

Figure 10

(a) $\text{Im}[\Omega_{\mathbf{B}}(p,q)]$ for $K^2(q) = K^2_{+\infty}H(q-l) + K^2_1 H(l^2-q^2) + K^2_{-\infty}H(-q-l)$, with $K_{-\infty} = 2, K_1 = 3, K_{+\infty} = 1, l = 5$, and $\bar{k} = 1$. (b) $\text{Im}[H(q-l)[K^2_{+\infty}-p^2]^{1/2} + H(l^2-q^2)[K^2_1-p^2]^{1/2} + H(-q-l)$ $[K^2_{-\infty}-p^2]^{1/2}]$. While many of the characteristics of the discontinuity surface apply in this case, the new feature is the oscillatory structure in the center layer, representative of the trapped modes for this focusing profile. The standard asymptotic results in (b) are clearly unable to capture the detailed oscillatory and correct singularity structure.

eigenvalues). As in the case of the discontinuity profile, Figures 9(b) and 10(b) display the real and imaginary parts, respectively, of the locally-homogeneous medium, operator symbol surfaces in each of the three regions, $H(q-l)[K^2_{+\infty}-p^2]^{1/2}+$

$H(l^2 - q^2)[K_1^2 - p^2]^{1/2} + H(-q - l)[K_{-\infty}^2 - p^2]^{1/2}$. Again, the figures readily illustrate that the elliptic pseudodifferential operator calculus cannot construct the detailed oscillatory and singularity structure associated with the three-layer composite profile.

The frequency dependence of the square-root Helmholtz operator symbol, ultimately, generates the frequency dependence of the total wavefield. In Figure 11, the real part of the operator symbol surface is displayed, for the specific case under consideration, for the sequence $\bar{k} = 2, 0.4, 0.1$, and 0.001, respectively. The sequence of figures illustrates the transition from the high- to the low-frequency limit, reflecting the decreasing number of trapped modes supported by this focusing profile with decreasing frequency. The high-frequency limit in Figure 11(a) is seen to be (to within the previously stated caveats) roughly a locally-homogeneous limit with an oscillatory component, which transitions in Figures 11(b) and (c) to intermediate regimes where the oscillatory features become quite prominent, finally approaching in Figure 11(d) the (near) q-independent, limiting, low-frequency form. More specifically, in the limit of zero frequency, the operator symbol surface in Figure 11(d) approaches the result given in Eqs. (36) and (37) for the case of $K_A = 3/2$ and $K_D = 1/2$, displaying a form similar to that illustrated in Figure 8.

Figures 12(a) and (b) compare the real part of the operator symbol surface for a focusing ($K_{+\infty} = 1$, $K_{-\infty} = 1$, $K_1 = \sqrt{7}/2$, $l = 5$, and $\bar{k} = 2$) and complementary defocusing ($K_{+\infty} = 1$, $K_{-\infty} = 1$, $K_1 = 1/2$, $l = 5$, and $\bar{k} = 2$), three-layer composite profile. Beyond the obvious difference in the "locally-homogeneous background structure," the principal difference lies in the nature of the oscillatory structure in the center layer. These two cases are connected by an analytic continuation in a manner similar to that connecting the previously-discussed focusing and defocusing quadratic profiles. Finally, for the symmetric, three-layer composite profile with $K_{+\infty} = K_{-\infty} = K_0$, in the limit of $l \to 0, K_1 \to \infty$, with $K_1^2 l$ constant, the delta function profile result in Eq. (35) is recovered.

4. Uniform Asymptotic Operator Symbol Constructions

While exact operator symbol constructions are quite valuable in illustrating the detailed mathematical theory and providing benchmark cases for approximate operator symbol constructions, the development of an exact, insightful, operator symbol representation for general profiles is quite unlikely. Ultimately, physical applications must depend upon asymptotic results. Since the theory of pseudodifferential operators was developed to provide a rigorous, mathematical foundation for "asymptotics", it is natural to ask why this vast machinery cannot simply be applied, in this case, to readily construct the symbol asymptotics for the square-root Helmholtz operator.

The theory of pseudodifferential operators expands the operator symbol in orders of smoothness, essentially resulting in a series of algebraic terms of increasing degree

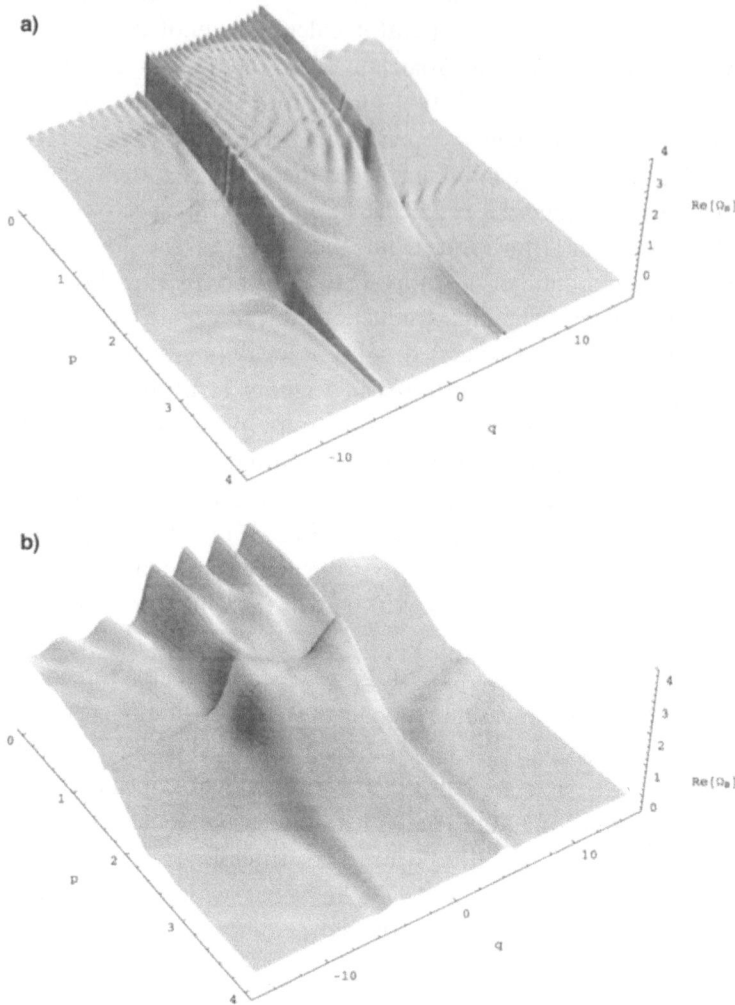

Figure 11

$\mathrm{Re}[\Omega_{\mathbf{B}}(p, q)]$ for $K^2(q) = K_{+\infty}^2 H(q - l) + K_1^2 H(l^2 - q^2) + K_{-\infty}^2 H(-q - l)$, with $K_{-\infty} = 2, K_1 = 3, K_{+\infty} = 1$, and $l = 5$. (a) $\bar{k} = 2$, (b) $\bar{k} = 0.4$, (c) $\bar{k} = 0.1$, and (d) $\bar{k} = 0.001$. This sequence of figures illustrates the transition from the high- to the low-frequency limit, reflecting the decreasing number of trapped modes in the process. The location of the square-root singularity in p remains constant throughout the transition. The low-frequency limit in (d) approaches the corresponding form illustrated in Figure 8.

of smoothness, valid in the ΨDO limit of $|p| \to \infty$. Nested within this smoothness expansion is a dual, implicit, "high-frequency" expansion in terms of a small, dimensionless parameter measuring the medium variation on the wavelength scale. Both asymptotic developments neglect the infinitely-smooth contributions of exponential order; that is to say, the pseudodifferential operator asymptotic analysis, inherently, only considers part of the solution. For hyperbolic wave propagation

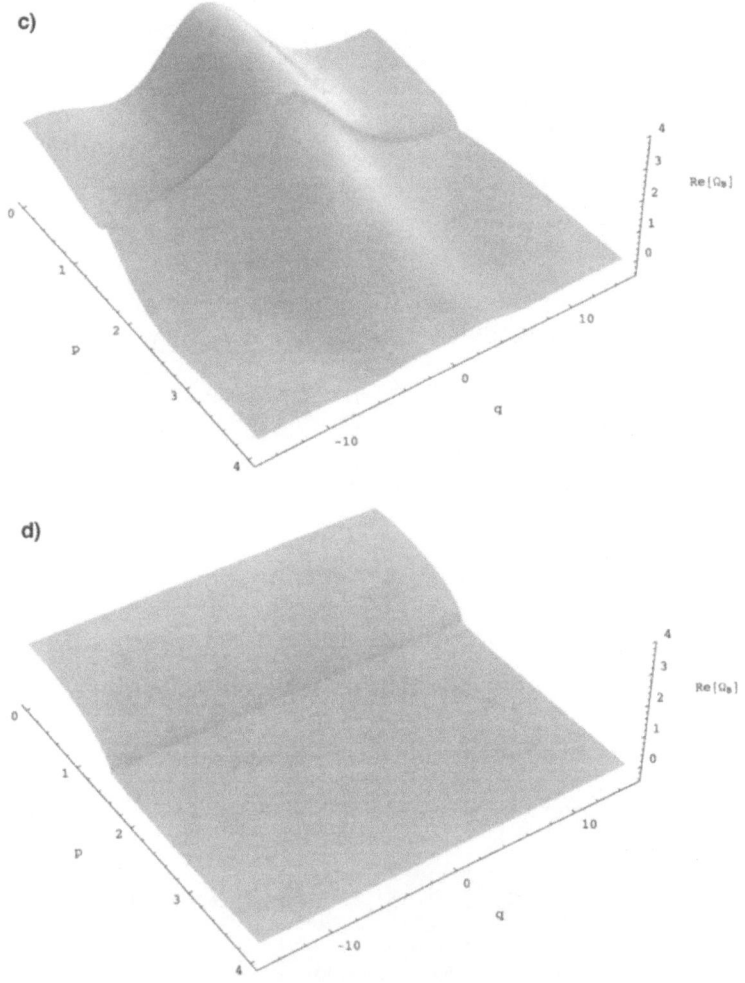

Figure 11c, d

models, such as the time-domain, plasma, and Klein-Gordon wave equations, which involve the square root of a positive operator, the smoothness-ordering, pseudodifferential operator symbol expansions can provide accurate symbol approximations over the entire phase space regime in an appropriate limit. In these cases, the neglected, infinitely-smooth contributions of exponential order are dominated by exponential decay, and can, indeed, be neglected. The elliptic Helmholtz equation, unlike the wave-front propagating, hyperbolic formulations, is a smoothing problem characterized by the nonpropagating, evanescent spectral components, the mathematical manifestation of which is the appearance of the square root of the indefinite, transverse Helmholtz operator. In this case, the pseudodifferential operator symbol expansion, while still valid in the ΨDO limit of $|p| \to \infty$, will no longer provide a uniformly valid, phase

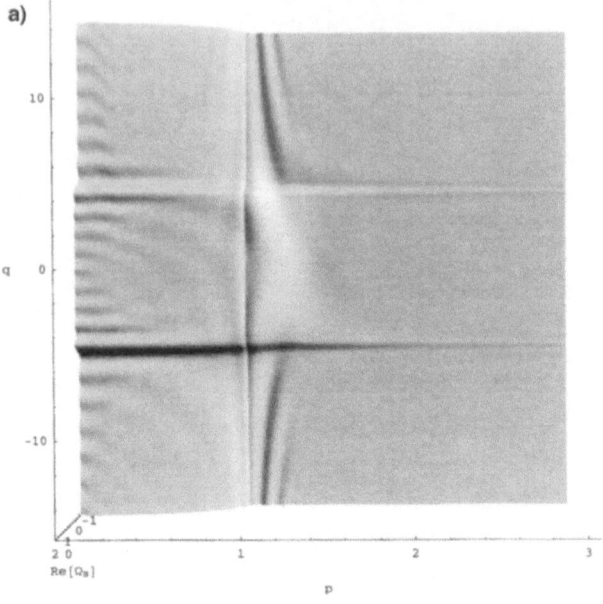

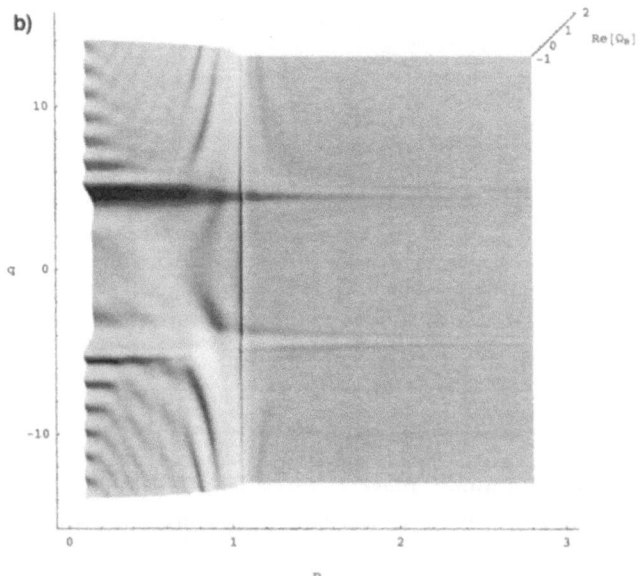

Figure 12

$\mathrm{Re}[\Omega_{\mathbf{B}}(p,q)]$ for $K^2(q) = K^2_{+\infty}H(q-l) + K^2_1 H(l^2 - q^2) + K^2_{-\infty}H(-q-l)$, with $K_{\pm\infty} = 1, l = 5$, and $\bar{k} = 2$. (a) $K_1 = \sqrt{7}/2(\approx 1.32)$ and (b) $K_1 = 1/2$. The complementary nature of the oscillatory structure in the center layer for the focusing (a) and defocusing (b) cases is to be noted.

space, operator symbol approximation in terms of a small, dimensionless parameter in an appropriate "high-frequency" regime. Rather, the expansion will be appropriate in a properly defined, "outer" phase space regime. In the corresponding "inner" phase space regime, the expansion will be singular, and the neglected terms of exponential order can be of comparable size to, or dominate, the leading, nonsingular, algebraic contribution. The ΨDO development cannot account for the oscillatory components of the operator symbol surfaces illustrated in the exact constructions in Section 3. Thus, a properly constructed, uniformly valid, high-frequency symbol approximation for the square-root Helmholtz operator must account for the appropriate contributions from the infinitely-smooth part of the operator kernel in just such a manner so as to simultaneously remove the singular, nonuniform behavior and incorporate the oscillatory features. Unlike the nonuniform, pseudodifferential operator symbol expansion, such a uniform approximation will be explicitly or implicitly characterized by a small, dimensionless parameter, and will not be universal, but, rather, will reflect the detailed scalings characterizing the refractive index field. Different expansions will apply in different asymptotic parameter regimes. Sometimes the consideration of part of the solution is sufficient (the hyperbolic wave propagation problems); sometimes it is not (the oscillatory, elliptic wave propagation problems).

The above points are readily illustrated. The nonuniform, singular, pseudodifferential operator symbol expansion for $d = 2$ takes the form

$$\Omega_{\mathbf{B}}(p, q) \sim [K^2(q) - p^2]^{1/2} - \frac{K^3(q)K''(q)}{8\bar{k}^2[K^2(q) - p^2]^{5/2}} + \cdots \quad , \tag{38}$$

where the primes denote differentiation with respect to q. While the expansion in Eq. (38) is appropriate in both the ΨDO smoothness and high-frequency limits, as a high-frequency expansion, the nonuniform and singular nature in phase space is quite evident. No matter how slowly the medium varies on the wavelength scale $(K''(q)/\bar{k}^2)$, a phase space neighborhood about the curve $K^2(q) = p^2$ always exists where the second term in Eq. (38) can be made arbitrarily large, with a singularity along the curve. The absence of any oscillatory character in Eq. (38) is also clear. Since the ultimate goal is to compute accurate wavefields, it is the high-frequency, as opposed to the pseudodifferential operator smoothness, expansion that provides the relevant asymptotic framework.

A uniform, high-frequency (UHF) approximation of the Weyl, square-root Helmholtz operator symbol can be constructed for sufficiently smooth profiles with a single scale of variation, and, for $d = 2$, is given by

$$\Omega_{\mathbf{B}}(p, q) \sim [K^2(q) - p^2]^{1/2} + \int_0^\infty du \, \cos[\bar{k}pu]K^2(q)$$
$$\times \left\{ A\left[\frac{H_1^{(1)}(\bar{k}I_0)}{I_0} + \frac{C}{\bar{k}}H_0^{(1)}(\bar{k}I_0) \right] - \frac{H_1^{(1)}(\bar{k}K(q)u)}{K(q)u} \right\} \quad , \tag{39}$$

where

$$A = \left(\frac{I_0}{I_1}\right)^{3/2} \left\{\frac{1}{K^2(q)} \left[K\left(q+\frac{u}{2}\right)K\left(q-\frac{u}{2}\right)\right]^{-1/2}\right\}, \tag{40}$$

$$C = \left(\frac{1}{8I_0}\right)\left\{\frac{15I_2}{I_1^2} - \frac{3}{I_0} - \frac{6}{I_1}\left[\frac{1}{K^2(q+u/2)} + \frac{1}{K^2(q-u/2)}\right]\right.$$

$$\left. + 2\left[\frac{K'(q-u/2)}{K^2(q-u/2)} - \frac{K'(q+u/2)}{K^2(q+u/2)}\right] - \tilde{I}\right\}, \tag{41}$$

$$I_m = \int_{q-u/2}^{q+u/2} dt[K(t)]^{1-2m} \quad (m = 0, 1, 2), \qquad \tilde{I} = \int_{q-u/2}^{q+u/2} dt\left[\frac{(K'(t))^2}{K^3(t)}\right], \tag{42}$$

and $H_v^{(1)}(\cdot)$ is the vth-order Hankel function of the first kind. The first term in Eq. (39) is the leading, nonsingular, algebraic contribution from the elliptic pseudodifferential operator calculus representing the locally-homogeneous result, while the second term replaces the first, nonuniform, singular term from the elliptic calculus with a uniform contribution which correctly incorporates the exponential contributions. The expression in Eq. (39) reproduces the high-frequency, pseudodifferential, and homogeneous medium limits, as well as the nonuniform result in Eq. (38) from a stationary phase evaluation of the integral about $u = 0$. It is most important to note that the Eq. (39) result represents the actual asymptotic behavior of the operator symbol, and is expressed solely in terms of the functions and parameters of the defining wave equation (11), as opposed to expansion coefficients, eigenvalues, eigenfunctions, and other numerically computed quantities, for example.

The Weyl ΨDO result in Eq. (38) and the uniform, high-frequency asymptotic result in Eq. (39) can be compared with exact operator symbol constructions to further illustrate the above points. While the Eq. (39) result does not apply to the exact, nonsmooth constructions in Section 3, comparisons with smooth profiles are appropriate. For example, for the defocusing quadratic profile in one transverse dimension (i.e., $K^2(q) = K_0^2 + \omega^2 q^2$), Figure 13 compares plots of the $q = 0$ slice of the $\Omega_B(p, q)$ surface generated from the exact (Eq. (34)), the Weyl pseudodifferential (ΨDO, Eq. (38)), and the uniform, high-frequency (UHF, Eq. (39)) operator symbol constructions. The nonuniform, singular nature of the pseudodifferential operator approximation, even well within the high-frequency regime, is clearly demonstrated, along with the absence of the oscillatory contributions of exponential order. The same comparison between the exact and uniform, high-frequency operator symbol constructions illustrates the near-perfect agreement, highlighting the full spectral range character of the uniform asymptotics. This single-slice comparison is typical of the operator symbol surface throughout the entire phase space.

The properties of the uniform, high-frequency operator symbol construction in Eq. (39) are further illustrated in Figure 14. In Figures 14(a) and (b), the $q = 0$ slice of the $\Omega_B(p, q)$ surface for the focusing and defocusing hyperbolic function profiles,

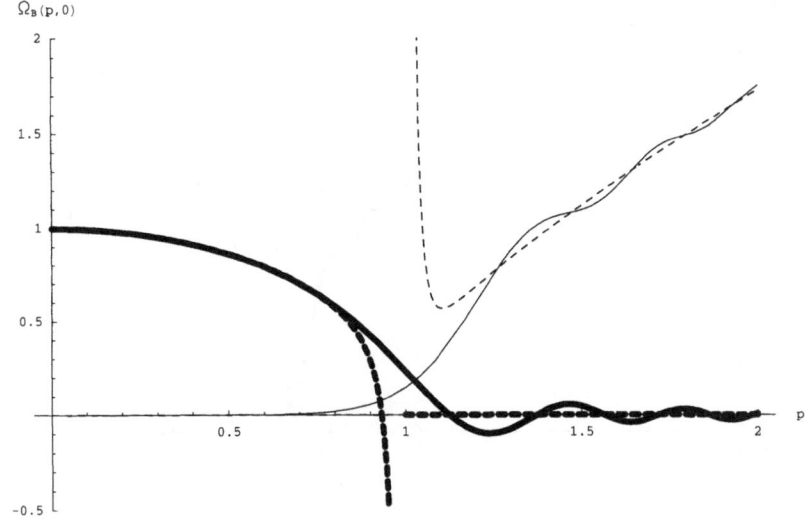

Figure 13

$\Omega_{\mathbf{B}}(p, q)$ vs. p for $K^2(q) = K_0^2 + \omega^2 q^2$, with $K_0 = \omega = 1, \bar{k} = 7.5$, and $q = 0$. Equation (34): $\text{Re}[\Omega_{\mathbf{B}}]$ (bold), $\text{Im}[\Omega_{\mathbf{B}}]$ (thin). ΨDO: $\text{Re}[\Omega_{\mathbf{B}}]$ (bold-dashed), $\text{Im}[\Omega_{\mathbf{B}}]$ (thin-dashed). UHF: $\text{Re}[\Omega_{\mathbf{B}}]$ (bold-dot-dashed), $\text{Im}[\Omega_{\mathbf{B}}]$ (thin-dot-dashed). The nonuniform, singular nature of the pseudodifferential operator (ΨDO) asymptotics is demonstrated, along with the absence of the oscillatory contributions. The uniform (UHF) asymptotic result is seen to capture all of the surface features throughout the phase space.

$K^2(q) = 1 + \text{sech}^2(q)$ and $K^2(q) = 1 - \frac{1}{2}\text{sech}^2(q)$, respectively, with $\bar{k} = 30$, illustrates the oscillatory nature of the surfaces and the square-root singularity at the background value of $p = K_0 = 1$ associated with the profiles. As previously mentioned in the discussion of the quadratic profiles, the oscillatory character of the operator symbol surface is confined to the locally-propagating portion of the phase space in the focusing case (Fig. 14(a)) and the locally-evanescent portion of the phase space in the defocusing case (Fig. 14(b)). Figure 14(c) illustrates the $q = 0$ slice of the $\Omega_{\mathbf{B}}(p, q)$ surface for the hyperbolic function profile, $K^2(q) = 1 + \frac{1}{2}\tanh(q)$, with $\bar{k} = 30$. Beyond the near-square-root, algebraic behaviour, the principal feature of interest is the square-root singularity occurring at the average value of the asymptotic refractive indices, $p = \frac{1}{2}\left[\sqrt{3/2} + \sqrt{1/2}\right] \approx 0.966$.

Finally, uniform low-frequency, operator symbol constructions can also be derived. A more detailed treatment of uniform asymptotic, operator symbol approximations can be found in the literature (FISHMAN *et al.*, 1997; DE HOOP and GAUTESEN, 2000).

5. Discussion

The exact construction cases for both the smooth and nonsmooth profiles exhibited in Section 3, along with the uniform, high-frequency, operator symbol

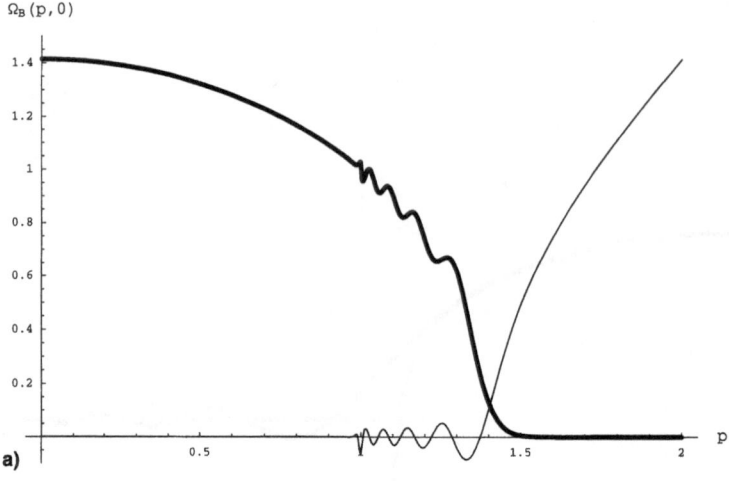

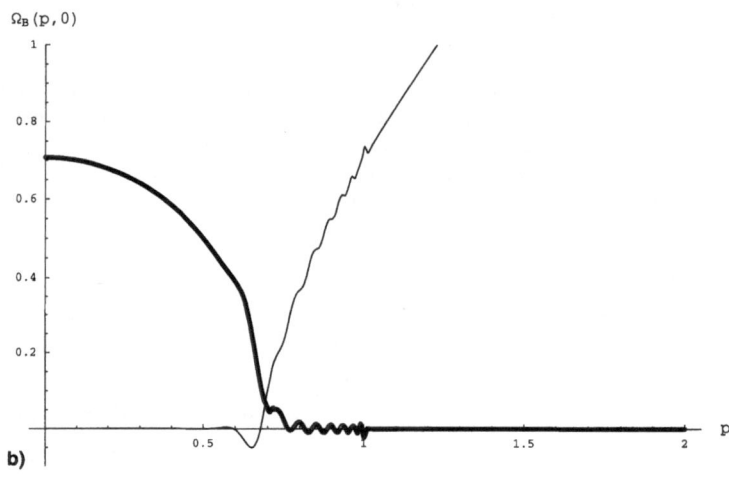

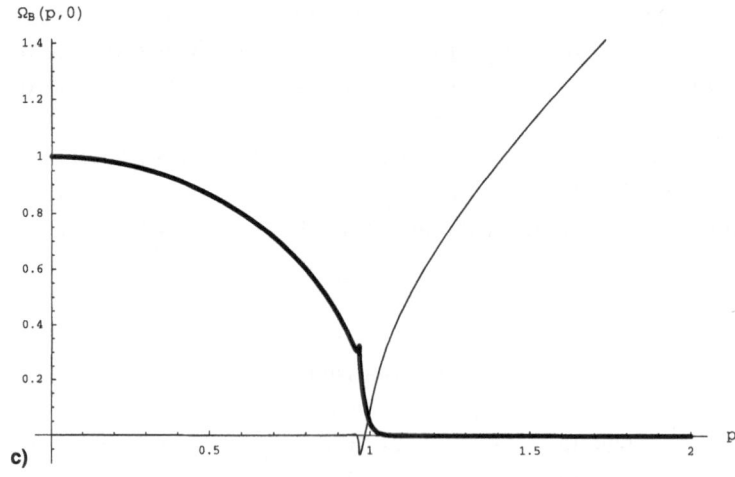

asymptotic approximation presented in Section 4, together, illustrate the character-istics of the square-root Helmholtz operator symbol. Starting in the high-frequency regime, an oscillatory structure is superimposed upon what is essentially a locally-homogeneous operator symbol. A distinction must be drawn between the smooth and nonsmooth cases regarding the details of this oscillatory component and the approach to the zero-wavelength limit, owing to the fact that there are no scales on which the jump discontinuity is slowly changing, unlike the case for the infinitely-differentiable profiles. In the low-frequency regime, the operator symbol surface is governed by Eq. (36), which, for the case of a uniform background profile, reduces to the homogeneous, background medium symbol. The singularity structure of the operator symbols is characterized, in general, by a square-root singularity occurring when p is equal to the average value to the two, asymptotic (half-space) refractive indicies. This result is also contained in the uniform, high-frequency expression in Eq. (39). The oscillatory structure is a fundamental part of the operator symbol surface, and cannot be viewed, in general, as a "perturbation" of the underlying algebraic structure. In addition to containing information about the trapped modes and the underlying profile, this oscillatory component is, ultimately, crucial for accurate, wide-angle, wavefield computations and the strict conservation of the integrated energy flux.

   Lying outside of the elliptic pseudodifferential operator calculus, the square root of the indefinite, transverse Helmholtz operator is quite distinct from the square roots of positive operators which occur in hyperbolic wave propagation problems, such as the time-domain and Klein-Gordon wave equations. The symbol surfaces corre-sponding to the hyperbolic wave propagation problems have no oscillatory compo-nent and can be uniformly approximated within the elliptic calculus by results analogous to Eq. (38), unlike the situation for the fixed-frequency, scalar Helmholtz equation. Moreover, by formulating the Helmholtz problem in the complex Laplace domain and employing integral representations developed in exact operator symbol construction cases, the connection between elliptic and hyperbolic wave propagation formulations can be examined in greater detail (FISHMAN et al., 2000). More specifically, Laplace transforming the time-domain wave equation and examining the results in the complex Laplace plane essentially establish that the hyperbolic and elliptic formulations are represented along the (positive) real and imaginary axes, respectively. While the elliptic pseudodifferential operator calculus applies along the

◄

Figure 14

UHF: $\mathrm{Re}[\Omega_\mathbf{B}(p,q)]$ vs. p (bold) and $\mathrm{Im}[\Omega_\mathbf{B}(p,q)]$ vs. p (thin). (a) $K^2(q) = 1 + \mathrm{sech}^2(q)$, (b) $K^2(q) = 1 - \frac{1}{2}\mathrm{sech}^2(q)$, and (c) $K^2(q) = 1 + \frac{1}{2}\tanh(q)$, all for $\bar{k} = 30$ and $q = 0$. The focusing (a) and defocusing (b) smooth profiles illustrate the location of the square-root singularity in p at the asymptotic, background value of the refractive index, and the confinement of the oscillatory structure to the locally-propagating and locally-evanescent parts of the spectrum in the phase space, respectively. The profile (c) has no connection to trapped modes, and, thus, an absence of oscillatory structure, and, further, illustrates the location of the square-root singularity in p at the average value of the asymptotic refractive indices.

(positive) real axis (and in an appropriate region about this axis), the neglect of the exponential-order contributions inherent in this asymptotic approximation would, ultimately, lead to singular, operator symbol approximations in a continuation process from the real to the imaginary axis in the complex Laplace plane. The exponentially small terms in the hyperbolic (time) domain become the oscillatory terms in the elliptic (frequency) domain, and cannot be neglected in a continuation process in the complex Laplace plane. Thus, establishing uniform asymptotic, operator symbol approximations for the elliptic formulation via continuation from the asymptotic, hyperbolic results is problematic, at best. The uniform, high-frequency result in Eq. (39) was established, on the other hand, directly in the frequency domain. It turns out that operator symbol, contour-integral representations, such as those given in Eqs. (31)–(33) for the quadratic profiles, allow for a complete treatment of the connection between the elliptic and hyperbolic formulations in the right-hand-half of the complex Laplace plane, providing the desired connection between the frequency- and time-domain problems (FISHMAN et al., 2000).

The high-frequency approximate wave theory, based on the uniform, high-frequency operator symbol construction in Eq. (39), is quite distinct from geometric, or semiclassical, approximations on the wavefield. This is an extremely important point for applications. Specifically, it can be briefly contrasted with the global, uniform, phase-space-based, asymptotic wavefield constructions associated with (1) the Maslov, or Lagrangian manifold, method, which effectively exploits the Fourier transform, and (2) the Klauder method based on the coherent-state transform (FISHMAN et al., 1997). Both the Maslov and Klauder approximate wavefield constructions are essentially nonuniform, high-frequency (geometric) wavefield asymptotics appropriately corrected for the variety of caustic structures inherent in the infinite-frequency approximation. The results are globally uniform wavefield representations, in terms of functions, which are still inherently "high-frequency." The wavefield approximation based on the uniform, high-frequency operator symbol construction, on the other hand, is a full path (functional) integral which, in retaining the "sum over paths" in the phase space, is a full-wave theory, free from caustic structures. Essentially, the phase functional in the path integral is uniformly approximated in the high-frequency limit. The difference between uniform, high-frequency asymptotics on the wavefield and the operator symbol is, roughly, that in the former case, the classical, high-frequency (geometric) asymptotic wavefield is considered as a global solution, while, in the latter case, it is essentially considered as an infinitesimal solution, repeatedly composed to produce a global solution. The key point is that high-frequency approximations on the square-root Helmholtz operator symbol can correspond to accurate wavefield regimes well outside of the geometric, or semiclassical, wavefield limit. This contention is well supported by numerical computations with the primitive, high-frequency, operator symbol approximation, essentially the leading term in Eq. (39) (FISHMAN et al., 1987; FISHMAN and WALES, 1987).

In addressing the correct phase space (singularity and oscillatory) structure of the operator symbols in an appropriate "inner" regime, the uniform asymptotic methods incorporate much more than the correct treatment of the evanescent wave contribution. This is an important point, and it is readily demonstrated by the comparisons of the exact, nonuniform asymptotic, and uniform asymptotic, operator symbol constructions illustrated in Figures 6, 7, 9, 10, and 13. These figures demonstrate that the nonuniform asymptotic treatment cannot properly represent both the high-angle propagation and near-evanescent, local phase space regimes. In a highly-variable environment, a necessary (global) p-component, for a given propagation calculation, could fall in both evanescent and propagating, local phase space regimes as q is varied over the relevant region of the phase space. (For example, imagine tracing the line $p = 2.5$ in Figures 9 and 10.) Moreover, the phase space and path integral approach, with its detailed asymptotic analysis of the operator symbols, provides a unified framework for the (generalized) Taylor series, operator rational approximation, and generalized phase screen representations of the square-root Helmholtz operator, so commonly found in the seismic literature. This follows since the phase space analysis provides the natural extension of the approximate, parabolic equation methods, as opposed to being a ray-based formulation, and the generalized phase screen can be viewed as a particular approximation of the operator symbol (FISHMAN, 1992; FISHMAN et al., 1997, 2000; DE HOOP et al., 2002).

The exact and approximate operator symbol constructions and corresponding wave equations have application in the construction of computational nonreflecting boundary conditions and the development of inverse algorithms for the multidimensional Helmholtz equation. Equation (12), specified to a fixed range point, provides the exact nonreflecting boundary condition appropriate for a transversely-inhomogeneous asymptotic regime. The Helmholtz, Weyl composition equation (13), on the other hand, provides the basis for the exact reconstruction of the square of the refractive index field from operator symbol data, supplying the first step in a layer-stripping procedure employing the complete Dirichlet-to-Neumann operator symbol and corresponding governing, composition-type equation. The high-frequency approximate reconstruction from the square-root Helmholtz operator symbol has been illustrated for several exact construction cases (FISHMAN, 1992; FISHMAN et al., 2000).

While the application of the previously outlined, directional wavefield decomposition, phase space, and path integral analysis to the reconstruction of the square of the refractive index field indicated above can be viewed as a "toy" problem, the application to the extension of the microlocal, asymptotic seismic inversion of real data is definitely not. This microlocal approach (tracking the propagation of singularities such as wavefronts) (DE HOOP et al., 1999; DE HOOP and BRANDSBERG-DAHL, 2000; STOLK and DE HOOP, 2000) basically extends the classical Kirchhoff-style imaging-inversion process in conjunction with migration-velocity analysis. The background medium is complex and elastic, with caustic formation taken into account. Given a smooth background medium (embedding), the singular medium component is reconstructed in a sequence of operations which can be summarized as (1) propagation (in the smooth embedding) at the level of the Maslov canonical operator, (2) modeling (scattering) in the Born/Kirchhoff approximation, (3) acquisition (restriction operator), (4) imaging (adjoint operator), (5) resolution (normal operator), and (6) inversion (left inverse operator). The above operations are mathematically represented by appropriate pseudodifferential and local and nonlocal Fourier integral operators. Artifacts arising in the resolution step can be estimated. Moreover, the reconstruction of the singular medium component can be iterated with the reconstruction of the background medium (smooth embedding) in what is effectively a "bootstrapping" approach. This inversion algorithm can be viewed as the global, uniform asymptotic extension of the geometrical method of "tracing rays" in conjunction with the Born/Kirchhoff approximation.

While the briefly outlined, microlocal, asymptotic seismic inversion approach can be quite successful in the inversion of real-world data sets, such as the multicomponent, ocean-bottom data from the Valhall field in Norway, which incorporate severe focusing effects, it is not able to account properly for waveguiding effects occurring in layers beneath the reservoir, for example, where the wave nature of the propagation process is crucial, and for the more rapidly varying medium properties (measured against a typical wavelength). This is a consequence of the microlocal approach only representing part of the wave solution, as emphasized in Section 4, in conjunction with the assumption of a sufficiently small and localized medium contrast. The functional integration and spectral analysis approach overcomes these limitations, in principle, by first applying the idea behind directional wavefield decomposition to construct the generalized Bremmer coupling series (DE HOOP, 1996) to account for all of the "multiples" in the original wavefield. This approach allows one to, in effect, "trace wave constituents"; hence, the concept of "wave tracing." In re-establishing the sequence: propagation — modeling — imaging — resolution — inversion, the (high-frequency) Maslov canonical operator is replaced by directional wavefield decomposition and the Born/Kirchhoff approximation by the generalized Bremmer coupling series. Mathematically, the cotangent vector in the microlocal analysis is replaced by an exact operator symbol at the level of the uniform asymptotic symbol calculus of Section 4, and the one-way wave propagator is written as a sequence of Trotter products (path integral), with the phase term being the square-root Helmholtz operator symbol. The remainder of the sequence: modeling — imaging — resolution — inversion is based on a decomposition of the Bremmer series, iterative methods, and the application of time-reversal mirrors (DE HOOP et al., 2002).

Acknowledgements

The research reported in this paper has been financially supported by the Naval Research Laboratory through an IPA with the University of New Orleans. Pedro

M. Jordan, of the Naval Research Laboratory at the Stennis Space Center, is gratefully acknowledged for the invaluable task of generating the figures in this paper with *Mathematica* 4.0. Finally, Guest Editor Ivan Pšenčík and the two referees are thanked for their efforts in the process to make this paper more accessible to the greater seismic community.

REFERENCES

DE HOOP, M. V. (1996), *Generalization of the Bremmer Coupling Series*, J. Math. Phys. *37*, 3246–3282.

DE HOOP, M. V., SPENCER, C., and BURRIDGE, R. (1999), *The Resolving Power of Seismic Amplitude Data: An Anisotropic Inversion/Migration Approach*, Geophys. *64*, 852–873.

DE HOOP, M. V. and BRANDSBERG-DAHL, S. (2000), *Maslov Asymptotic Extension of Generalized Radon Transform Inversion in Anisotropic Elastic Media: A Least-squares Approach*, Inverse Probs. *16*, 519–562.

DE HOOP, M. V. and GAUTESEN, A. K. (2000), *Uniform Asymptotic Expansion of the Generalized Bremmer Series*, SIAM J. Appl. Math. *60*, 1302–1329.

DE HOOP, M. V., FISHMAN, L., and JONSSON, B. L. G. (2002), *Acoustic Time-reversal Mirrors in the Framework of One-way Wave Theories*, manuscript submitted for publication.

FISHMAN, L., McCOY, J. J., and WALES, S. C. (1987), *Factorization and Path Integration of the Helmholtz Equation: Numerical Algorithms*, J. Acoust. Soc. Am. *81*, 1355–1376.

FISHMAN, L. and WALES, S. C. (1987), *Phase Space Methods and Path Integration: The Analysis and Computation of Scalar Wave Equations*, J. Comp. Appl. Math. *20*, 219–238.

FISHMAN, L. (1992), *Exact and Operator Rational Approximate Solutions of the Helmholtz, Weyl Composition Equation in Underwater Acoustics—The Quadratic Profile*, J. Math. Phys. *33*, 1887–1914.

FISHMAN, L. (1993), *One-way Wave Propagation Methods in Direct and Inverse Scalar Wave Propagation Modeling*, Radio Sci. *28*, 865–876.

FISHMAN, L., GAUTESEN, A. K., and SUN, Z. (1997), *Uniform High-frequency Approximations of the Square-root Helmholtz Operator Symbol*, Wave Motion *26*, 127–161.

FISHMAN, L., DE HOOP, M. V., and VAN STRALEN, M. J. N. (2000), *Exact Constructions of Square-root Helmholtz Operator Symbols: The Focusing Quadratic Profile*, J. Math. Phys. *41*, 4881–4938.

JORDAN, P. M. (2000), *Constructions of Helmholtz Operator Symbols for Three-layer Composite Media Using the MZB Rational Square-root Approximation*, unpublished.

JORDAN, P. M. (2001), *Comment on "Exact Constructions of Square-root Helmholtz Operator Symbols: The Focusing Quadratic Profile"* [J. Math. Phys. *41*, 4881 (2000)], J. Math. Phys. *42*, 4618–4623.

LU, Y. Y. and McLAUGHLIN, J. R. (1996), *The Riccati Method for the Helmholtz Equation*, J. Acoust. Soc. Am. *100*, 1432–1446.

LU, Y. Y. (1999), *One-way Large Range Step Methods for Helmholtz Waveguides*, J. Comp. Phys. *152*, 231–250.

LU, Y. Y., HUANG, J., and McLAUGHLIN, J. R. (2001), *Local Orthogonal Transformation and One-way Methods for Acoustic Waveguides*, Wave Motion, *34*, 193–207.

STOLK, C. C. and DE HOOP, M. V. (2000), *Microlocal Analysis of Seismic Inverse Scattering in Anisotropic, Elastic Media*, Center for Wave Phenomena, preprint.

(Received November 15, 2000, revised April 18, 2001, accepted April 24, 2001)

To access this journal online:
http://www.birkhauser.ch

Pure appl. geophys. 159 (2002) 1681–1689
0033–4553/02/081681–9 $ 1.50 + 0.20/0

❘ Pure and Applied Geophysics

Parabolic Equation Techniques for Seismic Waves

WAYNE JERZAK[1], MICHAEL D. COLLINS[2], RICHARD B. EVANS[3],
JOSEPH F. LINGEVITCH[2], and WILLIAM L. SIEGMANN[1]

Abstract—Parabolic equation techniques for solving seismic wave propagation problems are discussed. These techniques provide an excellent combination of accuracy and efficiency for problems involving laterally varying media. Due to recent improvements, the parabolic equation method is applicable to problems involving wide propagation angles, large variations in the properties of the medium, and all types of waves.

Key words: Parabolic equation, seismology, one-way wave equation, elastic waves.

1. Introduction

For many wave propagation problems in geophysics, the wave speeds and other properties of the medium vary strongly in the vertical but gradually in range (horizontal position). Such problems can be solved accurately and efficiently by factoring the operator in an elliptic wave equation to obtain a parabolic wave equation and then applying a rational function to approximate the square root of an operator (JENSEN *et al.*, 1994). Parabolic equation techniques have been applied to problems in seismology (LANDERS and CLAERBOUT, 1972), ocean acoustics (TAPPERT, 1977), surface gravity waves (RADDER, 1979), atmospheric waves (LINGEVITCH *et al.*, 1999), and other types of waves. In this paper, we discuss parabolic equation techniques, which have become highly developed for ocean acoustics problems, and their application to seismic problems. In Section 2, we derive a parabolic wave equation for the acoustic problem, which can be used to model seismic waves when the shear speed is low. In Section 3, we discuss parabolic equation techniques for generating an initial condition for the range marching solution, an efficient range marching algorithm, and an accurate approach for handling range dependence in the properties of the medium. In Section 4, we discuss the elastic parabolic equation and its state of development. The parabolic equation techniques discussed in this paper

---

[1] Rensselaer Polytechnic Institute, Troy, New York, U.S.A.
[2] Acoustics Division, Naval Research Laboratory, Washington, D.C., U.S.A.
[3] Science Applications International Corporation, New London, Connecticut, U.S.A.

have been implemented in computer codes that are available by anonymous ftp from ram.nrl.navy.mil.

2. *The Parabolic Wave Equation*

We derive the parabolic wave equation for a two-dimensional acoustics problem in Cartesian coordinates, where z is the depth and the range x is one of the horizontal directions. We assume that horizontal variations in the medium are gradual and approximate the medium in terms of a sequence of stratified regions. In each region, the depth dependence can be arbitrary and the acoustic pressure p satisfies the elliptic wave equation,

$$\frac{\partial^2 p}{\partial x^2} + \rho \frac{\partial}{\partial z}\left(\frac{1}{\rho}\frac{\partial p}{\partial z}\right) + k^2 p = 2i\delta(x)\delta(z - z_0) \ , \tag{1}$$

where k is the acoustic wave number, ρ is the density, and z_0 is the source depth. Rearranging Equation (1) in order to facilitate an expansion about the reference wave number k_0, we obtain

$$\frac{\partial^2 p}{\partial x^2} + k_0^2(1 + X)p = 0 \ , \tag{2}$$

$$X = k_0^{-2}\left(\rho \frac{\partial}{\partial z}\frac{1}{\rho}\frac{\partial}{\partial z} + k^2 - k_0^2\right) \ . \tag{3}$$

Factoring the operator in Equation (2), we obtain

$$\left(\frac{\partial}{\partial x} + ik_0(1 + X)^{1/2}\right)\left(\frac{\partial}{\partial x} - ik_0(1 + X)^{1/2}\right)p = 0 \ . \tag{4}$$

We neglect backscattered energy, which is assumed to be dominated by outgoing energy, and obtain the parabolic wave equation,

$$\frac{\partial p}{\partial x} = ik_0(1 + X)^{1/2}p \ . \tag{5}$$

In order to solve this equation numerically, it is necessary to approximate the square root of the operator. Applying a linear Taylor approximation, we obtain

$$\frac{\partial p}{\partial x} = ik_0\left(1 + \tfrac{1}{2}X\right)p \ . \tag{6}$$

This equation can be solved with standard numerical techniques. Greater accuracy can be achieved by approximating the square root with a rational function to obtain (BAMBERGER *et al.*, 1988).

$$\frac{\partial p}{\partial x} = ik_0 \left(1 + \sum_{j=1}^{N} \frac{a_{j,N}X}{1 + b_{j,N}X} \right) p \ . \tag{7}$$

This equation can also be solved with standard numerical techniques. By taking N sufficiently large, it can be used to obtain accurate solutions for problems involving wide propagation angles and large variations in k and ρ.

3. Parabolic Equation Techniques

The parabolic equation solution described in Section 2 is orders of magnitude faster than the direct numerical solution of Equation (1). It is possible to achieve even greater efficiency by using a special approach, which we describe along with other parabolic equation techniques in this section. The derivations of these techniques are based on the normal mode solution,

$$p(x,z) = \rho(z_0)^{-1} \sum_n \phi_n(z_0)\phi_n(z)k_n^{-1} \exp(ik_n x) \ , \tag{8}$$

where the normal modes ϕ_n and eigenvalues k_n^2 satisfy

$$k_0^2(1+X)\phi_n = k_n^2 \phi_n \ , \tag{9}$$

$$\int \rho^{-1} \phi_m \phi_n dz = \delta_{mn} \ . \tag{10}$$

Expanding the arbitrary analytic function f in a Taylor series and applying Equation (9), we obtain

$$f\big(k_0^2(1+X)\big)\phi_n = f\big(k_n^2\big)\phi_n \ . \tag{11}$$

Considering the case $f\big(k_n^2\big) = k_n^{-1} \exp(ik_n x)$ and applying the modal representation of the delta function,

$$\delta(z - z_0) = \rho(z_0)^{-1} \sum_n \phi_n(z_0)\phi_n(z) \ , \tag{12}$$

we obtain

$$p(x,z) = k_0^{-1}(1+X)^{-1/2}\exp\big(ik_0 x(1+X)^{1/2}\big)\delta(z - z_0) \ . \tag{13}$$

This equation can be applied to obtain an initial condition at $x = x_0$, where x_0 is comparable to a wavelength (COLLINS, 1992; 1999). To avoid encountering singular intermediate results in the implementation, we rearrange the initial condition to obtain

$$p(x_0,z) = k_0^{-1}(1+X)^{3/2}\exp\big(ik_0 x_0(1+X)^{1/2}\big)\sigma(z) \ , \tag{14}$$

$$\sigma(z) = (1 + X)^{-2} \delta(z - z_0) \ . \tag{15}$$

The intermediate solution $\sigma(z)$ is twice differentiable.

The initial condition can be advanced in range by integrating Equation (7). Greater efficiency can be achieved using the formal solution (COLLINS, 1993a),

$$p(x + \Delta x, z) = \exp\left(ik_0 \Delta x (1 + X)^{1/2}\right) p(x, z) \ , \tag{16}$$

which follows from Equation (13). Applying a rational approximation to the exponential operator, we obtain

$$p(x + \Delta x, z) = \exp(ik_0 \Delta x) \prod_{j=1}^{N} \frac{1 + \alpha_{j,N} X}{1 + \beta_{j,N} X} p(x, z) \ . \tag{17}$$

This solution provides higher-order accuracy in both X and $k_0 \Delta x$ and permits large range steps. An efficient approach for obtaining the coefficients has been developed by CEDERBERG and COLLINS (1997). In cylindrical geometry, the far-field counterpart to Equation (13) is

$$p(r, z) = (k_0 r)^{-1/2} (1 + X)^{-1/4} \exp\left(ik_0 r (1 + X)^{1/2}\right) \delta(z - z_0) \ . \tag{18}$$

Defining the new dependent variable $\tilde{p} = r^{1/2} p$, we obtain

$$\tilde{p}(r + \Delta r, z) = \exp\left(ik_0 \Delta r (1 + X)^{1/2}\right) \tilde{p}(r, z) \ . \tag{19}$$

We have described an initial condition and an approach for marching the solution through range-independent regions. It remains to specify a condition for handling the vertical interfaces between regions. Since the parabolic wave equation contains only one range derivative, it is possible to conserve only one quantity across these interfaces. Gradual range dependence can be handled accurately by conserving the energy flux (PORTER *et al.*, 1991; COLLINS and WESTWOOD, 1991), which is proportional to

$$E = \operatorname{Im} \int \rho^{-1} p^* \frac{\partial p}{\partial x} \, dz \ . \tag{20}$$

Applying Equations (5), (8), and (11) and the orthogonality of the modes, we obtain

$$E = k_0 \int \left| \rho^{-1/2} (1 + X)^{1/4} p \right|^2 dz \ . \tag{21}$$

We conclude that the energy flux can be conserved across an interface between range-independent regions A and B by applying the condition (COLLINS, 1993b),

$$\rho_B^{-1/2} (1 + X_B)^{1/4} p_B = \rho_A^{-1/2} (1 + X_A)^{1/4} p_A \ . \tag{22}$$

The small-angle limit of this condition,

$$(\rho_B c_B)^{-1/2} p_B = (\rho_A c_A)^{-1/2} p_A \ , \tag{23}$$

provides accurate solutions for many problems; the full condition in Equation (22) is seldom required.

We illustrate the parabolic equation solution for a problem involving a 25 Hz source that is located 50 m below the surface in a layer with undulating boundaries that overlies a basement. In the layer, the compressional speed is 1700 m/s, the density is 1.2 g/cm^3, and the attenuation is 0.05 dB/λ. In the basement, the compressional speed is 2000 m/s, the density is 1.5 g/cm ^{3}, and the attenuation is 0.1 dB/λ. The variable topography is handled by allowing the size of the computational grid to vary with range (COLLINS et al., 1995). A plot of transmission loss $-20\log_{10}|p|$ appears in Figure. 1. There are several guided modes in the layer near the source. Some of this energy escapes the layer at cutoff near r = 12 km .

4. The Elastic Parabolic Equation

The first factorization of the elastic wave equation was obtained by GREENE (1985). The first successful computations were obtained after stability problems

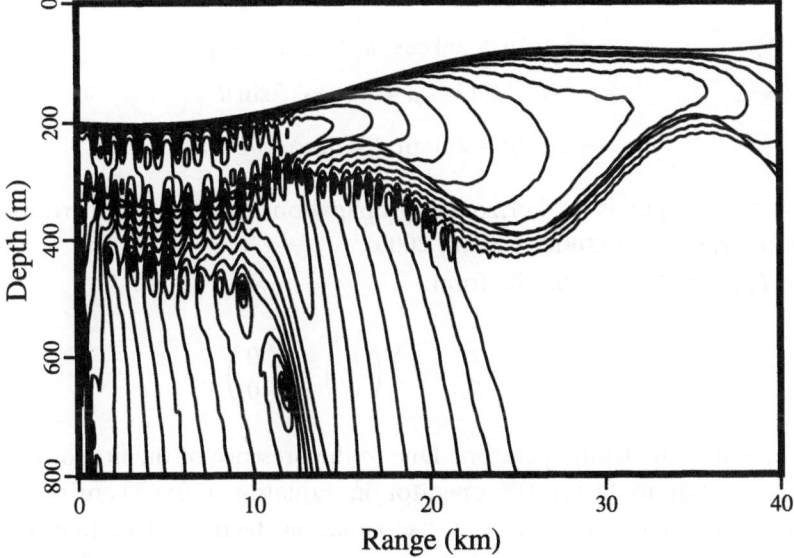

Figure 1

Transmission loss contours for an acoustics problem involving a 25 Hz source that is located 50 m below the surface in a layer with undulating boundaries that are shown in the figure. The model permits transmission into the basement, which has different sound speed, density, and attenuation than the layer, but does not permit transmission into the air above the top boundary. The contours are spaced by 5 dB.

associated with the evanescent modes were resolved (COLLINS, 1989; WETTON and BROOKE, 1990; COLLINS, 1991). The development of the elastic parabolic equation remains an active area of research. Some progress has been made in the development of approaches for accurately handling range dependence in the properties of the medium (COLLINS and SIEGMANN, 1999), but this is an area that requires further work. The elastic parabolic equation was originally developed for applications in ocean acoustics that involve elastic sediments but is readily adaptable to the seismic case.

We consider a two-dimensional problem involving a transversely isotropic medium (FREDRICKS *et al.*, 2000). In each range-independent region, the horizontal and vertical displacements u and w satisfy

$$(\lambda + 2\mu + 2v)\frac{\partial^2 u}{\partial x^2} + (\lambda + 4\kappa)\frac{\partial^2 w}{\partial x \partial z} + \frac{\partial}{\partial z}\left(\mu\frac{\partial w}{\partial x}\right) + \frac{\partial}{\partial z}\left(\mu\frac{\partial u}{\partial z}\right) + \rho\omega^2 u = 0 \;, \quad (24)$$

$$\mu\frac{\partial^2 w}{\partial x^2} + \mu\frac{\partial^2 u}{\partial x \partial z} + \frac{\partial}{\partial z}\left((\lambda + 4\kappa)\frac{\partial u}{\partial x}\right) + \frac{\partial}{\partial z}\left((\lambda + 2\mu - 2v)\frac{\partial w}{\partial z}\right) + \rho\omega^2 w = 0 \;, \quad (25)$$

where λ and μ are the Lamé constants, $\kappa = v = 0$ for an isotropic medium, and ω is the circular frequency. The anisotropic compressional and shear-wave speeds $c_p(\theta)$ and $c_s(\theta)$ correspond to the roots of the determinant,

$$a_{11}a_{22} - a_{12}a_{21} = 0 \;, \quad (26)$$

$$a_{11} = (\lambda + 2\mu + 2v)\cos^2\theta + \mu\sin^2\theta - \rho c^2 \;, \quad (27)$$

$$a_{12} = a_{21} = (\lambda + \mu + 4\kappa)\cos\theta\sin\theta \;, \quad (28)$$

$$a_{22} = (\lambda + 2\mu - 2v)\sin^2\theta + \mu\cos^2\theta - \rho c^2 \;, \quad (29)$$

where $\theta = 0$ corresponds to horizontal propagation and $\theta = \pi/2$ corresponds to vertical propagation (FREDRICKS *et al.*, 2000).

Equations (24) and (25) are in the form,

$$\left(R\frac{\partial^2}{\partial x^2} + S\frac{\partial}{\partial x} + T\right)\begin{pmatrix} u \\ w \end{pmatrix} = \begin{pmatrix} 0 \\ 0 \end{pmatrix} \;, \quad (30)$$

where R, S, and T are depth operators. Due to the presence of the first-order term in $\partial/\partial x$, it is difficult to factor the operator in Equation (30). There exist several formulations in which the elastic wave equation factors. The first successful implementations of the elastic parabolic equation were based on a formulation in terms of w and the dilatation $\Delta = \partial u/\partial x + \partial w/\partial z$. The elastic wave equation also factors in a formulation in terms of the variables w and $u_x = \partial u/\partial x$. One of the advantages of this formulation is that the dependent variables are continuous across horizontal interfaces. These variables satisfy the equations,

$$(\lambda + 2\mu + 2\nu)\frac{\partial^2 u_x}{\partial x^2} + (\lambda + 4\kappa)\frac{\partial^3 w}{\partial x^2 \partial z} + \frac{\partial}{\partial z}\left(\mu\frac{\partial^2 w}{\partial x^2}\right) + \frac{\partial}{\partial z}\left(\mu\frac{\partial u_x}{\partial z}\right) + \rho\omega^2 u_x = 0 \ ,$$

$$(31)$$

$$\mu\frac{\partial^2 w}{\partial x^2} + \mu\frac{\partial u_x}{\partial z} + \frac{\partial}{\partial z}((\lambda + 4\kappa)u_x) + \frac{\partial}{\partial z}\left((\lambda + 2\mu - 2\nu)\frac{\partial w}{\partial z}\right) + \rho\omega^2 w = 0 \ , \qquad (32)$$

which are in the form,

$$\left(L\frac{\partial^2}{\partial x^2} + M\right)\begin{pmatrix} u_x \\ w \end{pmatrix} = \begin{pmatrix} 0 \\ 0 \end{pmatrix} \ , \qquad (33)$$

where L and M are depth operators.

Multiplying Equation (33) by L^{-1} and rearranging, we obtain

$$\left(\frac{\partial^2}{\partial x^2} + k_0^2(1 + X)\right)\begin{pmatrix} u_x \\ w \end{pmatrix} = \begin{pmatrix} 0 \\ 0 \end{pmatrix} \ , \qquad (34)$$

$$X = k_0^{-2}(L^{-1}M - k_0^2) \ . \qquad (35)$$

Factoring the operator in Equation (34), we obtain

$$\left(\frac{\partial}{\partial x} + ik_0(1 + X)^{1/2}\right)\left(\frac{\partial}{\partial x} - ik_0(1 + X)^{1/2}\right)\begin{pmatrix} u_x \\ w \end{pmatrix} = \begin{pmatrix} 0 \\ 0 \end{pmatrix} \ . \qquad (36)$$

We neglect back-scattered energy, which is assumed to be dominated by outgoing energy, and obtain the parabolic wave equation,

$$\frac{\partial}{\partial x}\begin{pmatrix} u_x \\ w \end{pmatrix} = ik_0(1 + X)^{1/2}\begin{pmatrix} u_x \\ w \end{pmatrix} \ . \qquad (37)$$

This equation can be solved using generalizations of the approaches described in Section 3 for the acoustic parabolic equation, but special care must be taken for the elastic case in the design of the rational approximation to avoid instabilities associated with evanescent modes (WETTON and BROOKE, 1990; COLLINS, 1991).

We illustrate the elastic parabolic equation solution for an isotropic problem involving a 25 Hz source that is located 10 m below the surface in a 200-m thick layer overlying a basement. In the layer, the compressional speed is 2400 m/s, the shear speed is 1200 m/s, and the density is 1.2 g/cm³. In the basement, the compressional speed is 2800 m/s, the shear speed is 1400 m/s, and the density is 1.5 g/cm³. The compressional attenuation is 0.1 dB/λ and the shear attenuation is 0.2 dB/λ throughout the medium. A plot of the compressional transmission loss $-20\log_{10}|\Delta|$ appears in Figure 2. An interference pattern between a Rayleigh wave and guided waves appears near the surface.

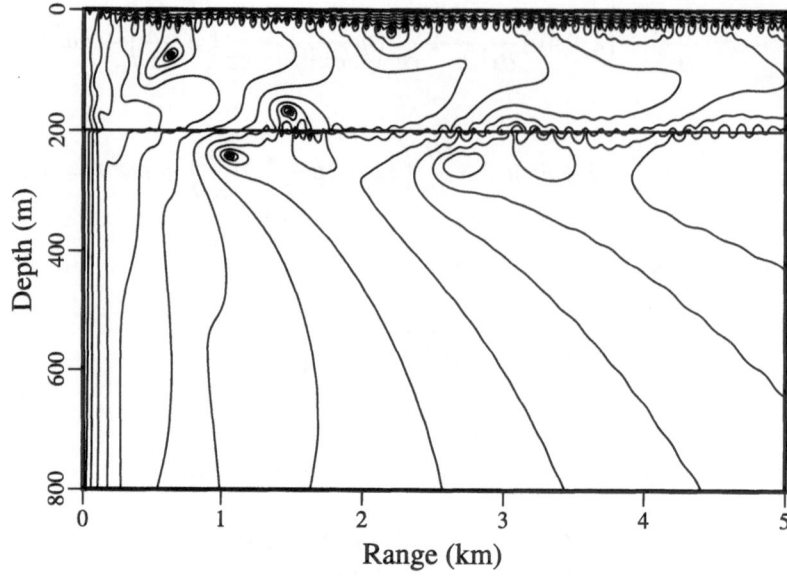

Figure 2
Compressional transmission loss contours for an elastic problem involving a 25 Hz source that is located 10 m below the surface in a 200 m thick layer. There is a contrast in wave speeds and density across the interface at $z = 200$ m. The interference pattern near the surface is due to a Rayleigh wave and guided waves in the layer. The contours are spaced by 5 dB.

5. Conclusion

Parabolic equation techniques that have become highly developed in ocean acoustics are adaptable to seismic problems. When shear waves can be ignored, the acoustic parabolic equation can be applied to seismic problems involving complex layering, variable topography, gradual range dependence, strong depth dependence, and wide propagation angles. The elastic parabolic equation handles all wave types accurately but requires further development for handling variable topography and other types of range dependence.

Acknowledgments

This work was supported by the Office of Naval Research.

REFERENCES

BAMBERGER, A., ENGQUIST, B., HALPERN, L., and JOLY, P. (1988), *Higher Order Paraxial Wave Equation Approximations in Heterogeneous Media*, SIAM J. Appl. Math. *48*, 129–154.

CEDERBERG, R. J., and COLLINS, M. D. (1997), *Application of an Improved Self-starter to Geoacoustic Inversion*, IEEE J. Ocean. Eng. *22*, 102–109.

COLLINS, M. D. (1989), *A Higher-order Parabolic Equation for Wave Propagation in an Ocean Overlying an Elastic Bottom*, J. Acoust. Soc. Am. *86*, 1459–1464.

COLLINS, M. D. (1991), *Higher-order Parabolic Approximations for Accurate and Stable Elastic Parabolic Equations with Application to Interface Wave Propagation*, J. Acoust. Soc. Am. *89*, 1050–1057.

COLLINS, M. D. (1992), *A Self-starter for the Parabolic Equation Method*, J. Acoust. Soc. Am. *92*, 2069–2074.

COLLINS, M. D. (1993a), *A Split-step Padé Solution for the Parabolic Equation Method*, J. Acoust. Soc. Am. *93*, 1736–1742.

COLLINS, M. D. (1993b), *An Energy-conserving Parabolic Equation for Elastic Media*, J. Acoust. Soc. Am. *94*, 975–982.

COLLINS, M. D. (1999), *The Stabilized Self Starter*, J. Acoust. Soc. Am. *106*, 1724–1726.

COLLINS, M. D., COURY, R. A., and SIEGMANN, W. L. (1995), *Beach Acoustics*, J. Acoust. Soc. Am. *97*, 2767–2770.

COLLINS, M. D., and SIEGMANN, W. L. (1999), *A Complete Energy Conservation Correction for the Elastic Parabolic Equation*, J. Acoust. Soc. Am. *105*, 687–692.

COLLINS, M. D., and WESTWOOD, E. K. (1991), *A Higher-order Energy-conserving Parabolic Equation for Range-dependent Ocean Depth, Sound Speed, and Density*, J. Acoust. Soc. Am. *89*, 1068–1075.

FREDRICKS, A. J., SIEGMANN, W. L., and COLLINS, M. D. (2000), *A Parabolic Equation for Anisotropic Elastic Media*, Wave Motion *31*, 139–146.

GREENE, R. R. (1985), *A High-angle One-way Wave Equation for Seismic Wave Propagation along Rough and Sloping Interfaces*, J. Acoust. Soc. Am. *77*, 1991–1998.

JENSEN, F. B., KUPERMAN, W. A., PORTER, M. B., and SCHMIDT, H., *Computational Ocean Acoustics* (American Institute of Physics, New York 1994), pp. 343–412.

LANDERS, T., and CLAERBOUT, J. F. (1972), *Numerical Calculations of Elastic Waves in Laterally Inhomogeneous Media*, J. Geophys. Res. *77*, 1476–1482.

LINGEVITCH, J. F., COLLINS, M. D., and SIEGMANN, W. L. (1999), *Parabolic Equations for Gravity and Acousto-gravity Waves*, J. Acoust. Soc. Am. *105*, 3049–3056.

PORTER , M. B., JENSEN, F. B., and FERLA, C. M. (1991), *The Problem of Energy Conservation in One-way Models*, J. Acoust. Soc. Am. *89*, 1058–1067.

RADDER, A. C. (1979), *On the Parabolic Equation Method for Water-wave Propagation*, J. Fluid Mech. *95*, 159–176.

TAPPERT, F. D., *The parabolic approximation method*, In *Wave Propagation and Underwater Acoustics* (eds. J. B. Keller and J. S. Papadakis) Lecture Notes in Physics, vol. 70 (Springer, New York 1977).

WETTON, B. T. R., and BROOKE, G. H. (1990), *One-way Wave Equations for Seismoacoustic Propagation in Elastic Waveguides*, J. Acoust. Soc. Am. *87*, 624–632.

(Received December 14, 2000, revised/accepted May 22, 2001)

Pure appl. geophys. 159 (2002) 1691–1706
0033–4553/02/081691–16 $ 1.50 + 0.20/0

❘ Pure and Applied Geophysics

A Full Waveform Test of the Southern California Velocity Model by the Reciprocity Method

LEO EISNER[1,2] and ROBERT W. CLAYTON[1]

Abstract—We apply the reciprocity method (EISNER and CLAYTON, 2001a) to compare the full waveform synthetic seismograms with a large number of observed seismograms. The reciprocity method used in the finite-difference modeling allows for the use of high quality data observed from the earthquakes distributed over the wide range of azimuths and depths. We have developed a methodology to facilitate the comparison between data and synthetics using a set of attributes to characterize the seismograms. These attributes are maximum amplitude, time delay and coda decay of the magnitude of the displacement vector. For the Southern California Velocity Model, Version 1 (MAGISTRALE *et al.*, 1996), we have found misfits between data and synthetics for paths traveling outside of the sedimentary basins and the western part of the Los Angeles and San Fernando basins.

Key words: Reciprocity, Southern California, finite difference, velocity model.

Introduction

With better numerical techniques to evaluate the seismic wave propagation in complex three-dimensional (3-D) heterogeneous media, the need for realistic velocity models arises. GRAVES and WALD (2001) show the necessity of well tested velocity models for source inversions. OLSEN and ARCHULETA (1996), EISNER and CLAYTON (2001c) apply finite-difference modeling in a 3-D velocity model to evaluate realistic long period site effects for Southern California. However, the outstanding issue is how well the 3-D models describe the earth properties important for the seismic wave propagation. The ultimate test of these models is how well they predict the observed full waveforms. Several studies have used the Southern California velocity models for simulations of Landers (OLSEN *et al.*, 1997; WALD and GRAVES, 1998) and Northridge (OLSEN and ARCHULETA, 1996) earthquakes and compared the full waveforms synthetics to the observed seismograms recorded during these earthquakes.

The current procedure is to evolve the wavefield outward from the source to a suite of observation points and compare the synthetics to the recorded data. This

[1] Seismological Laboratory, California Institute of Technology, Pasadena, CA, U.S.A.
[2] Present address: Schlumberger Cambridge Research, High Cross, Madingley Rd., Cambridge, CB30EL, U.K.

generally means one simulation for each source. We propose a method which will allow us to compare data and synthetics for a large number of source-receiver pairs with one run. By using reciprocal sources we can reduce the amount of calculations when determining synthetics for earthquakes at a few selected high quality stations. Furthermore, we can select stations which have available records for a multitude of earthquakes. EISNER and CLAYTON (2001a) show the reciprocity method for the above described application and discuss its numerical implementation and accuracy with the finite-difference technique.

We propose to use numerous small earthquakes computed with a finite-difference technique (GRAVES, 1996) to test the Southern California Velocity Model, Version 1.0, (MAGISTRALE *et al.*, 1996). The simulation of a large number of small earthquakes has several advantages; small earthquakes are generally distributed throughout the entire model, and the spatial distribution of sources enables us to illuminate the model from many azimuths. The depth variation of sources enables us to distinguish between the effects of the shallow and deep earthquakes. By including the weak motion data (small earthquakes) into this test of the velocity model, we are able to test regions where no large earthquake previously occurred, but which are potentially hazardous. Figure 1 illustrates the advantage of the proposed method. The velocity model is tested along many source-receiver paths which cross each other. The path crossing can be used to determine the sources of the discrepancies between the observed and synthetic seismograms.

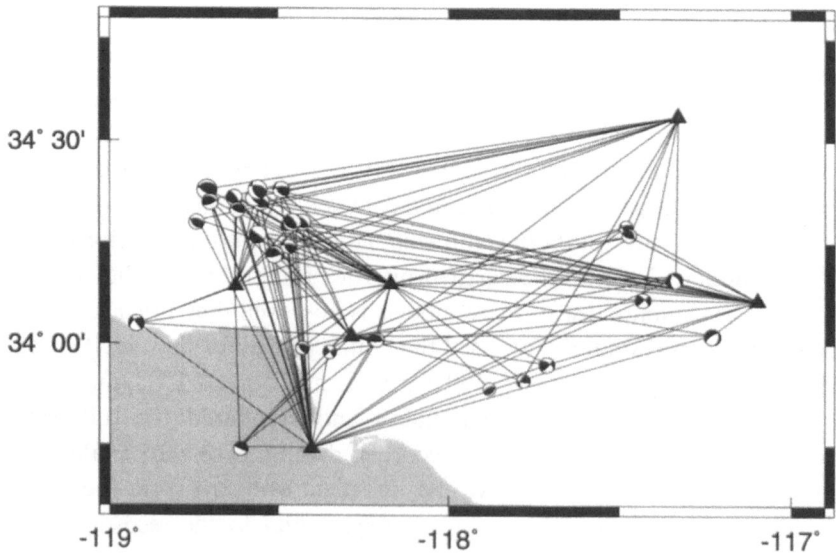

Figure 1

Map of Southern California showing the selected earthquakes and broadband stations (triangles) with straight lines connecting the epicenters and the receivers of the epicenter-receiver pairs used in the back-projection of the time shifts and coda decay.

WALD and GRAVES (1998) show that even for the long period data, observations and synthetics do not match in phase and amplitude. The real earth has more complexity than we are likely able to include in our model, and hence we do not expect an exact match of synthetics and data. In this study, we are interested in matching the main energy of the synthetic and observed seismograms; therefore the discrepancy between data and synthetics is measured by comparing simple attributes of the seismograms. These discrepancies in attributes can be used to determine the regions of the model that appear to be in error. We propose to determine these attributes from the time history of the displacement magnitude. The magnitude of the displacement provides a simple scalar quantity with which one can monitor timing, amplitude and coda of the seismograms. We chose to characterize the fit between synthetic seismograms and data by measuring the following attributes: the time shift (the shift of the synthetic seismogram for which it best matches the observed seismogram), maximum amplitude and coda decay (the rate at which the amplitude decays to zero). Therefore, a "good fit" in our study is a match of timing, coda decay, and maximum amplitude between the displacement magnitude of the observed and synthetic seismograms.

The previous studies (OLSEN et al., 1997; WALD and GRAVES, 1998; OLSEN and ARCHULETA, 1996) included triggered or incomplete seismograms in order to test the entire model. Some incomplete records from low quality stations are terminated before arrival of the later phases; this is a common problem for triggered seismograms. Since our study is based on weak motion data, we use broadband complete observed seismograms with absolute timing. We also develop a method of interpreting the differences between data and synthetics which is robust and uses the entire three-component seismograms. In this study, we only indicate the regions of the model that are inadequate. We do not attempt to update the model. This study of the Southern California Velocity Model, Version 1, includes the top low-velocity layers which are important for propagation of waves within a period range of interest.

The Testing Procedure

We apply the reciprocity method to simulate multiple sources recorded at a few receivers to reduce the amount of calculations. By invoking reciprocity, the number of simulations can be reduced to three times the number of receivers (EISNER and CLAYTON, 2001a). This method can also provide suites of source mechanisms and locations. For the example here, with 6 receivers and 32 sources (Fig. 2), we need only 18 simulations versus 32 simulations using the forward technique. If we had also wanted to include a variable double-couple mechanism for the point-source, the reciprocity method would still have required only 18 simulations versus 160 (5 moment tensor elements times 32 source locations) simulations with the direct

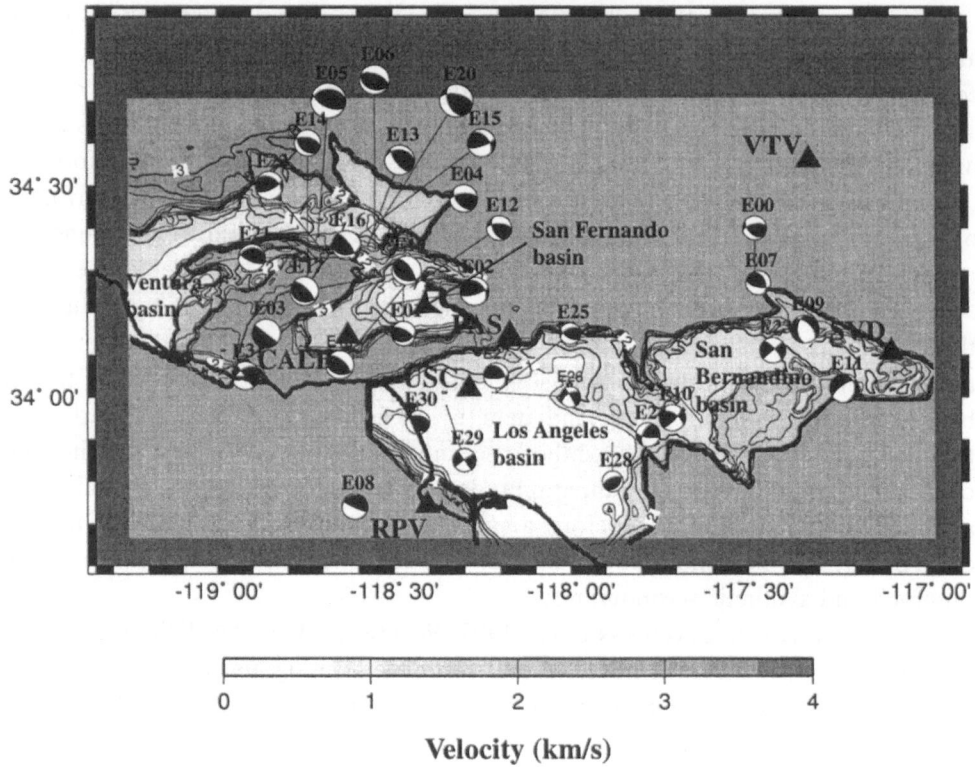

Figure 2

Map of Southern California showing the selected earthquakes (see Table 1), and broadband stations (triangles) for the test of the Southern California Velocity Model, Version 1 (MAGISTRALE et al., 1996). The shading and contours correspond to the Love wave group velocity of three seconds period. The contours are labeled at 1.0 and 2.0 km/sec and the contour interval is 0.25 km/sec. Four stations: Pasadena (PAS), Rancho Palos Verdes (RPV), Calabasas (CALB) and University of Southern California (USC) are situated in or near the deep parts of the Los Angeles and San Fernando basins. Two stations, Victorville (VTV) and Seven Oaks Dam (SVD), are outside of the major basins.

method. Source relocation would further increase the cost of the direct method but not of the reciprocal method.

We develop a new set of measurements to characterize the misfit. We use the magnitude of the displacement (MOD, vector length of all 3 components) as our basic measure

$$\mathrm{MOD}(t) = \sqrt{u_{\mathrm{rad}}^2(t) + u_{\mathrm{tra}}^2(t) + u_{\mathrm{up}}^2(t)} \ . \qquad (1)$$

Here $\mathrm{MOD}(t)$ is the time history of MOD and $u_{\mathrm{rad}}(t), u_{\mathrm{tra}}(t), u_{\mathrm{up}}(t)$ are the time histories of the individual components. Use of the MOD is convenient, because it allows a scalar quantity to represent the three components of a vector, and it is particularly useful in 3-D heterogeneous media where there is no simple decomposition of the seismogram into distinct phases, such as *SV* or *SH*, or surface waves such as Love or Rayleigh waves. It is a convenient measure of the first-order fit between data and synthetics that is sensitive to the travel time, amplitude and strength of coda. Furthermore, the MOD is not zero in the nodal direction of the radiation pattern, making it more suitable for comparison of amplitude ratio of data over synthetics. An entire three-component seismogram can be described by three time dependent spherical coordinates (an MOD, a particle motion's azimuth and a particle motion's declination). The azimuth and the declination depend on the direction from which a wave arrives and the type of the wave. In this study we are primarily interested in whether the model sufficiently represents the real earth so that we can reproduce main scattered waves in our numerical simulations. Consequently, we have chosen to base our comparison on measuring characteristics of the MOD, as it is more sensitive to the propagation effects of the large energy arrivals in a seismogram than the smaller arrivals. Beside MOD there are other variables suitable for measuring a fit between the data and synthetic seismograms in a 3-D space (e.g., measuring absolute distance between particle motion in the synthetic and the observed seismograms); however, MOD conveniently describes the criteria of the fit we were interested in: the timing, the coda decay and the maximum amplitude.

The MOD time histories of the real and synthetic data are compared by correlation to obtain the maximum cross-correlation and the time shift at which the maximum occurs. The value of the maximum cross-correlation determines the quality of the fit between data and synthetics. Since the cross-correlation of the MODs is a cross-correlation of the two positive functions, the mean value of the cross-correlation is 0.5. The time of the maximum correlation is a time shift of the synthetic seismogram for which it best matches the observed seismogram. Note that this definition of the time shift does not depend on an *a priori* selection of phases. Since the correlation is dominated by the maximum amplitude, we are likely determining the variations in surface-wave group velocities. However, this interpretation depends on the distance between the source and the receiver, source depth, source mechanism and several other parameters. The time shift between synthetics and the data is caused by the velocity deviation in the model between the source and the receiver. The time delays due to errors in horizontal locations of the earthquakes are not large enough to explain the observed time delays. A source mislocation should not appear as a consistent time delay in our model as the earthquakes are located by a very dense network of the stations, and time delays due to the mislocation are lower than time delays observed in this study. The noise in the observed seismograms or mismatch between the data and the synthetics may cause the cross-correlation to peak at a time shifted by a dominant period (cycle-skip). Therefore, the cross-correlation is tapered for time shifts longer than the dominant period of the signal to avoid this. We taper the cross-correlation function for times longer than the shortest period used in our signal (no shorter period can be a dominant period). We can invert the time shifts for a slowness variation with a tomographic method to show which parts of the model are most likely in error. Assuming most of the energy in the data and the synthetic seismograms travels along a straight line between the source and the receiver, we chose a simple back-projection method to map the time shifts into lines connecting corresponding epicenters and receivers. This is a simplification of the actual ray paths, however it gives us a good estimate regarding which regions of the model cause systematic time shifts.

To invert we divide the model into cells and the average slowness deviation of the i-th cell is determined by a simple inversion of the time shifts:

$$ds_i = \frac{\sum_j (L_{ij} dt_j)}{P_i + D} \ . \tag{2}$$

Here ds_i is a slowness deviation in i-th cell, dt_j is the time shift of the j-th epicenter-receiver pair, L_{ij} is the length of the straight line between the j-th epicenter-receiver pair in the i-th cell, D is damping, and $P_i = \sum_j L_{ij} L_{ji}$.

To measure the coda decay, we use a sliding window average (MONTALBETTI and KANASEWICH, 1970) of MOD. An exponential decay of the form

$$M(t) = M_0 \cdot e^{-b \cdot |t - t_0|}, \quad \text{for } t \geq t_0 \tag{3}$$

is fitted for the time $t \geq t_0$ by least-squares. Here t_0 is the time of the maximum of the $\text{MOD}(t)$, $M(t)$ is the sliding window averaged $\text{MOD}(t)$, b characterizes the decay of the coda, and M_0 is the maximum of the $M(t)$: $M_0 = M(t_0)$. The exponential decay of the $M(t)$ can be derived from the exponential decay of the energy at a seismogram computed for a random isotropic scattering medium (ZENG et al., 1991). Based on observation of the exponential decay of coda in data, we use this rate of decay as a first-order approximation for the coda decay of the long period signal.

We compare the coda of the synthetics and the data by comparing the decay of synthetics and the data if the maximum crosscorrelation is above 0.8. This level ensures we are looking at differences in coda decay and not simply misfit of entire seismograms. The coda is a measure of the complexity of the model, and we interpret it in the following manner: if b is larger for the synthetics than for the data, then our model does not generate enough coda and is lacking in complexity; if b is larger for the data than the synthetics, our model is too complex and generates excessive coda. Assuming small-angle scattering (WU and AKI, 1988) we may estimate that the sources of the observed scattering occur along a straight path between the epicenter and the receiver. This assumption is valid for

$$2\pi A \gg \lambda \ ,$$

where A is characteristic size of heterogeneity and λ dominating wavelength. An exact inversion for the scattering sources is beyond the scope of this study, however the proposed method identifies the regions of the model which consistently cause a discrepancy in the scattered energy between synthetics and data. An analogous back-projection can be used to identify these regions in the analogous manner as with the time shift anomalies:

$$dE_i = \frac{\sum_j (L_{ij} de_j)}{P_i + D} \quad \text{where } de_j = \left(\frac{b_j^{\text{data}} - b_j^{\text{synt}}}{\sqrt{b_j^{\text{synt}} b_j^{\text{data}}}} \right). \tag{4}$$

Here b_j^{synt} and b_j^{data} are determined from the fit of the synthetics and data (respectively) of equation (3), and dE_i is a relative error of the b value per distance in the i-th cell. The regions with positive dE_i are areas with too much scattering in the model and *vice versa*.

The last attribute we compare is M_0, the maximum of the MOD(t), which characterizes the overall source strength and model amplification. OLSEN and ARCHULETA (1996) and WALD and GRAVES (1998) have shown that the model amplification is well predicted by the 3-D velocity model for well constrained sources of large earthquakes. That is, they fit the maximum amplitudes within a factor of 2 between the data and the synthetics. As we have observed larger discrepancies of the maximum amplitude, we assume that the ratio of the maximum amplitude of the synthetics to data is not on average biased due to the 3-D velocity model, and we interpret it as a bias due to the strength (magnitude) of the source. For each earthquake we compute the ratio of the maximum the MOD(t) of the synthetics to data over all stations. We average these ratios over all stations used in a study. If the averaged ratio deviates significantly from 1.0, the estimated source magnitude is incorrect. Values larger than 1.0 can be interpreted as overestimated magnitude and values smaller than 1.0 can be interpreted as underestimated magnitude of an earthquake source.

Application to the Southern California Velocity Model

The reciprocity method and the measurement of the attributes discussed in the previous section are now applied to test the Southern California Velocity Model (SCVM), Version 1 (MAGISTRALE *et al.*, 1996). The model consists of sedimentary basins placed in a 1-D (HADLEY and KANAMORI, 1977) background medium. The sedimentary basin portions are based on geologic information of the surface geology and depth-to-basement rock, and other geological information. Note that the sedimentary basins have a very irregular shape and therefore the synthetic seismograms computed in this model are very sensitive to the source location.

Figure 2 and Table 1 show the selected earthquakes and their parameters (respectively) used in this study. The earthquakes represent the best spatial distribution of small earthquakes with a good signal-to-noise-ratio in periods of three seconds and longer. At shorter periods smaller velocity variations cause discrepancies between the data and synthetics, but this is compensated by having

Table 1

List of the selected earthquakes: Latitude (Lat.), Longitude (Lon.), Depth of the event, Strike, Dip and Rake use convention of AKI *and* RICHARDS *(1980), M is a magnitude of an earthquake, and ZH denotes source parameters determined by* ZHU *and* HELMBERGER *(1996); H by* HAUKSSON *(2000)*

Earth. name	Lat. (°)	Lon. (°)	Depth (km)	Strike (°)	Dip (°)	Rake (°)	M	Date yyyy/mm/dd	Source
E00	34.29	−117.48	11.1	258	61	52	3.4	1993/05/18	ZH
E01	34.29	−118.47	12.4	71	34	47	3.5	1994/01/19	ZH
E02	34.22	−118.51	14.8	74	72	61	4.1	1994/01/19	ZH
E03	34.27	−118.56	13.3	34	0	−4	4.2	1994/01/27	ZH
E04	34.38	−118.49	5.0	264	59	51	3.9	1994/01/28	ZH
E05	34.38	−118.71	13.3	266	34	65	5.0	1994/01/19	ZH
E06	34.35	−118.55	10.3	291	58	90	4.3	1994/01/24	ZH
E07	34.27	−117.47	11.3	104	62	58	3.5	1995/04/04	ZH
E08	33.74	−118.61	21.0	244	10	41	3.7	1995/03/01	ZH
E09	34.16	−117.34	9.8	181	48	−56	4.0	1997/06/28	ZH
E10	33.95	−117.71	9.9	43	75	32	3.8	1998/01/05	ZH
E11	34.02	−117.23	16.6	235	70	−68	4.1	1998/03/11	ZH
E12	34.24	−118.47	14.6	272	60	47	3.6	1994/01/18	ZH
E13	34.36	−118.57	8.9	279	30	51	4.2	1994/01/19	ZH
E14	34.30	−118.74	10.6	283	56	72	3.8	1994/01/19	ZH
E15	34.36	−118.63	13.6	71	72	49	4.0	1994/01/24	ZH
E16	34.36	−118.63	16.4	84	32	35	4.1	1994/01/24	ZH
E17	34.30	−118.45	11.1	103	24	61	4.1	1994/01/21	ZH
E18	34.30	−118.43	9.9	107	32	69	4.0	1994/01/23	ZH
E19	34.30	−118.46	10.7	111	29	60	4.2	1994/01/21	ZH
E20	34.38	−118.56	10.6	281	57	51	4.7	1994/01/18	ZH
E21	34.33	−118.62	15.8	257	27	57	3.9	1994/01/18	ZH
E22	34.35	−118.70	16.9	65	66	59	4.0	1996/05/01	ZH
E23	34.11	−117.43	5.6	48	90	16	3.7	1997/10/14	ZH
E24	33.91	−117.78	9.0	203	54	21	3.4	1997/01/31	ZH
E25	34.01	−118.21	12.9	120.0	55.0	130.0	3.1	1999/05/03	H
E26	34.01	−118.22	13.1	330.0	85.0	−150	3.0	1999/06/01	H
E27	34.01	−118.21	12.5	85.0	70.0	60.0	3.5	1999/05/30	H
E28	33.89	−117.88	5.1	45.0	35	70	3.0	1998/01/07	H
E29	33.98	−118.35	4.0	235	85	0	3.3	1997/04/04	H
E30	33.99	−118.43	13.8	145	60	120	3.4	1994/12/11	H
E31	34.05	−118.92	18.8	236	50	5	4.0	1995/02/19	ZH

more earthquakes that have a good signal-to-noise-ratio and thus improving the test of the velocity model. We use a triangular moment rate source time function of three seconds length in the modeling of the synthetic seismograms. The instrument response is removed from the observed data.

We have not inverted for source location or mechanism in our study because we use only a limited number of stations and therefore we would introduce an artificial bias due to station distribution. We use two catalogues of earthquake parameters: the primary catalogue of ZHU and HELMBERGER (1996) for 26 earthquakes with magnitude $5.5 > M_w > 3.4$ and a secondary catalog of earthquakes' parameters of

HAUKSSON (2000) for six earthquakes not listed by ZHU and HELMBERGER (1996). ZHU and HELMBERGER (1996) use surface waves and body waves to determine earthquake mechanisms and locations in a 1-D velocity model of Southern California. HAUKSSON (2000) uses direct first motion arrivals of P and S waves in a laterally heterogeneous model based on tomographic inversion. The catalogue of HAUKSSON (2000) lists also source parameters of the catalogue of ZHU and HELMBERGER (1996); therefore, we could compare the full waveform fit to the data for the source parameters listed in both catalogues. We found that the source parameters of ZHU and HELMBERGER (1996) fit the data better, especially the surface waves.

The lowest velocity in the model is clamped at 0.5 km/sec to allow the surface waves of three second and longer periods to maintain the same group velocities between the values in the velocity clamped and the original models. The velocity clamping at 0.5 km/sec replaces all velocities lower than 0.5 km/sec with 0.5 km/sec. For the SCVM, Version 1, only the S-wave velocities were clamped with the value of 0.5 km/sec in the top 600 meters. The simple velocity clamping is not the best method to preserve the surface-wave velocities for a velocity model (see EISNER and CLAYTON, 2001b for a detailed analysis); however, if the velocity clamping value (0.5 km/sec in this case) is sufficiently lower than the group velocities of the original model, it approximately maintains the same group velocities as in the original model. Figure 2 shows the Love wave group velocities in the SCVM, Version 1. The slowest regions of the Love wave group velocities are between 0.5 km/sec and 0.75 km/sec for the period of three seconds (the minimum is exactly 0.51 km/sec at 34.197°N latitude and 118.332°W longitude). Therefore, the wavelength of the surface waves propagating in the sedimentary basins is 1.5–2.1 km. We do not use attenuation in our modeling, since it is not part of the SCVM, Version 1. The attenuation would decrease the amount of the coda in the synthetic seismograms, which already tends to be underestimated by the model.

The Analyses of Individual Source-receiver Pairs

In this section we present examples of the fit between the synthetics and the data. Figure 3 shows an example of a good fit of synthetics to data in the presence of strong lateral heterogeneity between the earthquake E10 and the station PAS. The synthetic seismograms reproduce the late scattered arrivals fairly well on all components. The maximum crosscorrelation is at 0.92 and the time shift is 1.4 seconds, indicating that the model is slower on average than the data. The amplitude ratio of synthetics to data is 2.3, indicating the moment magnitude for this earthquake is overestimated (the average moment for the earthquake E10 is overestimated by the factor of 1.7). The coefficient of the decay is 30% larger for the synthetics.

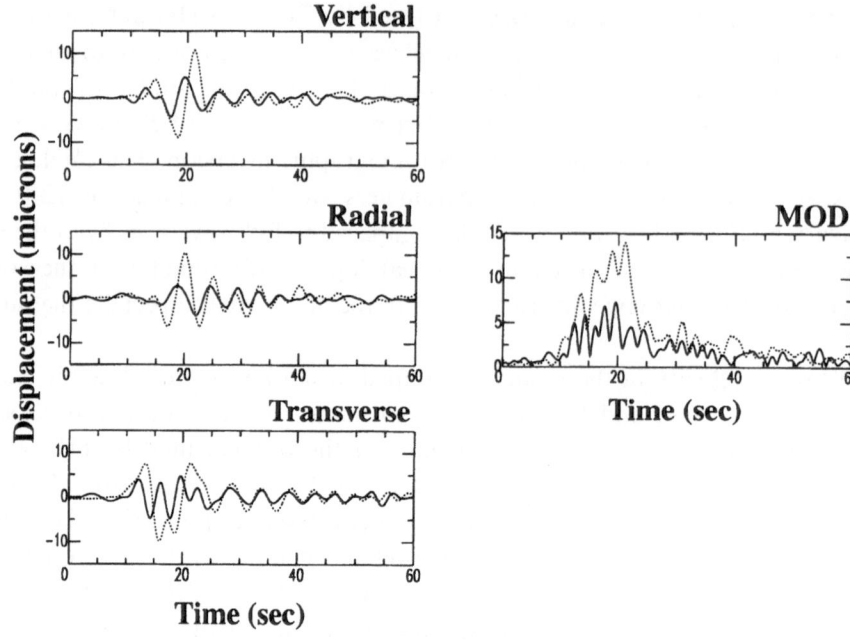

Figure 3

Example of a good fit between data and the synthetics for the earthquake E10 recorded at the station PAS (see Table 1). The seismograms show displacement in microns. Both synthetic seismograms and data are filtered between 3 and 20 seconds and the instrument response was removed. Data shown by solid line; synthetics by dashed line.

Figure 4 indicates another example of a good fit for the earthquake E24 at the station VTV. This example demonstrates the advantage of using the MOD to compare the synthetic seismograms and the data as the signal to noise ratio on the MOD component is better than on any of the vertical, radial or transverse components. This is also an example of the data with the worse signal-to-noise-ratio used for the inversion of the coda decay, maximum amplitude, or time shifts. The value of the maximum cross-correlation is 0.92 and the time shift is +0.25 sec. The amplitude ratio of synthetics to data is 1.25. The coefficient of decay is 10% larger for the synthetics. The synthetic seismograms match the timing and phase of the surface waves (time 30–40 seconds) well and no significant scattered energy arrives after the main pulse in either the observed or the synthetic seismograms.

Figure 5 shows the comparison of seismograms in which there are significant discrepancies between the synthetics and data for the station RPV and the earthquake E13. The first arrivals (15–30 sec) match in phase, timing and amplitude on all components extremely well; however, the later phases of data and synthetics diverge. The synthetic seismograms do not show large arrivals after 35 seconds, but the data show many large arrivals. The hypocenter of the earthquake E13 is at a

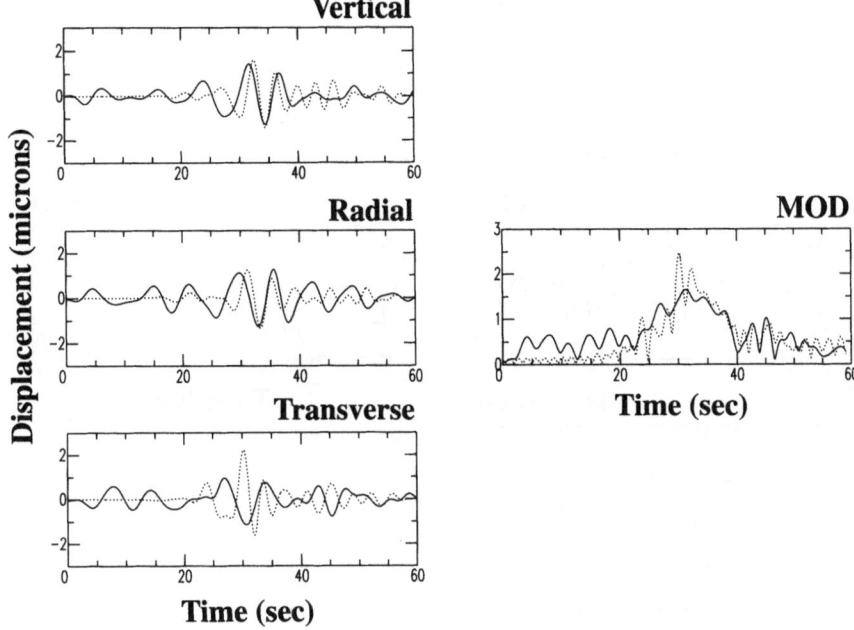

Figure 4

Example of a good fit between data and the synthetics for the earthquake E24 recorded at the station VTV (see Table 1). The seismograms show displacement in microns. Both synthetic seismograms and data are filtered between 3 and 20 seconds and the instrument response was removed. Data shown by solid line; synthetics by dashed line.

shallow depth (8.9 km) and therefore the earthquake excites surface waves. These waves propagate through the strongly heterogeneous San Fernando and western part of the Los Angeles basins before they are observed at the station RPV. The lack of scattered energy in the synthetic seismograms indicates these basins may be too simple in the velocity model. The maximum value of the cross-correlation is 0.87 and it is shifted by −1.5 seconds. The amplitude ratio is 0.84 and the coefficient of decay is 80% larger for the synthetics.

Figure 6 shows the comparison of data and synthetic seismograms at the station SVD for a shallow earthquake E04. The direct S wave and the surface waves (40 + sec) arrive ahead of the data. The latter arrivals observed in the data also exhibit more complexity not reproduced in the velocity model (50 + sec). A large portion of the path between the earthquake E04 and the station SVD is outside of the sedimentary basins and therefore the likely explanation of the timing shift is the fast background model (as was also observed by WALD and GRAVES, 1998). The lack of coda (50 + sec) in the synthetic seismograms indicates the background model should also have more complexity in order to explain the data. The maximum value of the cross-correlation is 0.81 and it is shifted by −1.8 sec. The amplitude ratio is 1.7 and the coefficient of decay is 400% larger for the synthetics.

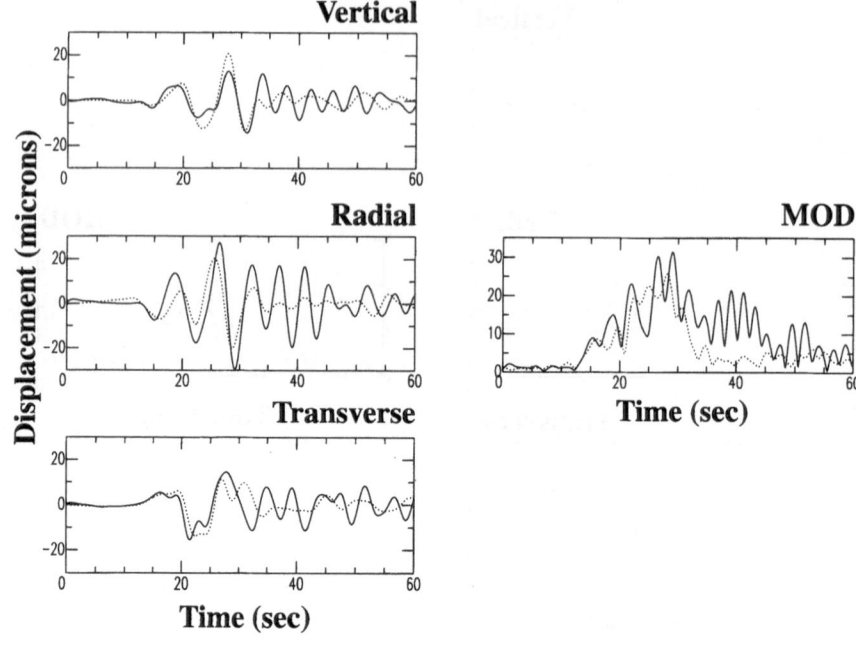

Figure 5

Example of a poor fit between the data and the synthetics for the earthquake E13 recorded at the station RPV (see Table 1). The seismograms show displacement in microns. Both synthetic seismograms and data are filtered between 3 and 20 seconds and the instrument response was removed. Data shown by solid line; synthetics by dashed line.

Errors of the Velocity Model

Finally we have used the back-projection techniques described earlier to summarize the comparison of all the synthetic and observed seismograms. We have used only seismograms with maximum cross-correlation higher than 0.8 within a maximum time shift of three seconds. The coda decay was measured for the sliding three seconds long window average of $MOD(t)$. The back-projections of equations (2) and (4) are damped for both the time shift inversion ($D = 8.1 10^{-5}\,\text{km}^{-2}$) and the coda inversion ($D = 8.1 10^{-6}\,\text{km}^{-2}$).

Figure 7 provides a summary of the comparison of maximum amplitude, coda decay and time shift between the observed and synthetic seismograms. The map A of Figure 7 shows the maximum amplitude comparison is dominated by the four underestimated earthquakes in the Los Angeles basin, however several factors may have biased the comparison of these amplitudes. The overestimated magnitude of the most western earthquake may have been caused by a complex 3-D coastal structure neglected in the 1-D velocity model used for the source magnitude inversion. There also seems to be a systematic bias to underestimate earthquakes to the north of the San Fernando basin and overestimate earthquakes with a hypocenter depth beneath

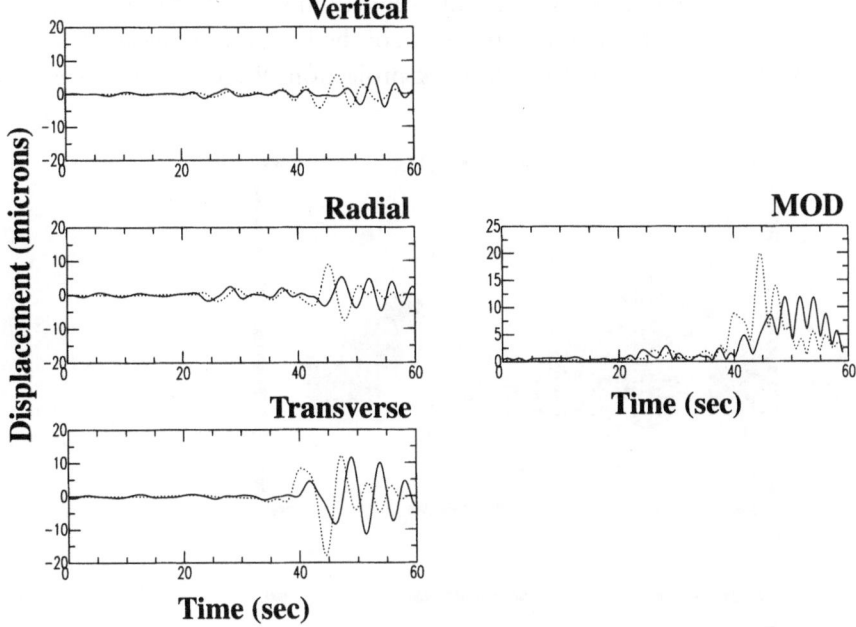

Figure 6

Example of a poor fit between the data and the synthetics for the earthquake E04 recorded at the station SVD (see Table 1). The seismograms show displacement in microns. Both synthetic seismograms and data are filtered between 3 and 20 seconds and the instrument response was removed. Data shown by solid line; synthetics by dashed line.

the San Fernando basin (earthquakes north of 34°N latitude and west of 118.5°W longitude). We attribute this effect to a discrepancy of the inversion for source parameters in the 1-D medium with a 3-D basin focusing (north of the San Fernando basin) and defocusing (below the San Fernando basin) of the energy.

The map B of Figure 7 shows results of the coda analysis. The map is dominated by the areas for which the tested model lacks coda. Including attenuation would further increase this discrepancy and hence our measurement is a lower bound. Therefore, the model would need even more complexity in order to explain the observed data. The lack of coda in the western part of Los Angeles and San Fernando basins and to the north of the Los Angeles basin reflects a lack of complexity in the velocity model. The small discrepancies in the model of the central Los Angeles basin and San Bernardino Basin indicate that on average the model is properly modeling the complexity observed in data. The coda discrepancy is most likely caused by the surface-wave scattering. However, some artifacts may be caused by a poor coverage of crossing paths, as can be seen in Figure 1. Also these results should not be considered as an inversion, but rather identification of regions which are sources of discrepancies between the observed and synthetic seismograms.

The map C of Figure 7 shows that the velocity model is too fast in the western part of the Los Angeles basin and to the north of the Los Angeles basin. This result is consistent with the results of the coda back-projection. We interpret this consistent

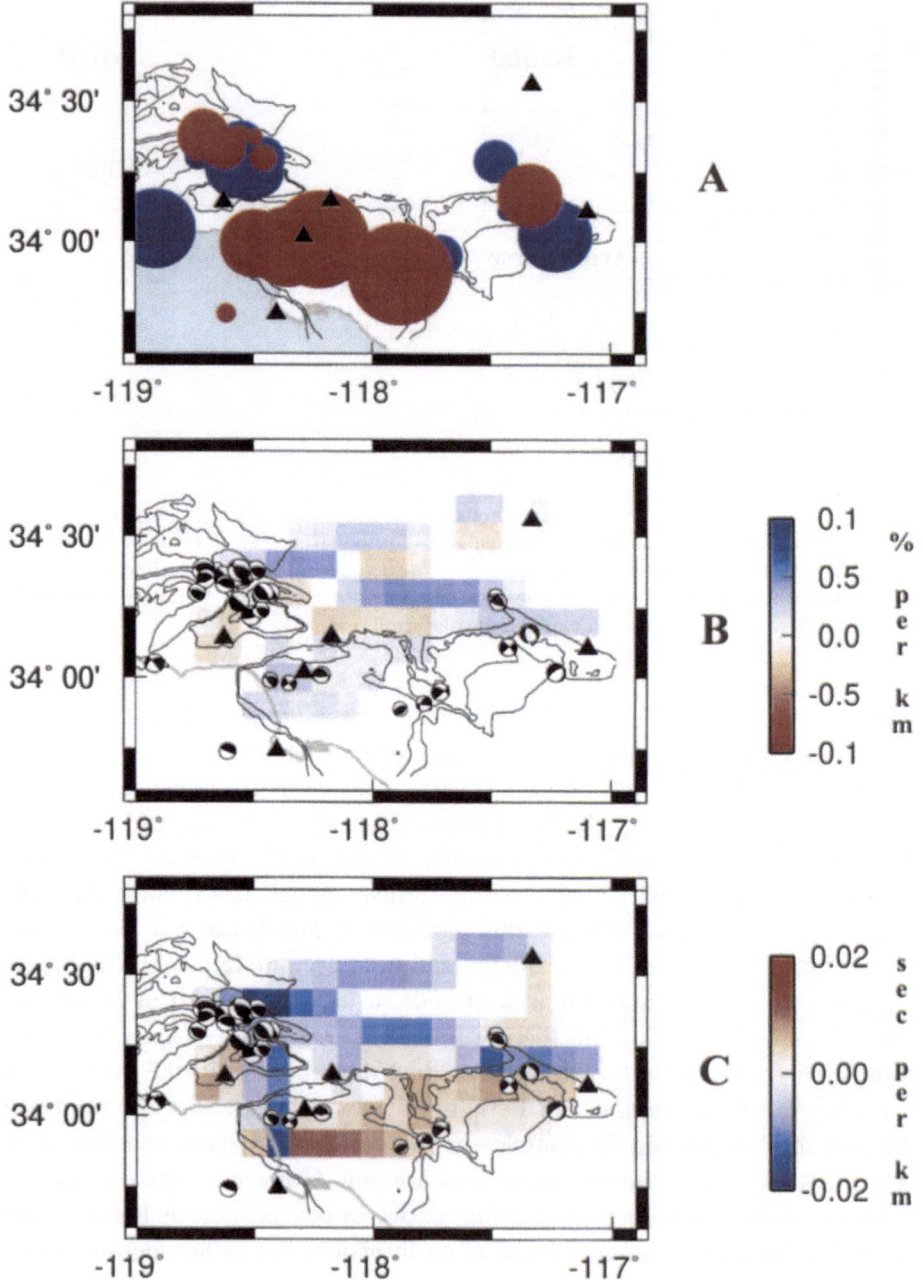

pattern to be a consequence of a too fast and too simple background 1-D velocity model (see the section Application to the Southern California velocity model) and that the western parts of the Los Angeles basin are also more complex than in the tested model. However, the central Los Angeles basin and San Bernardino basins seem to be too slow, which can be corrected with overall faster velocities in this part of the model.

Conclusions

We have shown that the reciprocal method provides a means for doing wave simulations that are otherwise expensive. In the example shown here, we are able to reduce the number of runs by a factor of two, and additional sources (when they become available) can be added with no extra computation.

We also developed a set of criteria for comparing data and synthetics computed in a complex 3-D media when exact matching of waveforms is not possible due to lack of model details or precision. The magnitude of displacement (MOD) measure has a number of advantages in this respect.

For the case study of the Southern California Velocity Model, Version 1, the characteristics of the fit between synthetic and recorded seismograms show consistent patterns, indicating regions which need to be improved to produce a better fit between data and synthetic seismograms. The bias would not be apparent unless numerous source receiver locations would be tested.

Acknowledgments

The authors would like to thank the reviewer Harold Magistrale and the two anonymous reviewers for their valuable suggestions. Special thanks go to Jascha Polet and Dr. Donald Helmberger for their input. Many of the figures were made with GMT (WESSEL and SMITH, 1991). This research was supported by the Southern

◄

Figure 7

Summary results of comparison of the synthetic and observed seismograms. The map A shows results of maximum amplitude comparison: The red circles correspond to the underestimated magnitude of an earthquake on average, the blue circles correspond to overestimated magnitude of an earthquake on average. The larger the circle, the larger the discrepancy. A circle of a radius zero, not printed, corresponds to a perfect fit. The largest circle corresponds to 4.2 times on average underestimated maximum amplitude. The contours correspond to the Love wave group velocity for a period of three seconds. The map B shows results of coda back-projection: the blue color corresponds to lack of the coda generated by the model, and the red color corresponds to too much coda generated by the synthetic model. The map C shows results of time shift back-projection: the blue color corresponds to overly fast parts of the model, and the red color corresponds to the slow parts of the model.

California Earthquake Center. SCEC is funded by NSF Cooperative Agreement EAR-8920136 and USGS Cooperative Agreements 14-08-0001-A0899 and 1434-HQ-97AG01718. The SCEC contribution number for this paper is 525.

REFERENCES

AKI, K. and RICHARDS, P.G., *Quantitative Seismology* (W.H. Freeman and Co., New York 1980).

EISNER, L. and CLAYTON, R.W. (2001a), *A Reciprocity Method for Multiple Source Simulations*, Bull. Seismol. Soc. Am., *91*, 553–560.

EISNER, L. and CLAYTON, R.W. (2001b), *Equivalent Medium Parameters for Numerical Modeling in Media with Near-surface Low-velocities*, submitted to Bull. Seismol. Soc. Am.

EISNER, L. and CLAYTON, R.W. (2001c), *Assessing Site and Path Effects by Full Waveform Modeling*, submitted to J. Geophys. Res.

GRAVES, R.W. (1996), *Simulating Seismic Wave Propagation in 3D Elastic Media Using Staggered-grid Finite Differences*, Bull. Seismol. Soc. Am. *86*, 1091–1106.

GRAVES, R.W. and WALD, D.J. (2001), *Resolution Analysis of Finite Fault Source Inversion Using 1D and 3D Green's Functions, Part I: Strong Motions*, J. Geophys. Res., *106*, 8745–8766.

HADLEY, D. and KANAMORI, H. (1977), *Seismic Structure of the Transverse Ranges, California*, Bull. Geol. Soc. Am. *88*, 1469–1478.

HAUKSSON, E. (2000), *Crustal Structure and Seismicity Distribution Adjacent to the Pacific and North America Plate Boundary in Southern California*, J. Geophys. Res. *105*, 13,875–13,903.

MAGISTRALE, H., MCLAUGHLIN, K., and Day, S. (1996), *A Geology-based 3D Velocity Model of the Los Angeles Basin Sediments*, Bull. Seismol. Soc. Am. *86*, 1161–1166. http://www.scecdc.scec.org/3Dvelocity/3Dvelocity.html

MONTALBETTI, J.F. and KANASEWICH, E.K. (1970), *Enhancement of Teleseismic Body Waves with a Polarization Filter*, Geophys. J. Roy. Astron. *21*, 119–129.

OLSEN, K.B. and ARCHULETA, R.J. (1996), *Three-dimensional Simulation of Earthquakes on the Los Angeles Fault System*, Bull. Seismol. Soc. Am. *86*, 575–596.

OLSEN, K.B., MADARIAGA, R., and ARCHULETA, R.J. (1997), *Three-dimensional Dynamic Simulation of the 1992 Landers Earthquake*, Science *278*, 834–838.

WALD, D.J. and GRAVES, R.W. (1998), *The Seismic Response of the Los Angeles Basin, California*, Bull. Seismol. Soc. Am. *88*, 337–356.

WESSEL, P. and SMITH, W.H.F. (1991), *Free Software Helps Map and Display Data*, EOS Trans. Am. Geophys. Union *72*, 441.

WU, R. and AKI, K. (1988), *Introduction: Seismic Wave Scattering in Three-dimensionally Heterogeneous Earth*, Pure appl. geophys. *128*, 1–6.

ZENG, Y. SU, F., and AKI, K. (1991), *Scattering Wave Energy Propagation in a Random Isotropic Scattering Medium 1. Theory*, J. Geophys. Res *96*, 607–619.

ZHU, L.P. and HELMBERGER, D.V. (1996), *Advancement in Source Estimation Techniques Using Broadband Regional Seismograms*, Bull. Seismol. Soc. Am. *86*, 1634–1641.

(Received March 3, 2001, revised February 5, 2001, accepted August 28, 2001)

To access this journal online:
http://www.birkhauser.ch

Pure appl. geophys. 159 (2002) 1707–1718
0033–4553/02/081707–12 $ 1.50 + 0.20/0

❙ Pure and Applied Geophysics

Constant Q and a Fractal, Stratified Earth

MIRKO VAN DER BAAN[1]

Abstract—Frequency-dependent measurements of the quality factor Q typically show a constant behaviour for low frequencies and a positive power-law dependence for higher frequencies. In particular, the constant Q pattern is usually explained using intrinsic attenuation models due to anelasticity with either a single or multiple superposed relaxation mechanisms — each with a particular resonance peak.

However, in this study, I show using wave localisation theory that a constant Q may also be due to apparent attenuation due to scattering losses. Namely, this phenomenon occurs if the earth displays fractal characteristics. Moreover, if fractal characteristics exist over a limited range of scales only, even an absorption band can be created—in accordance with observations. This indicates that it may be very difficult to distinguish between intrinsic and scattering attenuation on the basis of frequency-dependent measurements of the quality factor only.

Key words: Attenuation, fractals, inhomogeneous media, scattering, wave propagation.

Introduction

Frequency-dependent measurements of the quality factor Q have been made since at least the beginning of the century (see e.g., KNOPOFF, 1964 for historical references). Although Q is probably constant over a large frequency range in homogeneous materials, this is not the case in inhomogeneous media. Such frequency-dependent estimates are required, since they enable us in principle to infer more about the existing attenuation mechanisms in the considered medium (AKI, 1980) or to invert for the lateral and vertical distribution of Q (ROMANOWICZ, 1998). Both interpretations can then hopefully be related to geodynamical processes occurring in the earth.

In the earth, the obtained estimates typically show an approximately constant Q for periods between 10 s and 1 hour (0.28×10^{-4} to 0.1 Hz) and an increase with frequency for $f > 1$ Hz (DZIEWONSKI, 1979; SIPKIN and JORDAN, 1979; SATO and FEHLER, 1998). Namely, for S waves and frequencies between 1 and 25 Hz, Q is proportional to f^{α} with α ranging between 0.6 and 0.8 in the average (AKI, 1980; SATO and FEHLER, 1998). Low frequency estimates are usually obtained from normal mode and surface wave data, whereas higher frequencies are derived from body waves.

[1] School of Earth Sciences, University of Leeds, Leeds LS2 9JT, U.K.
E-mail: mvdbaan@earth.leeds.ac.uk

Combining the results of body-wave data with the lower and constant Q measurements of normal mode and surface wave data, AKI (1980) conjectured the existence of a maximum in attenuation around 0.5 to 1 Hz. Figure 1 displays a sketch of the situation. Although the existence of the peak has not yet unambiguously been proven (see SATO and FEHLER, 1998 for the latest review), considerable research has been done to explain its conjectured position and magnitude. Notwithstanding the fact that its position can be explained using a single relaxation model, this can be caused by two completely different mechanisms. Firstly, it may be due to thermoelastic effects involving grain sizes or crack lengths of the order of $1 \sim$ mm (ZENER, 1948; SAVAGE, 1966; AKI, 1980). Secondly, AKI (1980) proposed that apparent attenuation due to scattering losses may play a role. Subsequent research showed that, in particular, elastic scattering may cause the observed peak. Associated predictions indicate that heterogeneities with a typical scale-length of the order of 2 km may be involved (SATO and FEHLER, 1998).

In this study, however, I focus on the constant Q behaviour which has received less attention lately. The phenomenon has always been explained by anelastic intrinsic attenuation using either a single or multiple superposed relaxation mechanisms. However, as I will show, a constant Q can also be caused by scattering in a fractal, stratified earth. This is demonstrated using wave localisation theory

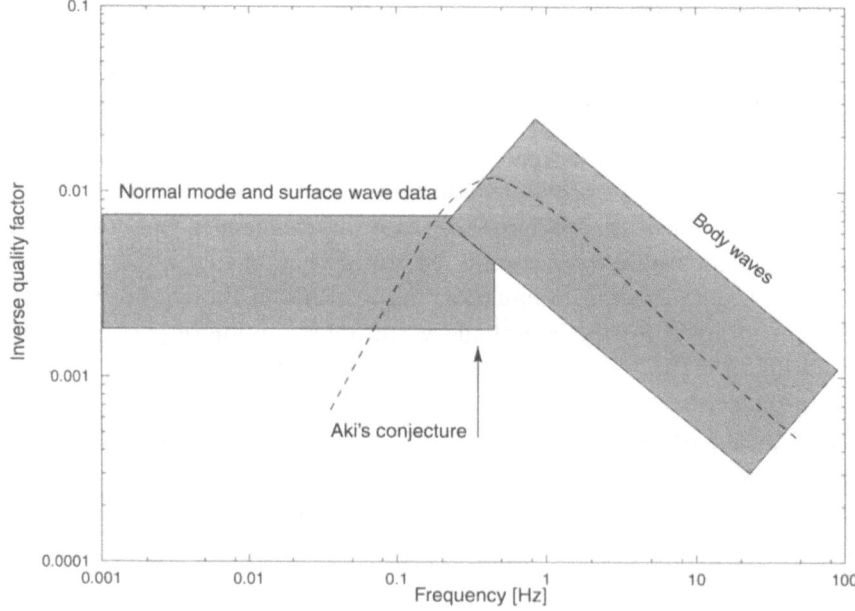

Figure 1

Schematic sketch of frequency dependent Q measurements. The height of the boxes indicates roughly the uncertainty in measurements. The position of the resonance peak conjectured by AKI (1980) is also shown. After SIPKIN and JORDAN (1979), AKI (1980) and SATO and FEHLER (1998).

(SHAPIRO and ZIEN, 1993; VAN DER BAAN, 2001). This multiple scattering theory is used since it describes the exact attenuation and dispersion of a plane-wave traversing a 1-D medium. Although an acoustic version of the theory is employed, similar predictions can be made using the full elastic theory.

First, I review some explanations for the occurrence of constant Q. Then, I briefly present wave localisation theory and its predictions for a fractal, stratified earth. Finally, I discuss some of its implications if Q measurements are to be interpreted in terms of scattering versus intrinsic attenuation mechanisms. I focus primarily on the physical implications of wave localisation theory, and in particular its predictions concerning scattering and apparent attenuation in fractal media in relation to constant Q behaviour. VAN DER BAAN (2001) describes the theory in more detail and illustrates its correctness using numerical simulations. Some of these numerical simulations are used here for different purposes.

Some Explanations for Constant Q

Constant Q behaviour is normally explained in two different ways. Either an anelastic mechanism is involved which attenuates waves independently of frequency or a multitude of relaxation mechanisms are required, each with a particular resonance peak. Their superposition is then assumed to yield a so-called absorption band resulting in a constant Q over a large frequency range.

Candidates for frequency-independent intrinsic attenuation mechanisms include hysteresis in static stress-strain curves of rocks (KRASILNIKOV, 1963; MCCALL and GUYER, 1994), frictional sliding on dry surfaces of thin cracks (WALSH, 1966), and nonlinear attenuation mechanisms (KNOPOFF and MACDONALD, 1960). However, attributing constant Q behaviour to hysteresis raises the question of which particular stress-strain model leads to a stress-strain cycle whose width is independent of frequency (KNOPOFF, 1964). On the other hand, hysteresis can be caused by cracks which open, close and slip during elastic loading (SATO and FEHLER, 1998), thereby pointing again to the study of WALSH (1966).

The second explanation uses a collection of intrinsic attenuation mechanisms, each with an individual resonance (Debye) peak (Fig. 2.a). Since the total inverse quality factor Q_{tot}^{-1} is simply a summation over each individual inverse quality factor Q_i^{-1}, a large collection of attenuation mechanisms can in principle cause a constant absorption band as displayed in Figure 2.b (LIU et al., 1976). These individual relaxation mechanisms can for instance be due to grain boundary sliding, the formation and movement of defects in crystal lattices and thermoelastic effects (ANDERSON, 1989).

However, it should be noted that a constant Q for all frequencies in combination with a non-zero effective phase velocity at zero frequency (i.e., in the long wavelength limit) is physically not possible. Namely, at zero frequency, the inverse quality factor

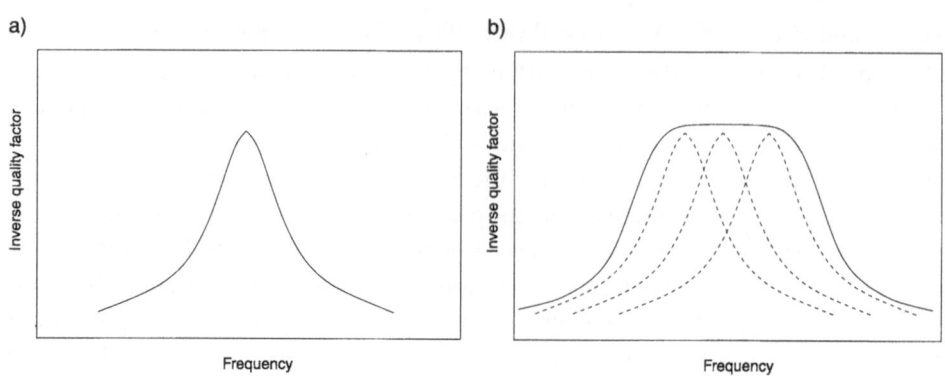

Figure 2
(a) A single relaxation mechanism with a Debye peak centred at a resonance frequency. (b) Several superposed mechanisms may form an absorption band. Adapted from LIU *et al.* (1976).

should approach zero ($Q^{-1} = 0$), thereby rendering the medium effectively homogeneous, and the phase velocity should converge to the effective medium velocity. These conditions are required, since otherwise causality is violated and the phase velocity becomes unbounded (FUTTERMAN, 1962). This can, e.g., be shown using the Kramers-Krönig relations which relate attenuation and dispersion (see also BELTZER, 1988). A constant Q alone is not in conflict with causality if the phase velocity is allowed to converge to zero in the long wavelength limit (KJARTANSSON, 1979). However, a zero phase velocity in the long wavelength limit conflicts with effective medium theories which state that it should converge to a finite constant.

Such a problem does not occur for the absorption band model of LIU *et al.* (1976). An additional advantage of their model is that it can deal with observations of Q being proportional to f^α for $f > 1$ Hz. It would simply indicate that the right edge, i.e., the corner frequency, starts at 1 Hz.

Unfortunately, however, the absorption band model is a phenomenological theory that does not give the faintest indication of which particular attenuation mechanisms are involved. To further complicate this problem, constant Q can also be caused by apparent attenuation due to scattering, as I will show using wave localisation theory.

Wave Localisation Theory

Wave localisation theory originally comes from the quantum theory of disordered solids (ANDERSON, 1958). It implies that the energy of a wave has an exponential fall-off for large distances from its maximum. This causes the envelope of a pulse to decay exponentially for long propagation distances in random, 1-D media. Hence, the transmission coefficient T behaves as $|T| = \lim_{L\to\infty} \exp(-\gamma L)$ with γ the so-called Lyapunov exponent and L the thickness of the random medium. This happens for almost any frequency and realisation of the medium. The mathematical proof of the exponential decay follows from random matrix theory and in particular the work of Fürstenberg (1963), Oseledec (1968) and Virster (1979). For more background, the interested reader is referred to Van der Baan (2001).

In order to determine the apparent attenuation of a wave propagating in a fractal, stratified earth, I consider a so-called matched-medium. That is, a random medium with thickness L is sandwiched between 2 homogeneous half-spaces with matching density ρ_0 and incompressibility κ_0^{-1}. The background density ρ_0 and incompressibility κ_0^{-1} are assumed to be constant and fluctuations occur in the incompressibility only. Thus the medium is described by

$$\kappa^{-1}(z) = \begin{cases} \kappa_0^{-1} & z < 0, \, z > L \\ \kappa_0^{-1}[1 + \sigma_\kappa(z)] & 0 \leq z \leq L \end{cases} \tag{1}$$

$$\rho(z) = \rho_0 \text{ everywhere}$$

with σ_κ the relative fluctuations of the incompressibility. For simplicity, I only treat the acoustic problem to prevent P-S and S-P conversions. In addition, the fluctuations are assumed to be stationary.

To determine the required Lyapunov exponent, Shapiro and Zien (1993) considered a plane-wave impinging from above on the inhomogeneous layer and used a second-order perturbation of the resulting wave equation. In this way they were able to show that, for vertical incidence, the Lyapunov exponent γ is given by

$$\gamma = \frac{1}{4} k_0^2 \int_0^\infty \mathrm{E}[\sigma_\kappa(z)\sigma_\kappa(z + \xi)] \cos(2k_0\xi) \, d\xi \tag{2}$$

with $\mathrm{E}[\cdot]$ the autocorrelation function of the random layer. The variable k_0 is equal to both the wavenumber in the background medium and the effective wavenumber in the random media, i.e., the wavenumber at zero frequency. The resulting expression for nonvertical incidence is only slightly more complicated.

Implications for a Fractal, Stratified Earth

To study the apparent attenuation in a fractal medium, I use the Von Kármán function given by

$$\mathrm{E}[\sigma_\kappa(z)\sigma_\kappa(z + \xi)] = \frac{2^{1-v}\langle\sigma_\kappa^2\rangle}{\Gamma(v)} \left(\frac{|\xi|}{a}\right)^v K_v\left(\frac{|\xi|}{a}\right), \tag{3}$$

where $K_v(\cdot)$ represents the modified Bessel function of the third kind of order v, also known as the MacDonald function. The constant a represents the typical scale-length of the heterogeneities, that is, of the fluctuations in the incompressibility.

For $v = 0$, this function has self-similar properties in the sense that it displays discontinuities on any scale. In addition, it has a constant variance for each octave interval of wavenumber for $k_0 a \gg 1$ (FRANKEL and CLAYTON, 1986). Moreover, its amplitude spectrum corresponds to a power law which is also indicative of fractal behaviour (MANDELBROT and WALLIS, 1969).

Calculating integral (2) for the Von Kármán function (3) with $v = 0$ results in

$$\gamma_{VK} = \frac{\pi\langle\sigma_\kappa^2\rangle}{4a\Gamma(0)} \frac{(k_0 a)^2}{(4(k_0 a)^2 + 1)^{1/2}}. \tag{4}$$

The reciprocal of the quality factor is obtained by means of

$$Q^{-1} = 2\gamma/k \approx 2\gamma/k_0, \tag{5}$$

yielding

$$Q_{VK}^{-1} \approx \frac{\pi\langle\sigma_\kappa^2\rangle}{2\Gamma(0)} \frac{k_0 a}{(4(k_0 a)^2 + 1)^{1/2}}. \tag{6}$$

Wave localisation theory also predicts the exact dispersion of the phase velocity and therefore the exact frequency dependence of the wavenumber $k(f)$ (VAN DER BAAN, 2001). However, for the present application, approximation (6) suffices since the phase velocity is a slowly monotonically rising function which also converges to a constant. Therefore, the overall trend of Q_{VK} is not changed.

The constant $\Gamma(0)$ is required to normalise the Von Kármán function for $v = 0$. However, in non-perfect fractals, i.e., in non-perfect realisations of the Von Kármán function, it is replaced by a finite constant.

Figure 3 displays the behaviour of Q^{-1} for a large range of scales $k_0 a$. It can clearly be seen that Q^{-1} becomes effectively a constant for $k_0 a > 2$. VAN DER BAAN (2001) showed that this phenomenon only happens for $v = 0$. Other functions, such as exponential and Gaussian functions, do not display such a behaviour either. In all these functions, a Debye peak occurs at $k_0 a \approx 1$. This indicates that, for such media, scattering is most efficient for wavelengths of the same order as the scale-length of the heterogeneities, i.e., due to Mie scattering. For fractal media, however, no such typical scale-length exists. Hence, any scale-length can be seen as a typical dimension and Mie scattering occurs at nearly all frequencies, thereby causing a constant Q.

The only exception occurs at zero frequency (long wavelength limit). In this case, the fractal medium becomes effectively homogeneous. Therefore, no scattering and thus attenuation occurs. Moreover, both the phase and group velocities converge to the expected effective medium and geometric optical velocities in respectively the low and high frequency limits (VAN DER BAAN, 2001). Hence, causality is respected and

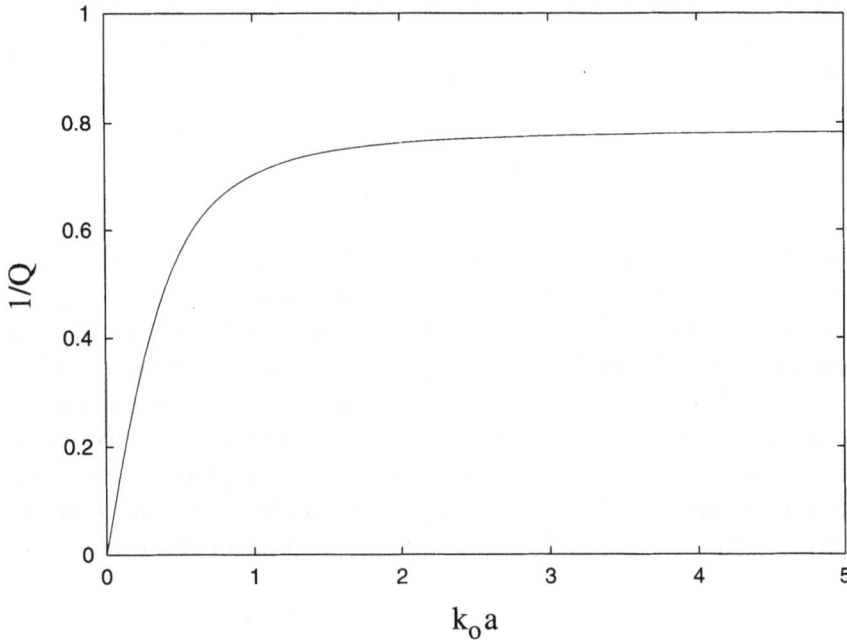

Figure 3

Inverse quality factor as predicted for a fractal, stratified medium by wave localisation theory. Vertical scale is expressed in units of $\langle \sigma_\kappa^2 \rangle / \Gamma(0)$.

the phase velocity is bounded, thereby preventing problems related to perfectly constant Q models (FUTTERMAN, 1962) as discussed before.

VAN DER BAAN (2001) also demonstrated the correctness of the predicted behaviour of Q^{-1} for different autocorrelation functions using an acoustic version of the reflectivity code of DIETRICH (1988). I use here some of these numerical simulations to show how effectively a constant absorption band can be created. In that study, a pulse propagated from the homogeneous halfspace below the inhomogeneous layer towards the surface. Its spectral content was compared to a reference wave traversing a homogeneous space and analysed using a wavelet transform. The quality factor was then obtained by means of a spectral ratio. The medium consisted of 4000 layers each 1 m thick with an average velocity of 4 km/s and a standard deviation of the fluctuations of the incompressibility σ_κ of 15%. This corresponds to a standard deviation of 7.5% of velocity. Sampling rate was 1 ms.

Figure 4 displays the measured Q^{-1} at normal incidence for the model described above. The results of 4 different numerical simulations are shown in addition to the theoretical predictions. The overall behaviour of both the numerical simulations and the theoretical predictions agree well except for the highest frequencies. Namely, for these frequencies, Q^{-1} decays again. This is probably due to the fact that a discrete representation of the medium is used which cannot simulate a perfect fractal. On the other hand, it clearly shows that, in non-perfect fractals, a constant Q only occurs within a given frequency band and that Q^{-1} becomes proportional to $f^{-\alpha}$ for higher frequencies in accordance with observations.

Discussion

Using different analysis methods, typical scale-lengths of heterogeneities have been established to exist in the earth over a very large range (see, e.g., WU and AKI, 1988a, for an overview). In this light, the existence of fractal distributions of heterogeneities within the earth becomes more likely (see also DOLAN *et al.*, 1998). And, as I have demonstrated, a fractal stratified earth may cause a constant Q because of apparent attenuation due to scattering. In addition, causality and a non-zero phase velocity at zero frequency are maintained. Moreover, as the numerical simulations have shown, in the case of non-perfect fractals, a constant Q only occurs over a certain frequency range, thereby effectively creating an absorption band.

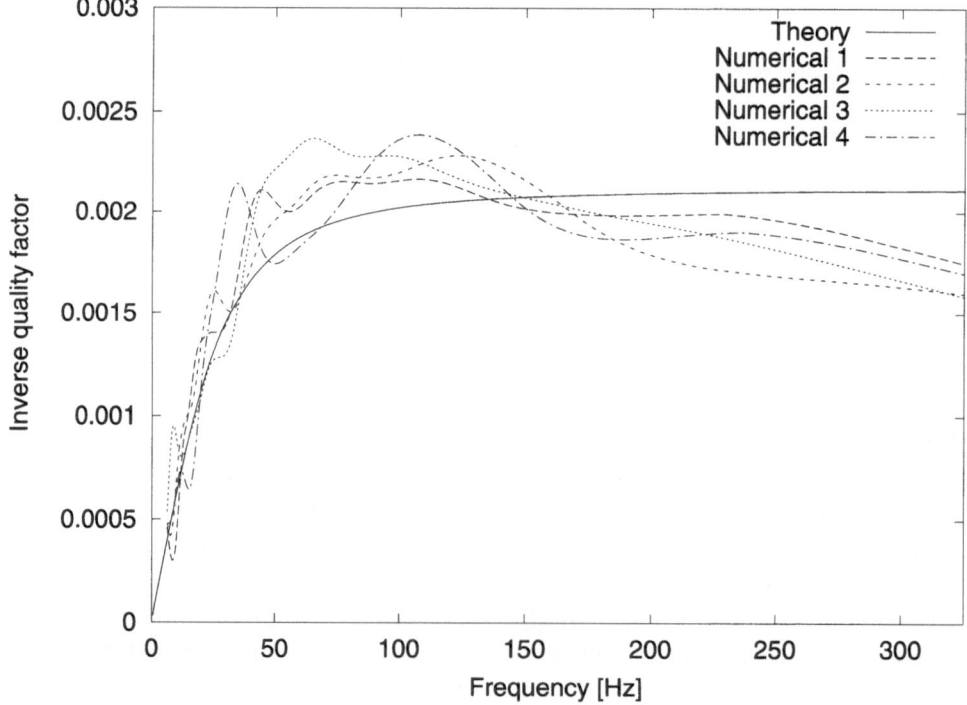

Figure 4
Numerical simulations for 4 different realisations of the Von Kármán function (broken lines) and the theoretical prediction (solid line). The factor $\Gamma(0)$ is replaced by the constant 13.0 to illustrate the general trend. After VAN DER BAAN (2001).

Furthermore, if we assume that 1 Hz forms the corner frequency for S waves ($v = 4$ km/s), and that Mie scattering is most efficient ($k_0a \approx 1$) then it is easily deduced that the heterogeneities of the lithosphere may have fractal characteristics for scales exceeding 1 km. In addition, wave localisation theory predicts a scattering Q of the order of 400 to 500 (Fig. 4) for vertical incidence and velocity fluctuations of 7.5%, whereas measured Q values range from 50 to 500 (Fig. 1), thereby indicating that intrinsic attenuation may play a dominant role (if wave localisation is the appropriate scattering mechanism). Conversely, scattering Q decreases with increasing angle of incidence towards 250 at 30° incidence, thereby reducing the role intrinsic attenuation plays (VAN DER BAAN, 2001). Naturally, this does not take into account that the heterogeneities in the earth are 3-D and elastic. However, the numerical simulations of FRANKEL and CLAYTON (1986) have shown that a constant Q also occurs for the same Von Kármán function in an elastic, 2-D medium.

More importantly, it also indicates that it may be even more complicated to distinguish between intrinsic attenuation due to anelasticity and apparent attenuation due to scattering than anticipated before. Namely, not only can an absorption band be mimicked by scattering mechanisms, but WENNERBERG and FRANKEL (1989) showed that even a single relaxation peak due to scattering attenuation is nearly indistinguishable from one created by intrinsic attenuation. This indicates that it may hardly be possible to distinguish between apparent and intrinsic attenuation by measuring Q only.

A possible way to circumvent this problem is to assume that either intrinsic attenuation or scattering is described by a particular physical mechanism with a known effect on, for instance, the energy density or Q. The non-explained part in the measurements is then solely due to the remaining mechanism.

Possible scattering mechanisms that can be invoked are wave localisation in finely layered media or radiative transfer of energy in 3-D structures. Approaches based on the latter mechanism describe the spatial (WU, 1985) or spatial and temporal distribution (ZHENG et al., 1991) of the energy density as a function of the contribution of scattering attenuation to total attenuation. Applications on real data can be found in WU and AKI (1988b), FEHLER et al. (1992) and JIN et al. (1994).

As an alternative, the phenomenological theory of FRANKEL and WENNERBERG (1987) can be considered which does not rely on the existence of a particular physical scattering mechanism. Their energy flux model simply assumes that scattering involves a transfer of energy from the incident wave to the coda with conservation of total energy, whereas intrinsic attenuation reduces the total amount of energy. For weak-scattering, their model is sufficiently accurate (ZHENG et al., 1991).

The drawback of applying such an approach lies, however, on the underlying implicit assumptions on which these mathematical theories are based. Namely, if their basic assumptions are violated, specific behaviour may be attributed to the wrong mechanism. In particular, these theories heavily rely on homogeneous and isotropic distributions of scatterers, and isotropic, pure-mode scattering. That is, scatterers are uniformly distributed in space, heterogeneities display identical horizontal and vertical scale-lengths, and no layering exists. In addition, scattering is angle independent, recorded energy consists entirely of either P or S waves, and no mode conversion occurs. Some of these basic assumptions are presently being questioned (MARGERIN et al., 1998).

Conclusions

The observed constant Q pattern in the earth for periods between 10 s and 1 hour has always been attributed to intrinsic attenuation mechanisms due to anelasticity. However, it has been demonstrated using wave localisation theory which takes multiple scattering into account that, in a fractal stratified earth, a constant Q can also be caused by apparent attenuation due to scattering. Moreover, if the earth displays fractal characteristics over a limited range of scales only then effectively an absorption band is created, as numerical simulations have shown. In this case, the inverse quality factor is only constant over a range of frequencies and tends to zero for other frequencies. This particular behaviour is in accordance with observations. This yields a further indication that it may be very difficult to distinguish between intrinsic and scattering attenuation using frequency-dependent Q measurements only. In particular, in most cases, the common assumption that scattering and intrinsic effects can be separated by allowing for frequency-dependent scattering only is not valid.

Acknowledgements

The author would like to thank Robert Nowack for his suggestion to write this short note, and José Carcione, Mike Kendall, Ivan Pšenčík and an anonymous reviewer for their comments on the original manuscript.

REFERENCES

AKI, K. (1980), *Attenuation of Shear-waves in the Lithosphere for Frequencies from 0.05 to 25 Hz*, Phys. Earth Planet. Inter. *21*, 50–60.

ANDERSON, P. W. (1958), *Absence of Diffusion in Certain Random Lattices*, Phys. Rev. *109*, 1492–1505.

ANDERSON, D. L. *Theory of the Earth* (Blackwell, Boston 1989).

BELTZER, A. I. (1988), *Dispersion of Seismic Waves by a Causal Approach*, Pure Appl. Geophys. *128*, 147–156.

DIETRICH, M. (1988), *Modeling of Marine Seismic Profiles in the $t - x$ and $\tau - p$ Domains*, Geophysics. *53*, 453–465.

DOLAN, S. S., BEAN, C. J., and RIOLLET, B. (1998), *The Broad-band Fractal Nature of Heterogeneities in the Upper Crust from Petrophysical Logs*, Geophys. J. Int. *132*, 489–507.

DZIEWONSKI, A. M. (1979), *Elastic and Anelastic Structure of the Earth*, Rev. Geophys. Space Phys. *17*, 303–312.

FEHLER, M., HOSHIBA, M., SATO, H., and OBARA, K. (1992), *Separation of Scattering and Intrinsic Attenuation for the Kanto-Tokai Region, Japan, Using Measurements of S-wave Energy versus Hypocentral Distance*, Geophys. J. Int. *108*, 787–800.

FRANKEL, A., and CLAYTON, R. W. (1986), *Finite-difference Simulations of Seismic Scattering: Implications for the Propagation of Short-period Seismic Waves in the Crust and Models of Crustal Heterogeneity*, J. Geophys. Res. *91*, 6465–6489.

FRANKEL, A., and WENNERBERG, L. (1987), *Energy-flux Model of Seismic Coda: Separation of Scattering and Intrinsic Attenuation*, Bull. Seismol. Soc. Am. *77*, 1223–1251.

FÜRSTENBERG, H. (1963), *Noncommuting Random Products*, Trans. Amer. Math. Soc. *108*, 377.

FUTTERMAN, W. I. (1962), *Dispersive Body Waves*, J. Geophys. Res. *67*, 5279–5291.

JIN, A., MAYEDA, K., ADAMS, D., and AKI, K. (1994), *Separation of Intrinsic and Scattering Attenuation in Southern California Using TERRAscope Data*, J. Geophys. Res. *99*, 17,835–17,848.

KJARTANSSON, E. (1979), *Constant Q-wave Propagation and Attenuation*, J. Geophys. Res. *84*, 4737–4748.

KNOPOFF, L., and MACDONALD, G. (1960), *Attenuation of Small Amplitude Stress Waves in Solids*, Rev. Mod. Phys. *30*, 1178–1192.

KNOPOFF, L. (1964), *Q*, Rev. Geophys. *2*, 625–660.

KRASILNIKOV, V. A., *Sound and Ultrasound Waves* (Israel program for Scientific Translations, Jerusalem, 3rd edition, pp. 302–303, 1963).

LIU, H.-P., KANAMORI, H., and ANDERSON, D. L. (1976), *Velocity Dispersion Due to Anelasticity; Implications for Seismology and Mantle Composition*, Geophys. J. R. Astron. Soc. *47*, 41–58.

MANDELBROT, B. B., and WALLIS, J. R. (1969), *Computer Experiments with Fractional Gaussian Noises*, Water Resour. Res. *5*, 228–267.

MARGERIN, L., CAMPILLO, M., and VAN TIGGELEN, B. (1998), *Radiative Transfer and Diffusion of Waves in a Layered Medium: New Insight into Coda Q*, Geophys. J. Int. *134*, 596–612.

McCALL, K., and GUYER, R. (1994), *Equation of State and Wave Propagation in Hysteretic Nonlinear Elastic Materials*, J. Geophys. Res. *99*, 23,887–23,897.

OSELEDEC, V. (1968), *A Multiplicative Ergodic Theorem. Lyapunov Characteristic Numbers for Dynamical Systems*, Trans. Moscow Math. Soc. *19*, 197–231.

ROMANOWICZ, B. (1998), *Attenuation Tomography of the Earth's Mantle: A Review of Current Status*, Pure appl. geophys. *153*, 257–272.

SATO, H., and FEHLER, M., *Seismic Wave Propagation and Scattering in the Heterogeneous Earth* (Springer-Verlag, New York 1998).

SAVAGE, J. (1966), *Thermoelastic Attenuation of Elastic Waves by Cracks*, J. Geophys. Res. *71*, 3929–3938.

SHAPIRO, S. A. and ZIEN, H. (1993), *The O'Doherty-Anstey Formula and the Localization of Seismic Waves*, Geophysics. *58*, 736–740.

SIPKIN, S. A. and JORDAN, T. H. (1979), *Frequency Dependence of Q_{ScS}*, Bull. Seismol. Soc. Am. *69*, 1055–1079.

VAN DER BAAN, M. (2001), *Acoustic Wave Propagation in one-dimensional Random Media: The Wave Localization Approach*, Geophys. J. Int., *145*, 631–646.

VIRSTER, A. D. (1979), *On the Products of Random Matrices and Operators*, Theor. Prob. Appl. *24*, 367.

WALSH, J. B. (1966), *Seismic Attenuation in Rock Due to Friction*, J. Geophys. Res. *71*, 2591–2599.

WENNERBERG, L. and FRANKEL, A. (1989), *On the Similarity of Theories of Anelastic and Scattering Attenuation*, Bull. Seismol. Soc. Am. *79*, 1287–1293.

WU, R.-S. and AKI, K. (1988a), *Introduction: Seismic Wave Scattering in Three-dimensionally Heterogeneous Earth*, Pure appl. geophys. *128*, 1–6.

WU, R.-S. and AKI, K. (1988b), *Multiple Scattering and Energy Transfer of Seismic Waves — Separation of Scattering Effect from Intrinsic Attenuation – II. Application to the Theory to the Hindu Kush Region*, Pure appl. geophys. *128*, 49–80.

WU, R.-S. (1985), *Multiple Scattering and Energy Transfer of Seismic Waves — Separation of Scattering Effect from Intrinsic Attenuation — I. Theoretical Modeling*, Geophys. J. R. astr. Soc. *82*, 57–80.

ZENER, C. M., *Elasticity and Anelasticity of Metals* (Univ. of Chicago Press, Chicago 1948).

ZHENG, Y., SU, F., and AKI, K. (1991), *Scattering Wave Energy Propagation in a Random Isotropic Scattering Medium. 1. Theory*, J. Geophys. Res. *96*, 607–619.

(Received September 29, 2000, revised January 1, 2001, accepted January 30, 2001)

 To access this journal online:
http://www.birkhauser.ch

Pure appl. geophys. 159 (2002) 1719–1736
0033–4553/02/081719–18 $ 1.50 + 0.20/0

┃Pure and Applied Geophysics

Time-domain Modeling of Constant-Q Seismic Waves Using Fractional Derivatives

José M. Carcione[1], Fabio Cavallini[1], Francesco Mainardi[2],
and Andrzej Hanyga[3]

Abstract—Kjartansson's constant-Q model is solved in the time-domain using a new modeling algorithm based on fractional derivatives. Instead of time derivatives of order 2, Kjartansson's model requires derivatives of order 2γ, with $0 < \gamma < 1/2$, in the dilatation-stress formulation. The derivatives are computed with the Grünwald-Letnikov and central-difference approximations, which are finite-difference extensions of the standard finite-difference operators for derivatives of integer order. The modeling uses the Fourier method to compute the spatial derivatives, and therefore can handle complex geometries. A synthetic cross-well seismic experiment illustrates the capabilities of this novel modeling algorithm.

Key words: Viscoelastic waves, fractional calculus, numerical modeling, seismology.

1. Introduction

Constant-Q models provide a good parameterization of seismic attenuation in rocks, in oil exploration and seismology. By reducing the number of parameters they allow an improvement of seismic inversion. Moreover, there is physical evidence that attenuation is almost linear with frequency (therefore Q is constant) in many frequency bands. BLAND (1960) and KJARTANSSON (1979) discuss a linear attenuation model with the required characteristics, but the idea is much older (SCOTT-BLAIR, 1949). Kjartansson's constant-Q model is based on a creep function of the form $t^{2\gamma}$, where t is time and $\gamma \ll 1$ for seismic applications. This model is completely specified by two parameters, i.e., phase velocity at a reference frequency and Q. Therefore, it is mathematically far simpler than any nearly constant Q, as for instance, a spectrum of Zener models (CARCIONE *et al.*, 1988). Due to its simplicity, Kjartansson's model is used in many seismic applications, mainly in its frequency-domain form. MAINARDI

[1] Istituto Nazionale di Oceanografia e di Geofisica Sperimentale – OGS, Borgo Grotta Gigante 42c, 34010 Sgonico, Trieste, Italy. E-mail: jcarcione@ogs.trieste.it
[2] Dipartimento di Fisica, Università di Bologna, via Irnerio 46, I-40126 Bologna, Italy. E-mail: mainardi@bo.infn.it
[3] Institute for Solid Earth Physics, Alle'gaten 41, N5007 Bergen, Norway. E-mail: andrzej@ifjf.uib.no

and TOMIROTTI (1997) interpreted the constant-Q model in terms of fractional derivatives and obtained its 1-D Green's function.

Seismic modeling in inhomogeneous media can, in principle, be performed in the frequency domain. However, the method is expensive when using differential formulations, since it involves solution of many Helmholtz equations. The alternative is to compute the solution through a time-convolution, although the resulting algorithm is relatively expensive. A purely differential – as opposed to integro-differential – formulation can be obtained by using fractional derivatives (CAPUTO and MAINARDI, 1971). Instead of time derivatives of order 2, Kjartansson's model requires derivatives of order $2 - 2\gamma$ with $0 < \gamma < 1/2$ in the dilatation formulation of the wave equation, and 2γ in the dilatation-stress formulation. The equation becomes parabolic since the phase velocity has no upper bound. Fractional derivatives appear also in Biot theory, related to memory effects in porous rocks at seismic frequencies with $\gamma = 1/4$ (GUREVICH and LOPATNIKOV, 1995) and at low- and high-frequency limits (FELLAH and DEPOLLIER, 2000). Fractional derivatives can be computed with the Grünwald-Letnikov and central-difference approximations, which are finite-difference extensions of the standard finite-difference approximation for derivatives of integer order (GRÜNWALD, 1867; LETNIKOV, 1868, GORENFLO, 1997). Unlike the standard operator of differentiation, the fractional operator increases in length as time increases, since it must keep the memory effects. However, after a given time period the operator can be truncated (short memory principle).

In the first part of this work we review the constant-Q model and calculate the complex modulus, phase velocity, and attenuation factor versus frequency. We then recast the acoustic wave equation in the time-domain in terms of fractional derivatives, and obtain the Grünwald-Letnikov and central-difference approximations. Then, we investigate the accuracy of the time discretization by comparing the exact and the finite-difference (FD) phase velocities and attenuation factors. The model is discretized on a mesh, and the spatial derivatives are calculated with the Fourier method by using the Fast Fourier Transform. This approximation is infinitely accurate for band-limited periodic functions with cutoff spatial wavenumbers smaller than the cutoff wavenumbers of the mesh. Finally, we test the modeling algorithms with an analytical solution for a 2-D homogeneous medium, and illustrate the method with seismic applications in inhomogeneous media.

2. Constant-Q Model

2.1. Stress-strain Relation

Stress σ and strain ϵ in a 1-D linear anelastic medium are related by a convolutional relation (BLAND, 1980),

$$\sigma(t) = \psi(t) * \dot{\epsilon}(t) \tag{1}$$

where ψ is the relaxation function, t is the time variable, the symbol "$*$" denotes time convolution, and a dot above a variable indicates time differentiation.

Let us define the relaxation function (KJARTANSSON, 1979)

$$\psi(t) = \frac{M_0}{\Gamma(1 - 2\gamma)} \left(\frac{t}{t_0}\right)^{-2\gamma} H(t) , \tag{2}$$

where M_0 is a bulk modulus, Γ is Euler's Gamma function, t_0 is a reference time, γ is a dimensionless parameter, and H is the Heaviside step function. The parameters M_0, t_0 and γ have precise physical meanings, which will become clear in the following analysis. Let us take the Fourier transform of equation (1). We obtain

$$\tilde{\sigma}(\omega) = \mathscr{F}(\dot{\psi}(t))\tilde{\epsilon}(\omega) \equiv M(\omega)\tilde{\epsilon}(\omega) , \tag{3}$$

where $\mathscr{F}$ is the Fourier transform operator, $M(\omega)$ is the complex modulus, and a tilde denotes the Fourier transform. After some calculations we gain

$$M(\omega) = M_0 \left(\frac{i\omega}{\omega_0}\right)^{2\gamma} , \tag{4}$$

where $\omega_0 = 1/t_0$ is the reference frequency.

2.2. Phase Velocity and Attenuation Factor

The complex velocity is

$$V = \sqrt{\frac{M}{\rho}} , \tag{5}$$

where ρ is the density. The phase velocity c is the frequency ω divided by the real part of the complex wavenumber. Then,

$$c = \left[\mathrm{Re}\left(\frac{1}{V}\right)\right]^{-1} . \tag{6}$$

Substituting equations (4) and (5) in (6) yields

$$c = c_0 \left|\frac{\omega}{\omega_0}\right|^{\gamma} \tag{7}$$

with

$$c_0 = \sqrt{\frac{M_0}{\rho}} \left[\cos\left(\frac{\pi\gamma}{2}\right)\right]^{-1} . \tag{8}$$

The attenuation factor is given by

$$\alpha = -\omega \, \mathrm{Im}\left(\frac{1}{V}\right) = \tan\left(\frac{\pi\gamma}{2}\right)\mathrm{sgn}(\omega)\frac{\omega}{c} \, . \tag{9}$$

The quality factor is defined as the peak energy stored during a cycle divided by the energy loss during the cycle. It is given by (e.g., CARCIONE and CAVALLINI, 1994)

$$Q = \frac{\mathrm{Re}(V^2)}{\mathrm{Im}(V^2)} = \frac{1}{\tan(\pi\gamma)} \, . \tag{10}$$

Firstly, we derive from equation (7) that c_0 is the phase velocity at $\omega = \omega_0$, the reference frequency, and that

$$M_0 = \rho c_0^2 \cos^2\left(\frac{\pi\gamma}{2}\right) \, . \tag{11}$$

Secondly, it follows from equation (10) that Q is independent of frequency, so that

$$\gamma = \frac{1}{\pi}\tan^{-1}\left(\frac{1}{Q}\right) \tag{12}$$

parameterizes the attenuation level. Hence we see that $Q > 0$ is equivalent to $0 < \gamma < 1/2$. Moreover, $c \to 0$ when $\omega \to 0$, and $c \to \infty$ when $\omega \to \infty$. It follows that very high frequencies of the signal propagate at almost infinite velocity, and the differential equation describing the wave motion is parabolic (e.g., PRÜSS, 1993).

2.3. Wave Equation in Differential Form

Let us consider a 2-D wave equation of the form

$$\frac{\partial^\beta w}{\partial t^\beta} = D\Delta w + f \, , \tag{13}$$

where $w(x, z, t)$ is a field variable, β is the order of the time derivative, D is a positive parameter, Δ is the 2-D Laplacian operator

$$\Delta = \frac{\partial^2}{\partial x^2} + \frac{\partial^2}{\partial z^2} \, , \tag{14}$$

and f is a forcing term. Consider a plane wave

$$\exp[i(\omega t - k_x x - k_z z)] \, , \tag{15}$$

with ω real and (k_x, k_z) the complex wavevector. Substituting the ansatz (15) in the wave equation (13) with $f = 0$ yields the dispersion equation

$$(i\omega)^\beta + Dk^2 = 0 \, , \tag{16}$$

where $k = (k_x^2 + k_z^2)^{1/2}$ is the complex wavenumber. Equation (16) is the Fourier transform of equation (13). The properties of the Fourier transform when it acts on fractional derivatives are well established, and a rigorous treatment is available in the literature (e.g., DATTOLI et al., 1998). Since $k^2 = \rho\omega^2/M$, comparison of equations (16) and (4) yields

$$\beta = 2 - 2\gamma, \quad \text{and} \quad D = \frac{M_0}{\rho}\omega_0^{-2\gamma} . \tag{17}$$

Equation (13), together with (17), is the wave equation corresponding to Kjartansson's stress-strain relation (KJARTANSSON, 1979). In order to obtain realistic values of the quality factor, corresponding to wave propagation in rocks, $\gamma \ll 1$ and the time derivative in equation (13) has a fractional order.

Kjartansson's wave equation (13) is a particular version of a more general wave equation for variable material properties. The convolutional constitutive equation (3) can be written in terms of fractional derivatives. In fact, it is easy to show, using equations (4) and (17), that it is equivalent to

$$\sigma = \rho D \frac{\partial^{2-\beta}\epsilon}{\partial t^{2-\beta}} . \tag{18}$$

Coupled with the constitutive equation (18) are the momentum equations

$$\frac{\partial\sigma}{\partial x} = \rho \frac{\partial^2 u_x}{\partial t^2} , \tag{19}$$

$$\frac{\partial\sigma}{\partial z} = \rho \frac{\partial^2 u_z}{\partial t^2} , \tag{20}$$

where u_x and u_z are the displacement components. Redefining

$$\epsilon = \frac{\partial u_x}{\partial x} + \frac{\partial u_z}{\partial z} \tag{21}$$

as the dilatation field, differentiating and adding equations (19) and (20), the substitution of equation (18) yields

$$\Delta_\rho \left(\rho D \frac{\partial^{2-\beta}\epsilon}{\partial t^{2-\beta}} \right) = \frac{\partial^2\epsilon}{\partial t^2} , \tag{22}$$

where

$$\Delta_\rho = \frac{\partial}{\partial x}\frac{1}{\rho}\frac{\partial}{\partial x} + \frac{\partial}{\partial z}\frac{1}{\rho}\frac{\partial}{\partial z} . \tag{23}$$

Multiplying by $(i\omega)^{\beta-2}$ the Fourier transform of equation (22) produces, after an inverse Fourier transform, the inhomogeneous wave equation

$$\frac{\partial^\beta\epsilon}{\partial t^\beta} = \Delta_\rho(\rho D\epsilon) + s , \tag{24}$$

where we included the seismic source s. This equation is of type (13) if the medium is homogeneous.

3. Numerical Algorithm

3.1. Grünwald-Letnikov and Central-difference Approximations to the Fractional Derivative

The Grünwald-Letnikov and central-difference approximations to the Riemann-Liouville fractional derivative of a function f are

$$\frac{\partial^v f(t)}{\partial t^v} \sim \frac{1}{h^v} \sum_{j=0}^{J} (-1)^j \binom{v}{j} f(t - jh) \tag{25}$$

and

$$\frac{\partial^v f(t)}{\partial t^v} \approx \frac{1}{h^v} \sum_{j=0}^{J} (-1)^j \binom{v}{j} f\left[t + \left(\frac{v}{2} - j\right)h\right] , \tag{26}$$

respectively, where h is the time step, and $J = t/h - 1$. These expressions are derived in Appendix A. They are first- and second-order accurate, respectively. The fractional derivative of f at time t depends on all the previous values of f. This is the memory property of the fractional derivative, related to field attenuation in our particular example. However, the binomial coefficients $\binom{v}{j}$ are negligible for j exceeding an integer J. This allows us to use the short-memory principle and hence to replace $\sum_{j=0}^{J}$ with $\sum_{j=0}^{L}$, where $L < J$ is the effective memory length (a small constant integer).

3.2. Dilatation-stress Formulation

3.2.1. Time discretization

Use of the Grünwald-Letnikov approximation (25) in equation (24) results in an implicit time-integration scheme, which can be expensive in terms of computer time and storage. An explicit scheme can be obtained if the wave equation is written in the dilatation-stress formulation. Using equations (18) and (22), and including the source term yields

$$\frac{\partial^2 \epsilon}{\partial t^2} = \Delta_\rho \sigma + s . \tag{27}$$

On the other hand, using (17), equation (18) becomes

$$\sigma = \rho D \frac{\partial^{2\gamma} \epsilon}{\partial t^{2\gamma}} . \tag{28}$$

Equations (27) and (28) are discretized at $t = (n - 1)h$ and $t = nh$, respectively, by using the central-difference and Grünwald-Letnikov approximations. We obtain

$$\epsilon^n = h^2(\Delta_\rho\sigma + s)^{n-1} + 2\epsilon^{n-1} - \epsilon^{n-2} \tag{29}$$

from (26)–(27), and

$$\sigma^n = \rho D h^{-2\gamma}\sum_{j=0}^{J}(-1)^j\binom{2\gamma}{j}\epsilon^{n-j} \tag{30}$$

from (25) and (28).

3.2.2. FD complex velocity

The dispersion relation relates the frequency with the wavenumber and allows the calculation of the phase velocity corresponding to each Fourier component. Time discretization implies an approximation of the dispersion relation.

Assuming constant material properties and substituting the ansatz (15) with $t = nh$ in equations (29) and (30) gives the following dispersion relation:

$$\sin\left(\frac{\omega h}{2}\right) = \frac{1}{2}\sqrt{D}kh^{1-\gamma}\left[\sum_{j=0}^{n-2}(-1)^j\binom{2\gamma}{j}\exp(-i\omega jh)\right]^{1/2}, \tag{31}$$

where k is the complex wavenumber. The FD approximation to the complex velocity is $\bar{V} = \omega/k$ where ω and k satisfy equation (31). If $\gamma = 0$, this velocity is real and we obtain the FD phase velocity

$$\bar{c} = \frac{c_0}{\mathrm{sinc}(\theta)}, \tag{32}$$

where $\mathrm{sinc}(\theta) = \sin(\theta)/\theta$ and $\theta = \omega h/2$. Equation (32) indicates that the FD velocity is greater than the true phase velocity. If $\gamma \neq 0$, the FD complex velocity can be written as

$$\bar{V} = \frac{\sqrt{D}h^{-\gamma}}{\mathrm{sinc}(\theta)}\left[\sum_{j=0}^{n-2}(-1)^j\binom{2\gamma}{j}\exp(-2i\theta j)\right]^{1/2}, \tag{33}$$

where equation (31) has been used.

3.3. Dilatation Formulation

3.3.1. Time discretization

An explicit scheme can be obtained with the central-difference approximation (26). In this case, equation (24) is discretized at $t = nh$. We obtain

$$\epsilon^{n+\beta/2} = h^\beta \Delta_\rho(\rho D \epsilon^n) - \sum_{j=1}^{J}(-1)^j \binom{\beta}{j}\epsilon^{n-j+\beta/2} + s^n \ . \tag{34}$$

In order to compute the spatial derivatives at nh we require an approximation for ϵ^n. Since $\beta \approx 2$, the simplest approximation is $\epsilon^n \approx \epsilon^{n-1+\beta/2}$. Similarly, the source can be introduced at times $t = (n - 1 + \beta/2)h$.

3.3.2. FD complex velocity

Substituting the ansatz (15) with $t = nh$ in equation (34) gives the FD complex velocity

$$\bar{V} = 2i\theta\sqrt{D}h^{\beta/2-1}\exp(-i\theta)\left[1 + \sum_{j=1}^{n-1}(-1)^j\binom{\beta}{j}\exp(-2i\theta j)\right]^{-1/2} . \tag{35}$$

In the above formula we have written $\exp(-i\theta)$ in place of $\exp(-i\beta\theta/2)$ because of the approximation $\epsilon^n \approx \epsilon^{n-1+\beta/2}$. Here lies, essentially, the difference between the Grünwald-Letnikov and central-difference approximations.

3.4. Accuracy. FD Phase Velocity and Attenuation Factor

The FD phase velocity is given by

$$\bar{c} = \left[\mathrm{Re}\left(\frac{1}{\bar{V}}\right)\right]^{-1} , \tag{36}$$

and the FD attenuation factor is

$$\bar{\alpha} = -\omega\,\mathrm{Im}\left(\frac{1}{\bar{V}}\right) . \tag{37}$$

If $h \to 0$ (i.e., $n \to \infty$), equation (33) becomes

$$\bar{V} \to \sqrt{D}\left[\frac{1 - \exp(-i\omega h)}{h}\right]^\gamma , \tag{38}$$

where we used the property

$$(1-z)^{2\gamma} = \sum_{j=0}^{\infty}(-1)^j\binom{2\gamma}{j}z^j, \quad z = \exp(-i\omega h) , \tag{39}$$

which is convergent if $|z| < 1$ (e.g., ITÔ, 1987). Using L'Hôpital rule as $h \to 0$ in equation (38) yields

$$\bar{V} = \sqrt{D}(i\omega)^\gamma , \tag{40}$$

which by virtue of equations (4), (5) and (17) gives the complex velocity V.

Similarly, when $h \rightarrow 0$, equation (35) becomes

$$\bar{V} = \sqrt{D}(i\omega)^{1-\beta/2} \, , \tag{41}$$

which is equivalent to (40).

Using the same arguments, the attenuation factor (9) is obtained from equation (37) if $h \rightarrow 0$.

4. Examples

Attenuation measurements in a relatively homogeneous medium (Pierre shale) were made by McDonal et al. (1958) near Limon, Colorado. They reported a constant-Q behavior with attenuation $\alpha = 0.12f$, where α is given in dB per 1000 ft and the frequency f in Hz. Conversion of units implies α (dB/1000 ft) = 8.686 α (nepers/1000 ft) = 2.6475 α (nepers/km). For low-loss solids, the quality factor is

$$Q = \frac{\pi f}{\alpha c} \, ,$$

with α given in nepers per unit length (Toksöz and Johnston, 1981). Since c is approximately 7000 ft/s (2133.6 m/s), the quality factor is $Q \simeq 32.5$. We consider a reference frequency $f_0 = \omega_0/(2\pi) = 250$ Hz, corresponding to the dominant frequency of the seismic source used in the experiments. Then, $\gamma = 0.0097955$, $\beta = 1.980409$, and $c_0 = \sqrt{M_0/\rho} = 2133.347$ m/s. The phase velocity (6) and attenuation factor (9) versus frequency $f = \omega/2\pi$ are shown in Figures 1a and 1b, respectively, where the open circles are the experimental points, the broken and dotted lines are the FD approximations (36) and (37), using the dilatation-stress and dilatation formulations, respectively. The memory lengths are 120 and 60, respectively. The curves correspond to a time step $h = 0.05$ ms. The dilatation formulation is more accurate because it is based on the central-difference approximation (26), and because the decay of the binomial coefficients in the series expansion is faster than in the dilatation-stress formulation. The latter fact is illustrated in Figure 2, which shows the logarithm of the absolute value of the binomial coefficients versus the summation index for the dilatation-stress formulation ($v = 2\gamma$, continuous line) and for the dilatation formulation ($v = \beta$, broken line).

The medium is discretized on a numerical mesh, with uniform vertical and horizontal grid spacings of 2 m, and 77×77 grid points. The spatial derivatives are calculated with the Fourier method by using the fast Fourier transform (FFT) (Kosloff and Baysal, 1982). The source, applied at the center of the mesh, is a Ricker-type wavelet, whose amplitude spectrum is a Gaussian function centered at 250 Hz. A band-limited source, such as a Butterworth filter with a low cut-off

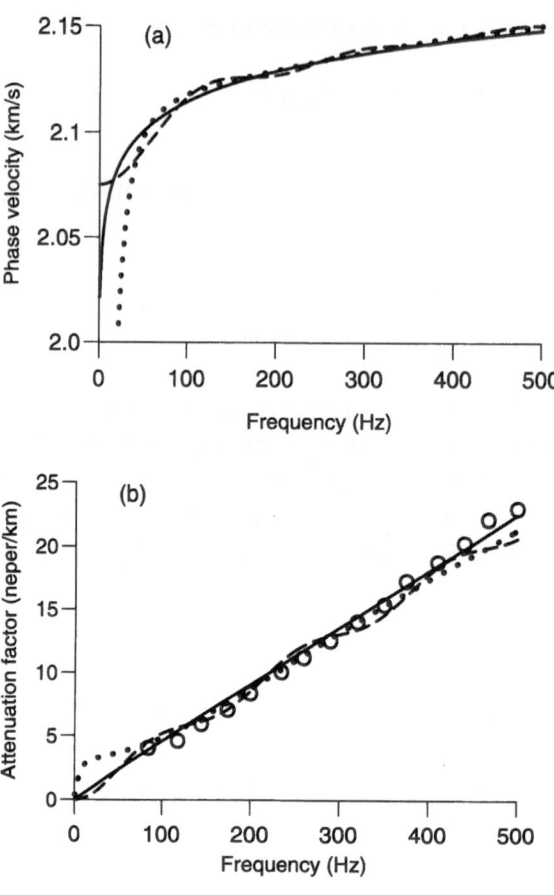

Figure 1

Phase velocity (a) and attenuation factor (b) versus frequency in Pierre shale (continuous line) as given by Eqs. (6) and (9), respectively. The broken and dotted lines are the FD approximations using the dilatation-stress (33) and the dilatation formulations (35), respectively. The memory lengths are, respectively, 120 and 60. The open circles are the experimental data reported by McDONAL *et al.* (1958).

frequency, should be used to avoid aliasing problems, since the phase velocity approaches zero at zero frequency. The maximum allowed frequency is $f_{max} = c_{min}/(2d)$, where d is the grid spacing and c_{min} is the phase velocity at the low cut-off frequency. However, this effect can be neglected, in virtue of the form of the phase-velocity curve (Fig. 1a) and because the energy of the Ricker wavelet is concentrated around its central frequency. The time step used in this simulation is 0.05 ms. Figure 3 compares two snapshots of the dilatation field computed at 36 ms, where (a) corresponds to the lossless case ($\gamma = 0$), and (b) to a lossy model of Pierre shale. The attenuation is evident in the latter case.

A 2-D analytical solution of equation (24) in a homogeneous medium can easily be obtained. The solution to the acoustic (lossless) equation is the zero-order Hankel

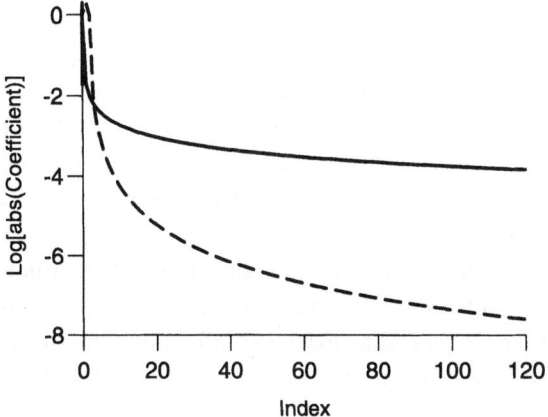

Figure 2

Decimal logarithm of the absolute value of the binomial coefficients versus the summation index for the dilatation-stress formulation ($v = 2\gamma = 2 \times 0.0097955$) (continuous line) and for the dilatation formulation ($v = \beta = 1.980409$) (broken line).

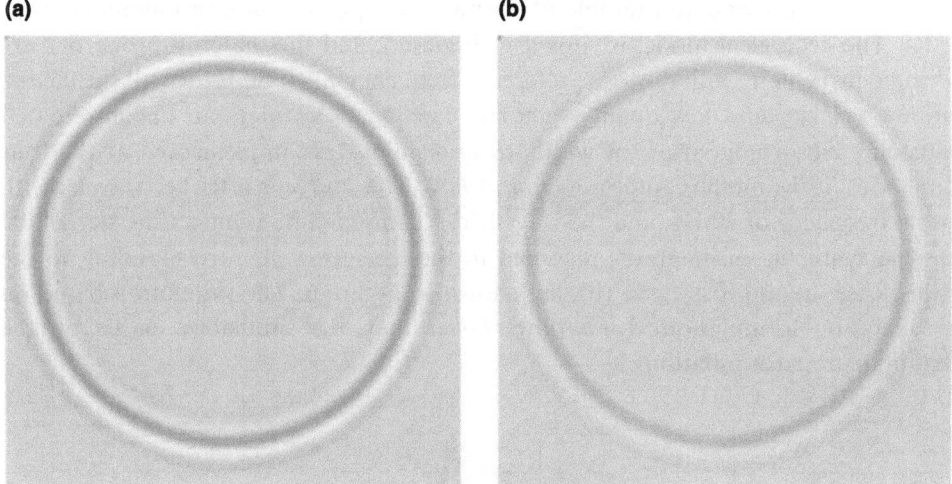

Figure 3

Snapshots of the dilatation field in a lossless medium equivalent to Pierre shale (a), and in a dissipative model of Pierre shale (b).

function of the second kind (MORSE and FESHBACH, 1953, sec. 11.2; CARCIONE *et al.*, 1988a),

$$G(x, z, x_0, z_0, \omega) = -i\pi H_0^{(2)}\left(\frac{\omega r}{c_0}\right) \tag{42}$$

where (x_0, z_0) is the source location, and

$$r = \left[(x - x_0)^2 + (z - z_0)^2\right]^{1/2} . \tag{43}$$

The viscoacoustic solution is obtained by invoking the correspondence principle (BLAND, 1960), i.e., by substituting the acoustic velocity c_0 with the complex velocity (5). We set $G(-\omega) = G^*(\omega)$, where the superscript $*$ denotes complex conjugation. This equation ensures that the inverse Fourier transform of the Green's function is real. The frequency-domain solution is then given by $w(\omega) = G(\omega)F(\omega)$, where F is the Fourier transform of the source. Because the Hankel function has a singularity at $\omega = 0$, we assume $G = 0$ for $\omega = 0$, an approximation that has no significant effect on the solution (note, moreover, that $F(0)$ is small). The time-domain solution $w(t)$ is obtained by a discrete inverse Fourier transform. We have tacitly assumed that w and dw/dt are zero at time $t = 0$.

Figure 4 compares numerical (dotted and broken lines) and analytical (solid line) solutions in the lossy case, at 40 m from the source. In this case, we used the dilatational formulation with memory lengths 20 (broken line) and 40 (solid line). The agreement is excellent for $L = 40$, while $L = 20$ yields a degraded numerical solution.

Finally, we provide an example of seismic wave propagation in inhomogeneous media. The geological model is shown in Figure 5, and the material properties are indicated in Table 1, with the same reference frequency $f_0 = 80$ Hz for all the media. The low velocity and low quality factor of medium 4 simulate an unconsolidated sandstone. Absorbing strips, of width 18 grid points, are implemented at the four boundaries of the mesh (CARCIONE *et al.*, 1988). The source is a Ricker wavelet with central frequency of 80 Hz, and the wavefield is computed by using a time step of 0.2 ms. The synthetic seismograms recorded in the receiver well, corresponding to the lossless case (a) and lossy case (b), are shown in Figure 6. The simulation based on the dilatation formulation is 4.5 times faster than the simulation based on the dilatation-stress formulation.

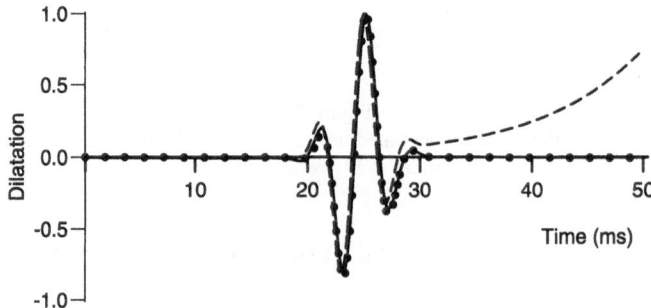

Figure 4

Comparison between numerical and analytical solutions at 400 m from the source. The dotted and broken lines correspond to memory lengths 40 and 20, respectively, and the solid line is the analytical solution. The medium is Pierre shale.

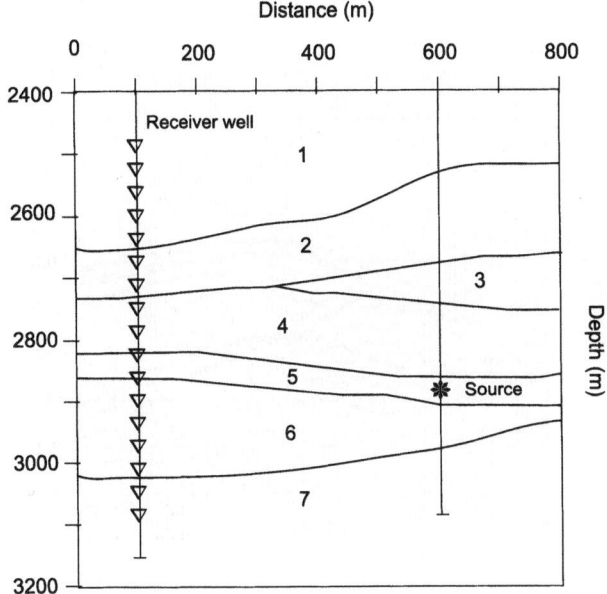

Figure 5
Geological model.

Table 1

Material properties

Medium	c_0 (km/s)	ρ (g/cm^3)	Q
1	3.2	2.5	100
2	3.3	2.52	110
3	3.6	2.58	120
4	2.9	2.4	30
5	3.6	2.7	140
6	3.7	2.71	150
7	3.85	2.72	165

5. Conclusions

The concept of fractional derivative has been used to simulate constant-Q wave propagation (Pierre shale). The equations were expressed in the dilatation-stress and dilatation formulations. The second approach is more accurate and efficient, however our numerical experiments indicate that the absorbing-boundary algorithm performs better with the first formulation. The validity and accuracy of the algorithms are verified by comparison with a novel 2-D analytical solution. The modeling is illustrated with a cross-well seismic experiment, using a Kjartansson's attenuation model, but this approach can provide important applications for porous media as

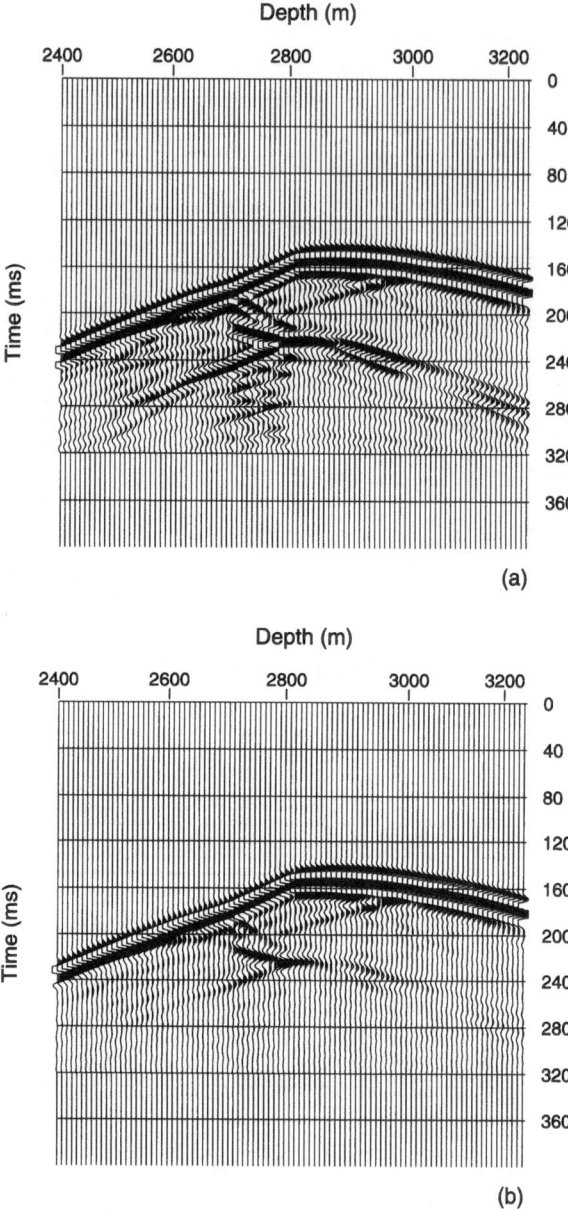

Figure 6
Acoustic (a) and viscoacoustic (b) synthetic seismograms of the dilatation field, corresponding to the model illustrated in Figure 5.

well, since fractional derivatives appear in Biot theory, related to viscodynamic effects at seismic frequencies.

Further research goals on this subject include: (i) an optimal finite-difference approximation of the fractional-derivative operator to reduce numerical dispersion

$[O(h^n), n \geq 2]$ and memory storage. The latter is closely related to the memory length, which in the present case exceeds the number of memory variables used in nearly constant-Q modeling algorithms based on mechanical models; (ii) to improve the absorbing-boundary algorithm in the dilatation formulation; (iii) the generalization to the elastic P-SV case; and (iv) applications of the method to wave propagation in porous media.

Appendix A. The Fractional Derivative

The notion of fractional derivative here adopted can be easily introduced via Fourier transform, since it is intended to generalize the rule of the Fourier transform for the common derivative of integer order of a well-behaved function of time by allowing noninteger powers of the frequency. If the time Fourier transform is defined as

$$[\mathscr{F}\phi](\omega) = \int\limits_{-\infty}^{+\infty} e^{-i\omega t}\phi(t)dt \ , \tag{44}$$

it is well known that

$$\left[\mathscr{F}\phi^{(n)}\right](\omega) = (i\omega)^n[\mathscr{F}\phi](\omega) \ , \tag{45}$$

where n is any positive *integer* number, and $\phi^{(n)}$ is the n-th derivative of ϕ. Our fractional derivative is defined in such a way that

$$\left[\mathscr{F}\phi^{(\alpha)}\right](\omega) = (i\omega)^\alpha[\mathscr{F}\phi](\omega) \ , \tag{46}$$

where now α is any positive *real* number. For α not integer, one can show that such a derivative is a special pseudo-differential operator that can be properly defined by introducing the integer m such that $m - 1 < \alpha < m$ and putting

$$\phi^{(\alpha)}(t) = \frac{1}{\Gamma(m - \alpha)}\frac{\mathrm{d}^m}{\mathrm{d}t^m}\int\limits_{-\infty}^{t}\phi(\tau)\frac{1}{(t - \tau)^{\alpha+1-m}}\,\mathrm{d}\tau \tag{47}$$

or, equivalently,

$$\phi^{(\alpha)}(t) = \frac{1}{\Gamma(m - \alpha)}\int\limits_{-\infty}^{t}\frac{\mathrm{d}^m\phi(\tau)}{\mathrm{d}\tau^m}\frac{1}{(t - \tau)^{\alpha+1-m}}\,\mathrm{d}\tau \ . \tag{48}$$

We note the possibility of interchanging the integer-order derivative with the integral, since the function $\phi(t)$ is assumed to decay sufficiently fast to zero for $t \to -\infty$ together with its (relevant) derivatives.

Appendix B. Grünwald-Letnikov and Central-difference
Approximations to the Fractional Derivative

Consider the backward first-order approximation of the first derivative,

$$\frac{\partial f(t)}{\partial t} \sim \frac{f(t) - f(t-h)}{h} \ . \tag{49}$$

This leads to the second derivative

$$\frac{\partial^2 f(t)}{\partial t^2} \sim \frac{1}{h}\left(\frac{\partial f(t)}{\partial t} - \frac{\partial f(t-h)}{\partial t}\right) \sim \frac{f(t) - 2f(t-h) + f(t-2h)}{h^2} \tag{50}$$

and to the third derivative

$$\frac{\partial^3 f(t)}{\partial t^3} \sim \frac{f(t) - 3f(t-h) + 3f(t-2h) - f(t-3h)}{h^3} \ . \tag{51}$$

The generalization is straightforward. The m-th derivative is

$$\frac{\partial^m f(t)}{\partial t^m} \sim \frac{1}{h^m}\sum_{j=0}^{m}(-1)^j\binom{m}{j}f(t-jh), \qquad m = 0, 1, 2, 3, \ldots . \tag{52}$$

A more accurate (second-order) approximation for the first derivative is

$$\frac{\partial f(t)}{\partial t} \approx \frac{1}{h}\left[f\left(t+\frac{h}{2}\right) - f\left(t-\frac{h}{2}\right)\right] \ . \tag{53}$$

This leads to the second-order accurate m-th derivative

$$\frac{\partial^m f(t)}{\partial t^m} \approx \frac{1}{h^m}\sum_{j=0}^{m}(-1)^j\binom{m}{j}f\left[t + h\left(\frac{m}{2} - j\right)h\right], \qquad m = 0, 1, 2, 3, \ldots . \tag{54}$$

The upper summation limit may be replaced by any integer larger than m, for example by $t/h - 1$, since

$$\binom{m}{j} = 0 \qquad \text{for } j > m.$$

There are no restrictions in the r.h.s. of equations (52) and (54) that require m to be an integer. Replacing m by any positive real number v in equations (52) and (54) gives the Grünwald-Letnikov approximation (25) and the central-difference approximation (26), respectively (GORENFLO, 1997). The fractional binomial coefficients can be defined in terms of Euler's Gamma function as

$$\binom{v}{j} = \frac{\Gamma(v+1)}{\Gamma(j+1)\Gamma(v-j+1)}$$

and can be calculated by a simple recursion formula

$$\binom{v}{j} = \frac{v-j+1}{j}\binom{v}{j-1}, \qquad \binom{v}{0} = 1 \ .$$

The extension of the upper limit from v to $t/h - 1$ has an important consequence. While in equations (52) and (54) the series has vanishing terms beyond $j = m$, in equations (25) and (26) these terms are different from zero. The approximations (25) and (26) are actually "differintegration" operators, since it can be shown that for negative v they represent the generalized Riemann sums.

For more details on the theory and applications of fractional calculus, the reader is referred to OLDHAM and SPANIER (1974), GORENFLO and MAINARDI (1997), and PODLUBNY (1999).

REFERENCES

BLAND, D. R., *The Theory of Linear Viscoelasticity* (Pergamon, New York 1960).
CAPUTO, M. and MAINARDI, F. (1971), *Linear Models of Dissipation in Anelastic Solids*, Riv. Nuovo Cimento (Ser. II) *1*, 161–198.
CARCIONE, J. M. and CAVALLINI, F. (1994), *A Rheological Model for Anelastic Anisotropic Media with Applications to Seismic Wave Propagation*, Geophys. J. Int. *119*, 338–348.
CARCIONE, J. M., KOSLOFF, D., and KOSLOFF, R. (1988a), *Wave Propagation Simulation in a Linear Viscoacoustic Medium*, Geophys. J. Roy. Astr. Soc. *93*, 393–407.
CARCIONE, J. M., KOSLOFF, D., and KOSLOFF, R. (1988b), *Wave Propagation Simulation in a Linear Viscoelastic Medium*, Geophys. J. Roy. Astr. Soc. *95*, 597–611.
DATTOLI, G., TORRE, A., and MAZZACURATI, G. (1998), *An Alternative Point of View to the Theory of Fractional Fourier Transform*, J. Appl. Math. *60*, 215–224.
FELLAH, Z. E. A. and DEPOLLIER, C. (2000), *Transient Acoustic Wave Propagation in Rigid Porous Media: A Time-domain Approach*, J. Acoust. Soc. Am. *107*, 683–688.
GORENFLO, R., *Fractional calculus, some numerical methods*. In *Fractals and Fractional Calculus in Continuum Mechanics* (eds. Carpinteri, A. and Mainardi, F.). (Springer Verlag, Wien 1997) pp. 277–290.
GORENFLO, R. and MAINARDI, F. (1997), *Fractional calculus: Integral and differential equations of fractional order*. In *Fractals and Fractional Calculus in Continuum Mechanics* (eds. Carpinteri, A. and Mainardi, F.). (Springer Verlag, Wien 1997) pp. 223–276.
GRÜNWALD, A. K. (1867), *Über "begrenzte" Derivationen und deren Anwendung*, Zeit. angew. Math. Physik *12*, 441–480.
GUREVICH, B. and LOPATNIKOV, S. L. (1995), *Velocity and Attenuation of Elastic Waves in Finely Layered Porous Rocks*, Geophys. J. Int. *121*, 933–947.
ITÔ, K., *Encyclopedic Dictionary of Mathematics*, (The M.I.T. Press, Cambridge, 1987).
KJARTANSSON, E. (1979), *Constant Q-wave Propagation and Attenuation*, J. Geophys. Res. *84*, 4737–4748.
KOSLOFF, D. and BAYSAL, E. (1982), *Forward Modeling by the Fourier Method*, Geophysics *47*, 1402–1412.
LETNIKOV, A. V. (1868), *Theory of Differentiation of Fractional Order*, Math. Sb. *3*, 1–68 (in Russian).

MAINARDI, F. and TOMIROTTI, M. (1997), *Seismic Pulse Propagation with Constant Q and Stable Probability Distributions*, Annali di Geofisica *40*, 1311–1328.

MCDONAL, F. J., ANGONA, F. A., MILSS, R. L., SENGBUSH, R. L., VAN NOSTRAND, R. G., and WHITE, J. E. (1958), *Attenuation of Shear and Compressional Waves in Pierre Shale*, Geophysics *23*, 421–439.

MORSE, P. M. and FESHBACH, H., *Methods of Theoretical Physics*, Vol. II (McGraw-Hill, New York 1953).

OLDHAM, K. B. and SPANIER, J., *The Fractional Calculus* (Academic Press, San Diego 1976).

PODLUBNY, I., *Fractional Differential Equations* (Academic Press, San Diego 1999).

PRÜSS, J., *Evolutionary Integral Equations and Applications* (Birkhäuser, Basel 1993).

SCOTT-BLAIR, G. W., *Survey of General and Applied Rheology* (Pitman, London 1949).

TOKSÖZ, M. N. and JOHNSTON, D. H., *Seismic wave attenuation*. In Geophysics, reprint series No. 2 (SEG, Tulsa 1981).

(Received September 21, 2000, revised December 4, 2000, accepted December 13, 2000)

 To access this journal online:
http://www.birkhauser.ch

Pure appl. geophys. 159 (2002) 1737–1748
0033–4553/02/081737–12 $ 1.50 + 0.20/0

▌Pure and Applied Geophysics

Numerical Computation of Nonlinear Inelastic Waves in Soil

WOLFGANG FELLIN[1]

Abstract—Godunov's method, a numerical method for solving conservation laws, is applied to nonlinear and inelastic wave propagation in soil. The solution is restricted to the one-dimensional case. An approximate Riemann solver for Godunov's method is presented. The capability of the numerical method is shown by a comparison with the analytical solution of a linear inelastic wave propagation. Finally the behaviour of the nonlinear inelastic soil is described by a hypoplastic constitutive law.

Key words: Nonlinear wave propagation, Godunov's method, Riemann solver, hypoplasticity.

1. Introduction

Generally seismologists study linear elastic wave propagation problems. This is a good approximation if the wave travels through intact rock. However, already in the case of strong motion the material behaviour of rock is no more linear, and the behaviour of weathered rock (soil) is nonlinear and inelastic even if the deformations are small. Sometimes the nonlinearity of the soil behaviour is not neglectable. In this case the solution must be found numerically because analytical solutions for nonlinear wave propagations are very rare.

Why use Godunov's method and not "simple" finite differences or finite elements for the numerical solution of soil waves? Because of the particulate stress-strain behaviour of soil (Fig. 1), shock fronts occur during wave propagation.

The material behaviour of soil is nonlinear and inelastic, as we see in the example of a confined compression test (Fig. 2). There a cylindrical, confined sample is compressed in an axial direction. The stress $\sigma = -F/A$ (F is the force acting on the top plate, A is the circular cross-section area of the sample) plotted versus the strain $\varepsilon = -s/h_0$ (s is the settlement of the top plate, h_0 is the initial height of the sample) is shown in Figure 1 for a compressive loading and subsequent unloading.

We see that the stiffness $E_s = \partial\sigma/\partial\varepsilon$ of the sample increases with increasing[2] stress and is higher at unloading than at loading. When we think in terms of

---

[1] Institute of Geotechnics and Tunnelling, University of Innsbruck, Technikerstraße 13, A-6020 Innsbruck, Austria. E-mail: wolfgang.fellin@uibk.ac.at

[2] "Increasing" is used in the sense that $|\sigma|$ increases.

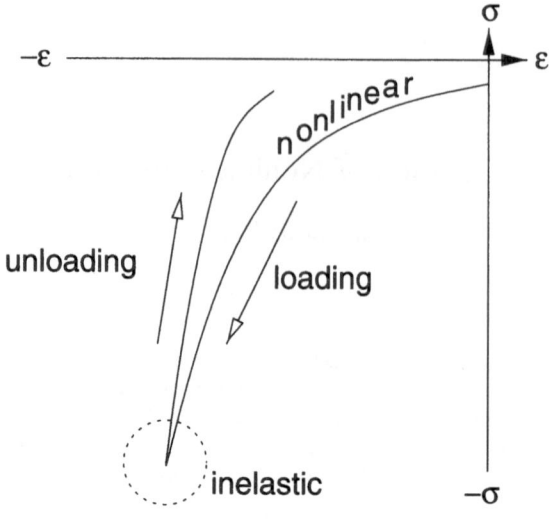

Figure 1
Soil behaviour in compression test.

wave propagation, the wave speed $c = \sqrt{E_s/\rho}$ increases with increasing stress during loading. Therefore the later wave parts (with higher stress levels) tend to overtake the earlier ones that started at a low stress level. Thus a shock front is formed.

A loading-unloading cycle will lead to a loading wave branch and a faster unloading wave branch (due to the higher unloading stiffness). Thus the unloading wave branch will "catch" the loading wave branch. At this moment the wave is not cancelled out, as we will see later. The wave front will become very steep (shock front).

Standard numerical methods cannot treat shock fronts without introducing some tricks, e.g., regularisation. Godunov's method (LE VEQUE, 1992, pp. 138–145) is based on the solution of propagation of shock waves, the so-called Riemann problem. Thus the method treats shocks adequately.

2. Introduction to One-dimensional Hypoplasticity

For all those who are not familiar with the concept of hypoplasticity, a short introduction is given. Roughly speaking, hypoplasticity is an advancement of the hypoelastic material law introduced by TRUESDELL (1956)

$$\overset{\circ}{\mathbf{T}} = \mathbf{f}(\mathbf{T}, \mathbf{D}) \ ,$$

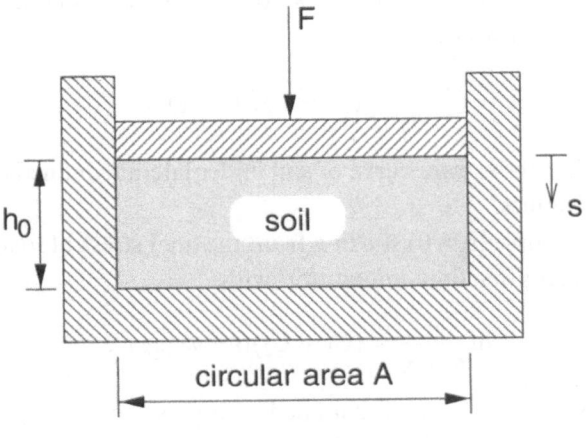

Figure 2
Compression test.

wherein $\overset{\circ}{\mathbf{T}}$ is an objective stress rate (TRUESDELL and NOLL, 1965) of the Cauchy stress. The tensorial hypoelastic function $\mathbf{f}$ is linear in the Cauchy stress $\mathbf{T}$ and the Euler's stretching $\mathbf{D}$.[3]

For our one-dimensional case (no rotations) the objective stress rate is equal to the material time derivative. The stretching tensor reduces to the time derivative of the strain

$$\dot{\sigma} = f(\sigma, \dot{\varepsilon}) \ .$$

With such a hypoelastic function f the nonlinear behaviour of the soil can be described. But loading and unloading will follow the same stress-strain path, because the material law is elastic.

The main idea to obtain an inelastic behaviour is to create a function h

$$\dot{\sigma} = h(\sigma, \dot{\varepsilon}) \ ,$$

which is linear in σ but nonlinear in $\dot{\varepsilon}$. This was first proposed by KOLYMBAS (1977). Hypoplastic laws are mainly developed to describe the behaviour of sands (granular materials). As the behaviour of sand is assumed to be approximately time-independent, the function h is time-scale invariant.

To give an initial sense of the functionality of the hypoplastic material law, a very simple one-dimensional version is presented now to describe the soil behaviour in a compression test (FELLIN, 2000a)

$$\dot{\sigma} = C_1 \sigma \dot{\varepsilon} + C_2 \sigma |\dot{\varepsilon}| \ , \tag{1}$$

---

[3] A hypoplastic material law is linear in $\mathbf{T}$ but nonlinear in $\mathbf{D}$.

with $C_1 < C_2 < 0$. If we integrate this law for compressive loading ($\dot{\varepsilon} < 0$) starting from an initial stress σ_0 we obtain

$$\ln \frac{\sigma}{\sigma_0} = (C_1 - C_2)(\varepsilon - \varepsilon_0) \ .$$

The logarithmic stress-strain curve of soil under lateral confined conditions is well known in soil mechanics.

Unloading the sample ($\dot{\varepsilon} > 0$) starting from the final stress of loading σ_{max} at the strain ε_{max} we finish after time integration with

$$\ln \frac{\sigma}{\sigma_{max}} = (C_1 + C_2)(\varepsilon - \varepsilon_{max}) \ .$$

The unloading path of a lateral confined compression test follows a different logarithmic curve than the loading path, which is also well known in soil mechanics. The loading and unloading curve for a confined compression test simulated with the simple version of the hypoplastic law Eq. (1) is plotted in Figure 3 in comparison with a laboratory experiment with sand.

The first hypoplastic law of KOLYMBAS (1985) has been further developed. In a more recent version the void ratio e of the soil has been added as an additional state parameter (VON WOLFFERSDORFF, 1996)

$$\dot{\mathbf{T}} = \mathbf{h}(\mathbf{T}, \mathbf{D}, e) \ .$$

The calibration of the material parameters of this law can easily be done with the help of standard laboratory tests (HERLE, 1997). For further introduction into three-dimensional hypoplasticity see KOLYMBAS (2000).

3. Problem Specification

In this paper we consider the numerical solution to the one-dimensional wave equation in a hypoplastic material for small deformations and velocities

$$\partial_t v(x, t) = \frac{1}{\rho(x, t)} \partial_x \sigma(x, t) \ , \tag{2}$$

where the velocity $v = \partial_t u(x, t)$ is the partial derivative[4] of the displacement $u(x, t)$ with respect to the time, σ is the Cauchy stress and ρ is the density of the soil. The spatial coordinate is $x \in [0, \infty)$ and the time is $t \in [0, \infty)$.

The material behaviour is prescribed by the hypoplastic law presented by VON WOLFFERSDORFF (1996). This reads in a general form for one dimension and for small deformations and velocities

---

[4] $\partial_t v(x, t) := \dfrac{\partial v}{\partial t}(x, t), \partial_x \sigma(x, t) := \dfrac{\partial \sigma}{\partial x}(x, t).$

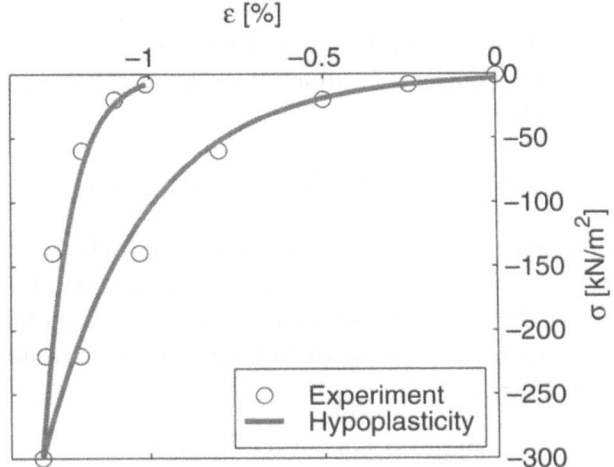

Figure 3
Hypoplastic approximation Eq. (1) to a laboratory test: $C_1 = -775$, $C_2 = -433$ ($\sigma_0 = -3,4$ kN/m^2).

$$\partial_t \sigma(x,t) = h\Big(\sigma(x,t), \partial_t \varepsilon(x,t), e(x,t)\Big) \tag{3}$$

with the strain $\varepsilon = \partial_x u(x,t)$ and e being the void ratio of the soil. The function h is later specified in Eq. (8).

We also use the compatibility condition

$$\partial_t \varepsilon(x,t) = \partial_x v(x,t) \ . \tag{4}$$

4. Numerical Solution

The numerical method of Godunov is very well presented in LE VEQUE (1992) and is not repeated here. Also the overall Riemann problem is shown in LE VEQUE (1992).

The existence of the solution to the Riemann problem of the system of nonlinear first-order differential equations (2), (3) and (4) has not yet been shown. SCHEARER and SCHAEFFER (1996) demonstrated the existence of the Riemann solution using the very restricting assumption that the hypoplastic law does not depend on the stress state. Also for an elasto-plastic constitutive law, which has in principle similar properties as the hypoplastic one, the existence and solution of the Riemann problem has been shown (TRANGENSTEIN and PEMBER, 1991).

To find an approximate Riemann solver for the nonlinear inelastic wave propagation, we have to linearize the problem.

4.1 Linearization

The nonlinear function h is linearized at each time step t^n and in each cell $[x_{j-1/2}, x_{j+1/2}]$.[5] We use the tangential stiffness $E = \partial\sigma/\partial\varepsilon$ at the time t^n and obtain from Eq. (3) with Eq. (4)

$$\partial_t \sigma(x, t) = E\Big(\sigma(x, t), \partial_x v(x, t), e(x, t)\Big)\partial_x v(x, t) . \qquad (5)$$

In this linearization the very important incrementally inelastic property of the hypoplastic law is sustained. The density ρ in each cell is held constant over the time step. The stiffness and the density are updated at the end of each time step.

Herewith we can write the system of nonlinear first-order differential equations (2) and (5) in the form

$$\partial_t \mathbf{q}(x, t) + \mathbf{A}(\mathbf{q}, x, t)\partial_x \mathbf{q}(x, t) = \mathbf{0} \qquad (6)$$

with

$$\mathbf{q}(x, t) = \begin{bmatrix} v(x, t) \\ \sigma(x, t) \end{bmatrix}, \quad \mathbf{A}(\mathbf{q}, x, t) = \begin{bmatrix} 0 & -1/\rho(x, t) \\ -E(\mathbf{q}, x, t) & 0 \end{bmatrix} .$$

Discretization of Eq. (6) leads to a piecewise linear system, for which the solution to the Riemann problem exists.

4.2 Riemann Problem in the Discretization

We follow the main idea of the Riemann solver proposed by LE VEQUE (CLAWPACK User Note #13) for acoustics in heterogeneous media. We consider the cells $j - 1$ and j at time t_n. The cell j has at time t_n material properties ρ_j^n and $E_j^n(\mathbf{Q}_j^n)$.[6] Similar notation applies to cell $j - 1$

$$\mathbf{A}_j^n = \begin{bmatrix} 0 & -1/\rho_j^n \\ -E_j^n & 0 \end{bmatrix}, \quad \mathbf{A}_{j-1}^n = \begin{bmatrix} 0 & -1/\rho_{j-1}^n \\ -E_{j-1}^n & 0 \end{bmatrix} .$$

The solution to the Riemann problem between the states $\mathbf{Q}_{j-1}^n$ in cell $j - 1$ and $\mathbf{Q}_j^n$ in cell j consists of two waves. The right-going wave moves into cell j with velocity $c_j^n = \sqrt{E_j^n/\rho_j^n}$, which is the positive eigenvalue λ_2 of $\mathbf{A}_j^n$. The left-going wave moves into cell $j - 1$ with velocity $-c_{j-1}^n = -\sqrt{E_{j-1}^n/\rho_{j-1}^n}$, which is the negative eigenvalue λ_1 of $\mathbf{A}_{j-1}^n$.

We decompose the jump in the numerical solution $\Delta\mathbf{Q}_{j-1,j}^n = \mathbf{Q}_j^n - \mathbf{Q}_{j-1}^n$ at the interface between the cells into the eigenvectors $\mathbf{r}_1$ of $\mathbf{A}_{j-1}^n$ and $\mathbf{r}_2$ of $\mathbf{A}_j^n$. Consequently the right-going wave is $\alpha_{2j}^n \mathbf{r}_{2j}^n$ and the left-going wave is $\alpha_{1j-1}^n \mathbf{r}_{1j-1}^n$. The sum of these

---

[5] For details to the discretization see LE VEQUE (1992, pp. 97 ff), and FELLIN (1999, pp. 89–100).

[6] The numerical solution $\mathbf{Q}$ used here is the mean value of the solution $\mathbf{q}$ over a cell.

two waves must be equal to the jump in the numerical solution (the proper physical jump conditions)

$$\alpha_{1j-1}^n \mathbf{r}_{1j-1}^n + \alpha_{2j}^n \mathbf{r}_{2j}^n = \Delta \mathbf{Q}_{j-1,j}^n \ . \tag{7}$$

From Eq. (7) we obtain a linear system which can be solved for α_{1j-1}^n and α_{1j}^n, yielding:

$$\alpha_{1j-1}^n = \frac{c_j^n \Delta \sigma_{j-1,j}^n + E_j^n \Delta v_{j-1,j}^n}{c_{j-1}^n E_j^n + c_j^n E_{j-1}^n}, \quad \alpha_{2j}^n = \frac{-c_{j-1}^n \Delta \sigma_{j-1,j}^n + E_{j-1}^n \Delta v_{j-1,j}^n}{c_{j-1}^n E_j^n + c_j^n E_{j-1}^n} \ .$$

The numerical fluxes[7] in Godunov's method are

$$\mathbf{A}^-(\mathbf{Q}_j^n - \mathbf{Q}_{j-1}^n) = \alpha_{1j-1}^n c_{j-1}^n \begin{bmatrix} c_{j-1}^n \\ E_{j-1}^n \end{bmatrix}, \quad \mathbf{A}^+(\mathbf{Q}_j^n - \mathbf{Q}_{j-1}^n) = \alpha_{2j}^n c_j^n \begin{bmatrix} c_j^n \\ E_j^n \end{bmatrix} \ .$$

This Riemann solver must be implemented into the software package CLAWPACK (LE VEQUE, 1994).

5. Comparison with Analytical Solution

An analytical solution for the following simple conditions is obtained by NIKITIN (1998) and FELLIN (1999). The one-dimensional continuum is a semi-infinite thin rod (Fig. 4). Strain and inertia orthogonal to the rod axis x are neglected as well as density changes due to wave propagation. The rod is loaded at its left end ($x = 0$) by a stress jump at time $t = 0$ from zero to σ_1 and at $t = t_1$ back to zero (Fig. 5).

The constitutive relation is very simple (linear inelastic)

$$\dot{\sigma} = \begin{cases} E_1 \dot{\varepsilon} : & \dot{\varepsilon} < 0 \text{ first loading,} \\ E_2 \dot{\varepsilon} : & \dot{\varepsilon} > 0 \text{ unloading and } \dot{\varepsilon} < 0 \text{ reloading} \ . \end{cases}$$

The results of a calculation with $\sigma_1 = -1000 \text{ kN/m}^2$, $t_1 = 0,01$ s and typical material properties for sand $E_1 = 10 \text{ MN/m}^2$, $E_2 = 80 \text{ MN/m}^2$ and $\rho = 1528 \text{ kg/m}^3$ are plotted in Figures 6 and 7. The analytical solution in these figures is (FELLIN, 2000b, pp. 95–103)

$$v_2 = \frac{\sigma_1}{\rho c_1} \frac{c_1 - c_2}{c_2}, \ v_3 = \frac{\sigma_1}{\rho c_1} \frac{c_1 - c_2}{c_2 + c_1}, \ \sigma_3 = \sigma_1 \frac{c_2 - c_1}{c_2 + c_1} \quad \text{with} \quad c_1 = \sqrt{E_1/\rho}, \ c_2 = \sqrt{E_2/\rho} \ ,$$

$$\varepsilon_2 = \sigma_1 \frac{E_2 - E_1}{E_2 E_1}, \ \varepsilon_{3,a} = \frac{\sigma_1}{E_1} \frac{c_2 - c_1}{c_2 + c_1}, \ \varepsilon_{3,b} = \frac{\sigma_1}{E_1} \frac{c_2 - c_1}{c_2 + c_1} \frac{2c_1^2 + c_2^2 + 2c_1 c_2}{c_2^2} \ .$$

---

[7] A numerical flux is the mean value of the physical flux over a time step. For further explanation see LE VEQUE (1992, page 141).

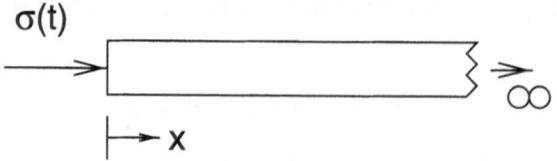

Figure 4
Semi-infinite rod.

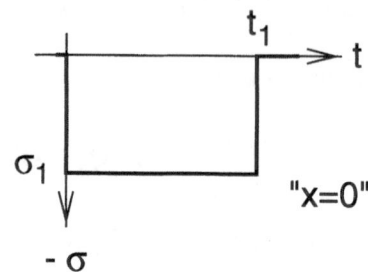

Figure 5
Loading on the left end, $x = 0$.

The numerical solutions were obtained with 800 cells and first order Godunov in Figure 6. The shock fronts are smoothed due to the internal numerical damping of the first order Godunov method, however, the values in between the shocks are exact. Increasing the number of cells would not give decidedly better results.

The shock front can be better predicted with the same number of cells but using a higher order method[8] with flux limiter[9] (LE VEQUE, 1992, pp. 173 ff) (see Fig. 7). Here the so called superbee[10] flux limiter gives the best result. The computation time for the higher order method is 2.25 times higher for this example.

6. Example

A laterally confined rod of soil is not a one-dimensional problem. The strains $\varepsilon_y = \varepsilon_z$ orthogonal to the compression direction x are zero, although the stresses

---

[8] This method adds a correction to the numerical fluxes, to overcome the numerical damping of the first-order Godunov's method.

[9] The flux correction depends on the solution. It is nearly zero if the solution is smooth. If the solution is discontinuous the sum of the Godunov's flux and the correction are nearly a higher order flux with lower damping.

[10] Numerical solutions obtained by the use of higher order fluxes tend to oscillate near a shock front. Some higher order methods give oscillations in front of a shock, others oscillate behind the shock. A good flux limiter method chooses the correction as a mixture of such two methods. The superbee flux limiter is a special way to combine two higher order fluxes to avoid oscillations.

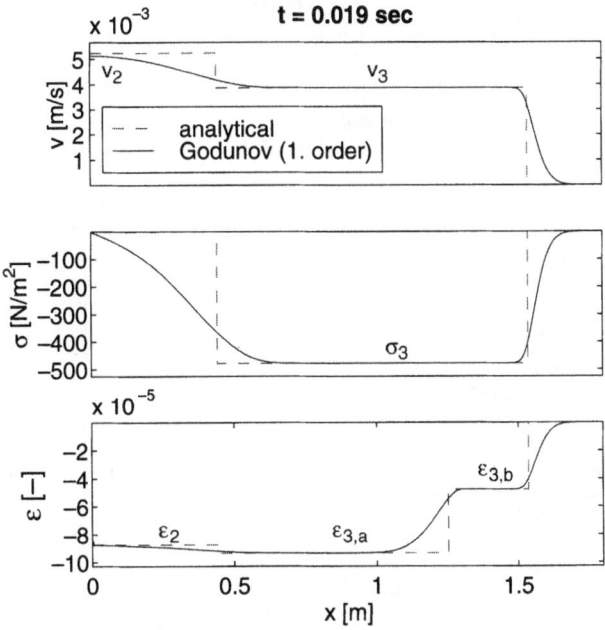

Figure 6
Solution with first-order Godunov.

$\sigma_y = \sigma_z$ are not equal to zero. To stay in our one-dimensional world, we must make a further assumption for the lateral stresses

$$\sigma_y = \sigma_z = K_0 \sigma_x$$

which is quite accurate loading and reasonably accurate during unloading.[11] The coefficient of lateral stress at rest can be estimated $K_0 \approx 1 - \sin \varphi_c$ with the critical friction angle of the soil φ_c. With this simplification the hypoplastic law (VON WOLFFERSDORFF, 1996) is very simple

$$\dot{\sigma} = \frac{f_b f_e}{1 + 4K_0^2} \left[(1 + 2K_0)^2 \dot{\varepsilon} + a^2 \dot{\varepsilon} + f_d \frac{a}{3} (5 + 12K_0 - 4K_0^2)|\dot{\varepsilon}| \right] \qquad (8)$$

with

$$f_b := \frac{h_s}{n} \left(\frac{e_{i0}}{e_{c0}} \right)^\beta \frac{1 - e_i}{e_i} \left(\frac{3p_s}{h_s} \right)^{1-n} \left[3 + a^2 - a\sqrt{3} \left(\frac{e_{i0} - e_{d0}}{e_{c0} - e_{d0}} \right)^\alpha \right]^{-1}, \quad f_d := \left(\frac{e - e_d}{e_c - e_d} \right)^\alpha,$$

$$f_e := \left(\frac{e_c}{e} \right)^\beta, \quad a := \frac{\sqrt{3}(3 - \sin \varphi_c)}{2\sqrt{2} \sin \varphi_c}, \quad \frac{e_i}{e_{i0}} = \frac{e_c}{e_{c0}} = \frac{e_d}{e_{d0}} = \exp \left[-\left(\frac{3p_s}{h_s} \right)^n \right], \quad p_s = (1 + 2K_0)\sigma.$$

---

[11] One can also add a third equation to Eq. (6) which describes the evolution of $\sigma_y = \sigma_z$. But this will not change the results significantly.

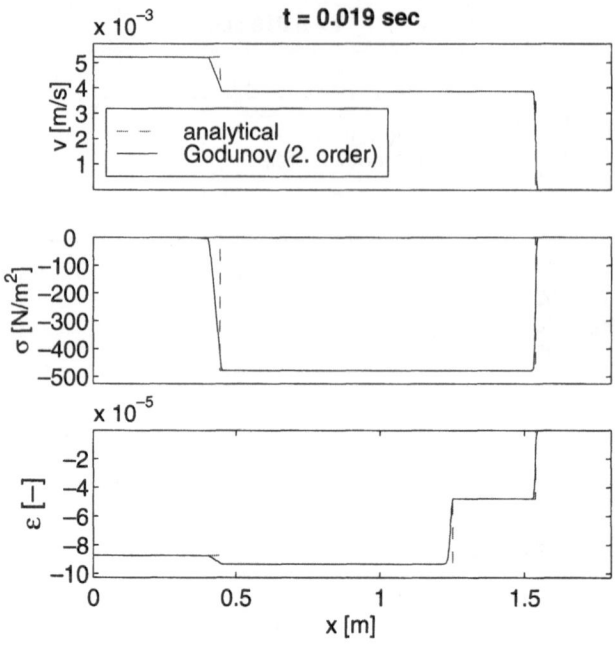

Figure 7
Solution with Godunov's method, second-order corrections, superbee flux limiters.

The results for a hypoplastic wave computed with second-order Godunov and superbee flux limiters (400 cells) are shown in Figure 8. The material parameters used are for Karlsruhe sand (HERLE, 1997): critical friction angle $\varphi_c = 30°$, granular hardness $h_s = 5800$ MPa, exponent $n = 0.28$, void ratio at highest density $e_{d0} = 0.53$, void ratio at critical density $e_{c0} = 0.84$, void ratio at lowest density $e_{i0} = 1.00$, coefficients $\alpha = 0.13$ and $\beta = 1.05$. The initial conditions are: $\sigma_0 = -100$ N/m^2, void ratio $e_0 = 0.735$, density $\rho_0 = 1528$ kg/m^3. The load at the left end is $\sigma(0,t) = \hat{\sigma}\sin(\Omega t) + \sigma_0$ with $\hat{\sigma} = -1000$ N/m^2 and $\Omega = 314.16$ 1/s acting from $t = 0$ to $t = \pi/\Omega$ (half a period).

As we see in Figure 7 the second-order Godunov with flux limiters can predict the wave propagation very well. Thus the results in Figure 8 approach exactness. Only the shock front is not as steep as it should be, nonetheless its position is correct.

7. Conclusions

During wave propagation in nonlinear inelastic materials, shock fronts occur. It is necessary to use a numerical method which can treat shock fronts.

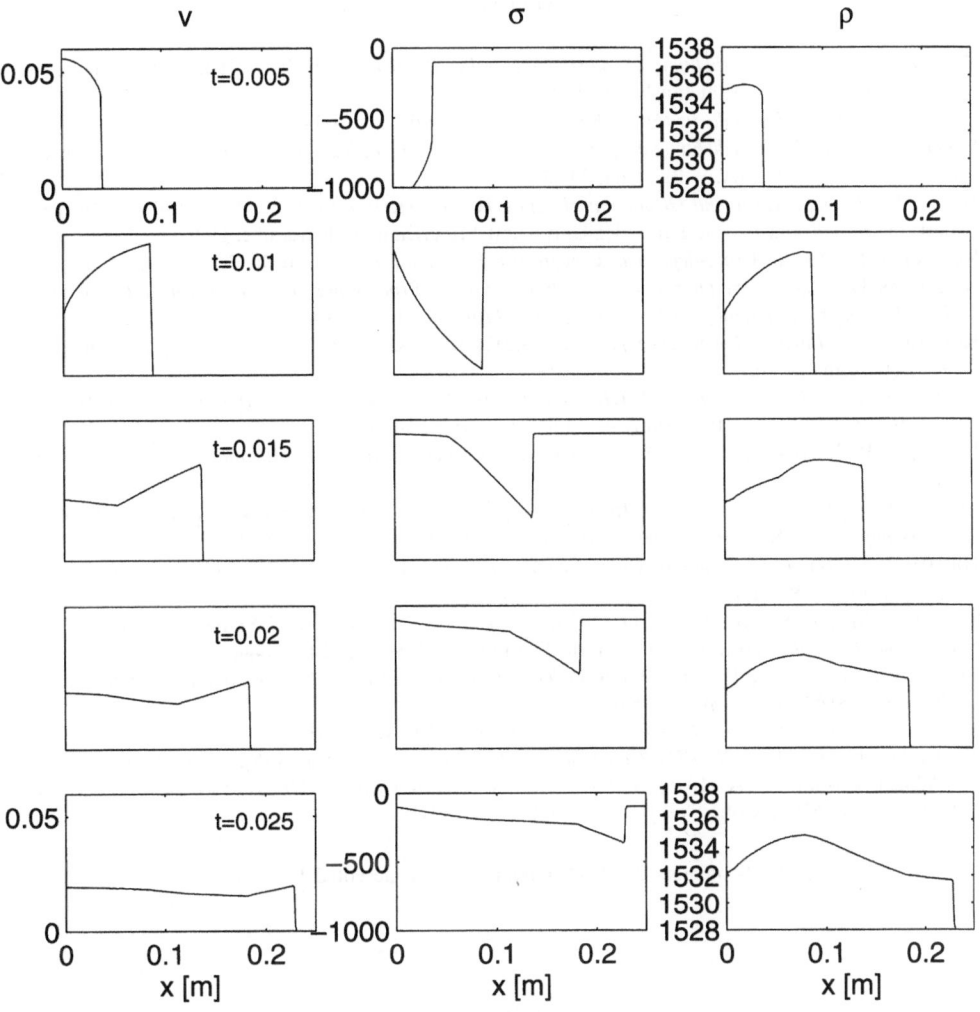

Figure 8
One-dimensional hypoplastic wave; second-order Godunov with superbee flux limiters.

Godunov's method is a good choice. Geotechnical engineering problems such as foundation oscillating, vibration compaction and others are usually two- or three-dimensional. Therefore it is necessary to develop multidimensional methods.[12] A first attempt of calculating two-dimensional hypoplastic waves with Godunov's method is given by FELLIN (2000b, pp. 132–145).

---

[12] See also LE VEQUE (1992, p. 207).

REFERENCES

FELLIN, W., *Rütteldruckverdichtung als plastodynamisches Problem*. Ph.D. Thesis (Institut für Geotechnik und Tunnelbau, Universität Innsbruck, 1999).

FELLIN, W. (2000a), *Hypoplastizität für Einsteiger*, Bautechnik *77*, 10–14.

FELLIN, W., *Rütteldruckverdichtung als plastodynamisches Problem*, Advances in Geotechnical Engineering and Tunnelling, vol. 2 (A.A. Balkema, 2000b).

HERLE, I., *Hypoplastizität und Granulometrie einfacher Korngerüste* (Veröffentlichung des Institutes für Bodenmechanik und Felsmechanik der Universität Fridericiana in Karlsruhe, Heft 142, 1997).

KOLYMBAS, D. (1977), *A rate-dependent Constitutive Equation for Soils*, Mech. Res. Comm. *4*, 367–372.

KOLYMBAS, D., *A generalized hypoelastic constitutive law*. In *Proceedings of XI International Conference on Soil Mechanics and Foundation Engineering, San Francisco, vol. 5.* (Balkema 1985) p. 2626.

KOLYMBAS, D., *Introduction to Hypoplasticity*, Advances in Geotechnical Engineering and Tunnelling, vol. 1 (Balkema 2000).

LE VEQUE, R. J. (online) *CLAWPACK, A Software Package for Conservation Laws and Hyperbolic Systems*, http://www.amath.washington.edu/~rjl/clawpack.html.

LE VEQUE, R. J., *Numerical Methods for Conservation Laws, Lectures in Mathematics* (Birkhäuser Verlag 1992).

LE VEQUE, R. J., *Clawpack — A software package for solving multi-dimensional conservation laws*. In *Proceedings of the 5th International Conference on Hyperbolic Problems* (1994).

NIKITIN, L. (1994), *The Problem of Unloading Materials of the KGB (Kolymbas, Gudehus, Bauer) Form*, personal note in Russian.

SCHEARER, M. and SCHAEFFER, G. D. (1996), *Riemann Problem for 5 × 5 Systems of Fully Nonlinear Equations Related to Hypoplasticity*, Math. Methods Appl. Sci. *19*, 1433–1444.

TRANGENSTEIN, J. A. and PEMBER, B. (1991), *The Riemann Problem for Longitudinal Motion in an Elastic-plastic Bar*, SIAM J. Sci. Stat. Comp. *12*, 180–207.

TRUESDELL, C. (1956), *Hypo-elasticity*, J. Rational Mech. Analysis *4*, 83–133.

TRUESDELL, C. and NOLL, W., *The Non-Linear Field Theories of Mechanics* (Springer 1965).

VON WOLFFERSDORFF, P.-A. (1996), *A Hypoplastic Relation for Granular Materials with a Predefined Limit State Surface*, Mech. Cohesive-Frictional Mat. *1*, 251–271.

(Received September 29, 2000, January 5, 2001, accepted January 30, 2001)

 To access this journal online:
http://www.birkhauser.ch

Pure appl. geophys. 159 (2002) 1749–1769
0033–4553/02/081749–21 $ 1.50 + 0.20/0

© Birkhäuser Verlag, Basel, 2002

❙Pure and Applied Geophysics

Propagation of Pulses in Viscoelastic Media

ANDRZEJ HANYGA[1]

Abstract — It is shown that in hereditary models of media with regular convolution kernels discontinuities propagate with the wavefronts. A singularity always remains sharp and propagates with the wavefront. Oscillatory signals lag behind the wavefronts and are preceded by distortions. In media with singular memory, signals lag behind wavefronts. Singular signals are immediately smoothed and spread out.

Key words: Viscoelasticity, wave propagation.

1. Introduction

We shall examine a few relatively simple examples of partial differential equations which exhibit dispersion, dissipation and attenuation and can be considered as simple models of viscoelastic or poroelastic wave propagation. In general such equations involve differential and convolution operators. There are significant differences between equations involving only regular convolution kernels and equations involving at least one singular convolution kernel. Poroelastic media and an overwhelming majority of specific physical models of viscoelastic materials involve singular convolution kernels (HANYGA, 2001b). The Cole–Cole dispersion law for electromagnetic waves in dielectric media, also known in polymer rheology as the Bagley-Torvik model, belongs to this class.

The case of regular memory is more complex and has a long history.

In the 1880s Lord Rayleigh introduced the concept of group velocity in order to characterize the speed of propagation of non-monochromatic signals in purely dispersive media. Its rigorous derivation for wave packets (or finite duration pulses) was possible when Lord Kelvin developed the stationary phase method. The derivation depends on the absence of frequency-dependent attenuation. Consequently it applies to dispersive modes but it does not extend to dispersion associated with attenuation.

In the beginning of the last century it was realized that dispersion can be anomalous: the group velocity can exceed the phase speed and its high-frequency

---

[1] Institute of Solid Earth Physics, University of Bergen, N5007 Allégaten 41, Norway.

limit. In particular, the Lorentz-Kramers model of dielectric dispersion (BRILLOUIN, 1960) exhibits anomalous dispersion in some frequency ranges. For electromagnetic waves in a dispersive dielectric, the high-frequency limit is the speed of light *in vacuo*. In relativity signals, whatever they are, are considered as a tool for synchronizing clocks. It is a fundamental assumption that they cannot travel faster than light. Assuming that signals, interpreted in the sense of propagating distinctive patterns of the electromagnetic field, play this role, it was thought unacceptable that they might propagate with group speed. This argument does not sound convincing since the very structure of the Maxwell equations ensures that the electromagnetic field cannot advance at a speed exceeding the speed of light *in vacuo* and in this sense the basic principle of relativity is never violated.

This apparent contradiction between relativity and anomalous dispersion prompted VON LAUE (1905) and SOMMERFELD (1914) to pose the problem of signal velocity in a dispersive dielectric. Sommerfeld provided the first answer in 1914, however, Brillouin, working with Sommerfeld in Munich at that time, produced a more elaborate analysis. Since in dissipative media either frequencies or wave number vectors are complex, Brillouin applied saddle point analysis, which was a relatively novel technique at that time. Rigorous foundations of the saddle point method would be laid later. Brillouin's analysis was further clarified by others, including MAINARDI (1983b), who applied it in viscoelasticity. Independently, THAU (1974) applied a similar analysis to the Klein-Gordon and Boussinesq equations, considered as prime examples of dispersive-dissipative behavior. As pointed out by Mainardi, viscoelastic media[1] also exhibit anomalous dispersion, hence anomalous dispersion seemed to be in conflict with the upper limit for the propagation speed implied by the hyperbolic character of the equations.

Sommerfeld and Brillouin tried to demonstrate that oscillatory signals propagate with a velocity which they called signal velocity. The signals they considered were sinusoidal wave trains windowed by the Heaviside function. According to their definition after a time r/S, where S is the signal velocity and r is the propagation distance, the signal amplitude attains more than one-half of its final value. Signal velocity of electromagnetic waves in a dispersive dielectric, as defined by Brillouin, never exceeds the speed of light. In a more general context, including hyperbolic models of viscoelasticity, signal velocity does not exceed the wavefront speed.

An important by-product of the analysis undertaken by Sommerfeld and Brillouin was the discovery of Sommerfeld and Brillouin forerunners, often dominating the main signal.

The behavior of both stress waves in viscoelastic media and electromagnetic waves in Lorentz-Kramers dielectric media is qualitatively well represented by a single scalar integro-differential equation with a regular convolution kernel. As noted

---

[1] We assume throughout this paper the hereditary model of viscoelasticity.

by MAINARDI (1983a,b), the high-frequency behavior of solutions of such equations can be expressed in terms of a three-dimensional version of the telegraph equation. The telegraph equation has two parameters. One of them controls dispersion; the other one introduces both attenuation and dispersion. It turns out that an inequality discriminates between the predominantly dispersive case and the predominantly attenuating case. It will be shown in Section 3 that Green's functions for the three-dimensional telegraph equation can be determined explicitly. In both cases Green's functions consist of an exponentially damped delta singularity propagating with the wavefront and a tail which is oscillatory in the predominantly dispersive case and monotone in the other. In the oscillatory case the Green's function has a dense oscillation near the wavefront.

The solution for a finite-bandwidth signal is given by the convolution of the signal with Green's function of the medium. The zero phase point of the main signal clearly propagates with the wavefront. The main signal is accompanied by distorting signals. These are formed from the low and high frequencies of the input signal and can dominate over the main signal before it reaches a full amplitude (OUGHSTUN and SHERMAN, 1994). For media with resonant properties (Lorentz-Kramers dispersion) the distortions appear as Sommerfeld and Brillouin forerunners. The Brillouin forerunner typically has a higher amplitude. At higher distances the main signal can be negligible in comparison with the Brillouin forerunner. Except for the frequency ranges exhibiting anomalous dispersion, signal velocity is equal to group velocity.

Wave propagation in hereditary viscoelastic media with singular memory is discussed in Section 6. Hereditary viscoelastic media with singular memory are considered in more detail by HANYGA and SEREDYŃSKA (1999a,b; 2001) (sharp signals) and HANYGA and ROK (2000) (oscillatory finite band-width finite-duration signals). The first results in this direction were obtained in BUCHEN and MAINARDI (1975) as well as LOKSHIN and ROK (1978a,b). The analysis presented in this paper is less rigorous and more elementary. The most important conclusion is a delay of the pulse with respect to the wavefront.

The evolution of the signal in a viscoelastic medium with singular memory, including poroelastic and poroacoustic media, differs from the regular memory case. In a medium with nonsingular memory, a discontinuity always remains sharp, albeit its amplitude decreases exponentially in time. In media with singular memory, an initial discontinuity is immediately smoothed and it gradually flattens out. These remarks show that the concept of travel time might require revision in the case of viscoelastic and poroelastic media.

In a homogeneous medium with singular memory, pulses do not propagate with constant speeds. Instead, they lag behind the wavefronts. The pulse delay is not a linear function of propagation time and the effective signal speed decreases in time. The time lag can be explicitly calculated in some models for initially sharp signals. It has also been demonstrated numerically for oscillatory signals resembling the Ricker wavelet (HANYGA and ROK, 2000). The dependence of the time lag on the dominant frequency is only slight. Pulse delay has been observed in seismic prospection as a delay of seismic pulses with respect to their travel times estimated from acoustic log data (Steen Petersen, priv. comm.).

Wave propagation problems in media with regular memory and singular memory can be solved numerically by finite-difference schemes provided simplifications are made. Thus, in the regular case it is usually assumed that the relaxation function can be represented in terms of a finite spectrum of exponential relaxations with relaxation times equally spaced on the logarithmic scale (DAY and MINSTER, 1984; EMMERICH and KORN, 1987; CARCIONE et al., 1998a,b). This allows elimination of integral operators at the cost of supplementing the equations of motion with additional differential equations representing internal relaxation. The most important flaw of this approximation is the appearance of arbitrarily chosen relaxation times, which therefore lack physical meaning. The relaxation strengths depend on the choice of the relaxation times and hence they are also non-physical.

On the other hand, for viscoelastic models involving singular memory formulation in terms of generalized fractional derivatives provides a fairly efficient discretization of the non-local integro-differential operator (HANYGA, 1999b).

In the context of ray tracing methods, singular memory models are more convenient than regular memory models. For a general dissipative-dispersive medium with hereditary effects, complex rays are indispensable to account for the dispersion and attenuation of the signal (HANYGA and SEREDYŃSKA, 2000a; HANYGA, 1999a). In the case of a singular hereditary medium, the singular part of the memory controls the behavior of the wavefield at the wavefront, while the regular part merely modifies the amplitude decay rate and the signal tail. Instead of applying complex ray tracing, it is now possible to express the complex-valued phase function in terms of a few real-valued functions which can be calculated by real ray tracing. The first of these functions is travel time, which determines the propagation of the wavefront. The other functions determine the time lag of the pulse and its smoothing.

Following MAINARDI (1983b) we shall study simple scalar models of dispersive media, avoiding the irrelevant complexities of Maxwell and viscoelastic equations.

In Section 2 a simple scalar model of poroelasticity and viscoelasticity is chosen for further analysis. As shown by Mainardi, it also represents some properties of Lorentz-Kramers dispersion in dielectrics. Two cases, representing predominantly dispersive and predominantly dissipative behavior, are distinguished. In Section 3 propagation of singularities represented by Green's functions is discussed. In Section 4 propagation of pulses is discussed. The complicated structure of the solution is discussed for the dispersive dielectrics in Section 5. In Section 6 the viscoelastic model with singular memory is briefly recapitulated along with its implications.

2. Simple Model Equations

Hereditary effects can enter the equations of motion in two ways: convolution operators can appear either in the part of the differential operator involving time derivatives (such as in poro-elasticity (NORRIS, 1986)):

$$\rho_\infty K(t) * u_{tt} = \nabla \cdot [C \nabla u] \tag{1}$$

or in the spatial part, through a generalization of Hooke's law to a Boltzmann hereditary model of viscoelasticity:

$$\rho u_{tt} = \nabla \cdot [C * \nabla u]. \tag{2}$$

In poro-acoustics and visco-poro-elasticity, memory effects appear on both sides of the equation of motion.

In the context of the familiar viscoelastic constitutive relation (CHRISTENSEN, 1971; GROSS, 1953; FABRIZIO and MORRO, 1992):

$$\sigma(t) = G_0\, e(t) + \int_0^\infty \dot{G}(\xi) e(t - \xi)\, d\xi$$

$$\equiv G_\infty\, e(0) + \int_0^\infty G(\xi) \dot{e}(t - \xi)\, d\xi \tag{3}$$

the weakly singular case is defined by $\dot{G}(\xi) \sim \text{const} \times t^{-\alpha}$, $0 < \alpha, 1$ (note that $v = \sqrt{G_0/\rho}$ is the wavefront speed, or the limit of phase speed for infinite frequency).

The memory kernels in either case have the form $K(t) = K_0 \delta(t) + K_1(t)$, $C(t) = C_0 \delta(t) + C_1(t)$ with locally integrable functions $K_1(t), C_1(t)$ and hence the associated convolution operators are invertible in the convolution algebra.

In a homogeneous isotropic medium, the convolution kernels are scalar functions independent of $\mathbf{x}$ and the associated convolution operators commute with the gradient operator. Consequently the convolution operators can be moved to one side of the equation, either from the left-hand side of the equation to the right-hand side or *vice versa*. Moreover, if $R(t) = R_0 \delta(t) + R_1(t)$ is the inverse of $K(t)$ in the convolution algebra, $R(t) * K(t) = \delta(t)$, then and $K_1(t) \sim \text{const} \times t^{-\alpha}$ for $t \to 0$, then $R_1(t) \sim \text{const} \times t^{-\alpha}$ for $t \to 0$. Indeed, the Fourier transform

$$\hat{K}_1(\omega) := \int_{-\infty}^{\infty} \exp(i\omega t) K(t)\, dt \sim D\Gamma(1 - \alpha)(-i\omega)^{\alpha - 1}$$

entails

$$1 + \hat{R}_1(\omega)/R_0 = 1/\left[1 + D\Gamma(1 - \alpha)(-i\omega)^{\alpha-1}/K_0\right] \sim 1 - D\Gamma(1 - \alpha)(-i\omega)^{\alpha-1}/K_0$$

and $R_1(t) \sim -d(R_0/K_0)t^{-\alpha}$. It follows that the basic properties of solutions of eqs. (1) and (2) are the same, provided the kernels have the same singularities at $t = 0$.

For our purposes both forms of hereditary theories are equivalent. Equation (1) has however simpler solutions for some representative kernels. Its equivalent (2) has the same solutions although a complicated kernel. We shall therefore focus on eq. (1).

Viscoelastic models encompass a wide variety of constitutive equations and resulting equations of motion. In particular, Newtonian viscosity $\sigma = N\dot{e}$ and various hereditary models interpolating between elastic Hooke's law and Newtonian viscosity (PIPKIN, 1986) lead to parabolic equations of motion and propagation speeds which are unbounded in the high-frequency limit. Some current models of marine sediments (DEANE, 1997; BUCKINGHAM, 1997; BUCKINGHAM, 1998) are of this type. In seismology constant-Q equations (CAPUTO, 1969; CAPUTO and MAINARDI, 1976; KJARTANSSON, 1979; CARCIONE, et al., 2000) fall into this class. In viscoelasticity the assumption of $\dot{G} \sim \mathrm{const} \times t^{-1-\alpha}$, with an infinite G_0, has also been considered. Infinite speeds of propagation and a parabolic behavior also arise in hereditary viscoelasticity with strongly singular kernels (RENARDY, 1982).

Parabolic equations are inconvenient for use in seismological applications as they do not allow for a rigorous definition of travel time. We are thus left with hereditary viscoelastic models with regular or weakly singular kernels. For eq. (1) this means that $K_1(t)$ is either regular at $t = 0$ or $K_1(t) \sim \mathrm{const} \times t^{-\alpha}$ with $0 < \alpha < 1$. The inequality $\alpha \geq 1$ defines the strongly singular case.

We shall focus on a simple but representative case of a scalar equation

$$K(t) * u_{tt} - c^2 \nabla^2 u = f(t, \mathbf{x}) \tag{4}$$

with $K(t) = \delta(t) + K_1(t)$. Mainardi noted that truncating the asymptotic expansion of the Fourier transformed convolution kernel $K(t)$

$$\hat{K}(\omega) \sim 1 + a(-i\omega)^{-1} + b(-i\omega)^{-2} + \cdots \tag{5}$$

one obtains a three-dimensional version of the telegraph equation

$$u_{tt} + a u_t + b u - c^2 \nabla^2 u = f(t, \mathbf{x}). \tag{6}$$

The predominantly dispersive and predominantly dissipative cases are now defined by the inequalities $\Delta > 0$ and $\Delta < 0$, with

$$\Delta := b - \frac{a^2}{4}. \tag{7}$$

ainardi (1983b) also demonstrated that viscoelasticity (and, in particular, the Standard Linear Solid model) corresponds to $\Delta < 0$.

3. Propagation of Singularities: Green's Functions

We shall now derive an explicit Green's function for eq. (6).

Let $\tilde{f}$ denote the one-sided Laplace transform of an arbitrary causal function $f(t)$.

The initial-value problem for eq. (6) with $f(t, \mathbf{x}) = F(t)\delta(\mathbf{x})$

$$
\begin{aligned}
u(0) &= u_0\delta(\mathbf{x}) \\
u_t(0) &= v_0\delta(\mathbf{x})
\end{aligned}
\tag{8}
$$

can be solved explicitly.

Let $\gamma(s) := s^2 + as + b$, $h(s) = \tilde{F}(s) + (s + a)u_0 + v_0$. The Laplace transformation of (6), (8) yields the equation

$$
s^2\tilde{u} + as\tilde{u} + b\tilde{u} - c^2\nabla^2\tilde{u} = h(s)\delta(\mathbf{x}).
\tag{9}
$$

The spatial Fourier transform $U(s, \mathbf{k})$ of $\tilde{u}(t, \mathbf{x})$ satisfies the equation

$$
(s^2 + as + b + c^2\mathbf{k}^2)U = h(s),
\tag{10}
$$

whence, after a routine inversion of the Fourier transformation

$$
\tilde{u}(s, \mathbf{x}) = \frac{1}{(2\pi)^2 irc^2} \int_{-\infty}^{\infty} \frac{1}{\gamma(s)/c^2 + k^2} e^{ikr} k \, dk
\tag{11}
$$

where $k := |\mathbf{k}|$, $r := |\mathbf{x}|$. Integral (11) can be calculated by the residue method

$$
\tilde{u}(s, \mathbf{x}) = \frac{h(s)}{(2\pi)^2 irc^2} e^{-\gamma(s)^{1/2} r/c},
\tag{12}
$$

where the square root $\gamma(s)^{1/2}$ is defined in such a way that

$$
\operatorname{Re} \gamma(s)^{1/2} \geq 0.
\tag{13}
$$

Let $s = \sigma - a/2$, so that $\gamma = \sigma^2 + \Delta$. The Green function $g(t, \mathbf{x})$ of eq. (6) is defined by $u_0 = v_0 = 0$ and $f(t, \mathbf{x}) = \delta(t)\delta(\mathbf{x})$ we have $A(s) = 1$. Applying simple transformations and referencing a table of Laplace transforms in ABRAMOWITZ and STEGUN (1970), eqs. (29.3.92–93), the following results are obtained: $g(t, \mathbf{x}) = 0$ for $t < r/c$, while for $t \geq r/c$

$$
g(t, \mathbf{x}) = \frac{1}{4\pi rc^2} e^{-at/2} \times
$$
$$
\begin{cases}
\left[\delta(t - r/c) - J_1\left(\sqrt{\Delta(t^2 - r^2/c^2)}\right)\sqrt{\Delta}(r/c)\sqrt{t^2 - r^2/c^2}\right] & \text{if } \Delta > 0 \\
\left[\delta(t - r/c) + I_1\left(\sqrt{|\Delta|(t^2 - r^2/c^2)}\right)\sqrt{|\Delta|}(r/c)/\sqrt{t^2 - r^2/c^2}\right] & \text{if } \Delta < 0
\end{cases}
\tag{14}
$$

where I_1 denotes the Macdonald function of the first kind (ABRAMOWITZ and STEGUN, 1970).

For either of the two cases Green's function involves an exponentially damped delta spike, followed by an oscillatory tail in the case of $\Delta > 0$ and a monotone tail

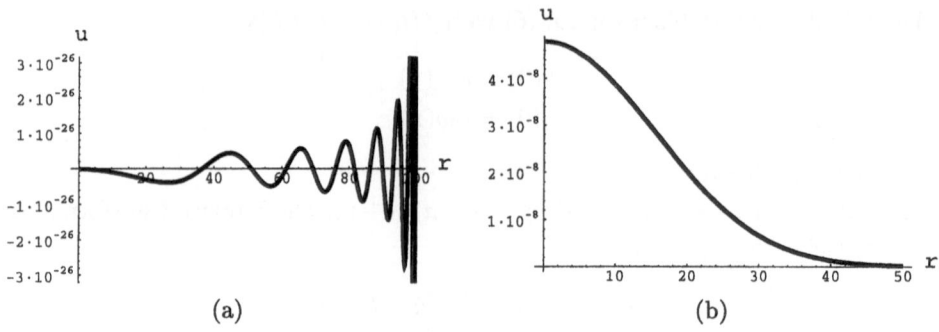

Figure 1

Green function for eq. (6) as a function of r at $t = 10$; (a) for $\Delta > 0$, (b) for $\Delta < 0$.

decaying from the source toward the wavefront if $\Delta < 0$. In the first case the oscillations are dense at the wavefront. Figure 1 shows $g(t, \mathbf{x})$ as a function of r. Note that in both cases the wavefront is at $r = 100$.

For $a = 0$ and $b > 0$ the energy, if properly defined (Section A), is conserved, and Green's function involves an undamped oscillation.

4. Propagation of Localized Oscillatory Pulses

The solution of eq. (6) for an arbitrary pulse $F(t)$ emitted by the point source is given by the convolution $u(t, \mathbf{x}) = F(t) * g(t, \mathbf{x})$. This representation is not very helpful for the study of the pulse propagation in a dispersive medium. It is therefore more convenient to use asymptotic methods for this purpose.

The Sommerfeld-Brillouin theory of pulse propagation is based on the saddle point method of asymptotic evaluation of the inverse Laplace transform (12) for large r, at fixed $\zeta = ct/r$. The Bromwich contour is replaced by an equivalent steepest descent path (Fig. 2) and pole contributions are added if the contour deformation sweeps over them (see Appendix for detailed arguments).

The pole is defined in terms of the signal parameters, while the saddle point is defined in terms of the medium parameters. The pole contribution has the appearance of the propagating signal while the saddle contribution is mainly dependent on the medium parameters. The steepest descent path depends on the parameter ζ. For $\zeta = 1$, i.e., at the passage of the wavefront, the saddle points lie at infinity. For $\zeta \to \infty$, or $t \to \infty$, the saddle points tend to the branching points and the steepest descent path contracts to the cut defined by eq. (13).

For some intermediate value of ζ, which we denote by $1/v_S$, the steepest descent path crosses the pole. For $\zeta > 1/v_S$ the integral over the Bromwich contour is equivalent to the integral over the steepest descent path *plus* the pole contribution. Consequently the parameter v_S is interpreted as the signal velocity and $t = r/v_S$ as the arrival time of the signal. The steepest descent path is identified here rather formally

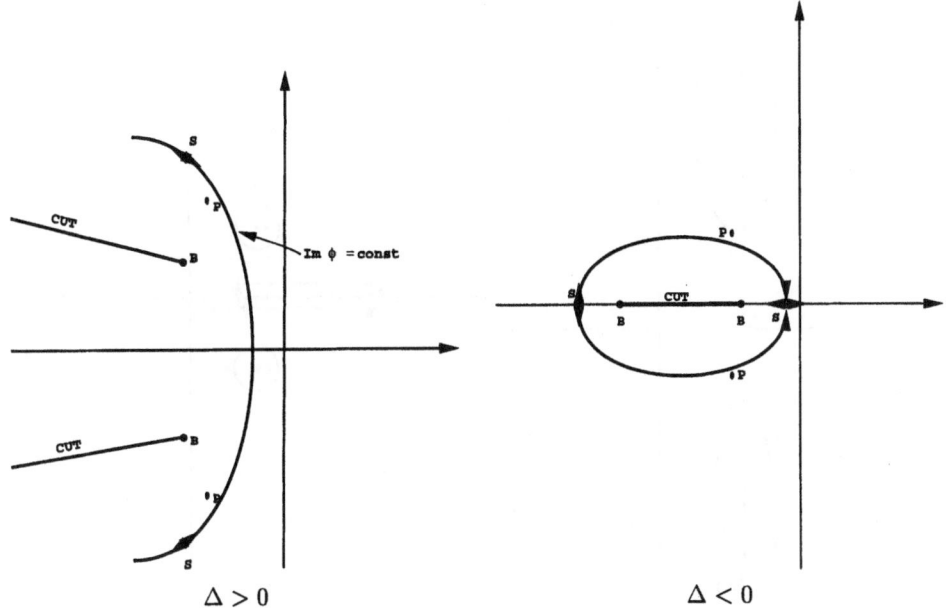

Figure 2
The steepest descent contour.

with the curve Im[phase] = const. The contribution of the saddle point following the wavefront is interpreted as a forerunner.

There is a weak point in this argument. The saddle point method is an asymptotic method and, as such, it determines the arrival of the main signals as opposed to the tails. The exact shape of the steepest descent path, except for its direction in a proximity of the saddle points, is irrelevant for the asymptotic solution and *eo ipso* for the signals. The SDP is determined in terms of the steepest descent directions at the saddles and its equivalence with the Bromwich contour according to the Jordan lemma. Consequently, the value of ζ at which the steepest descent path crosses the pole is not uniquely defined. More correctly, the definition of the signal velocity should be based on dominance of the pole over the saddle contribution (OUGHSTUN and SHERMAN, 1994). Dominance however does not constitute a sharp criterion and hence it does not determine any sharp transition time.[2]

A rigorous splitting of the solution into the signal and a transient requires a more precise definition of the latter, e.g., as the contribution of the cut given by the Hankel contour encircling the cut shown in Figure 3. The possibility of such a contour deformation is discussed in the Appendix. With this definition however the pole contribution appears immediately after the passage of the wavefront.

---

[2] Additional complications arise as regard pseudo-oscillatory signals commonly used in seismology, such as the Ricker wavelet. Such signals are not represented by pole contributions; instead, they modify the phase of the integrand.

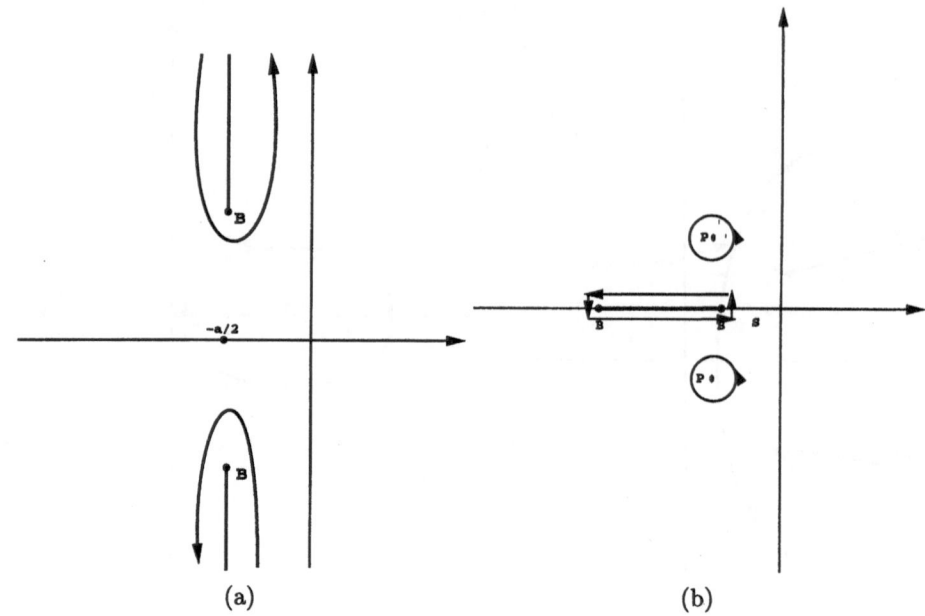

Figure 3
Contours for the main signal and the transient. (a) $\Delta > 0$; (b) $\Delta < 0$.

Let the source signal be a damped sinusoidal wave train

$$F(t) = e^{-At} \sin(Bt + C)\theta(t) \tag{15}$$

with the Laplace transform

$$\tilde{F}(s) = \cos C \frac{B}{(A+s)^2 + B^2} + \sin C \frac{A+s}{(A+s)^2 + B^2}. \tag{16}$$

Without loss of generality we can set $C = 0$.

The solution of eq. (6) consists of the residue contribution u_R and the contribution of the cuts u_C:

$$u(t, \mathbf{x}) = u_R(t, \mathbf{x}) + u_C(t, \mathbf{x}). \tag{17}$$

The pole contribution

$$u_R(t, \mathbf{x}) = \frac{1}{4\pi r c^2} e^{-At} \sin(Bt - k_0 r/c) e^{-\alpha_0 r} \tag{18}$$

with

$$k_0 = \operatorname{Im} \gamma(-A + iB)^{1/2} \tag{19}$$

$$\alpha_0 = \operatorname{Re} \gamma(-A + iB)^{1/2} \tag{20}$$

represents the main signal with apparent effects of spatial attenuation and dispersion. Its speed of propagation cB/k_0 depends on the signal frequency B. Note that the wavenumber $k_0 + i\alpha_0$ is complex-valued even if the frequency $\omega = B + iA$ is real.

For $\Delta < 0$ the contribution of the cut is

$$u_C(t, \mathbf{x}) = \frac{B}{4\pi^2 rc^2} \int_{s_-}^{s_+} \frac{1}{(A+s)^2 + B^2} e^{st} \sin(\Gamma(s)r/c)ds \tag{21}$$

with $s_\pm = -(a/2) \pm \sqrt{-\Delta}$, $\Gamma(s) := \sqrt{(s - s_-)(s_+ - s)}$. Note that at every distance r the term $u_C(t, \mathbf{x})$ decays in time faster than $e^{-s_+ t}$. The decay rate of the transient is a property of the medium, in contrast to the decay rate of the main signal.

The case of $\Delta > 0$ is more complicated but the results are qualitatively similar. The cuts run vertically from the branching points to infinity (Appendix B). Since the cuts lie to the left of the vertical line $\mathrm{Re}\, s = -a/2$, the contribution of the cuts decays faster than $e^{-at/2}$. The steepest descent path crosses the lines $\mathrm{Re}\, \gamma^{1/2} = 0$. The Bromwich contour can however be replaced by any contour in the left-hand complex plane which runs to infinity in such a way that $t\,\mathrm{Re}\, s - |s|r/c < 0$ asymptotically. We can therefore imagine that the cuts are pushed out of the way as shown in Figure 3.

Numerical experiments of MAINARDI (1984) show that the total solution, calculated by the formula $u = g * F$, tends to u_R with time. The transient $u_C(t, \mathbf{x})$ and the main pulse $u_R(t, \mathbf{x})$ for $\Delta > 0$ and $\Delta < 0$ with r are shown in Figures 4 and 5, respectively.

5. Propagation in Media with Lorentz-Kramers dispersion

The effects of excitation of atoms in a dielectric are represented by Lorentz dispersion, with the index of refraction given by the formula

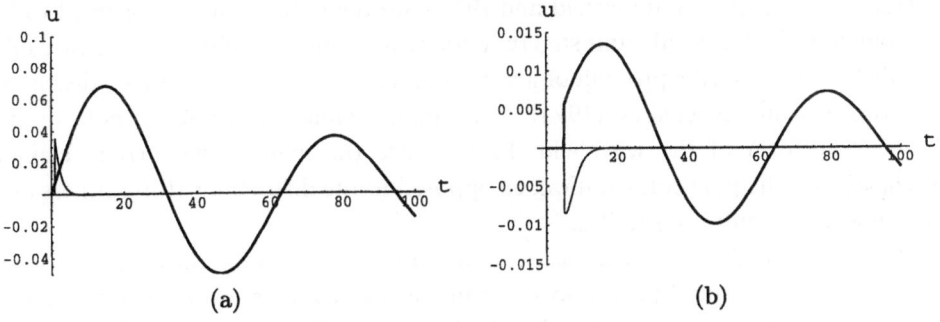

(a)   (b)

Figure 4

The main pulse and the transient for $\Delta > 0$: (a) at $r = 1\,\mathrm{km}$, (b) at $r = 5\,\mathrm{km}$. The transient at $r = 5\,\mathrm{km}$ is scaled vertically by a factor of 500.

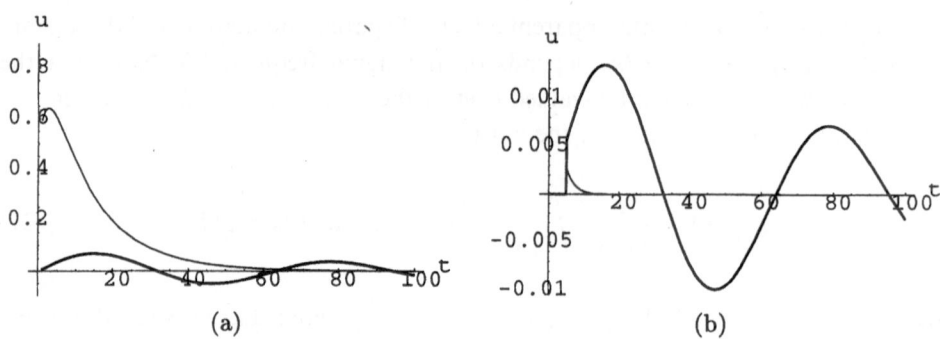

Figure 5
The main signal (thick line) and the transient (thin line) for $\Delta < 0$. (a): at $r = 1$ (b): at $r = 5$.

$$n(\omega) = \left[\frac{\omega^2 - \omega_1^2 + 2i\delta\omega}{\omega^2 - \omega_0^2 + 2i\delta\omega}\right]^{1/2} \qquad (22)$$

with $\omega_1^2 > \omega_0^2$, $\delta > 0$. The resulting phase function for one-dimensional propagation is $\Phi = i(x\omega/c)[n(\omega) - ct/x]$. A detailed analysis of the singularities of $\Phi(\omega)$ for complex ω and asymptotic behavior of the solution for large $x\omega/c$ is presented in OUGHSTUN and SHERMAN (1994). We shall only wrap up the most important results of their analysis.

The function $\Phi(\omega)$ is defined on a complex plane with two horizontal branch cuts joining the points $\pm\sqrt{\omega_0^2 - \delta^2} - i\delta$ with $\pm\sqrt{\omega_1^2 - \delta^2} - i\delta$. Two distant saddles $S_D^{\pm}$ approach the outer branch points starting at infinity for $\theta := ct/x = 1$. Two other saddles $S_N^{\pm}$ lie on the imaginary axis for θ sufficiently close to 1. As θ increases, the imaginary saddles collide for $\theta = \theta_0$ and move apart along the line $\Im\omega = -\delta$, approaching the inner branching points of $\Phi(\omega)$.

For $\theta < \theta_0$ a contour can be constructed along the steepest descent paths through the steepest descent directions of the upper imaginary saddle S_N^+ and the two distant saddles. For $\theta > \theta_0$ it passes through all the saddles.

Traditionally, since Sommerfeld and Brillouin, the input signal is assumed to be the product of the Heaviside unit step function with a sinusoid, although the case of a finite-duration signal (the product of a boxcar and a sinusoid) is also briefly discussed in OUGHSTUN and SHERMAN (1994). Such input signals introduce a pole in the Laplace transform of the wavefield. In this case the main signal arrival can be associated with the pole contribution, as opposed to the distortions (forerunners and tails), associated with the saddles.

The Sommerfeld forerunner, associated with the distant saddles, appears as a strongly oscillating initial part, appearing immediately after the wavefront passage. The Brillouin forerunner, associated with the saddles $S_N^{\pm}$, appears as a bump some time later, when the contribution of the saddles $S_N^{\pm}$ becomes comparable and then larger than the contribution of the distant saddles, at some $\theta_B > \theta_0$.

Formally, the pole contribution appears when the pole lies between the saddle contour and the original contour (the real axis). Its contribution becomes significant when it is comparable or larger than the saddle contributions, at $\theta = \theta_1$, say. This defines an estimated signal velocity $v = c/\theta_1$. The Brillouin forerunner often has a higher amplitude than the main signal. More complicated patterns arise if the input signal has finite duration, with the Sommerfeld and Brillouin forerunners following and preceding the arrivals associated with the main signal.

Note that in the case of an input signal in the form of a delta spike, the Sommerfeld "forerunner" is also present, but it is preceded by the main signal. This is due to the extremely broadband aspect of singular signals. The large and small frequencies propagating with higher speeds which play a marginal role in the formation of a narrow band signal, but their participation in the formation of a broadband signal is on a par with the intermediate frequencies. Therefore a broadband signal is no longer preceded by marginal frequency components.

Brillouin and Sommerfeld precursors for TEM waves in a dispersive dielectric were also investigated by a more sophisticated method (wave splitting in time domain) in KARLSSON and RIKTE (1998).

These results shows that in a dispersive medium signals are no longer associated with the wavefronts. The appearance of strong forerunners can lead to a wrong identification of the input signal and travel time.

Asymptotic analysis (HANYGA and SEREDYŃSKA, 2000a; HANYGA, 1999a) shows that in general dispersive and attenuating media complex ray tracing is required in order to account for the variation of the pulse shape. In the case of singular memory it is however still possible to describe the signal evolution in the context of real ray asymptotics, as discussed in the next section.

6. Propagation of Signals in Viscoelastic Media with Singular Memory

The solutions of an integro-differential equation with a singular memory kernel are $\mathscr{C}^\infty$-smooth at the wavefronts (and everywhere except at the source (LOKSHIN and SUVOROVA, 1982; DAUTRAY and LIONS, 1992; RENARDY, 1982; HRUSA and RENARDY, 1985; HANYGA, 2001b). Signal peaks are delayed with respect to their arrival times implied by the travel time.

We shall give here an elementary derivation of the delay of an originally delta-spiked pulse for eq. (1) with $K(t) \sim \text{const} \times t^{-1/2}$ for $t \to 0$. For a rigorous derivation of exact and asymptotic solutions see HANYGA and SEREDYŃSKA (1999b).

Our hypothesis is compatible with the following asymptotic expansion of the Laplace transform $\tilde{K}(s)$ of $K(t)$:

$$\tilde{K}(s) = 1 + K_1 s^{-1/2} + K_2 s^{-1} + \cdots \tag{23}$$

for $s \to \infty$. The above expression and an asymptotic expansion of the Laplace-transformed solution $\tilde{u}(s, \mathbf{x})$ will be substituted in the Laplace-transformed eq. (1):

$$\rho_\infty \tilde{K}(s)\left[s^2\tilde{u} - su(0) - u_t(0)\right] = \nabla \cdot [\nabla \tilde{u}] + F(s). \tag{24}$$

The half-integer powers of s from (23) must be balanced by the corresponding expressions in the asymptotic expansion of $\tilde{u}(s, \mathbf{x})$, hence it is assumed that

$$\tilde{u}(s, \mathbf{x}) \sim e^{-\phi(s,\mathbf{x})} \sum_{n=0}^{\infty} s^{-n/2} u_n(\mathbf{x}). \tag{25}$$

Let us begin with the hypothesis that $\phi(s, \mathbf{x}) = sT(\mathbf{x})$. The eikonal equation is obtained by substituting both asymptotic expansions in eq. (24) and setting the coefficients of s^2 equal to zero, with $u_0 \neq 0$,

$$C(\nabla T)^2 - \rho_\infty = 0. \tag{26}$$

Setting the coefficient of $s^{3/2}$ equal to zero and using eq. (26) yields an equation $K_1 u_0 = 0$ which cannot be satisfied. A work-around is provided by the assumption that $\phi(s, \mathbf{x}) = sT(\mathbf{x}) + s^{1/2}\alpha(\mathbf{x}) + O[1]$. The coefficients of $s^{3/2}$ now yield a transport equation for the auxiliary parameter α:

$$\nabla T \cdot \nabla \alpha = \frac{1}{2}\rho_\infty K_1 \tag{27}$$

along the rays $d\mathbf{x}/d\tau = \nabla T$ of the eikonal $T(\mathbf{x})$.

In order to ensure the existence of the inverse Laplace transform and causality we assume that $\alpha \geq 0$ and $s^{1/2}$ is the principal value of the complex square root, with the complex s-plane cut along the negative real axis. Note that the inequality $\alpha \geq 0$ is satisfied for all $\tau \geq 0$ if it is satisfied at the source $\tau = 0$ and $K_1 \geq 0$. The last inequality is a property of the medium. It can be deduced from the assumption that the time-averaged dissipation in a periodic motion is non-negative (HANYGA and SEREDYŃSKA, 1999b).

Without loss of generality it can be assumed that

$$\phi(s, \mathbf{x}) = sT(\mathbf{x}) + s^{1/2}\alpha(\mathbf{x}). \tag{28}$$

Indeed, in the general case $\phi = sT + s^{1/2}\alpha + \beta + s^{-1/2}\gamma + \cdots$

$$\exp(-\phi) \sim \exp(-sT - s^{1/2}\alpha)\exp(-\beta)\left\{1 + O\left[s^{-1/2}\right]\right\} \tag{29}$$

and the two last factors can be included in the amplitudes.

In the Fourier domain $s = -i\omega$ (we define the Fourier transformation by $\hat{u}(\omega, \mathbf{x}) = (1/2\pi)\int_{-\infty}^{\infty} e^{i\omega t}\, dt u(t, \mathbf{x})dt)$ and the exponential factor assumes the following form:

$$\exp(i\omega T)\exp\left(i\sqrt{|\omega/2|}\alpha\right)\exp\left(-\sqrt{|\omega/2|}\alpha\right). \tag{30}$$

For a point source in a homogeneous medium $T = r/c$, $\alpha = Ar$, where $c = \sqrt{C/\rho_\infty}$ and A is a constant. For $r > 0$ the spectrum of the signal decays exponentially even if the signal emitted by the source is a singularity with an algebraically decaying spectrum. Consequently the signal is immediately smoothed out. The equations are however hyperbolic and for a delta-spiked signal $(F(s) = 1)$ the solution $\tilde{u}(s, \mathbf{x}) = \exp\left(-sT(\mathbf{x}) - s^{1/2}\alpha(\mathbf{x})\right)\left[1 + O\left[s^{-1/2}\right]\right]$ has an explicit asymptotic inverse Laplace transform

$$u(t, \mathbf{x}) \cong \frac{\alpha}{2\sqrt{\pi}}(t - T)^{-3/2}\exp\left(\alpha^2/(4(t - T))\right) \tag{31}$$

for $t > 0$ and $u(t, \mathbf{x}) = 0$ for $t < 0$.

The expression on the right-hand side of eq. (31) is a unimodal function with a unique maximum at $t = T(\mathbf{x}) + \alpha^2/6$ (Fig. 6). It vanishes with all its derivatives at the wavefront $T(\mathbf{x}) = t$. For a point source in a homogeneous medium the maximum lies at $t = r/c + A^2r^2$, hence it lags behind the wavefront and the delay is proportional to r^2.

An exact Green's function for $K_2 = K_1^2/4$ is shown in Figure 7.

Propagation of oscillatory signals was investigated numerically by HANYGA and ROK (2000). Two phenomena are apparent: (1) delay of the signal with respect to the wavefront, (2) exponential attenuation of signal peaks. There is only a slight dependence of the delay on frequency.

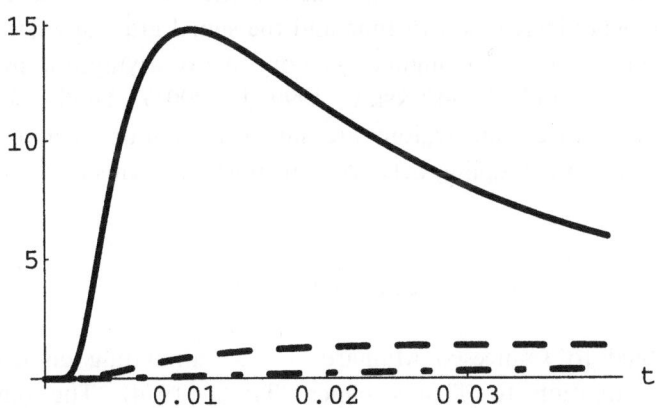

Figure 6
Smoothed delta spike for $K(t) \sim t^{-1/2}$.

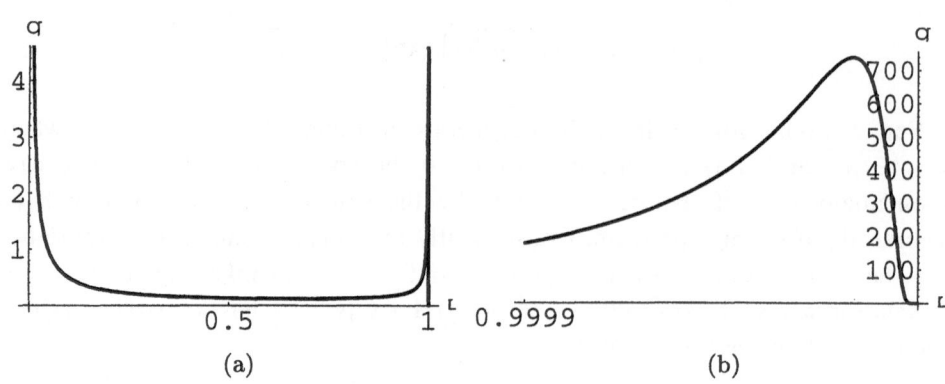

Figure 7
(a) A snapshot of the Green function; (b): detailed behavior at the wavefront.

7. Conclusions

In a hereditary medium with a regular memory kernel, singular signals (delta spike, discontinuities) travel with the wavefront, i.e., with the high-frequency limiting phase speed. The amplitude of a singularity is exponentially attenuated but the singularity remains sharp. An oscillatory signal propagates with the phase speed corresponding to its dominating frequency. The main signal is accompanied by distortions. The onset of each signal coincides with the wavefront. The distortions can however dominate the wavefield before the main signal attains its full amplitude. In this case the main signal is preceded by forerunners. The amplitude of a forerunner can exceed the final amplitude of the main signal. At large propagation distances the main signal can be negligible in comparison with the forerunners.

In a hereditary medium with singular memory kernels, singularities are immediately smoothed out. A pulse propagates with a delay with respect to the wavefront. The delay increases with time and the signal gradually flattens out.

For hereditary media with singular memory, a ray asymptotic method can be developed HANYGA and SEREDYŃSKA, 1999a,b, 2001). Uniformly asymptotic methods applicable in caustic regions are significantly more complicated than in the elastic case due to a coupling between transport equations (HANYGA, 2001a).

Acknowledgments

I am indebted to Francesco Mainardi for a stimulating discussion and for attracting my attention to Thau's paper (THAU, 1974). The remarks of an anonymous reviewer allowed improvements in the final draft. Financial support by Norsk Hydro under the contract NH-B44-5003496-00 is acknowledged.

Appendix A

Thermodynamic Analysis of the Equations

For a better understanding of the physical meaning of various parameters we shall briefly discuss the dissipation in the media represented by the equations of Section 3.

Multiplying eq. (6) by u_t and rearranging terms we derive the following energy balance

$$\frac{d}{dt}\left(\frac{1}{2}u_t^2 + c^2\frac{1}{2}(\nabla u)^2 + \frac{1}{2}bu^2\right) + au_t^2 = c^2\nabla\cdot(u_t\nabla u). \tag{32}$$

It is thus clear that the term bu, responsible for a part of the dispersion, is not dissipative. On the other hand the term au_t introduces both dispersion and dissipation. It is clear that

$$a \geq 0 \tag{33}$$

in an attenuating medium. Well-posedness of the initial-value problems also requires that the energy should have a lower bound, hence

$$b \geq 0. \tag{34}$$

For $a = 0$, $b > 0$ the energy is conserved, although the system is dispersive.

Equation 1 with a singular kernel requires more sophisticated analysis (HANYGA, 2001b).

Appendix B

Cuts, Saddles, Steepest Descent Paths and Hankel Loops

Let $x := \operatorname{Re} s$, $y := \operatorname{Im} s$.

The phase function for the inverse Laplace transform of (12) is $\Phi(s) := st - \gamma(s)^{1/2}r/c$. The branching points are the roots of $\gamma(s) \equiv s^2 + as + b$.

The cuts are primarily defined by the equation $\operatorname{Re}\gamma(s)^{1/2} = 0$ or, equivalently, $\operatorname{Im}\gamma(s) = 0$ and $\operatorname{Re}\gamma(s) < 0$. More explicitly,

$$y(2x + a) = 0 \tag{35}$$

$$x^2 + ax + b < y^2. \tag{36}$$

Equation (35) implies that either (i) $y = 0$, or (ii) $x = -a/2$.

In case (i) eq. (36) implies that $\Delta > 0$ and the cut joins the two branching points.

In case (ii) we must assume that $\Delta < 0$, or else the cut would cut the Riemann surface

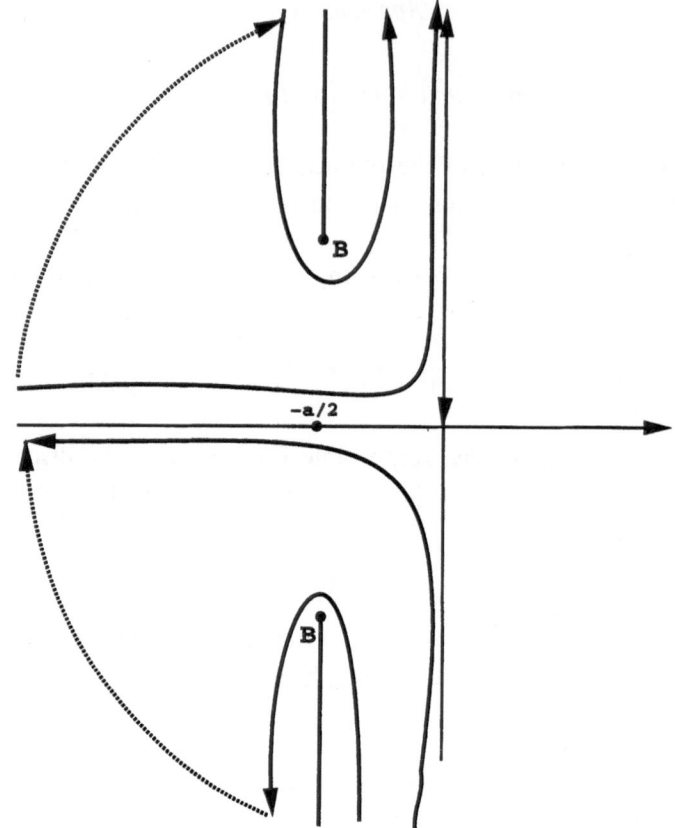

Figure 8
Deformation of the Bromwich contour for $\Delta > 0$.

in two. Equation (36) shows that the cuts run vertically from either branching point to infinity.

The saddles are given by $\Phi'(s) = 0$, or

$$\frac{s + a/2}{\gamma(s)^{1/2}} = \zeta. \tag{37}$$

This can be transformed to the equation

$$s^2 + as + b + \Delta/[4(\zeta^2 - 1)] = 0. \tag{38}$$

Assume that $\Delta \neq 0$.

For $\zeta \to 1$ it is clear that the saddles escape to infinity. For $\zeta \to \infty$ the saddles collapse onto the branching points.

The direction of the steepest descent is determined by $\text{Im}\,\Phi = 0$ and the condition that on the SDP $\text{Re}\,\Phi$ achieves a local maximum at the saddle. It is easy to see that

$$\Phi''(s) = -(r/c)\gamma(s)^{-3/2}\Delta. \tag{39}$$

For $\Delta < 0$ we have $\gamma^{3/2} > 0$ at the saddles and the steepest descent direction is vertical. For $\Delta > 0$ we note that $\arg\gamma = \arg(s - s_+)(s - s_-) = \pm\pi$ at the upper/lower saddle and $\gamma^{-3/2}$ is accordingly positive/negative imaginary. Hence the steepest descent directions are at $\mp\pi/4$ from the horizontal direction.

By the Jordan lemma, the Bromwich contour can be replaced by any topologically equivalent contour approaching infinity in a direction such that $\text{Re}\,\Phi \sim xt - |s|(r/c)\text{sgn}\,\text{Re}\,\gamma(s)^{1/2} < 0$, provided the contribution of the pole is taken into account if appropriate. For $\Delta > 0$ and $t > r/c$ this condition defines two sectors in the left half of the complex plane contiguous to the imaginary axis. For $t > r/c$ one can always construct an SDP that will be contained in these sectors. In order to construct a contour encircling the cuts, one can add to the Bromwich contour, a contour running from zero to infinity along the negative real axis and a contour running from infinity back to zero. In view of the condition $\text{Re}\,\gamma^{1/2} \geq 0$, Jordan's lemma allows replacement the additional contours by the contours running to infinity along the upper cut on its left side and from infinity along the lower cut on its left side (Fig. 8).

For $\Delta < 0$ the inequality $\text{Re}\,\gamma^{1/2} \geq 0$ holds in the left half of the complex plane as long as the cut is not intersected. Consequently, the Bromwich contour can be replaced by any closed contour encircling the cut.

REFERENCES

ABRAMOWITZ, M., and STEGUN, I., *Mathematical Tables.* (Dover, New York, 1970).

BRILLOUIN, L., *Wave Propagation and Group Velocity.* (Academic Press, New York, 1960).

BUCHEN, P. W., and MAINARDI, F. (1975), *Asymptotic Expansions for Transient Viscoelastic Waves*, J. de Mécanique *14*, 597–608.

BUCKINGHAM, M. J. (1997), *Theory of Acoustic Attenuation, Dispersion and Pulse Propagation in Unconsolidated Granulated Materials Including Marine Sediments*, J. Acoust. Soc. Am. *102*, 2579–2596.

BUCKINGHAM, M. J. (1998), *Theory of Compressional and Shear Waves in Fluidlike Marine Sediments*, J. Acoust. Soc. Am. *103*, 288–299.

CAPUTO, M., *Elasticità e dissipazione.* (Zanichelli, Bologna, 1969).

CAPUTO, M., and MAINARDI, F. (1976), *New Dissipation Model based on Memory Mechanism*, Pure appl. geophys. *91*, 134–147.

CARCIONE, J. M., CAVALLINI, F., MAINARDI, F., and HANYGA, A. (2000), *Time-domain Seismic Modeling of Constant-Q Wave Propagation Using Fractional Derivatives*, Pure appl. geophys., this issue.

CARCIONE, J. M., KOSLOFF, D., and KOSLOFF, R. (1998a), *Wave Propagation Simulation in a Linear Viscoacoustic Medium*, Geophys. J. R. astr. Soc. *93*, 393–407.

CARCIONE, J. M., KOSLOFF, D., and KOSLOFF, R. (1998b), *Wave Propagation Simulation in a Linear Viscoacoustic Medium*, Geophys. J. R. astr. Soc. *95*, 597–611.

CHRISTENSEN, R. M., *Theory of Viscoelasticity: An Introduction* (Academic Press, New York, 1971).

DAUTRAY, R., and LIONS, J.-L., *Mathematical Analysis and Numerical Methods for Science and Technology.* (Springer-Verlag, Berlin, 1992), vol. 5.

DAY, S. M., and MINSTER, J. B. (1984), *Numerical Simulation of Wavefields Using a Padé Approximant Method*, Geophys. J. R. astr. Soc. *78*, 105–118.

DEANE, G. B. (1997), *Internal Friction and Boundary Conditions in Lossy Fluid Seabeds*, J. Acoust. Soc. Am. *101*, 233–240.

EMMERICH, M., and KORN, M. (1987), *Incorporation of Attenuation into Time-domain Computation of Seismic Wavefields*, Geophysics *52*, 1252–1264.

FABRIZIO, M., and MORRO, A., *Mathematical Problems in Linear Viscoelasticity* (SIAM, Philadelphia, 1992).

GROSS, B., *Mathematical Structure of the Theories of Viscoelasticity* (Hermann, Paris, 1953).

HANYGA, A. (1999a), *Asymptotic theory of wave propagation in viscoporoelastic media*. In *Theoretical and Computational Acoustics '97*, (eds. Y.-C. Teng, E.-C. Shang, Y.-H. Pao, M. H. Schultz, and A. D. Pierce) (World-Scientific, Singapore, 1999) pp. 429–448.

HANYGA, A. (1999b), *Simple memory models of attenuation in complex viscoporous media*. In *Proceedings of the 1st Canadian Conference on Nonlinear Solid Mechanics, Victoria, BC, June 16–20, 1999*, Volume 2, pp. 420–436.

HANYGA, A. (2001a), *Uniformly Asymptotic Solutions of Wave Equations with Singular Memory Function*, J. Comput. Acoustics *9*, 1–18.

HANYGA, A. (2001b), *Wave Propagation in Media with singular Memory*, Math. and Comput. Mech.

HANYGA, A., and ROK, V. E. (2000), *Wave Propagation in Micro-heterogeneous Porous Media: A Model Based on an Integro-differential Equation*, J. Acoust. Soc. Amer. *107*, 2965–2972.

HANYGA, A., and SEREDYŃSKA, M. (1999a), *Asymptotic Ray Theory in Poro- and Viscoelastic Media*, Wave Motion *30*, 175–195.

HANYGA, A., and SEREDYŃSKA, M. (1999b), *Some Effects of the Memory Kernel Singularity on Wave Propagation and Inversion in Poroelastic Media, I: Forward Modeling*, Geophys. J. Int. *137*, 319–335.

HANYGA, A., and SEREDYŃSKA, M. (2000), *Ray Tracing in Elastic and Viscoelastic Media*, Pure appl. geophys. *157*, 679–717.

HANYGA, A., and SEREDYŃSKA, M. (2002), *Asymptotic Wavefront Expansions in Hereditary Media with Singular Memory Kernels*, Quart. Appl. Math., in press.

HRUSA, W., and RENARDY, M. (1985), *On Wave Propagation in Linear Viscoelasticity*, Quart. Appl. Math. *43*, 237–253.

KARLSSON, A., and RIKTE, S. (1998), *The Time-domain Theory of Forerunners*, J. Opt. Soc. Amer. A*15*, 487–502.

KJARTANSSON, E. (1979), *Constant Q-wave Propagation and Attenuation*, J. Geophys. Res. *84*, 4737–4748.

LOKSHIN, A. A., and ROK, V. E. (1978a), *Automodel Solutions of Wave Equations with Time Lag*, Russian Math. Surveys *33*, 243–244.

LOKSHIN, A. A., and ROK, V. E. (1978b), *Fundamental Solutions of the Wave Equation with Delayed Time*, Doklady AN SSSR *239*, 1305–1308.

LOKSHIN, A. A., and SUVOROVA, Y. V., *Mathematical Theory of Wave Propagation in Media with Memory* (Moscow University Publ., Moscow, 1982), in Russian.

MAINARDI, F. (1983a), *On Signal Velocity for Anomalous Dispersive Waves*, Nuovo Cimento B *74*, 52–58.

MAINARDI, F. (1983b), *Signal Velocity for Transient Waves in Linear Dissipative Media*, Wave Motion *5*, 33–41.

MAINARDI, F. (1984), *On linear dispersive waves with dissipation*, In *Wave Phenomena: Modern Theory and Applications* (eds. Rogers, C. and Moodie, T. B.) (Elsevier Science Publishers B. V., Amsterdam, 1984) pp. 307–317.

NORRIS, A. N. (1986), *On the Viscodynamic Operator in Biot's Theory*, J. Wave-Material Interaction *1*, 365–380.

OUGHSTUN, K. E., and SHERMAN, G. C., *Electromagnetic Pulse Propagation in Causal Dielectrics* (Springer-Verlag, Berlin, 1994).

PIPKIN, A. C., *Lectures on Viscoelasticity Theory* (Springer-Verlag, Berlin, 1986), 2nd ed.

RENARDY, M. (1982), *Some Remarks on the Propagation and Non-propagation of Discontinuities in Linearly Viscoelastic Liquids*, Rheol. Acta *21*, 251–254.

SOMMERFELD, A. (1914), *Über die Fortpflanzung des Lichtes in dispergierenden Medien*, Ann. Physik (4) *44*, 177–202.

THAU, S. A., *Linear dispersive waves*, In *Nonlinear Waves* (eds. Leibovich, S. and Seebass, A. R.) (Cornell University Press, Ithaca, NY, 1974) pp. 44–81.
VON LAUE, M. (1905), Ann. Physik (4) *18*, 523.

(Received October 3, 2000, revised/accepted March 8, 2001)

 To access this journal online:
http://www.birkhauser.ch

Pure appl. geophys. 159 (2002) 1771–1789
0033–4553/02/081771–19 $ 1.50 + 0.20/0

© Birkhäuser Verlag, Basel, 2002

Γ **Pure and Applied Geophysics**

Scattering of Seismic Waves by Cracks with the Boundary Integral Method

KIYOSHI YOMOGIDA,[1] and RAFAEL BENITES[2]

Abstract—We develop a new scheme to compute 2-D SH seismograms for media with many flat cracks, based on the boundary integral method. A dry or traction-free boundary condition is applied to crack surfaces although other kinds of cracks such as wet or fluid-saturated cracks can be treated simply by assigning different boundary conditions. While body forces are distributed for cavities or inclusions to express scattered wave, dislocations (or displacement discontinuities between the top and the bottom surfaces of each crack) are used as fictitious sources along crack surfaces. With these dislocations as unknown coefficients, the scattered wave is expressed by the normal derivative of Green's function along the crack surface, which is called "double-layer potentials" in the boundary integral method, while we used "single-layer potentials" for cavities or inclusions. These unknowns are determined so that boundary conditions or crack surfaces are satisfied in the least-squared sense, for example, traction-free for dry cracks. Seismograms with plane-wave incidence are synthesized for homogeneous media with many cracks. First, we check the accuracy of our scheme for a medium with one long crack. All the predicted phases such as reflected wave, diffraction from a crack tip and shadow behind the crack are simulated quite accurately, under the same criterion as in the case for cavities or inclusions. Next, we compute seismograms for 50 randomly distributed cracks and compare them with those for circular cavities. When cracks are randomly oriented, waveforms and the strength of scattering attenuation are similar to the cavity case in a frequency range higher than $kd \simeq 2$ where the size of scatterers d (i.e., crack length or cavity diameter) is comparable with the wavelength considered (k is the wavenumber). On the other hand, the scattering attenuation for cracks becomes much smaller in a lower frequency range ($kd < 2$) because only the volume but not detail geometry of scatterers becomes important with wavelength much longer than each scatterer. When all the cracks are oriented in a fixed direction, the scattering attenuation depends strongly on the incident angle to the crack surface as frequency increases ($kd > 2$): scattering becomes weak for cracks oriented parallel to the direction of the incident wave, while it gets close to the cavity case for cracks aligned perpendicular to the incident wave.

Key words: Scattering, crack media, attenuation, boundary method.

1. Introduction

Small-scale heterogeneities in the earth recently have been studied in detail, mainly by scattering and attenuation of high-frequency seismic waves (e.g., SATO and

[1] Division of Earth and Planetary Sciences, Graduate School of Science, Hokkaido University, Sapporo 060-0810, Japan. E-mail: yomo@ep.sci.hokudai.ac.jp
[2] Institute of Geological and Nuclear Sciences, P.O. Box 30368, Lower Hutt, New Zealand.

FEHLER, 1998). Most of these studies have adopted the random medium model of CHERNOV (1960): velocities fluctuate randomly in space, and their characters are described by a small number of model parameters such as an autocorrelation function of velocity fluctuations. Although this type of random media may be suitable to explain the fluctuation of direct P or S waves from teleseismic events observed by an array (e.g., AKI, 1973), coda as well as late arrival phases used in most studies of small-scale heterogeneities are composed mainly of backward scattering shear waves (e.g., SATO and FEHLER, 1998). These waves are controlled largely by sharp impedance contrasts such as irregular interfaces rather than Chernov-type smooth velocity fluctuations, as shown by our previous studies (e.g., YOMOGIDA and BENITES, 1995; YOMOGIDA et al., 1997).

As for Chernov-type random media, analytic approaches are applied to seismic waves in media with randomly distributed impedance contrasts only under very restricted conditions such as small concentrations of heterogeneities and the wavelengths involved to be considerably larger than the size of heterogeneities (e.g., VARADAN et al., 1978; HUDSON, 1981; YAMASHITA, 1990; KAWAHARA and YAMASHITA, 1992). Furthermore, these studies have taken into account only the scattering and attenuation of direct waves but not late arriving coda waves. Besides laboratory experiments using rock samples (e.g., MATSUNAMI, 1990; NISHIZAWA et al., 1997), numerical approaches can deal with the scattering, of not only direct waves but also coda waves, in media with large concentrations of heterogeneities and in a wide frequency range, including the case in which the considered wavelengths are similar to the size of heterogeneities. Using the boundary integral method, we have studied scattering and attenuation of high-frequency seismic waves in a deterministic manner (e.g., BENITES et al., 1992; YOMOGIDA et al., 1997). We have used two-dimensional circular cavities or inclusions in addition to elliptical cavities as heterogeneities. In these studies, we emphasized how important the effect of multiple scattering is in our adopted models. For example, YOMOGIDA and BENITES (1995) compared synthetic seismograms including multiple scattering with those without it (i.e., single scattering only), and showed that the omission of multiple scattering in most cases is not valid. With randomly distributed cracks of the same concentrations as the cavities, the models in this study cannot be handled correctly only with single scattering or based on the Born approximation.

Some researchers have insisted that heterogeneities in the earth, particularly in the crust, are mainly attributed to cracks or fractures whose thickness is much smaller than their length (e.g., CRAMPIN and LOVELL, 1991). In fact, cracks are abundant in active fault zones and their effects on seismic wave propagation have been occasionally observed clearly (e.g., LI et al., 1994). Assuming that cracks are responsible for scattered waves in seismic records of high frequency, some numerical studies have been conducted to simulate waveforms or frequency dependency of scattering and attenuation of seismic waves, such as BOUCHON (1987), COUTANT (1989), and MURAI et al. (1995).

Since we still do not know what kinds of heterogeneities are indeed dominant in the observed scattering characteristics of high-frequency seismic waves, it should be useful to understand the difference in the scattering of seismic waves between flat cracks and circular cavities or inclusions by numerical methods applied to similar models. In other words, it is useful to study the effect of the shape of heterogeneities on seismic wave scattering. Such quantitative comparisons have not been done in the above previous studies, therefore in this study we establish a new numerical scheme for seismic wave scattering in a medium with randomly distributed cracks, extending our boundary integral method for cavities or inclusions (BENITES *et al.*, 1992). Although we consider only two-dimensional *SH*-wave problems for cracks in this study, our numerical approach can be easily extended to 2-D *P-SV* waves or even three-dimensional problems, as we have developed for cavity problems (BENITES *et al.*, 1997). Dry cracks whose interfaces are flat are the only example in this study, but other kinds of cracks can be handled naturally, simply by assigning appropriate boundary conditions on the crack surface (see examples in KAWAHARA, 1998). More realistic models such as 3-D cracks may be ideal but presently requires tremendously large computations, although our preliminary study of 3-D simulations of seismic scattering is in progress. As in the previous studies with cavities, many important scattering characteristics obtained from 2-D simulations can be used in order to understand realistic 3-D problems to the extent we analyze their physical implications. The *SH*-wave models in this study, which does not include the effect of any conversions between *P* and *S* waves, can be widely applied to the observations of *S* waves and coda for local earthquakes because the *S*-to-*S* scattering is highly dominant over the *S*-to-*P* scattering from both theoretical (e.g., SATO and FEHLER, 1998) and numerical (e.g., YOMOGIDA *et al.*, 1997) studies.

First we show our computational scheme for seismic waves in media with randomly distributed cracks by comparing it to the one for cavities. This new scheme is then tested for a medium with one long crack in comparison with analytical solutions. Seismograms or waveforms are synthesized for media with many cracks that replace circular cavities in our previous study (BENITES *et al.*, 1992). Attenuation due to scattering is measured by the amplitude decay of direct waves, emphasizing its frequency-dependent characteristics. Changing the orientation angle of cracks with respect to the incident wave, both waveforms and frequency-dependent scattering attenuation are quantitatively investigated as seismic anisotropy due to the preferred orientation of cracks.

2. Method

In this section we present formulations and computational scheme for two-dimensional *SH* waves in media with many cracks, extending our boundary integral method for cavities (BENITES *et al.*, 1992). Here we shall consider wavefields in the frequency domain with a single angular frequency ω. In general, the total seismic wave at a given point P is expressed by the summation of the incident wave $u_0(P)$ and the scattered wave by all the cracks $u_s(P)$, as shown in Figure 1:

$$u(P) = u_0(P) + u_s(P). \qquad (1)$$

The representation theorem in elastodynamics (e.g., AKI and RICHARDS, 1980) gives the n-th component of the scattered wave by the j-th crack as a function of displacement discontinuities $[u_i(\xi)]$ ($\equiv u_i^+(\xi) - u_i^-(\xi)$ where u_i^+ and u_i^- are the displacements of the upper and lower surfaces $\Sigma_j(\xi)$ of the j-th crack, respectively) on the cracks:

$$u_n(\mathbf{x}) = \int_{\Sigma_j} [u_i] v_k C_{ikpq} \frac{\partial G_{np}}{\partial \xi_q} d\Sigma_j(\xi) \qquad (2)$$

where v_k is the normal unit vector of the crack surface, C_{ikpq} is the elastic constant tensor and G_{np} is the Green's function. For an isotropic medium with M cracks, the scattering wave u_s for 2-D SH waves in (1) and (2) is reduced to

$$u_s(P) = \sum_{j=1}^{M} \int_{\Sigma_j} \mu \Delta v \frac{\partial G(P, \xi_j)}{\partial \xi_j} d\Sigma_j(\xi), \qquad (3)$$

where μ is the shear modulus, Δv is the dislocation on the crack surface and $G(P, \xi_j)$ is the 2-D SH-wave Green's function with source at ξ_j and receiver at P

$$G(P, \xi_j) = \frac{i}{4} H_0^{(2)} \left(\frac{\omega}{\beta} |P - \xi_j| \right), \qquad (4)$$

where $H_0^{(2)}$ is the zeroth order Hankel function of the second kind and β is the shear-wave velocity. Unknown quantities $\mu \Delta v$ on the crack surface are discretized or

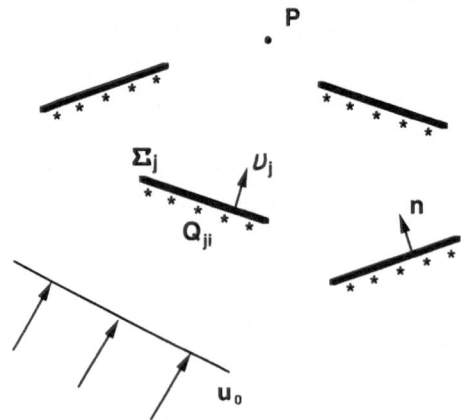

Figure 1
Boundary integral method for 2-D SH wave in a medium with many cracks.

expressed as fictitious line sources ($i = 1, 2, \ldots, N$ for each crack) with coefficients a_{ji} located at Q_{ji}. The total wavefield at P is then expressed by

$$u(P) = u_0(P) + \sum_{j=1}^{M} \sum_{i=1}^{N} a_{ji} \frac{\partial G(P, Q_{ji})}{\partial Q_{ji}}. \tag{5}$$

In boundary integral methods, this representation of fictitious sources is called "double-layer potentials" by URSELL (1973) while we used "single-layer potentials" for models with cavities or inclusions where the spatial derivatives of Green's function are replaced by Green's functions themselves in (5) (see BENITES et al., 1992).

Similar to a cavity or inclusion model, the unknown coefficients a_{ji} are determined to satisfy boundary conditions on crack surface. For dry cracks, the traction component normal to each crack surface must be free, thus the condition becomes

$$\frac{\partial u(P)}{\partial n} = \frac{\partial u_0(P)}{\partial n} + \sum_{j=1}^{M} \sum_{i=1}^{N} a_{ji} \frac{\partial^2 G(P, Q_{ji})}{\partial n \partial Q_{ji}} = 0 \tag{6}$$

on every crack surface (i.e., the observation point P is now located on a crack) whose unit normal vector is n. The unknown parameters a_{ji} are determined so that the boundary conditions (6) are satisfied over all the crack surfaces with an element of length along the crack ds in the least-squares sense

$$\min \mathscr{L} \equiv \sum_{j=1}^{M} \int_{\Sigma_j} \left| \frac{\partial u}{\partial n} \right|^2 ds. \tag{7}$$

That is,

$$\frac{\partial \mathscr{L}}{\partial a_{ji}^*} = 0, \tag{8}$$

which yields a set of $N \times M$ simultaneous linear equations for a_{ji}. Solving this set of equations, we obtain a_{ji}, followed by the calculation of the wavefield at point P by (5). In order to synthesize waves in the time domain, we only need the inverse Fourier transform of (5) obtained for many values of angular frequency ω.

In this paper we consider only 2-D *SH* waves in media with many dry cracks, however but the present scheme can be naturally extended to more complicated problems. Formulations in two-dimensional *P-SV* and three-dimensional cases are explicitly included in the representation theorem (2). This equation also can be used for tensile cracks in addition to the present shear cracks. We have adopted dry cracks by assigning the boundary condition (6) on each crack surface, nonetheless other kinds of cracks (e.g., water-saturated or with fluid of no viscosity) can be simulated with corresponding boundary conditions. Since the general consensus has not been yet obtained on appropriate boundary conditions for cracks in the crust (e.g., CRAMPIN and LOVELL, 1989; KAWAHARA, 1998), we do not investigate the difference in scattering by crack types but only show results for dry cracks.

3. Test

In this section, we shall observe how the present boundary integral method gives accurate and reliable results with simple models for which analytical solutions can be compared. First, we compare the analytical solution for one crack in a homogeneous full space with a vertically incident plane wave (MAL, 1970). Figure 2 shows displacement discontinuities or dislocations on the crack surface (Δv in (3)) for the analytical solution of MAL (1970) and our result by the boundary integral method with 50 fictitious sources in several frequencies (k_2 is same as ka where k is the wavenumber and a is the half crack length). Displacement discontinuities are normalized by the value at the center of the crack in the limit of zero frequency. Our numerical result agrees well with the analytical solution, and the wavefield expressed by these displacement discontinuities or fictitious sources is sufficiently accurate.

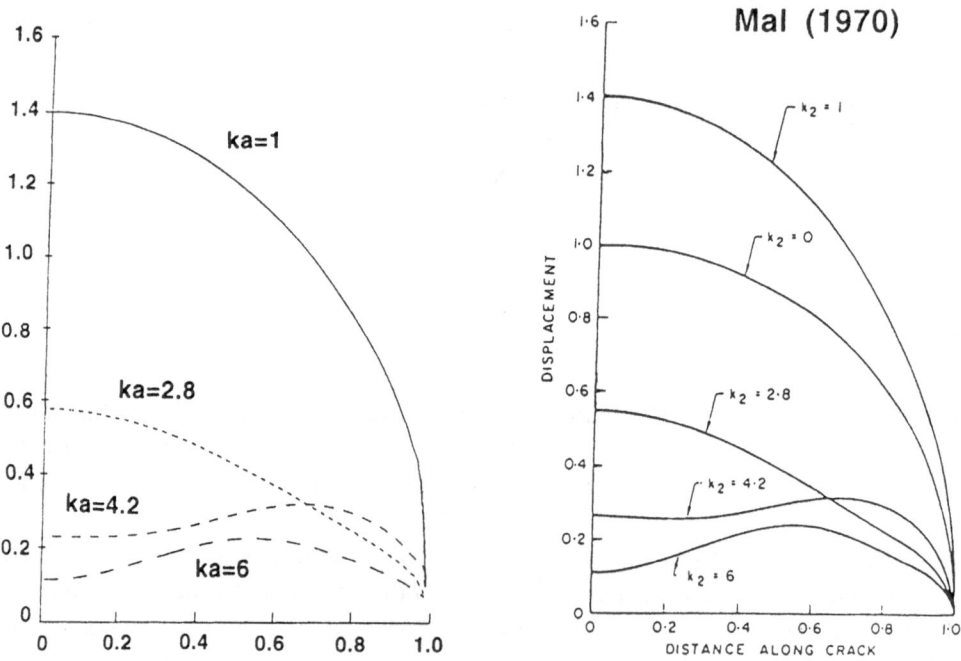

Figure 2

One crack in a homogeneous full space with a vertically incident plane wave. Dislocations along the crack surface Δv for an analytical solution (MAL, 1970) and those by our boundary integral method with 50 sources. Scale normalized by the center of the crack in the limit of zero frequency. k_2 of MAL (1970) corresponds to ka (k: wavenumber, a: half crack length).

The distance between two adjacent fictitious sources should be less than one quarter of the wavelength considered in the case of the single-layer potentials of BENITES et al. (1992). By changing the number of fictitious sources, we estimate the lower limit of the number of fictitious sources in the discretization process in (5), and find a criterion similar to the previous case can be applied for the present double-layer potential case. Another numerical trick is required in our boundary integral method: in order to avoid the singularity of the Green's function (5) when a source is very close to a receiver (i.e., $|P - Q_{ji}| \simeq 0$), we need to shift the fictitious source slightly away from the crack surface. The distance of this shift is also confirmed to be similar to that in our previous single-potential case for stable numerical results.

The successful comparison in Figure 2 is, nevertheless, only for a relatively low-frequency range ($kd < 10$, where $d\ (= 2a)$ is the crack length) where the accuracy of the numerical results is easily attained in general. For realistic seismic scattering problems, we would like to compute accurate time-domain seismograms or waveforms with a central frequency as large as $kd \simeq 10$ or a wavelength much smaller than each crack length. Figure 3 shows synthetic seismograms at a receiver array behind a long crack in a homogeneous full space for a vertically incident plane wave. As for the remaining examples, model parameters are given as normalized or non-dimensionalized values so that we can translate our results into models of different scale lengths. In this case, the crack length is 20, the distance between receivers and the crack is 2.5, and the shear-wave velocity is 1. Waveforms are convolved with a Ricker wavelet of central frequency equal to 1.5. The corresponding wavelength of 0.67 is much smaller than the crack length. The whole time window of seismograms in Figure 3 is 50, and we show seismograms only at receivers above one half of the crack because of the symmetry of this model.

Since the seismograms in Figure 3 are of sufficiently high frequency, we can distinguish several phases, which is also revealed by the analytical solution in the time domain. In Figure 3, we can clearly observe three phases: the direct incident wave, the wave scattered from a tip of the crack (denoted by "t"), and the diffracted or creeping wave behind the shadow of the crack ("C"), as also observed in the cavity case (BENITES et al., 1992). It might not be surprising that the direct incident wave is not seen above the crack or that there is a shadow zone because the crack surface does not allow the incident wave to transmit and all the energy is completely reflected. This clear shadow zone is, however, one of the most critical tests for the numerical accuracy in our boundary integral method. Wavefield in the shadow zone (i.e., the zero-energy field) is the summation of the incident wave u_0 and the wave u_s scattered by the crack as shown in (1), and the scattered wave is expressed by discretized fictitious sources on the crack, as in (5). In Figure 4 with the incident wave coming from above, the incident wave and the wave reflected by the crack surface "r" arrive at receivers with some time delay while they arrive simultaneously in Figure 3.

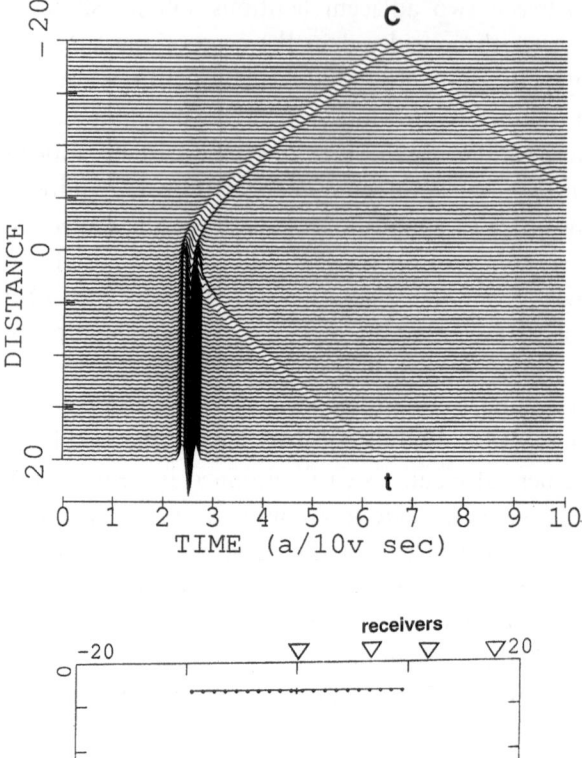

Figure 3

Waveform for a long crack with a vertically incident plane wave from below. Receivers are located on the surface only in the right half of the model. The depth of the crack is 2.5, the crack length d is 20, and the shear-wave velocity β is 1. Seismograms are convolved with a Ricker wavelet of the central frequency f_c to be 1.5, and their total record length is 50. The incident wave is observed, together with a scattered wave from a crack tip "t" and a diffracted or creeping wave behind the crack "C".

The shadow in Figure 3 shows that the scattered wave u_s has the opposite polarity of the phase "r" in Figure 4 and cancels out the incident wave u_0 perfectly. We can consider the shadow in Figure 3 as one of the most pathological tests against our scheme, guaranteeing its high numerical accuracy.

Besides the shadow, the crack produces scattered waves as two clear phases denoted by "C" and "t". These phases are also well explained in both amplitude and arrival time, compared with the analytical solution for a half-infinite crack. The accuracy of these phases is also confirmed by the result in Figure 5 for an obliquely (45 degrees) incident plane wave. In a more strict sense, we may compare these

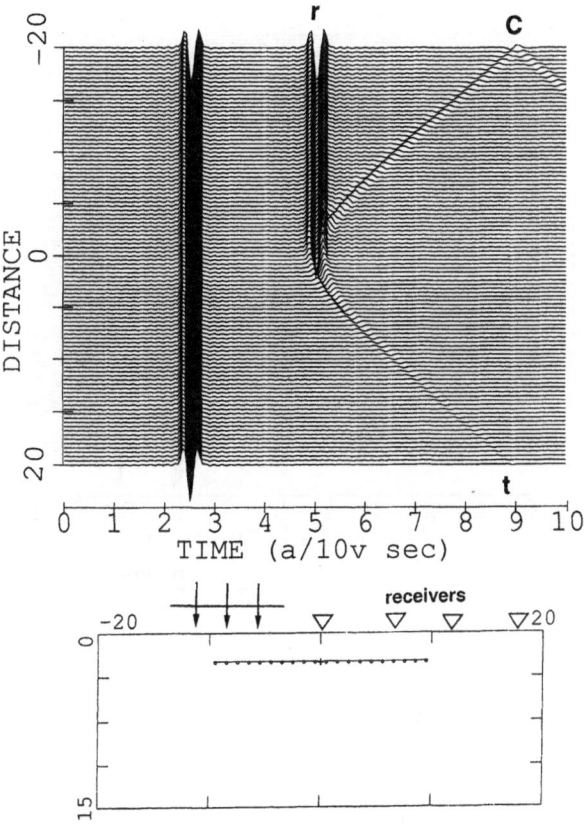

Figure 4
Same as Figure 3 except for a vertically incident plane wave from above. The reflected wave from the crack
"*r*" is additionally observed.

seismograms with the analytical solutions obtained by the Wiener–Hopf method for a model with a semi-infinite crack (e.g., ACHENBACH, 1973).

The above results are obtained in a very high-frequency range where each phase is clearly separated and ray-theoretical interpretations are possible. In many seismic scattering problems, the wavelength considered is comparable to or longer than the size of heterogeneities or cracks. Figure 6 shows synthetic seismograms for a finite-length crack with a vertically incident plane wave. The frequency range is now much lower than in Figure 3. The non-dimensional frequency with respect to the crack length, kd, is 188 in Figure 3 while it is 7.5 in Figure 6.

Waves diffracted from both ends of the crack into the shadow zone behind the crack obscure the shadow so that amplitude is not zero but small and arrival time is delayed at receivers behind the crack. Both the creeping wave "*C*" and the diffracted wave from the crack tips "*t*" are also seen but not as clearly as in Figure 3. These seismograms do not appear to differ significantly from those for the cavity case presented in BENITES *et al.* (1992). The seismograms in Figure 6 cannot be checked

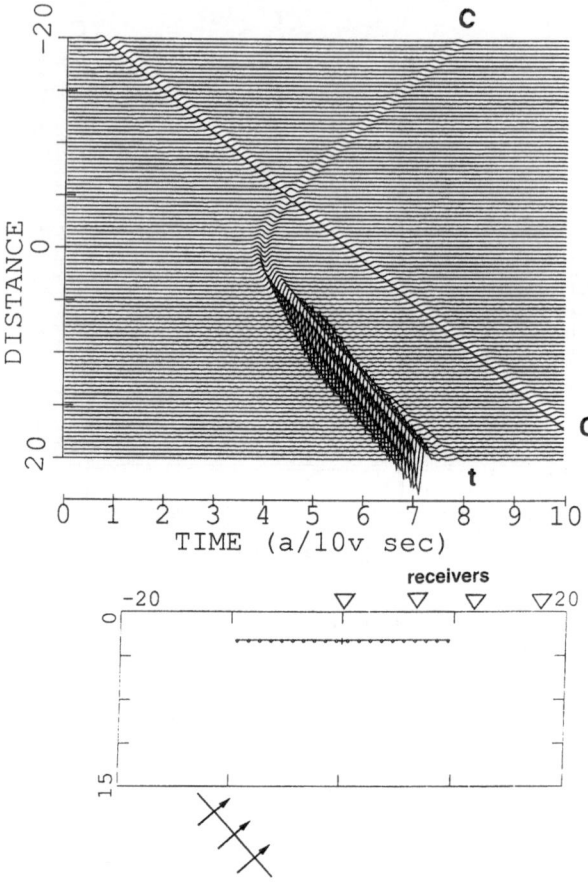

Figure 5

Same as Figure 3 except for an obliquely (45 degrees) incident plane wave from below.

by analytical solutions, but their accuracy should be better or more easily attained, compared with the previous examples in Figures 3–5, because our numerical scheme requires more careful selection of parameters involved with the discretization of fictitious sources, as frequency increases.

4. Attenuation Due to Scattering

Our boundary integral method can deal with the scattering by many scatterers (cracks in this study) in the same manner as one scatterer, including every degree of multiple scattering. As described in the introduction, the main purpose is the comparison of the nature of seismic scattering between randomly distributed cracks in this study and cavities in our previous studies. Figure 7 shows synthetic seismograms for a homogeneous full space with 50 randomly distributed circular

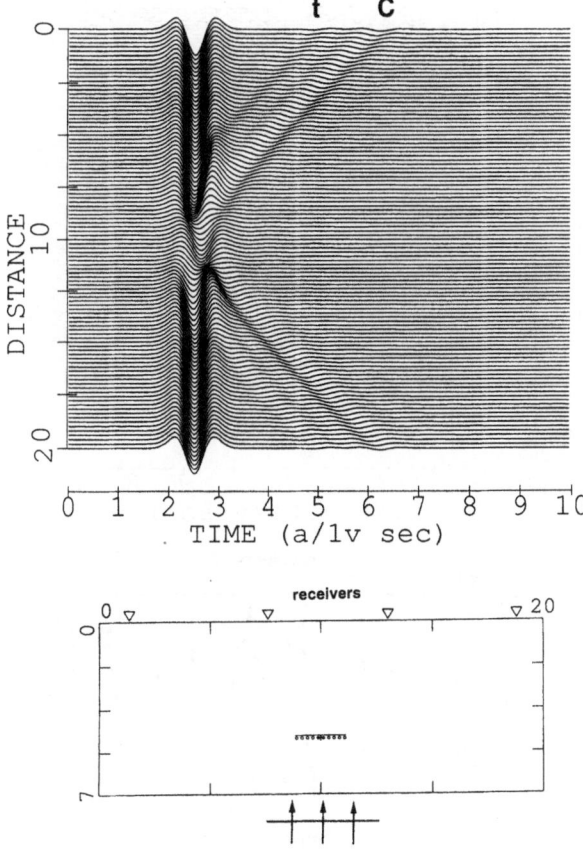

Figure 6
Waveform for a finite-length crack ($d = 2.4$) with a vertically incident plane wave. The distance of receivers from the crack is 5, and $f_c = 0.5$ (or $kd = 7.5$). Both amplitude decay and phase delay behind the crack are observed.

cavities with a vertically incident plane wave (BENITES et al., 1992). Direct waves are attenuated due to the scattering by cavities, and coda-like late arrival waves are observed. Measuring the amplitude attenuation of direct waves for bandpassed synthetic seismograms, the scattering attenuation factor Q is obtained in different frequency ranges. Figure 8 shows the frequency dependency of Q^{-1} for the scattering attenuation with the model of Figure 7.

YOMOGIDA and BENITES (1995) demonstrated that the strength of multiple scattered waves is not negligible compared with that of single scattering in this model. For this reason, analytic approaches for media with randomly distributed cracks or cavities cannot be applied. Figure 8 also depicts three curves based on a single scattering theory of WU (1982). Although it deals with Chernov-type random media, its overall result agrees well with previous studies on crack models (e.g., VARADAN et al., 1978; YAMASHITA, 1990) if the concentration of heterogeneities is

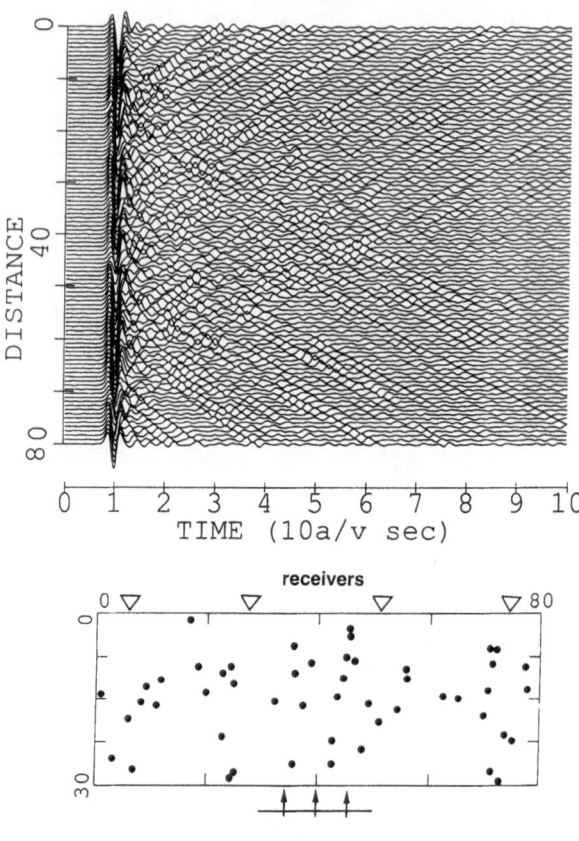

Figure 7
Waveform for a homogeneous full space with 50 randomly distributed circular cavities with a vertically incident plane wave. The diameter of cavities d is 1, and $f_c = 0.6$ ($kd = 3.8$) in whose frequency the degree of scattering is nearly the maximum (after BENITES et al., 1992).

small. Since the simple expression for the frequency dependency of Q^{-1} by WU (1982) helps us to understand the involved physical processes of scattering easily, as explained below, we refer it here even though the heterogeneities in our model are too strong to be applied and WU (1982) does not deal with crack models. A peak value of Q^{-1} occurs around the non-dimensional frequency kd around 2, where d is the diameter of the cavities. Similar to WU (1982), Q^{-1} at higher frequency decreases nearly proportional to $(kd)^{-1}$, which agrees with the prediction of the geometrical ray theory. At lower frequency, it should gradually follow the Rayleigh scattering with $Q^{-1} \propto (kd)^2$ although our example cannot measure this character precisely due to the very small attenuation in such a case.

Analogous to this cavity model, we set a crack model: a homogeneous full space with 50 randomly distributed cracks at the same locations as the cavities in Figure 7, as shown in Figure 9. The length of the cracks is the same as the diameter of the cavities, and the orientation of all the cracks is horizontal or perpendicular to the

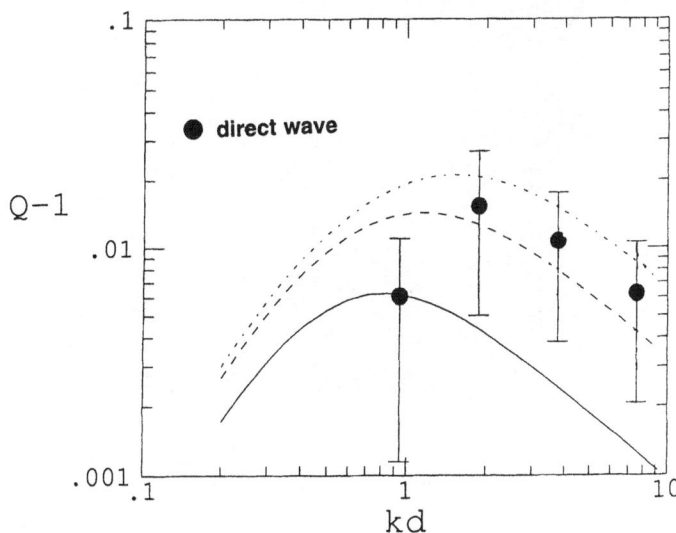

Figure 8

Frequency dependency of Q^{-1} for scattering of direct waves . Three curves are predictions based on a single scattering theory of WU (1982) (modified from YOMOGIDA and BENITES, 1995).

vertically incident plane wave. Synthetic seismograms with the same central frequency of the Ricker wavelet as that in Figure 7, $f_c = 0.6$ or $kd = 3.8$, are shown in Figure 9. The overall features of these two sets of seismograms (i.e., Figs. 7 and 9) appear to be very similar for both direct waves and coda. In other words, the scattering character should not be so different between the two models, at least in this frequency range.

Next, the length and spatial distribution of cracks are kept the same as in Figure 9, but their orientation is changed. Synthetic seismograms are shown in Figures 10 and 11 for the orientations of cracks of 45 and 75 degrees with respect to the incident wave, respectively. As the crack orientation becomes parallel to the propagating direction of the incident wave, the direct wave attenuates less and the excitation of coda becomes weaker. This phenomenon can be described as the apparent seismic anisotropy of a medium with cracks of preferred orientation, as studied before (e.g., KAWAHARA and YAMASHITA, 1992).

In Figure 12, the frequency dependency of scattering attenuation Q^{-1} is shown for media with cracks with three different orientations, together with the same three curves as in Figure 8, based on the single scattering theory (WU, 1982), in order to compare the present results for cracks with those for cavities (Fig. 8). In all the cases, the scattering attenuation has a peak value around $kd = 2$, similar to the cavity case in Figure 8. At lower frequencies, Q^{-1} or the strength of scattering attenuation by cracks is significantly lower than in the cavity case, and the difference among the

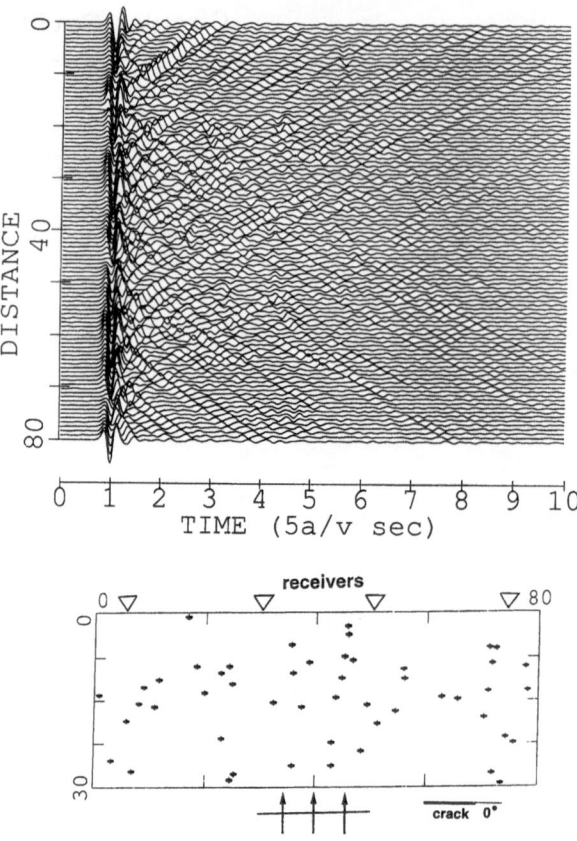

Figure 9

Waveform for a homogeneous full space with 50 randomly distributed cracks with a vertically incident plane wave. The length of cracks d is 1, and $f_c = 0.6$ ($kd = 3.8$). Cracks are aligned horizontally or perpendicular to the incident wave. The locations of the cracks are the same as those of the cavities in Figure 7.

three crack orientations is relatively small. If the considered wavelength becomes larger than the size of heterogeneities (i.e., the crack length in this case), the scattering is no longer affected by detailed geometry of heterogeneities but only by the entire volume of heterogeneities (e.g., SATO and FEHLER, 1998). Cavities have finite volumes while the volume of cracks is ideally zero. In a low frequency range, cracks in the crust may not play an important role in observed scattering phenomena of seismic waves.

In contrast, the crack orientation becomes very important in a frequency range higher than $kd \simeq 2$ or where the wavelength considered is shorter than the crack length. Compared with the result for the cavity model (Fig. 8), Q^{-1} values are grossly similar to the crack case if cracks are oriented perpendicularly to the propagating direction of the incident wave. The more obliquely cracks are oriented, the less the scattering attenuation. This feature is more noticeable as frequency increases. The

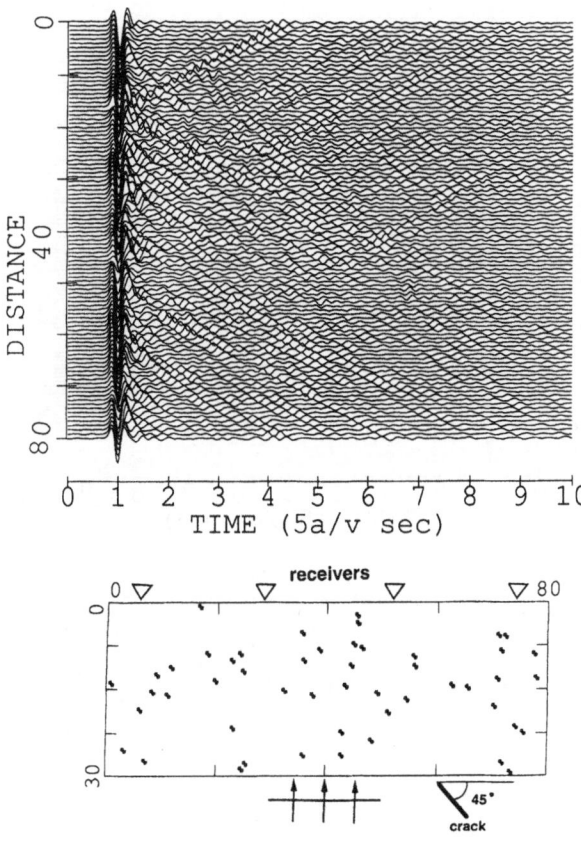

Figure 10
Same as Figure 9 except for cracks inclined by 45 degrees to the incident wave.

above results have been investigated in several previous studies (e.g., MURAI *et al.*, 1995). In these high frequency ranges, we can apply the geometrical ray theory to the interpretation of scattering characters, and the total cross section for scattering becomes close to the actual cross section of all the heterogeneities with respect to the incident wave (e.g., SATO and FEHLER, 1998). Due to the common crack length and cavity diameter as well as their spatial distribution, the total cross section of cavities in Figure 7 is same as that of cracks in Figure 9, resulting in very similar values of Q^{-1} in both cases. The total cross section becomes smaller for oblique cracks, which can explain smaller Q^{-1} values for the other two crack models in Figure 12. The above result in a high frequency range of $kd > 2$ gives quantitative features of apparent seismic anisotropy in media with preferred oriented cracks, as discussed in many previous crack models (e.g., CRAMPIN and LOVELL, 1991). Our scheme takes the effect of multiple scattering into full consideration as a decided advantage over many previous studies in order to evaluate this anisotropy quantitatively in a wide frequency range.

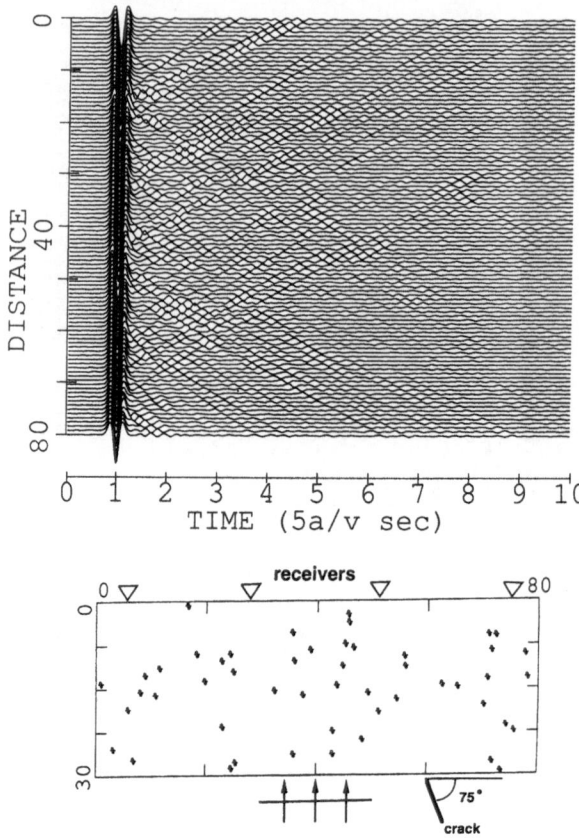

Figure 11
Same as Figure 9 except for cracks inclined by 75 degrees to the incident wave.

5. Conclusions

In this study, we present the boundary integral method to calculate seismic waves in media with many cracks. In contrast with the previous cavity case of BENITES *et al.* (1992) investigated with single-layer potentials, we introduce double-layer potentials or displacement discontinuities to represent the wave scattered by cracks. Explicit formulations with discrete fictitious sources are presented for two-dimensional *SH* waves in a medium with dry cracks, although our formulations can be easily extended to more complicated problems such as 2-D *P-SV* waves or 3-D cases, by changing the equation (3) and the Green's function (4) into appropriate ones from the general equation (2). Assigning an appropriate boundary condition over the crack surface, other types of cracks such as fluid-saturated ones can be simulated. As in the previous cavity case, the present method includes every degree of higher-order multiple scattering among cracks.

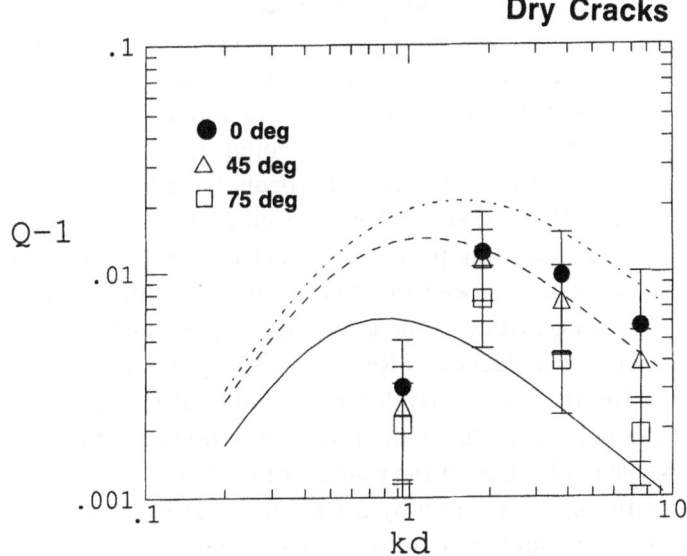

Figure 12
Frequency dependency of Q^{-1} for crack models with various orientations. For comparison with Figure 8, the three curves of Wu (1982) are also plotted.

For a homogeneous full space with one long crack, we check the accuracy and stability of our boundary integral method by comparison with analytical solutions. We succeed in obtaining seismograms with a shadow zone behind the crack, a phase scattered from a crack tip and a diffracted wave creeping into the shadow. As long as the interval of fictitious sources is less than a quarter of the wavelength considered, our results are proved to be accurate and stable.

In order to make a quantitative comparison of the frequency dependency of scattering between circular cavities and cracks, we calculate seismograms in a homogeneous full space with 50 randomly distributed cracks whose distribution and size are the same as the previous cavity model. Similar to the cavity case, scattering attenuation or Q^{-1} peaks around $kd = 2$, that is, the wavelength considered (k is the wavenumber) is comparable to the size of heterogeneities (or crack length d). In a lower frequency range (i.e., wavelength much longer than the crack length), Q^{-1} values become considerably smaller than those in the cavity case. Not detailed shape but only the volume of heterogeneities becomes the essential factor in scattering, and circular cavities should attenuate seismic waves far more efficiently than cracks at low frequency. In a frequency range higher than $kd \simeq 2$, Q^{-1} values are very similar to those in the cavity case if cracks are oriented perpendicularly to the propagating direction of the incident wave. As cracks are aligned more obliquely to the incident wave, Q^{-1} values become smaller, particularly as the frequency increases. If the wavelength is shorter than the representative size of heterogeneities, we can apply the geometrical ray theory to the interpretation of scattering, and the cross section for scattering approaches the actual cross section of heterogeneities for the incident wave. This change of effective cross section with crack orientation causes the apparent anisotropy in seismic scattering only at high frequency.

As clearly shown in this study, cavities or volumetric heterogeneities are more effective than cracks in seismic scattering, particularly in the low frequency range with wavelength longer than the size of heterogeneities. Recent studies in seismology have repeatedly emphasized the importance of cracks in the crust, however as regards the scattering of seismic waves, we should take a more careful attitude towards the interpretation of heterogeneities in the crust. More quantitative comparisons with cracks and volumetric heterogeneities for the scattering of seismic waves should be conducted in the future, particularly for other kinds of cracks and cavities (i.e., inclusions), for example, saturated with viscous or non-viscous fluid.

Another possible implication of the results in this study is the stronger frequency dependency of scattering attenuation by cracks than that by cavities. The scattering attenuation decreases rapidly at frequencies lower than $kd = 2$ (Fig. 12), and that might be extremely useful in certain important seismic observations. For example, we may be able to distinguish a fracture with a few single cracks from that with many short ones, based on the observation of the frequency dependency of scattering attenuation, because the transition from strong to weak scattering attenuation should occur at a lower frequency in the latter case, due to the smaller value of d.

Acknowledgments

The launch of the present study was financially supported by the Earthquake Research Institute Cooperative Research Program (1995-G-02). The discussions with Prof. Teruo Yamashita and Drs. Jun Kawahara and Yoshio Murai as well as useful comments by Dr. Michel Bouchon, one anonymous reviewer and the guest editor (Dr. Ivan Pšenčík) were greatly appreciated. This study was also partially supported by Grant-in-Aid for Scientific Research (C) (No. 10640409) under the auspice of the Ministry of Education, Science and Culture of Japan.

REFERENCES

ACHENBACH, J. D., *Wave Propagation in Elastic Solids* (North-Holland, Amsterdam 1973).

AKI, K. (1973), *Scattering of P Waves under the Montana Lasa*, J. Geophys. Res., *78*, 1334–1346.

AKI, K., and RICHARDS, P. G., *Quantitative Seismology, Theory and Methods*, vol. 1 (W. H. Freeman and Company, San Francisco 1980).

BENITES, R., AKI, K., and YOMOGIDA, K. (1992), *Multiple Scattering of SH Waves in 2-D Media with Many Cavities*, Pure appl. geophys. *138*, 353–390.

BENITES, R., YOMOGIDA, K., ROBERTS, P. M., and FEHLER, M. (1997), *Scattering of Elastic Waves in 2-D Composite Media I: Theory and Test*, Phys. Earth Planet. Inter. *104*, 161–173.

BOUCHON, M. (1987), *Diffraction to Elastic Waves by Cracks or Cavities Using the Discrete Wave Number Method*, J. Acoust. Soc. Am. *81*, 1671–1676.

CHERNOV, J. D., *Wave Propagation in a Random Medium* (McGraw-Hill, New York 1960).

COUTANT, O. (1989), *Numerical Study of the Diffraction of Elastic Waves by Fluid-filled Cracks*, J. Geophys. Res. *94*, 17,805–17,818.

CRAMPIN, S. and LOVELL, H. (1991), *A Decade of Shear-wave Splitting in the Earth's Crust: What Does it Mean? What Use Can we Make of it? and What Should we Do Next*, Geophys. J. Int. *107*, 387–407.

HUDSON, J. A. (1981), *Wave Speeds and Attenuation of Elastic Waves in Material Containing Cracks*, Geophys. J. Roy. Astr. Soc. *64*, 133–150.

KAWAHARA, J. (1998), *Seismic Scattering by Cracks Containing Liquids*, Proc. 4th SEGJ Inter. Symp., 161–167.

KAWAHARA, J., and YAMASHITA, T. (1992), *Scattering of Elastic Waves by a Fracture Zone Containing Randomly Distributed Cracks*, Pure appl. geophys. *139*, 121–144.

LI, Y.-G., LEARY, P. C., ADAMS, D., and HASEMI, A. (1994), *Seismic Guided Waves Trapped in the Fault Zone of the Landers, California, Earthquake of 1992*, J. Geophys. Res. *99*, 11,705–11,722.

MAL, A. K. (1970), *Interaction of Elastic Waves with a Griffith Crack*, Int. J. Eng. Sci. *8*, 763–776.

MATSUNAMI, K. (1990), *Laboratory Measurements of Spatial Fluctuations and Attenuation of Elastic Waves by Scattering due to Random Heterogeneities*, Pure appl. geophys. *132*, 197–220.

MURAI, Y., KAWAHARA, J., and YAMASHITA, T. (1995), *Multiple Scattering of SH Waves in 2-D Elastic Media with Distributed Cracks*, Geophys. J. Int. *122*, 925–937.

NISHIZAWA, O., SATOH, T., LEI, X., and KUWAHARA, Y. (1997), *Laboratory Studies of Seismic Wave Propagation in Inhomogeneous Media Using a Laser Doppler Vibrometer*, Bull. Seismol. Soc. Am. *87*, 809–823.

SATO, H., and FEHLER, M. C. *Seismic Wave Propagation and Scattering in the Heterogeneous Earth* (Springer-Verlag, New York 1998).

URSELL, F. (1973), *On the Exterior Problems of Acoustics*, Proc. Camb. Phil. Soc. *74*, 117–125.

VARADAN, V. K., VARADAN, V. V., and PAO, Y.-H. (1978), *Multiple Scattering of Elastic Waves by Cylinders of Arbitrary Cross Section. I. SH Waves*, J. Acoust. Soc. Am. *63*, 1310–1319.

WU, R. S. (1982), *Attenuation of Short-period Seismic Waves due to Scattering*, Geophys. Res. Lett. *9*, 9–12.

YAMASHITA, T. (1990), *Attenuation and Dispersion of SH Waves due to Scattering by Randomly Distributed Cracks*, Pure appl. geophys. *132*, 545–568.

YOMOGIDA, K., and BENITES, R. (1995), *Relation between Direct Wave Q and Coda Q: A Numerical Approach*, Geophys. J. Int. *123*, 471–483.

YOMOGIDA, K., BENITES, R., ROBERTS, P. M., and FEHLER, M. (1997), *Scattering of Elastic Waves in 2-D Composite Media II: Waveforms and Spectra*, Phys. Earth Planet. Inter. *104*, 175–192.

(Received September 25, 2000, December 5, 2000, accepted February 9, 2001)

Pure appl. geophys. 159 (2002) 1791–1810
0033–4553/02/081791–20 $ 1.50 + 0.20/0

▌Pure and Applied Geophysics

Application of the Medium Covariance Functions to Travel-time Tomography

LUDĚK KLIMEŠ[1]

Summary—The slowness describing the geological structure is assumed to be a representation of a random medium described in terms of the first two statistical moments: the mean value and the medium covariance function. The proposed approach to seismic travel-time tomography enables the equations for the inversion of travel times to be derived without additional subjective *a priori* information, and the accuracy and resolution of the resulting seismic model of the geological structure to be estimated.

Key words: Seismic model, travel-time tomographic inversion, covariance functions and matrices, smoothing, resolution.

1. Introduction

The seismic model (macro model) approximates the spatial distribution of the material properties of the geological structure.

It is obvious that geological structures cannot be definitely described in terms of a finite number of model parameters: proceeding from large to small scales, increasingly more details of the medium may be discovered. From this point of view, the seismic model can never represent all details of the medium, no matter how numerous the seismic data are.

On the other hand, we cannot consider the individual small details of the medium to be mutually independent. Such a medium, with statistical properties of white noise, would be unpredictable from a finite number of seismic data and seismic inversion would be nonsense. We must assume that the probability of the same or similar material properties at two points increases with decreasing distance of the points. This property of the medium may, in the roughest approximation, be described in terms of medium covariance functions.

This paper, concentrating on the inversion of travel times, follows the approach of FRANKLIN (1970), TARANTOLA and VALETTE (1982), TARANTOLA and NERCESSIAN (1984) and TARANTOLA (1987), similarly as the related paper of

[1] Department of Geophysics, Charles University, Ke Karlovu 3, 121 16 Praha 2, Czech Republic.
E-mail: klimes@seis.karlov.mff.cuni.cz

MAURER *et al.* (1998). Consideration of the medium covariance functions in the travel-time inversion is inevitable if the ray coverage of the medium is sparse and uneven. Since travel times contain only the information on the average slowness along the corresponding two-point rays, we would have no information on the behavior of slowness between the rays without the medium covariance functions. The medium covariance functions provide statistically correct interpolation of slowness between the rays, especially over distances larger than the dimensions of Fresnel volumes. The properties of this inversion have numerically been demonstrated by MAURER *et al.* (1998) by a very presentable 2-D synthetic example. Analogous behavior of the inversion for 1-D electrical resistivities has been shown by PILKINGTON and TODOESCHUCK (1990). The medium covariance functions simultaneously enable the information, contained in travel times picked in several solitary stations, to be protected from the superior force of travel times picked along dense receiver profiles, whereas the still often used "naive" least squares, with incorrect travel-time weights, might result in substantial loss of information and considerably bad models. Different inverse methods are designed for different kinds of data and should not be applied wrongly. For example, in contrast with this paper, the inverse theory developed by BACKUS and GILBERT (1970) is designed for the inversion of gross earth data but not for the inversion of data which are considerably sparse and uneven with respect to their resolution. For comparison refer to Box 7.2 of TARANTOLA (1987).

In this paper, the slowness describing the geological structure is assumed to be a representation of a random medium with known statistical properties. We assume that some travel times corresponding to the geological structure (i.e., to the representation of the random medium) are known, and determine the seismic model as the "best" approximation to the (possibly smoothed) slowness describing the geological structure. The "best" approximation means the smallest Lebesgue L_2 norm of the standard deviations between the model and the (possibly smoothed) slowness describing the geological structure.

This approach enables us to arrive at the final formulae for linearized travel-time tomography and to separate the smoothing effects of (a) the application of the covariance functions in the inversion for the non-smoothed medium, (b) the inversion for the smoothed medium compared to the inversion for the non-smoothed medium, (c) the finite model parameterization, and (d) the additional *a priori* smoothing information specified in terms of the Sobolev norm, see Section 3.5.

The derived equations apply to the final model resulting from an iterative inversion. The final model should be independent of the iterations, depending only on the mean model, the medium covariance function, and the data. Yet the equations still may be applied to individual linearized iterations in such a way that the ray characteristic functions are approximated by the ones calculated in the initial model of the iteration. The Einstein summation over repetitive indices is used throughout the paper.

2. Formulation

2.1. Random Slowness

Assume that slowness $u(\mathbf{x})$ in the geological structure is a realization of the random quantity with the first statistical moment (*mean value*)

$$u^0(\mathbf{x}) = \langle u(\mathbf{x}) \rangle \tag{1}$$

and the second statistical moment

$$\langle u(\mathbf{x}_1)\, u(\mathbf{x}_2) \rangle \;. \tag{2}$$

We shall call *a priori* covariance function

$$C(\mathbf{x}_1, \mathbf{x}_2) = \langle u(\mathbf{x}_1)\, u(\mathbf{x}_2) \rangle - u^0(\mathbf{x}_1) u^0(\mathbf{x}_2) \tag{3}$$

of the slowness the *medium covariance function*. Here $u^0(\mathbf{x})$ is the *a priori* known deterministic component of the random medium serving as the background slowness in estimating the medium covariance function $C(\mathbf{x}_1, \mathbf{x}_2)$, which should be determined prior to the proposed tomographic inversion.

We shall see that the determination of $u^0(\mathbf{x})$ by means of very strongly overdetermined least squares is consistent with the proposed tomographic inversion. The least-square fitting of $u^0(\mathbf{x})$ must be very strongly overdetermined in order to leave a sufficient amount of information to estimate medium covariance function $C(\mathbf{x}_1, \mathbf{x}_2)$.

The determination of $u^0(\mathbf{x})$, parameterized by several parameters, and $C(\mathbf{x}_1, \mathbf{x}_2)$, parameterized by one or a few parameters, should be carried out prior to the proposed tomographic inversion. If the medium covariance function depends only on vectorial distance $\mathbf{x}_1 - \mathbf{x}_2$, it may be expressed in terms of *medium correlation function* $c(\mathbf{x})$,

$$C(\mathbf{x}_1, \mathbf{x}_2) = c(\mathbf{x}_1 - \mathbf{x}_2) \tag{4}$$

(TARANTOLA, 1987). One possibility of determining $u^0(\mathbf{x})$ and $c(\mathbf{x})$ has been demonstrated by KLIMEŠ (2002c). Various types of the medium correlation functions $c(\mathbf{x})$ used in geophysics have been systematically listed by KLIMEŠ (2002b).

2.2. Travel Times and Rays

Field travel times T_I are the integrals of slowness $u(\mathbf{x})$ along rays Ω_I plus the *picking errors*,

$$T_I = \int G_I(\mathbf{x}) u(\mathbf{x}) \mathrm{d}\mathbf{x} + \delta T_I \;, \tag{5}$$

where the integral with ray characteristic function $G_I(\mathbf{x})$ is taken over the model volume. *Uppercase subscripts* are used hereinafter to denote travel times and the corresponding rays. We assume zero mean values of the picking errors,

$$\langle \delta T_I \rangle = 0 \ , \tag{6}$$

and introduce the covariance matrix of picking errors (*data covariance matrix*)

$$T_{IJ} = \langle \delta T_I \ \delta T_J \rangle \tag{7}$$

describing errors δT_I. We assume no correlation between picking errors δT_I and the exact slowness $u(\mathbf{x})$,

$$\langle \delta T_I \ u(\mathbf{x}) \rangle = 0 \ . \tag{8}$$

Ray characteristic functions

$$G_I(\mathbf{x}) = \frac{\partial T_I}{\partial u(\mathbf{x})} \ , \tag{9}$$

also called "travel-time sensitivity functions" or "sampling functions", characterize the sensitivity of travel times to slowness perturbations. Ray characteristic functions $G_I(\mathbf{x})$ should ideally be calculated for the exact medium, but during linearized inversion they are approximated by the ones calculated in the best model of the last iteration.

The exact ray characteristic functions, describing "thick rays," depend on the definition of the arrival times picked for the inversion. For examples refer, e.g., to WOODWARD and ROCCA (1988), COATES and CHAPMAN (1990), WOODWARD (1992), PRATT et al. (1996) or MARQUERING et al. (1998).

From the point of view of the high-frequency asymptotic ray theory, the ray characteristic function $G_I(\mathbf{x})$ of "thin" ray Ω_I has the form of the 2-D Dirac distribution $\delta(q^1, q^2)$ in ray-centered coordinates (q^1, q^2, q^3), with q^1 and q^2 being the Cartesian coordinates in the plane perpendicular to the ray parameterized by arclength q_3,

$$G_I(\mathbf{x}(q^1, q^2, q^3)) = \delta(q^1, q^2) \tag{10}$$

for the values of q^3 between the source and the receiver, zero for other q_3. For the two-point ray tracing refer, e.g., to BULANT (1996, 1999).

If the spacing between two-point rays exceeds the dimensions of the relevant Fresnel volumes considerably, the sensitivity of the resulting seismic model to the shape of the ray characteristic functions vanishes and the differences between "thick" and "thin" rays practically disappear.

2.3. Seismic Model

The slowness distribution $\bar{u}(\mathbf{x})$ in the seismic model is considered to be an approximation to the smoothed version $\bar{u}(\mathbf{x})$ of the slowness $u(\mathbf{x})$ of the medium,

$$\bar{u}(\mathbf{x}) = \int D(\mathbf{x}, \mathbf{x}') u(\mathbf{x}') d\mathbf{x}' \ , \tag{11}$$

where $D(\mathbf{x}_1, \mathbf{x}_2)$ is the given *smoothing function*. The smoothing function is included here for the sake of generality. It may express *a priori* ideas on the resolution of the model and may strongly influence the local standard deviation between the model and the geological structure, because the small details in the medium are much worse resolved relative to larger inhomogeneities. The *smoothing function may be taken in the form of the Dirac distribution*, $D(\mathbf{x}_1, \mathbf{x}_2) = \delta(\mathbf{x}_1 - \mathbf{x}_2)$. In such a case, no *a priori* smoothing is applied during the tomographic travel-time inversion, and the slowness distribution $\tilde{u}(\mathbf{x})$ in the seismic model is an approximation to the slowness $u(\mathbf{x})$ of the medium. Smoothing may then be applied to the model *a posteriori*. This approach may be, to some extent, limited only by an insufficiently rough parameterization of the model or other functions.

Model slowness $\tilde{u}(\mathbf{x})$ depends on the exact slowness $u(\mathbf{x})$ through travel times T_I. We assume the model to be linearly dependent on travel times,

$$\tilde{u}(\mathbf{x}) = R^0(\mathbf{x}) + R_K(\mathbf{x})[T_K - T_K^0] \ , \tag{12}$$

where

$$T_I^0 = \int G_I(\mathbf{x})u^0(\mathbf{x})\mathrm{d}\mathbf{x} \tag{13}$$

is used to denote the integrals with the ray characteristic functions $G_I(\mathbf{x})$ corresponding to exact rays Ω_I (*reference travel times*), although we know that this linearity may be violated by inaccurate approximations to $G_I(\mathbf{x})$ during the iterative linearized inversion. Note that, because of (5) and (13) with (1) and (6),

$$\langle T_K - T_K^0 \rangle = 0 \ . \tag{14}$$

We assume here that the differences between $u(\mathbf{x})$, $\bar{u}(\mathbf{x})$ and $\tilde{u}(\mathbf{x})$ are sufficiently small enabling calculation of the corresponding travel times along the same rays within an accuracy which would not distort the complete covariance matrix S_{IJ} of travel times, introduced later on by equation (34). In other words, we assume that the *linearization of travel times with respect to medium perturbations may be applied in the vicinity of the exact model*.

After inserting (5) and (13), equation (12) reads

$$\tilde{u}(\mathbf{x}) = R^0(\mathbf{x}) + \int R(\mathbf{x}, \mathbf{x}')[u(\mathbf{x}') - u^0(\mathbf{x}')]\mathrm{d}\mathbf{x}' + R_K(\mathbf{x})\delta T_K \ , \tag{15}$$

where

$$R(\mathbf{x}_1, \mathbf{x}_2) = R_K(\mathbf{x}_1)G_K(\mathbf{x}_2) \ , \tag{16}$$

is the *resolving kernel* of the *resolving operator*, linearly projecting the exact slowness $u(\mathbf{x})$ onto model $\tilde{u}(\mathbf{x})$.

For the sake of generality, we allow the resolving kernel to be constrained by zero to several *a priori* conditions indexed here with α,

$$\Delta^\alpha R^0(\mathbf{x}) + \int R(\mathbf{x},\mathbf{x}')\,u^\alpha(\mathbf{x}')\mathrm{d}\mathbf{x}' = \int D(\mathbf{x},\mathbf{x}')[\Delta^\alpha u^0(\mathbf{x}') + u^\alpha(\mathbf{x}')]\mathrm{d}\mathbf{x}' , \quad (17)$$

where $0 \le \Delta^\alpha \le 1$ are the weighting coefficients. For weighting coefficients $\Delta^\alpha = 1$, these projection constraints express the condition that the travel-time inversion yields exact results in the structures whose differences from $u^0(\mathbf{x})$ lie in a linear subspace generated by a few functions $u^\alpha(\mathbf{x})$. For weighting coefficients $\Delta^\alpha = 0$, we do not require the travel-time inversion to yield exact results in the structure described by $u^0(\mathbf{x})$, but enforce the exact results for perturbations $u^\alpha(\mathbf{x})$ of a structure from $u^0(\mathbf{x})$, allowing $u^0(\mathbf{x})$ to be smoothed by means of the Sobolev scalar products. We shall see later in Section 3.4 that these conditions are closely related to adjusting mean slowness $u^0(\mathbf{x})$ by a linear combination of functions $u^\alpha(\mathbf{x})$, fitted by least squares.

2.4. Parameterization of the Model and Resolving Kernel

Assume that the slowness distribution $\tilde{u}(\mathbf{x})$ in the seismic model, which is the best possible approximation to $\bar{u}(\mathbf{x})$, is composed of a finite set of *basis functions* $B_i(\mathbf{x})$ indexed hereinafter with *lowercase subscripts*,

$$\tilde{u}(\mathbf{x}) = B_i(\mathbf{x})u_i . \quad (18)$$

For example, tricubic B-splines may be employed for the basis functions $B_i(\mathbf{x})$. Comparing now (18) with (12), we see that functions $R^0(\mathbf{x})$ and $R_K(\mathbf{x})$ should also be composed of the basis functions,

$$R^0(\mathbf{x}) = B_i(\mathbf{x})R_i^0 , \quad (19)$$

$$R_K(\mathbf{x}) = B_i(\mathbf{x})R_{iK} , \quad (20)$$

that the resolving kernel (16) takes the form

$$R(\mathbf{x}_1,\mathbf{x}_2) = B_i(\mathbf{x}_1)R_{iI}G_I(\mathbf{x}_2) , \quad (21)$$

and that equation (12) can be expressed in component form

$$u_i = R_i^0 + R_{iI}[T_I - T_I^0] . \quad (22)$$

Assume that optional functions $u^\alpha(\mathbf{x})$ may also be composed of the same basis functions,

$$u^{\alpha}(\mathbf{x}) = u_i^{\alpha}\, B_i(\mathbf{x}) \ . \tag{23}$$

Optional constraints (17) with (19), (21) and (23), L_2-projected on the basis functions, then take the form

$$\Delta^{\alpha} R_i^0 + R_{iI} T_I^{\alpha} = \Delta^{\alpha} \bar{u}_i^0 + \bar{u}_i^{\alpha} \ , \tag{24}$$

where

$$T_I^{\alpha} = \int G_I(\mathbf{x}) u^{\alpha}(\mathbf{x}) d\mathbf{x} \ , \tag{25}$$

are the travel-time variations with respect to functions $u^{\alpha}(\mathbf{x})$,

$$\bar{u}_i^{\alpha} = B_{ik}^{-1} \iint B_k(\mathbf{x}_1) D(\mathbf{x}_1, \mathbf{x}_2) u^{\alpha}(\mathbf{x}_2) d\mathbf{x}_1\, d\mathbf{x}_2 \tag{26}$$

are the components of the Lebesgue L_2 projections of smoothed perturbations $u^{\alpha}(\mathbf{x})$ onto the basis functions, and

$$\bar{u}_i^0 = B_{ik}^{-1} \iint B_k(\mathbf{x}_1) D(\mathbf{x}_1, \mathbf{x}_2) u^0(\mathbf{x}_2) d\mathbf{x}_1 d\mathbf{x}_2 \tag{27}$$

are the components of the Lebesgue L_2 projections of the smoothed mean slowness onto the basis functions.

2.5. Model Covariance Function

The posterior covariance function describing the deviation of model slowness $\tilde{u}(\mathbf{x})$ from the smoothed exact slowness $\bar{u}(\mathbf{x})$ is

$$\tilde{C}(\mathbf{x}_1, \mathbf{x}_2) = \langle [\tilde{u}(\mathbf{x}_1) - \bar{u}(\mathbf{x}_1)][\tilde{u}(\mathbf{x}_2) - \bar{u}(\mathbf{x}_2)] \rangle \ . \tag{28}$$

We shall call it the *model covariance function*. Although the model covariance function is frequently neglected in the geophysical literature, it represents together with the model the most important result of the inversion. The model without the model covariance function is worthless from the scientific point of view. We thus derive here the equations for the full posterior covariance function $\tilde{C}(\mathbf{x}_1, \mathbf{x}_2)$, although only the posterior variance function $\tilde{C}(\mathbf{x}, \mathbf{x})$ appears in the objective function for the inversion.

Inserting (11) and (15) into (28), and considering (1), (6) and (8), we arrive at

$$\begin{aligned}
\tilde{C}(\mathbf{x}_1, \mathbf{x}_2) = \iint &[R(\mathbf{x}_1, \mathbf{x}_1') - D(\mathbf{x}_1, \mathbf{x}_1')][R(\mathbf{x}_2, \mathbf{x}_2') - D(\mathbf{x}_2, \mathbf{x}_2')] \\
&\times \langle [u(\mathbf{x}_1') - u^0(\mathbf{x}_1')][u(\mathbf{x}_2') - u^0(\mathbf{x}_2')] \rangle d\mathbf{x}_1'\, d\mathbf{x}_2' + R_I(\mathbf{x}_1) R_J(\mathbf{x}_2) \langle \delta T_I\, \delta T_J \rangle \\
&+ \left[R^0(\mathbf{x}_1) - \int D(\mathbf{x}_1, \mathbf{x}_1') u^0(\mathbf{x}_1') d\mathbf{x}_1' \right] \left[R^0(\mathbf{x}_2) - \int D(\mathbf{x}_2, \mathbf{x}_2') u^0(\mathbf{x}_2') d\mathbf{x}_2' \right] .
\end{aligned} \tag{29}$$

Using then (3), (7), and (20), we arrive at

$$\tilde{C}(\mathbf{x}_1, \mathbf{x}_2) = \iint [R(\mathbf{x}_1, \mathbf{x}_1') - D(\mathbf{x}_1, \mathbf{x}_1')][R(\mathbf{x}_2, \mathbf{x}_2') - D(\mathbf{x}_2, \mathbf{x}_2')]C(\mathbf{x}_1', \mathbf{x}_2')\mathrm{d}\mathbf{x}_1' \, \mathrm{d}\mathbf{x}_2'$$

$$+ B_i(\mathbf{x}_1)B_j(\mathbf{x}_2)R_{iI}R_{jJ}T_{IJ} + \left[R^0(\mathbf{x}_1) - \int D(\mathbf{x}_1, \mathbf{x}_1')u^0(\mathbf{x}_1')\mathrm{d}\mathbf{x}_1'\right]$$

$$\times \left[R^0(\mathbf{x}_2) - \int D(\mathbf{x}_2, \mathbf{x}_2')u^0(\mathbf{x}_2')\mathrm{d}\mathbf{x}_2'\right] . \tag{30}$$

Insertion of (19) and (21) into (30) yields

$$\tilde{C}(\mathbf{x}_1, \mathbf{x}_2) = \iint D(\mathbf{x}_1, \mathbf{x}_1')D(\mathbf{x}_2, \mathbf{x}_2')C(\mathbf{x}_1', \mathbf{x}_2')\mathrm{d}\mathbf{x}_1' \, \mathrm{d}\mathbf{x}_2'$$

$$- B_i(\mathbf{x}_1)R_{iI} \iint D(\mathbf{x}_2, \mathbf{x}_2')C(\mathbf{x}_2', \mathbf{x}_1')G_I(\mathbf{x}_1')\mathrm{d}\mathbf{x}_1' \, \mathrm{d}\mathbf{x}_2'$$

$$- B_j(\mathbf{x}_2)R_{jJ} \iint D(\mathbf{x}_1, \mathbf{x}_1')C(\mathbf{x}_1', \mathbf{x}_2')G_J(\mathbf{x}_2')\mathrm{d}\mathbf{x}_1' \, \mathrm{d}\mathbf{x}_2'$$

$$+ B_i(\mathbf{x}_1)B_j(\mathbf{x}_2)R_{iI}R_{jJ}\left[\iint G_I(\mathbf{x}_1')C(\mathbf{x}_1', \mathbf{x}_2')G_J(\mathbf{x}_2')\mathrm{d}\mathbf{x}_1' \, \mathrm{d}\mathbf{x}_2' + T_{IJ}\right]$$

$$+ \left[B_i(\mathbf{x}_1)R_i^0 - \int D(\mathbf{x}_1, \mathbf{x}_1')u^0(\mathbf{x}_1')\mathrm{d}\mathbf{x}_1'\right]$$

$$\times \left[B_j(\mathbf{x}_1)R_j^0 - \int D(\mathbf{x}_2, \mathbf{x}_2')u^0(\mathbf{x}_2')\mathrm{d}\mathbf{x}_2'\right] . \tag{31}$$

Introducing the *ray back-projective functions* (or ray resolution functions)

$$\Psi_I(\mathbf{x}) = \langle T_I\left[u(\mathbf{x}) - u^0(\mathbf{x})\right]\rangle = \int C(\mathbf{x}, \mathbf{x}')G_I(\mathbf{x}')\mathrm{d}\mathbf{x}' , \tag{32}$$

representing the covariances between the slowness and field travel times, the *a priori geometrical travel-time covariance matrix*

$$\Theta_{IJ} = \iint G_I(\mathbf{x}_1)C(\mathbf{x}_1, \mathbf{x}_2)G_J(\mathbf{x}_2)\mathrm{d}\mathbf{x}_1 \, \mathrm{d}\mathbf{x}_2 = \int G_I(\mathbf{x}_1)\Psi_J(\mathbf{x}_1)\mathrm{d}\mathbf{x}_1 \tag{33}$$

(CHERNOV, 1960; TARANTOLA, 1987), and the *complete travel-time covariance matrix*

$$S_{MN} = \langle [T_M - T_M^0][T_N - T_N^0]\rangle = T_{MN} + \Theta_{MN} , \tag{34}$$

equation (31) for the model covariance function becomes

$$\tilde{C}(\mathbf{x}_1, \mathbf{x}_2) = \iint D(\mathbf{x}_1, \mathbf{x}_1') D(\mathbf{x}_2, \mathbf{x}_2') C(\mathbf{x}_1', \mathbf{x}_2') d\mathbf{x}_1' \, d\mathbf{x}_2'$$

$$- B_i(\mathbf{x}_1) R_{iI} \int D(\mathbf{x}_2, \mathbf{x}_2') \Psi_I(\mathbf{x}_2') d\mathbf{x}_2' - B_j(\mathbf{x}_2) R_{jJ} \int D(\mathbf{x}_1, \mathbf{x}_1') \Psi_J(\mathbf{x}_1') d\mathbf{x}_1'$$

$$+ B_i(\mathbf{x}_1) B_j(\mathbf{x}_2) R_{iI} R_{jJ} S_{IJ} + \left[B_i(\mathbf{x}_1) R_i^0 - \int D(\mathbf{x}_1, \mathbf{x}_1') u^0(\mathbf{x}_1') d\mathbf{x}_1' \right]$$

$$\times \left[B_j(\mathbf{x}_2) R_j^0 - \int D(\mathbf{x}_2, \mathbf{x}_2') u^0(\mathbf{x}_2') d\mathbf{x}_2' \right] . \tag{35}$$

3. Inversion

3.1. Sobolev Scalar Product

Model covariance function (35) describes the deviation of the resulting model slowness $\tilde{u}(\mathbf{x})$ from the smoothed exact slowness $\bar{u}(\mathbf{x})$. The deviation causes the differences between the synthetic wavefields calculated in the model and exact wavefields measured in geological structures. To minimize the deviation, the objective function minimizing the squared standard deviation $\tilde{C}(\mathbf{x}, \mathbf{x})$ may be considered.

However, there is also another source of the differences between the synthetic and exact wavefields: inaccuracy of the wavefield modeling methods in overly complex models. The accuracy of all numerical methods increases with the smoothness of the model. The smoothness of the model may be conveniently expressed in terms of the square

$$(\|\tilde{u} - u^\Phi\|^\Phi)^2 = ((\tilde{u} - u^\Phi, \tilde{u} - u^\Phi))^\Phi \tag{36}$$

of the *Sobolev norm* $\| \bullet \|^\Phi$ of the difference between $\tilde{u}(\mathbf{x})$ and given function

$$u^\Phi(\mathbf{x}) = B_i(\mathbf{x}) u_i^\Phi . \tag{37}$$

We may combine several smoothing expressions (36), indexed by superscript Φ. Here

$$((f, g))^\Phi = \sum_{\alpha_1} \sum_{\alpha_2} \sum_{\alpha_3} \sum_{\beta_1} \sum_{\beta_2} \sum_{\beta_3} \int W_{\alpha_1 \alpha_2 \alpha_3 \beta_1 \beta_2 \beta_3}^\Phi(\mathbf{x})$$

$$\times \left[\left(\frac{\partial}{\partial x^1} \right)^{\alpha_1} \left(\frac{\partial}{\partial x^2} \right)^{\alpha_2} \left(\frac{\partial}{\partial x^3} \right)^{\alpha_3} f(\mathbf{x}) \right] \left[\left(\frac{\partial}{\partial x^1} \right)^{\beta_1} \left(\frac{\partial}{\partial x^2} \right)^{\beta_2} \left(\frac{\partial}{\partial x^3} \right)^{\beta_3} g(\mathbf{x}) \right] d\mathbf{x} \tag{38}$$

is the *Sobolev scalar product* of functions $f(\mathbf{x})$ and $g(\mathbf{x})$, defined in terms of weighting functions $W_{\alpha_1 \alpha_2 \alpha_3 \beta_1 \beta_2 \beta_3}^\Phi(\mathbf{x})$, see TARANTOLA (1987). The order of the Sobolev scalar product is determined by non-zero coefficients $W_{\alpha_1 \alpha_2 \alpha_3 \beta_1 \beta_2 \beta_3}^\Phi$. The coefficients are chosen according to the application of the velocity model being constructed. The minimization of the derivatives has been suggested by MENKE (1984), but his equations for the first derivatives are applicable only to the properly normalized 1-D triangular basis functions on a regular grid, and his equations for the second derivatives only to the properly normalized 1-D quadratic B-splines on a regular grid.

As the relative standard deviation $\sqrt{\tilde{C}(\mathbf{x}, \mathbf{x})}$ quantifies the local travel-time error per the small distance of wave propagation due to the inaccuracy of the model, it is desirable to adjust weighting functions $W^{\Phi}_{\alpha_1 \alpha_2 \alpha_3 \beta_1 \beta_2 \beta_3}(\mathbf{x})$ in such a way that the Sobolev norm $\|\tilde{u} - u^{\Phi}\|^{\Phi}$ approximates the analogous local travel-time errors due to the inaccuracies of the numerical methods expected to be applied in the resulting model. For examples of constructing particular Sobolev norms for various numerical forward-modeling methods refer to KLIMEŠ (2000), BULANT (2002) and ŽÁČEK (2002). We only mention here two Sobolev scalar products relevant to the ray tracing. The complicated behavior of rays depends on the second velocity derivatives (KLIMEŠ, 2002a) and may thus be restricted by minimizing the second-order Sobolev norm $\| \bullet \|^{\Phi}$ given by a completely symmetric tensor of coefficients (KLIMEŠ, 2000), with $u^{\Phi}(\mathbf{x}) = 0$. The second-order travel-time perturbation depends on the perturbation of the first velocity derivatives (FARRA, 1999; ČERVENÝ, 2001; KLIMEŠ, 2002d). It may thus be useful to limit the velocity model updates within individual iterations of the linearized inversion by means of the isotropic first-order Sobolev norm $\| \bullet \|^{\Phi}$, with $u^{\Phi}(\mathbf{x})$ equal to the initial model of the respective iteration of the linearized inversion.

In order to improve the efficiency and accuracy of numerical forward-modeling algorithms to be applied in the resulting model, it may thus be desirable to constrain the spatial slowness variations by means of adding squared Sobolev norm $(\|\tilde{u} - u^{\Phi}\|^{\Phi})^2$, quantifying the roughness of the resulting slowness distribution $\tilde{u}(\mathbf{x})$, to the objective function y minimizing the squared standard deviation $\tilde{C}(\mathbf{x}, \mathbf{x})$ integrated over the model volume.

Unfortunately, the addition of $((\tilde{u} - u^{\Phi}, \tilde{u} - u^{\Phi}))^{\Phi}$ would violate the assumption (15) of the linear dependence of the model on travel times. Since the squared standard deviation $\tilde{C}(\mathbf{x}, \mathbf{x})$ between the medium and its model is minimized, instead of minimizing the difference corresponding to a particular representation, we should treat the Sobolev norm consistently and minimize the mean value $\langle ((\tilde{u} - u^{\Phi}, \tilde{u} - u^{\Phi}))^{\Phi} \rangle$ instead of $((\tilde{u} - u^{\Phi}, \tilde{u} - u^{\Phi}))^{\Phi}$.

3.2. The Objective Function

The objective function is thus assumed in the form

$$ y = \int \tilde{C}(\mathbf{x}, \mathbf{x}) d\mathbf{x} + \sum_{\Phi} \langle ((\tilde{u} - u^{\Phi}, \tilde{u} - u^{\Phi}))^{\Phi} \rangle , \qquad (39) $$

where $((\bullet, \bullet))^\Phi$ are zero to several Sobolev scalar products, quantifying the roughness of the resulting slowness distribution. Inserting (35) and (18) with (22) into (39), we arrive at

$$y = \iiint D(\mathbf{x}, \mathbf{x}_1)D(\mathbf{x}, \mathbf{x}_2)C(\mathbf{x}_1, \mathbf{x}_2)d\mathbf{x}_1\,d\mathbf{x}_2\,d\mathbf{x}$$

$$- 2R_{iI} \iint B_i(\mathbf{x}_1)D(\mathbf{x}_1, \mathbf{x}_2)\Psi_I(\mathbf{x}_2)d\mathbf{x}_1\,d\mathbf{x}_2 + B_{ij}R_{iI}R_{jJ}S_{IJ}$$

$$+ B_{ij}R_i^0 R_j^0 - 2R_i^0 \iint B_i(\mathbf{x}_1)D(\mathbf{x}_1, \mathbf{x}_2)u^0(\mathbf{x}_2)d\mathbf{x}_1\,d\mathbf{x}_2$$

$$+ \iiint D(\mathbf{x}, \mathbf{x}_1)D(\mathbf{x}, \mathbf{x}_2)u^0(\mathbf{x}_1)u^0(\mathbf{x}_2)d\mathbf{x}_1\,d\mathbf{x}_2\,d\mathbf{x}$$

$$+ \sum_\Phi ((B_i, B_j))^\Phi \langle [R_i^0 - u_i^\Phi + R_{iI}(T_I - T_I^0)][R_j^0 - u_j^\Phi + R_{jJ}(T_J - T_J^0)]\rangle\ , \quad (40)$$

where

$$\boxed{B_{ij} = \int B_i(\mathbf{x})B_j(\mathbf{x})d\mathbf{x} \qquad (41)}$$

is the L_2 scalar product of the basis functions, and $((B_i, B_j))^\Phi$ are the Sobolev scalar products of basis functions $B_i(\mathbf{x})$ and $B_j(\mathbf{x})$.

Considering (14) and (34), equation (40) reads

$$y = \iiint D(\mathbf{x}, \mathbf{x}_1)D(\mathbf{x}, \mathbf{x}_2)C(\mathbf{x}_1, \mathbf{x}_2)d\mathbf{x}_1\,d\mathbf{x}_2\,d\mathbf{x} - 2\bar{\Psi}_{iI}B_{ij}R_{jI}$$

$$+ B_{ij}[R_{iI}R_{jJ}S_{IJ} + (R_i^0 - \bar{u}_i^0)(R_j^0 - \bar{u}_j^0)] - \bar{u}_i^0 B_{ij}\bar{u}_j^0$$

$$+ \sum_\Phi ((B_i, B_j))^\Phi [R_{iI}R_{jJ}S_{IJ} + (R_i^0 - u_i^\Phi)(R_j^0 - u_j^\Phi)]$$

$$+ \iiint D(\mathbf{x}, \mathbf{x}_1)D(\mathbf{x}, \mathbf{x}_2)u^0(\mathbf{x}_1)u^0(\mathbf{x}_2)d\mathbf{x}_1\,d\mathbf{x}_2\,d\mathbf{x}\ , \qquad (42)$$

where components $\bar{u}_i^0$ of the Lebesgue L_2 projections of the smoothed mean slowness onto the basis functions are defined by (27), and

$$\boxed{\begin{aligned}\bar{\Psi}_{iJ} &= B_{ik}^{-1} \iint B_k(\mathbf{x}_1)D(\mathbf{x}_1, \mathbf{x}_2)\Psi_J(\mathbf{x}_2)d\mathbf{x}_1\,d\mathbf{x}_2 \\ &= B_{ik}^{-1} \iiint B_k(\mathbf{x}_1)\,D(\mathbf{x}_1, \mathbf{x}_1')\,C(\mathbf{x}_1', \mathbf{x}_2)\,G_J(\mathbf{x}_2)d\mathbf{x}_1\,d\mathbf{x}_1'\,d\mathbf{x}_2 \qquad (43)\end{aligned}}$$

are the components of the Lebesgue L_2 projections of the smoothed ray back-projective functions (covariances between the smoothed slowness and travel times) onto the basis functions.

3.3. The Minimum of the Objective Function with Respect to the Resolving Kernel

Differentiating (42), under constraints (24), with respect to R_i^0, we obtain the minimum objective function for

$$B_{ij}[R_j^0 - \bar{u}_j^0] + \sum_\Phi ((B_i, B_j))^\Phi [R_j^0 - u_j^\Phi] = \Lambda_i^\alpha \Delta^\alpha , \tag{44}$$

where Λ_i^α are Lagrange multipliers corresponding to optional constraints (24). Note that the Einstein summation over repetitive indices applies also to superscripts $^\alpha$. The coefficients of the inverted mean value $u^0(\mathbf{x})$ thus have the form

$$R_i^0 = \left[B_{ij} + \sum_\Phi ((B_i, B_j))^\Phi \right]^{-1} \left[B_{jk}\bar{u}_k^0 + \sum_\Phi ((B_i, B_j))^\Phi u_j^\Phi + \Lambda_i^\alpha \Delta^\alpha \right] , \tag{45}$$

where $[B_{ji} + \sum_\Phi ((B_j B_i))^\Phi]^{-1}$ are the components of the matrix inverse to matrix $[B_{ij} + \sum_\Phi ((B_i B_j))^\Phi]$.

Differentiating (42), under constraints (24), with respect to R_{iI}, we see that the minimum objective function is obtained for

$$\left[B_{ij} + \sum_\Phi ((B_i, B_j))^\Phi \right] R_{jJ} S_{JI} - B_{ik}\bar{\Psi}_{kI} = \Lambda_i^\alpha T_I^\alpha \tag{46}$$

with the same Lagrange multipliers Λ_i^α as in (44). The coefficients of the resolving kernel thus have the form

$$R_{jJ} = \left[B_{ji} + \sum_\Phi ((B_j B_i))^\Phi \right]^{-1} [B_{ik}\bar{\Psi}_{kI} + \Lambda_i^\alpha T_I^\alpha] S_{IJ}^{-1} , \tag{47}$$

where S_{NM}^{-1} are the components of the matrix inverse to the complete travel-time covariance matrix S_{MN} defined by (34).

Without constraints (24), there are no Lagrange multipliers Λ_i^α in equations (45) and (47). The model coefficients then depend on travel times through equation (22) with

$$R_i^0 = \left[B_{ij} + \sum_\Phi ((B_i, B_j))^\Phi \right]^{-1} \left[B_{jk}\bar{u}_k^0 + \sum_\Phi ((B_i, B_j))^\Phi u_j^\Phi \right] \tag{48}$$

and

$$R_{jJ} = \left[B_{ji} + \sum_\Phi ((B_j B_i))^\Phi \right]^{-1} B_{ik}\bar{\Psi}_{kI} S_{IJ}^{-1} . \tag{49}$$

In this case of no constraints (24), Section 3.4 may be skipped, continuing with Section 3.5.

3.4. Constraints Adjusting the Mean Slowness

Let us now discuss the meaning of constraints (24). Inserting (45) and (47) into constraints (24), we obtain equations

$$
\left[B_{ij} + \sum_\Phi ((B_i, B_j))^\Phi \right]^{-1}
$$

$$
\times \left[\left(B_{jk} \bar{u}_k^0 + \sum_\Phi ((B_i, B_j))^\Phi u_j^\Phi + \Lambda_i^\alpha \Delta^\alpha \right) \Delta^\beta + \left(B_{ik} \bar{\Psi}_{kI} + \Lambda_i^\alpha T_I^\alpha \right) S_{IJ}^{-1} T_J^\beta \right]
$$

$$
= \Delta^\alpha \bar{u}_i^0 + \bar{u}_i^\alpha \tag{50}
$$

for Lagrange multipliers Λ_i^α. The Lagrange multipliers are then

$$
\Lambda_i^\alpha = \left[T_M^\alpha S_{MN}^{-1} T_N^\beta + \Delta^\alpha \Delta^\beta \right]^{-1}
$$

$$
\times \left[B_{ij} \bar{u}_j^\beta + \sum_\Phi ((B_i, B_j))^\Phi \left[\bar{u}_j^\beta + (\bar{u}_j^0 - u_j^\Phi) \Delta^\beta \right] - B_{ij} \bar{\Psi}_{jJ} S_{JK}^{-1} T_K^\beta \right] . \tag{51}
$$

Inserting (45) and (47) with Lagrange multipliers (51) into (22) and subtracting (22) with (48) and (49), we see that the increment in u_i due to constraints (24) is

$$
\Delta u_i = \left[B_{ij} \bar{u}_j^\beta + \sum_\Phi ((B_i, B_j))^\Phi \left[\bar{u}_j^\beta + (\bar{u}_j^0 - u_j^\Phi) \Delta^\beta \right] - B_{ij} \bar{\Psi}_{jJ} S_{JK}^{-1} T_K^\beta \right] U^\beta \tag{52}
$$

with

$$
U^\beta = \left[T_M^\beta S_{MN}^{-1} T_N^\alpha + \Delta^\beta \Delta^\alpha \right]^{-1} \left[T_P^\alpha S_{PR}^{-1} (T_R - T_R^0) + \Delta^\alpha \right] . \tag{53}
$$

The same increment in u_i can be achieved by updating the mean medium to

$$
u^{0\text{new}}(\mathbf{x}) = u^0(\mathbf{x}) + u^\alpha(\mathbf{x}) U^\alpha , \tag{54}
$$

replacing functions (37) by

$$
u_i^{\Phi\text{new}} = u_i^\Phi + [\bar{u}_i^\alpha + (\bar{u}_i^0 - u_i^\Phi) \Delta^\alpha] U^\alpha , \tag{55}
$$

and then performing inversion (22) without constraints (24). Note that coefficients (53) can be obtained by minimizing objective function

$$
y^{(U)} = [T_I - T_I^0 - T_I^\alpha U^\alpha] S_{IJ}^{-1} [T_J - T_J^0 - T_I^\beta U^\beta] + [U^\alpha \Delta^\alpha - 1]^2 , \tag{56}
$$

where the perturbed reference travel times $T_J^{0\text{new}} = T_J^0 + T_J^\alpha U^\alpha$ coincide with the reference travel times (13) calculated in the perturbed mean medium (54).

The first term of addition in objective function (56) corresponds to constraints (24) with $\Delta^\alpha = 0$, and its interpretation is straightforward. It represents the least-square fitting of given travel times T_I by perturbed mean medium (54). Conversely, the second term in objective function (56), present if at least one Δ^α is non-zero, is far more difficult to interpret. The second term tends to keep smoothed perturbation (54) of the mean medium close to the linear subspace generated by the right-hand sides of constraints (24). It then seems natural to choose $\Delta^\alpha = 0$ when constraints (24) are applied.

The impact of the application of constraints (24) on model $\tilde{u}(x)$ resulting from the inversion is equivalent to adjusting the *a priori* mean slowness $u^0(\mathbf{x})$ through a linear combination (54) of functions $u^\alpha(\mathbf{x})$, fitted by least squares according to equation (53) prior to the inversion, and then performing inversion (22) with perturbed mean slowness (54) but without constraints (24). The least-square fitting (56) should be very strongly overdetermined. We may consider this procedure as a part of the determination of the statistical moments $u^0(\mathbf{x})$ and $C(\mathbf{x}, \mathbf{x}')$ of the medium and assume no constraints (24) during the travel-time inversion.

For example, imagine that we have no idea about the mean slowness and thus choose $u^0(\mathbf{x}) = 0$, but we wish the inversion to be exact in a homogeneous space. We may thus introduce one constraint (24) with $u^1(\mathbf{x}) = 1$ and $\Delta^1 = 0$. Inversion with such a constraint yields the same model as the determination of $u^0(\mathbf{x}) = constant$ by means of least squares (54) and its use in the inversion without constraints.

We shall not discuss the impact of constraints (24) on the model covariance matrix (35) in this paper.

3.5. Discussion

Let us now decompose equation (22) with coefficients (48) and (49) into several steps to gain a better insight into the proposed travel-time tomographic inversion.

The travel-time differences with respect to the reference travel times in mean slowness are back-projected onto the model by the operator

$$[T_I - T_I^0] \rightarrow f(\bullet) = u^0(\bullet) + r_J(\bullet)[T_J - T_J^0] \ , \tag{57}$$

where

$$r_J(\mathbf{x}) = \Psi_K(\mathbf{x}) S_{KJ}^{-1} \ , \tag{58}$$

with ray back-projective functions $\Psi_K(\mathbf{x})$ given by (32). Considering either exact ray characteristic functions or their approximations from the last iteration of the linearized inversion, equation (57) is identical to equation (7.127) of Section 7.5.2 [or equation (18) of Problem 7.1] by TARANTOLA (1987). The back-projection by means of functions $\Psi_K(\mathbf{x})$ is not only statistically correct, but is also much smoother than the back-projection by means of the ray characteristic functions $G_K(\mathbf{x})$, especially if the ray characteristic functions are narrow in comparison with the distances between two-point rays. The application of the medium covariance functions has thus a significant smoothing effect.

The resulting back-projection of the weighted travel-time differences is then optionally smoothed with the smoothing function,

$$f(\bullet) \rightarrow \bar{f}(\bullet) = \int D(\bullet, \mathbf{x}) f(\mathbf{x}) \mathrm{d}\mathbf{x} \ . \tag{59}$$

There is no reason to apply this smoothing if the ray characteristic functions $G_K(\mathbf{x})$ correctly account for the sensitivity of travel times to the medium at finite frequencies.

The optionally smoothed back-projection is then projected onto the basis functions

$$\bar{f}(\bullet) \rightarrow B_i(\bullet) \, \bar{f}_i = B_i(\bullet) \, B_{ij}^{-1} \int B_j(\mathbf{x}) \, \bar{f}(\mathbf{x}) \mathrm{d}\mathbf{x} \ . \tag{60}$$

This projection should modify the velocity model as little as possible. The projection onto the basis functions should never substitute smoothing by means of the Sobolev scalar products.

The projection onto the basis functions is then smoothed with the optional Sobolev scalar products,

$$B_i(\bullet) \, \bar{f}_i \rightarrow \tilde{u}(\bullet) = B_i(\bullet) \left[B_{ij} + \sum_{\Phi} ((B_i, B_j))^{\Phi} \right]^{-1} \left[B_{jk} \bar{f}_k + \sum_{\Phi} ((B_j, B_k))^{\Phi} u_k^{\Phi} \right] \ . \tag{61}$$

Step (57) is independent of the *a priori* information contained within the smoothing function $D(\mathbf{x}_1, \mathbf{x}_2)$, model parameterization, and smoothing expressed in terms of the Sobolev scalar products $((\bullet, \bullet))^{\Phi}$. These subjective *a priori* parts of information are applied step-by-step *a posteriori* as expressed in equations (59), (60) and (61).

Equation (61) shows that the *a priori* additional smoothing term $\sum_{\Phi} \langle ((\tilde{u} - u^{\Phi}, \tilde{u} - u^{\Phi}))^{\Phi} \rangle$ to the objective function (39) acts just as a convolutional low-pass filter applied *a posteriori* to the result of the tomographic travel-time inversion performed without the Sobolev scalar products. It is thus possible to perform the inversion without the Sobolev scalar products. *A posteriori* smoothing is then at the discretion of the seismologist.

4. Special Cases

4.1. Special Choice of the Smoothing Function

We discuss here the very important special case of the smoothing function $D(\mathbf{x}_1, \mathbf{x}_2)$ coinciding with the integral kernel of the projection operator onto the model subspace generated by the basis functions,

$$D(\mathbf{x}_1, \mathbf{x}_2) = B_i(\mathbf{x}_1)B_{ij}^{-1}B_j(\mathbf{x}_1) \ . \tag{62}$$

Equation (43) then simplifies to

$$\bar{\Psi}_{iJ} = B_{ij}^{-1} \iint B_j(\mathbf{x}_1)C(\mathbf{x}_1, \mathbf{x}_2)G_J(\mathbf{x}_2)\mathrm{d}\mathbf{x}_1\,\mathrm{d}\mathbf{x}_2 \ . \tag{63}$$

Note that the same applies if $D(\mathbf{x}_1, \mathbf{x}_2)$ takes the form of the Dirac distribution $\delta(\mathbf{x}_1 - \mathbf{x}_2)$ or, more generally, if $D(\mathbf{x}_1, \mathbf{x}_2)$ coincides with the Dirac distribution on the model subspace.

In a self-affine random medium, the squared standard deviation $\tilde{C}(\mathbf{x}, \mathbf{x})$ of the resulting slowness $\tilde{u}(\mathbf{x})$ may be very large for very narrow smoothing function $D(\mathbf{x}_1, \mathbf{x}_2)$, and may go to infinity for $D(\mathbf{x}_1, \mathbf{x}_2)$ approaching the Dirac distribution $\delta(\mathbf{x}_1 - \mathbf{x}_2)$. On the other hand, for the smoothing function in the form of (62), the model covariance function (35) takes the form

$$\tilde{C}(\mathbf{x}_1, \mathbf{x}_2) = B_i(\mathbf{x}_1)\tilde{C}_{ij}B_j(\mathbf{x}_1) \tag{64}$$

with

$$\tilde{C}_{ij} = C_{ij} - R_{iK}\bar{\Psi}_{jK} - \bar{\Psi}_{iK}R_{jK} + R_{iI}S_{IJ}R_{jJ} + [R_i^0 - \bar{u}_i^0][R_j^0 - \bar{u}_j^0] \ , \tag{65}$$

where

$$C_{ij} = B_{ik}^{-1}B_{jl}^{-1} \iint B_k(\mathbf{x}_1)C(\mathbf{x}_1, \mathbf{x}_2)B_l(\mathbf{x}_2)\,\mathrm{d}\mathbf{x}_1\,\mathrm{d}\mathbf{x}_2 \tag{66}$$

are the components of the projection

$$C^{(BB)}(\mathbf{x}_1, \mathbf{x}_2) = B_i(\mathbf{x}_1)C_{ij}B_j(\mathbf{x}_2) \tag{67}$$

of the medium covariance function $C(\mathbf{x}_1, \mathbf{x}_2)$ onto the model space generated by the basis functions.

4.2. Special Choice of the Ray Characteristic Functions

Assume here that the ray characteristic functions $G_I(\mathbf{x})$ are replaced by their projections $G_I^{(B)}(\mathbf{x})$ onto the model subspace generated by basis functions $B_i(\mathbf{x})$,

$$G_J^{(B)}(\mathbf{x}) = B_i(\mathbf{x})B_{ik}^{-1} \int B_k(\mathbf{x})G_J(\mathbf{x})d\mathbf{x} = B_i(\mathbf{x})B_{ik}^{-1}G_{kJ} \ , \tag{68}$$

where

$$G_{iJ} = \int B_i(\mathbf{x})G_J(\mathbf{x})d\mathbf{x} \tag{69}$$

is the variation of the Jth synthetic travel time with respect to the ith model slowness parameter. This simplification may be justified, e.g., if the model parameterization is so fine that the "thick rays" (68) are mostly situated within the Fresnel volumes corresponding to frequencies significant for travel-time picking.

Integral (33) may then be expressed in terms of integrals (66) and (69),

$$\Theta_{IJ} = G_{iI}C_{ij}G_{jJ} \ , \tag{70}$$

and integration along rays in integral (43) may be replaced with the projection onto the basis functions,

$$\bar{\Psi}_{iJ} = B_{ij}^{-1} \iiint B_j(\mathbf{x}_1)D(\mathbf{x}_1,\mathbf{x}_1')C(\mathbf{x}_1',\mathbf{x}_2)B_l(\mathbf{x}_2)d\mathbf{x}_1\,d\mathbf{x}_1'\,d\mathbf{x}_2\,B_{lj}^{-1}G_{jJ} \ . \tag{71}$$

Note that (70) and (71) are also valid for general $G_{iI}(\mathbf{x})$ if $C(\mathbf{x}_1,\mathbf{x}_2)$ is replaced by $C^{(BB)}(\mathbf{x}_1,\mathbf{x}_2)$, see (67).

4.3. Special Choice of the Smoothing Function and Ray Characteristic Functions

In the special case of both options (62) and (68), equation (70) may be used to calculate integral (33). In addition, integral (43) takes the simple form

$$\bar{\Psi}_{iJ} = C_{ij}G_{jJ} \ , \tag{72}$$

compared with (63) and (71). In consequence, equation (65) simplifies to

$$\tilde{C}_{ij} = [\delta_{ik} - R_{il}G_{kl}]C_{kl}[\delta_{jl} - R_{jJ}G_{lJ}] + R_{iI}T_{IJ}R_{jJ} + [R_i^0 - \bar{u}_i^0][R_j^0 - \bar{u}_j^0] \ . \tag{73}$$

5. Some Consequences

5.1. Residual Travel-time Covariance Matrix

The covariance matrix of the travel-time residuals is

$$\tilde{S}_{IJ} = \langle (\tilde{T}_I - T_I)(\tilde{T}_J - T_J) \rangle \ , \tag{74}$$

where

$$\tilde{T}_I = \int G_I(\mathbf{x})\tilde{u}(\mathbf{x})\mathrm{d}\mathbf{x} \tag{75}$$

and T_I are given by (5). Inserting (5) and (75), equation (74) reads

$$\tilde{S}_{IJ} = \iint G_I(\mathbf{x}_1)G_J(\mathbf{x}_1)\langle[\tilde{u}(\mathbf{x}_1) - u(\mathbf{x}_1)][\tilde{u}(\mathbf{x}_2) - u(\mathbf{x}_2)]\rangle\mathrm{d}\mathbf{x}_1\mathrm{d}\mathbf{x}_2$$
$$- \int G_I(\mathbf{x})\langle\tilde{u}(\mathbf{x})\delta T_J\rangle\mathrm{d}\mathbf{x} - \int\langle\delta T_I\tilde{u}(\mathbf{x})\rangle G_J(\mathbf{x})\mathrm{d}\mathbf{x} + \langle\delta T_I\ \delta T_J\rangle \ . \tag{76}$$

Here $\langle[\tilde{u}(\mathbf{x}_1) - u(\mathbf{x}_1)][\tilde{u}(\mathbf{x}_2) - u(\mathbf{x}_2)]\rangle$ coincides with (28) for $\bar{u}(\mathbf{x}) = u(\mathbf{x})$, i.e., for smoothing function $D(\mathbf{x}_1, \mathbf{x}_2)$ in the form of the Dirac distribution $\delta(\mathbf{x}_1 - \mathbf{x}_2)$ (no smoothing). Inserting (35) with $D(\mathbf{x}_1, \mathbf{x}_2) = \delta(\mathbf{x}_1 - \mathbf{x}_2)$ for $\langle[\tilde{u}(\mathbf{x}_1) - u(\mathbf{x}_1)][\tilde{u}(\mathbf{x}_2) - u(\mathbf{x}_2)]\rangle$, and

$$\langle\tilde{u}(\mathbf{x})\delta T_J\rangle = B_i(\mathbf{x})R_{iK}\langle\delta T_K\ \delta T_J\rangle = B_i(\mathbf{x})R_{iK}T_{KJ} \tag{77}$$

into (76), we arrive at

$$\tilde{S}_{IJ} = \Theta_{IJ} - G_{iI}R_{iK}\Theta_{KJ} - \Theta_{IL}R_{jL}G_{jJ} + G_{iI}R_{iK}S_{KL}R_{jL}G_{jJ}$$
$$+ [R_i^0 G_{iI} - T_I^0][R_j^0 G_{jJ} - T_J^0] - G_{iI}R_{iK}T_{KJ} - T_{IK}R_{jK}G_{jJ} + T_{IJ} \ . \tag{78}$$

Thus the statistical expectation of the travel-time residuals is described by the covariance matrix

$$\tilde{S}_{IJ} = [\delta_{IK} - G_{iI}R_{iK}]S_{KL}[\delta_{LJ} - R_{jL}G_{jJ}] + [R_i^0 G_{iI} - T_I^0][R_j^0 G_{jJ} - T_J^0] \ , \tag{79}$$

where S_{KL} is given by (34).

6. Conclusions

In this paper, we have discussed the approach to seismic travel-time tomography enabling the accuracy and resolution of the resulting model of a geological structure (macro model) to be estimated. The medium covariance functions yield correct travel-time weights for uneven ray coverage of the medium and provide statistically correct interpolation of slowness between two-point rays.

The presented approach already enables us to consider *a priori* information such as the obviously much greater probability of slow materials close to the earth's surface than deeper in the earth, or greater probability of slow materials in the low and concave parts of the earth's periphery than in the high-elevated and convex parts.

On the other hand, we have still made various simplifications. We have considered only smooth models without structural interfaces, and have restricted the study to a single material parameter – slowness. However, we believe that this approach to seismic travel-time tomography may further be generalized.

Acknowledgements

The author is indebted to Sergei Shapiro for fruitful discussions essential for this study.

This research has been supported by the Grant Agency of the Czech Republic under Contracts 205/95/1465, 205/01/0927 and 205/01/D097, by the Grant Agency of the Charles University under Contract 237/2001/B-GEO/MFF, by the Ministry of Education of the Czech Republic within Research Project J13/98 113200004, and by the members of the consortium "Seismic Waves in Complex 3-D Structures" (see "http://sw3d.mff.cuni.cz").

REFERENCES

BACKUS, G.E., and GILBERT, F. (1970), *Uniqueness in the Inversion of Inaccurate Gross Earth Data*, Phil. Trans. R. Soc. London *266A*, 123–192.

BULANT, P. (1996), *Two-point Ray Tracing in 3-D*, Pure appl. geophys. *148*, 421–447.

BULANT, P. (1999), *Two-point Ray Tracing and Controlled Initial-value Ray Tracing in 3-D Heterogeneous Block Structures*, J. seism. Explor. *8*, 57–75.

BULANT, P. (2002), *Sobolev Scalar Products in the Construction of Velocity Models – Application to Model Hess and to SEG/EAGE Salt Model*, Pure appl. geophys. *159*, 1487–1506.

ČERVENÝ, V. (2001), *Seismic Ray Theory* (Cambridge Univ. Press, Cambridge).

CHERNOV, L.A. (1960), *Wave Propagation in a Random Medium* (McGraw-Hill, New York).

COATES, R.T., and CHAPMAN, C.H. (1990), *Ray Perturbation Theory and the Born Approximation*, Geophys. J. int. *100*, 379–392.

FARRA, V. (1999), *Computation of 2nd-order Travel-time Perturbation by Hamiltonian Ray Theory*, Geophys. J. int. *136*, 205–217.

FRANKLIN, J.N. (1970), *Well-posed Stochastic Extensions of Ill-posed Linear Problems*, J. math. Anal. Appl. *31*, 682–716.

KLIMEŠ, L. (2000), *Sobolev scalar products in the construction of velocity models*. In *Seismic Waves in Complex 3-D Structures, Report 10* (Dep. Geophys. Charles Univ., Prague) pp. 15–40 (online at "http://sw3d.mff.cuni.cz").

KLIMEŠ, L. (2002a), *Lyapunov Exponents for 2-D Ray Tracing without Interfaces*, Pure appl. geophys. *159*, 1465–1485.

KLIMEŠ, L. (2002b), *Correlation Functions of Random Media*, Pure appl. geophys. *159*, 1811–1831.

KLIMEŠ, L. (2002c), *Estimating the Correlation Function of a Self-affine Random Medium*, Pure appl. geophys. *159*, 1833–1853.

KLIMEŠ, L. (2002d), *Second-order and Higher-order Perturbations of Travel Time in Isotropic and Anisotropic Media*, Stud. geophys. geod. *46*, 213–248.

MARQUERING, H., NOLET, G., and DAHLEN, F.A. (1998), *Three-dimensional Waveform Sensitivity Kernels*, Geophys. J. int. *132*, 521–534.

MAURER, H., HOLLIGER, K., and BOERNER, D.E. (1998), *Stochastic Regularization: Smoothness or Similarity?*, Geophys. Res. Lett. *25*, 2889–2892.

MENKE, W. (1984), *Geophysical Data Analysis: Discrete Inverse Theory* (Academic Press, Orlando).

PILKINGTON, M., and TODOESCHUCK, J.P. (1990), *Stochastic Inversion for Scaling Geology*, Geophys. J. int. *102*, 205–217.

PRATT, R.G., SONG, Z.-M., WILLIAMSON, P., and WARNER, M. (1996), *Two-dimensional Velocity Models from Wide-angle Seismic Data by Wavefield Inversion*, Geophys. J. int. *124*, 323–340.

TARANTOLA, A. (1987), *Inverse Problem Theory* (Elsevier, Amsterdam).

TARANTOLA, A., and NERCESSIAN, A. (1984), *Three-dimensional Inversion without Blocks*, Geophys. J. R. astr. Soc. *76*, 299–306.

TARANTOLA, A., and VALETTE, B. (1982), *Generalized Nonlinear Inverse Problems Solved Using the Least Squares Criterion*, Rev. Geophys. Space Phys. *20*, 219–232.

WOODWARD, M.J. (1992), *Wave-equation Tomography*, Geophysics *57*, 15–26.

WOODWARD, M.J., and ROCCA, F. (1988), *Wave-equation tomography*. In *Expanded Abstracts of 58th Annual Meeting (Anaheim)* (Soc. Explor. Geophysicists, Tulsa) pp. 1232–1235.

ŽÁČEK, K. (2002), *Smoothing the Marmousi Model*, Pure appl. geophys. *159*, 1507–1526.

(Received October 10, 2000, revised April 2, 2001, accepted May 17, 2001)

 To access this journal online:
http://www.birkhauser.ch

Pure appl. geophys. 159 (2002) 1811–1831
0033–4553/02/081811–21 $ 1.50 + 0.20/0

▌Pure and Applied Geophysics

Correlation Functions of Random Media

LUDĚK KLIMEŠ[1]

Summary — In geophysics, the correlation functions of random media are of principal importance for understanding and inverting the properties of seismic waves propagating in geological structures. Unfortunately, the kinds of correlation functions inappropriate for the description of geological structures are often assumed and applied. The most frequently used types of correlation functions are thus summarized and reviewed in this paper, together with an explanation of the physical meaning of their parameters.

A stationary random medium is assumed to be realized in terms of a white noise filtered by a spectral filter. The spectral filter is considered isotropic, in a simple general form enabling the random media used in geophysics to be specified. The medium correlation functions, corresponding to the individual special cases of the general random medium (Gaussian, exponential, von Kármán, self-affine, Kummer), are then derived and briefly discussed. The corresponding elliptically anisotropic correlation functions can simply be obtained by linear coordinate transforms.

Key words: Random medium, covariance function, correlation function, scaling geology, fractal geology.

1. Introduction

In order to estimate the relation of the seismic model to the geological structure and, in consequence, to estimate the relation of the synthetic quantities, calculated in the seismic model, to reality, it is important to obtain an estimate of the medium covariance function. On large scales, the medium covariance function is of principal importance in refraction travel-time tomographic inversion (TARANTOLA, 1987; MAURER et al., 1998; KLIMEŠ, 2002a), especially when estimating the accuracy of the seismic model, its relation to the geological structure, or the covariance matrix describing the statistics of synthetic travel times. On smaller scales, correlation functions make it possible to estimate the scattering attenuation (e.g., CHERNOV, 1960; WU, 1982) or to approximately upscale a rough distribution of material parameters and estimate the relation to a smoother distribution of effective material parameters (e.g., SHAPIRO and KNEIB, 1993; SHAPIRO et al., 1996; MÜLLER and SHAPIRO, 2001).

[1] Department of Geophysics, Charles University, Ke Karlovu 3, 121 16 Praha 2, Czech Republic.
E-mail: klimes@seis.karlov.mff.cuni.cz

In contrast to the importance of the correlation functions, the kinds of correlation functions inappropriate for the description of geological structures are applied in too many geophysical papers, whereas the kinds of correlation functions suitable for the description are applied rarely in the geophysical literature. For example, the physically reasonable values of the Hurst exponent N, describing the scaling properties of geological structures, are $-1/2 < N < 0$, whereas the two values most popular in the geophysical literature are $N = -3/2$ and $N = 1/2$. The author thus has come to the conclusion that it is reasonable to briefly summarize the most frequently used types of correlation functions, together with an explanation of the physical meaning of their parameters in geophysical applications.

We assume each material parameter of the geological structure to be composed of a heterogeneous mean value and a particular representation of the stationary (statistically homogeneous) random medium in this paper. The stationary random medium is assumed to be realized in terms of a white noise filtered by a spectral filter. In this way, the medium covariance function may be expressed in terms of the correlation function.

The aim of this paper is to summarize systematically and discuss briefly the types of random media used in geophysics. The generic spectral filter is considered isotropic, in a simple general form described by four constants, enabling the random media used in geophysics to be specified. The medium correlation functions, corresponding to the individual special cases of the general random medium, are then derived and briefly discussed. The corresponding elliptically anisotropic correlation functions can simply be obtained from their generic isotropic counterparts by linear coordinate transforms, see Section 3.2.

2. Correlation Functions of Material Parameters

The material parameters of the *geological structure* are assumed to be realizations of random media with known statistical characteristics. This approach enables us to account satisfactorily for the differences between the geological structure and the *seismic model* of the structure, based on our quite incomplete information regarding the structure. The statistical characteristics represent our knowledge of the deterministic features of the structure and may change with more information regarding the structure. The randomness corresponds to the unknown details in the geological structure and may decrease with more information regarding the structure.

2.1. Material parameters

The geological structure is assumed to be expressed in terms of spatial distribution $u = u(\mathbf{x})$ of several *material parameters* u. For example, the material parameters may represent the density, P- and S-wave velocities or slownesses, or the elements of the stiffness matrix.

The *microscale heterogeneities* of these parameters (heterogeneities on scales shorter than the wavelength) may influence the propagation of seismic waves of a particular frequency in such a way that the propagation corresponds to some frequency-dependent *effective material parameters* $\bar{u}(\mathbf{x}, \omega)$ "locally homogeneous" on microscales. For the study of *macroscale heterogeneities* (e.g., seismic travel-time tomography) material parameters $u(\mathbf{x})$ may appropriately be replaced with the effective material parameters $\bar{u}(\mathbf{x}, \omega)$. The corresponding seismic model is then often called the *macromodel*, particularly in seismic prospecting. Hereinafter, "material parameter $u(\mathbf{x})$" may represent either any of the material parameters $u(\mathbf{x})$ or any of the effective material parameters $\bar{u}(\mathbf{x}, \omega)$ at a given frequency.

2.2. Mean Value of the Material Parameter

Mean value

$$u_0(\mathbf{x}) = \langle u(\mathbf{x}) \rangle \tag{1}$$

of material parameter $u(\mathbf{x})$ describes the most probable value of the material parameter. The mean value $u_0(\mathbf{x})$ of the material parameter is a very smooth function in most cases.

The unknown difference

$$U(\mathbf{x}) = u(\mathbf{x}) - u_0(\mathbf{x}) \tag{2}$$

between the material parameter and its mean value $u_0(\mathbf{x})$ is considered to be a realization of the random quantity.

2.3. Generalized Correlation Function of the Material Parameter

The unknown difference between the material parameter and its mean value may be characterized in terms of the *medium covariance function*

$$C(\mathbf{x}_1, \mathbf{x}_2) = \langle U^*(\mathbf{x}_1) U(\mathbf{x}_2) \rangle . \tag{3}$$

The asterisk * denoting complex conjugacy is included here for completeness only. We shall consider real-valued material parameters in this paper.

To characterize the properties of the medium covariance function better, we introduce the *generalized correlation function*

$$c^{\mathrm{G}}(\mathbf{y}, \mathbf{x}) = C(\mathbf{y} - \tfrac{1}{2}\mathbf{x}, \mathbf{y} + \tfrac{1}{2}\mathbf{x}) = \langle U^*(\mathbf{y} - \tfrac{1}{2}\mathbf{x}) \ U(\mathbf{y} + \tfrac{1}{2}\mathbf{x}) \rangle . \tag{4}$$

The first argument of $c^{\mathrm{G}}(\mathbf{y}, \mathbf{x})$ is the midpoint of the pair of points $\mathbf{x}_1 = \mathbf{y} - \tfrac{1}{2}\mathbf{x}$ and $\mathbf{x}_2 = \mathbf{y} + \tfrac{1}{2}\mathbf{x}$ in which the correlation of difference U is studied, whereas the second argument is the distance of the points. The generalized correlation function should vary very moderately with respect to the first argument, whereas it may vary very rapidly with respect to the second argument (distance of points), especially in the vicinity of $\mathbf{x} = \mathbf{0}$.

The Fourier transform of the generalized correlation function with respect to the second argument may be introduced, for instance, in the form

$$\widehat{c^G}(\mathbf{y}, \mathbf{k}) = (2\pi)^{-d} \int\limits_{-\infty}^{+\infty} \mathbf{dx} \, \exp(-i\mathbf{k}^T\mathbf{x}) \, c^G(\mathbf{y}, \mathbf{x}) \ , \tag{5}$$

where d is the Euclidean dimension of the space, e.g., $d = 3$ in 3-D. Here we have denoted

$$\int\limits_{-\infty}^{+\infty} \mathbf{dk} = \int\limits_{-\infty}^{+\infty} dk_1 \int\limits_{-\infty}^{+\infty} dk_2 \cdots \int\limits_{-\infty}^{+\infty} dk_d \ . \tag{6}$$

The inverse Fourier transform of (5) is

$$c^G(\mathbf{y}, \mathbf{x}) = \int\limits_{-\infty}^{+\infty} \mathbf{dk} \, \exp(i\mathbf{k}^T\mathbf{x}) \, \widehat{c^G}(\mathbf{y}, \mathbf{k}) \ . \tag{7}$$

Relation to the Wigner distribution
The *Wigner distribution* of function $\varphi(\mathbf{y})$ is

$$\varphi^W(\mathbf{y}, \mathbf{k}) = (2\pi)^{-d} \int\limits_{-\infty}^{+\infty} \mathbf{dx} \, \exp(-i\mathbf{k}^T\mathbf{x}) \, \varphi^*(\mathbf{y} - \tfrac{1}{2}\mathbf{x}) \, \varphi(\mathbf{y} + \tfrac{1}{2}\mathbf{x}) \ . \tag{8}$$

It may also be expressed in terms of the Fourier transform $\hat{\varphi}(\mathbf{p})$ of $\varphi(\mathbf{y})$,

$$\varphi^W(\mathbf{y}, \mathbf{k}) = \int\limits_{-\infty}^{+\infty} \mathbf{dp} \, \exp(-i\mathbf{p}^T\mathbf{y}) \, \hat{\varphi}^*(-\mathbf{k} - \tfrac{1}{2}\mathbf{p}) \, \hat{\varphi}(-\mathbf{k} + \tfrac{1}{2}\mathbf{p}) \ , \tag{9}$$

see, e.g., BASTIAANS (1981) or RYZHIK *et al.* (1996).

The Fourier transform (5) of the generalized correlation function is thus the mean value of the Wigner distribution of the unknown difference between the material parameter and its mean value,

$$\widehat{c^G}(\mathbf{y}, \mathbf{k}) = \langle U^W(\mathbf{y}, \mathbf{k}) \rangle \ , \tag{10}$$

and the generalized correlation function is its inverse Fourier transform,

$$c^G(\mathbf{y}, \mathbf{x}) = \int\limits_{-\infty}^{+\infty} \mathbf{dk} \, \exp(i\mathbf{k}^T\mathbf{x}) \langle U^W(\mathbf{y}, \mathbf{k}) \rangle \ . \tag{11}$$

2.4. Correlation Function of the Stationary Random Medium

Random medium $u(\mathbf{x})$ is *stationary* if mean value $u_0(\mathbf{x})$ is constant and generalized correlation function $c^G(\mathbf{y}, \mathbf{x})$ is independent of the first argument, depending only on distance $\mathbf{x} = \mathbf{x}_1 - \mathbf{x}_2$,

$$c^G(\mathbf{y}, \mathbf{x}) = c(\mathbf{x}) \ , \tag{12}$$

where $c(\mathbf{x})$ is the *correlation function* of the stationary random medium (TARANTOLA, 1987). Note that we do not assume mean value $u_0(\mathbf{x})$ to be constant in this paper.

2.5. Synthetic realizations of a stationary random medium

(a) On a regular rectangular grid, we generate a pseudorandom realization $W(\mathbf{x})$ of the *white noise* of the desired statistical distribution of the functional values, with the *unit standard deviation*.

(b) Calculate the d-dimensional Fourier transform $\hat{W}(\mathbf{k})$ of the white noise.

(c) Multiply the Fourier transform with the *spectral filter* $\hat{F}(\mathbf{k})$,

$$\hat{U}(\mathbf{k}) = \hat{F}(\mathbf{k})\hat{W}(\mathbf{k}) \ . \tag{13}$$

(d) Calculate the d-dimensional inverse Fourier transform $U(\mathbf{x})$ of product $\hat{U}(\mathbf{k})$. The multiplication factors of the forward and inverse Fourier transform may, of course, be different from equations (5) and (7).

(e) Add the appropriate mean value $u_0(\mathbf{x})$ to arrive at the desired pseudorandom realization,

$$u(\mathbf{x}) = u_0(\mathbf{x}) + U(\mathbf{x}) \ . \tag{14}$$

The Fourier transform of the correlation function is then

$$\hat{c}(\mathbf{k}) = (2\pi)^{-d} \hat{F}^*(\mathbf{k}) \hat{F}(\mathbf{k}) \ , \tag{15}$$

where $\hat{F}^*(\mathbf{k})\hat{F}(\mathbf{k})$ is the power spectrum of the filter. The corresponding correlation function is

$$c(\mathbf{x}) = (2\pi)^{-d} \int_{-\infty}^{+\infty} d\mathbf{k} \ \exp(i\mathbf{k}^T\mathbf{x})\hat{F}^*(\mathbf{k})\hat{F}(\mathbf{k}) \ . \tag{16}$$

3. Isotropic and Elliptically Anisotropic Correlation Functions

3.1. Isotropic Correlation Functions

The realizations $U(\mathbf{x})$ of a random medium with the *isotropic correlation function* may be obtained by multiplying the Fourier transform of the realizations of a white noise of unit standard deviation by *isotropic spectral filter*

$$\hat{F}(\mathbf{k}) = \hat{f}(k) \tag{17}$$

with

$$k = (\mathbf{k}^{\mathrm{T}}\mathbf{k})^{\frac{1}{2}} \ , \tag{18}$$

and inversely Fourier transforming the products back into the space domain. Remember that $U(\mathbf{x})$ denotes just the random component taken with respect to the mean value, i.e., that $U(\mathbf{x})$ has zero mean value.

The power spectrum of the filter is then the function of k only,

$$P(k) = \hat{f}^{*}(k)\,\hat{f}(k) \ . \tag{19}$$

Assuming that a white noise has the unit standard deviation, the corresponding *isotropic correlation function* is then, see (16),

$$c(\mathbf{x}) = (2\pi)^{-d} \int_{-\infty}^{+\infty} dk_1 \cos(k_1 x_1) \cdots \int_{-\infty}^{+\infty} dk_d \cos(k_d x_d) P(k) \ , \tag{20}$$

where d is the Euclidean dimension of the space.

We may rotate, before integrating, the k_1 axis into the direction of vector $\mathbf{x}$ to arrive at

$$c(\mathbf{x}) = (2\pi)^{-d} \int_{-\infty}^{+\infty} dk_1 \cos(k_1 x) \int_{-\infty}^{+\infty} dk_2 \cdots \int_{-\infty}^{+\infty} dk_d \, P(k) \ , \tag{21}$$

where

$$x = (\mathbf{x}^{\mathrm{T}}\mathbf{x})^{\frac{1}{2}} \ . \tag{22}$$

For $d > 1$, we introduce the distance r from the k_1 axis,

$$r = [(k_2)^2 + \cdots + (k_d)^2]^{\frac{1}{2}} \ , \tag{23}$$

and recall the equation

$$V_d(r) = \frac{\pi^{\frac{d}{2}}}{\Gamma(\frac{d}{2}+1)} r^d \tag{24}$$

for the volume of the d-dimensional sphere of radius r. Differentiating the volume with respect to the radius, we obtain the surface of the sphere,

$$S_d(r) = \frac{d\pi^{\frac{d}{2}}}{\Gamma(\frac{d}{2}+1)} r^{d-1} = \frac{2\pi^{\frac{d}{2}}}{\Gamma(\frac{d}{2})} r^{d-1} \ . \tag{25}$$

Integrating (21) over the surface of the $(d-1)$-dimensional sphere of radius r in subspace $k_1 = \mathit{constant}$, and taking into account that the integrands are constant along the surface, we arrive at

$$c(\mathbf{x}) = (2\pi)^{-d} \frac{4\pi^{\frac{d-1}{2}}}{\Gamma(\frac{d-1}{2})} \int\limits_0^{+\infty} dk_1 \cos(k_1 x) \int\limits_0^{+\infty} dr\, r^{d-2} P\left(\sqrt{(r)^2 + (k_1)^2} \right) . \tag{26}$$

We now switch from the integration over k_1 and r to the integration over k_1 and $k = \sqrt{(r)^2 + (k_1)^2}$,

$$c(\mathbf{x}) = (2\pi)^{-d} \frac{4\pi^{\frac{d-1}{2}}}{\Gamma(\frac{d-1}{2})} \int\limits_0^{+\infty} dk\, kP(k) \int\limits_0^{k} dk_1 \cos(k_1 x)[k^2 - (k_1)^2]^{\frac{d-3}{2}} . \tag{27}$$

For $d > 1$, the integral with respect to k_1 may be calculated,

$$\int\limits_0^{k} dk_1 \cos(k_1 x)[k^2 - (k_1)^2]^{\frac{d-3}{2}} = 2^{\frac{d-4}{2}}\, \pi^{\frac{1}{2}}\Gamma(\frac{d-1}{2})k^{\frac{d-2}{2}}x^{\frac{2-d}{2}} J_{\frac{d-2}{2}}(kx) , \tag{28}$$

where $J_\nu(x)$ is the Bessel function and $\Gamma(x)$ is the Gamma function. Inserting (28) into (27), the correlation function may be expressed in terms of the Hankel transform of $k^{\frac{d-1}{2}}P(k) = k^{\frac{d-1}{2}}\hat{f}^*(k)\hat{f}(k)$,

$$\boxed{c(\mathbf{x}) = (2\pi)^{-\frac{d}{2}} x^{\frac{1-d}{2}} \int\limits_0^{+\infty} dk\sqrt{kx}\, J_{\frac{d-2}{2}}(kx)k^{\frac{d-1}{2}}\, \hat{f}^*(k)\hat{f}(k) .} \tag{29}$$

Taking into account that

$$J_{-\frac{1}{2}}(\xi) = \sqrt{\frac{2}{\pi}} \frac{\cos(\xi)}{\sqrt{\xi}} , \qquad J_{\frac{1}{2}}(\xi) = \sqrt{\frac{2}{\pi}} \frac{\sin(\xi)}{\sqrt{\xi}} , \tag{30}$$

equation (29) may be used for $d \geq 1$, see (21) for $d = 1$. On the other hand, relations (30) enable the expression of Hankel transform (29) in the form of Fourier transform in 1-D,

$$c(\mathbf{x}) = \pi^{-1} \int\limits_0^{+\infty} dk \cos(kx)\, \hat{f}^*(k)\hat{f}(k) , \tag{31}$$

and 3-D,

$$c(\mathbf{x}) = \frac{1}{2\pi^2 x} \int\limits_0^{+\infty} dk \sin(kx)k\, \hat{f}^*(k)\hat{f}(k) . \tag{32}$$

3.2. Elliptically Anisotropic Correlation Functions

Equation (29) remains valid if we generalize norms (18) and (22) to

$$k = (\mathbf{k}^T \mathbf{L} \mathbf{k})^{\frac{1}{2}}, \quad x = (\mathbf{x}^T \mathbf{L}^{-1} \mathbf{x})^{\frac{1}{2}}, \tag{33}$$

where the positive-definite symmetric scaling matrix $\mathbf{L}$ may account for the elliptical anisotropy of the correlation functions (GOFF and JORDAN, 1988).

4. Examples of Isotropic and Elliptically Anisotropic Correlation Functions

Most of the isotropic or elliptically anisotropic spectral filters (17) recently considered in the geophysical literature are special cases of filter

$$\hat{f}(k) = \kappa[a^{-2} + k^2]^{-\frac{d}{4} - \frac{N}{2}} \exp\left(-\frac{a_G^2 k^2}{8}\right), \tag{34}$$

described by four constants κ, N, a_G and a. The individual special filters may be fitted by the relevant choice of *Hurst exponent* (Hurst parameter, Hurst number) N and two correlation lengths a_G (Gaussian correlation length, see Sections 4.1 and 4.4) and a (von Kármán correlation length, see Section 4.2).

Function (34) is the product of spectral filter

$$\hat{f}_{S_N}(k) = \kappa k^{-\frac{d}{2} - N}, \tag{35}$$

corresponding to the self-affine random medium, with high-pass wavenumber filter

$$\hat{f}_a(k) = [1 + (ak)^{-2}]^{-\frac{d}{4} - \frac{N}{2}} \tag{36}$$

and low-pass wavenumber filter

$$\hat{f}_{a_G}(k) = \exp\left(-\frac{a_G^2 k^2}{8}\right), \tag{37}$$

see Figure 1.

Note that, with regards to the discussion in Section 4.3, the self-affine random medium with Hurst exponents

$$-1/2 < N < 0 \tag{38}$$

seems to be suitable for the description of the material properties of the geological structures. For the illustration of the random medium given by spectral filter (34) see Figure 2.

For example, one possibility of estimating the mean value $u_0(\mathbf{x})$ and parameters κ and N of the correlation function from travel times in the Western Bohemia region, under the assumption of a self-affine random medium, has been demon-

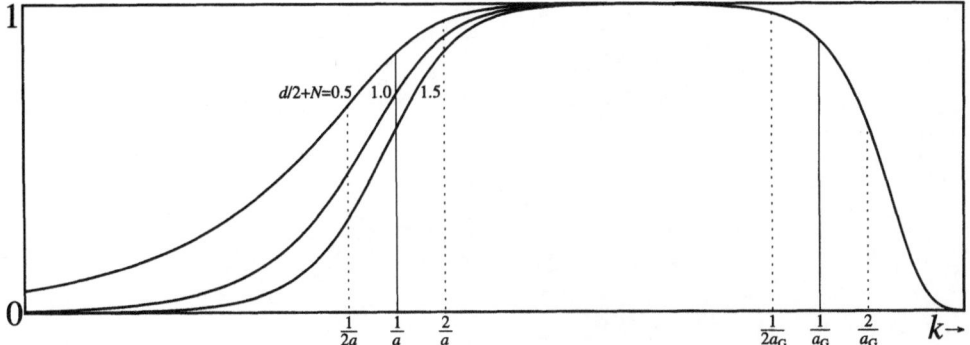

Figure 1

High-pass filter $\hat{f}_a(k)$ controlled by correlation length a and low-pass Gaussian filter $\hat{f}_{a_G}(k)$ controlled by correlation length a_G. The horizontal wavenumber axis has a logarithmic scale.

strated by KLIMEŠ (1996, 2002b). The results suggest that the value of Hurst exponent N, chosen in Figures 2, 4, 7 and 8, is acceptable, e.g., for the Western Bohemia region.

The application of the low-pass filter of correlation length a_G removes small-scale heterogeneities. The resulting random medium remains self-affine at distances considerably longer than correlation length a_G and describes the features of the geological structure seen by seismic waves of limited bandwidth rather than the entire geological structure.

The application of the high-pass filter of correlation length a suppresses large-scale heterogeneities. The resulting random medium remains self-affine at distances considerably shorter than correlation length a, and corresponds to the difference between the material parameter and its mean value rather than to the geological structure itself, provided that the mean value fits the heterogeneities considerably larger than correlation length a.

If both the low-pass and high-pass filters are applied, the random medium is self-affine at distances x between the correlation lengths,

$$a_G \ll x \ll a \ . \tag{39}$$

Correlation lengths a_G and a thus serve as the *inner and outer cutoff scales* of the self-affine random medium (MANDELBROT, 1977).

4.1. Gaussian Correlation Function ($N = -d/2$)

Choosing Hurst exponent $N = -d/2$, we obtain a white noise filtered by low-pass filter (37). Spectral filter (34) then reads

$$\hat{f}_G(k) = \kappa \, \exp\left(-\frac{a_G^2 k^2}{8}\right) \ . \tag{40}$$

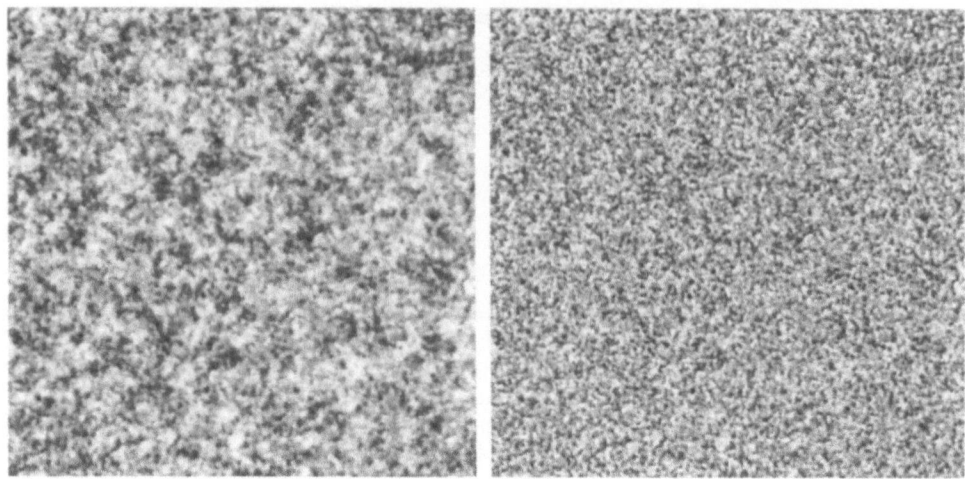

Figure 2

A 2-D representation of the random medium described by Hurst exponent $N = -0.2$ and correlation lengths $a_G = 0.005H$ and $a = 0.020H$, where H is the dimension of the square sample. Compare with Figure 7 imagined without both small and large heterogeneities.

Figure 3

A 2-D representation of the Gaussian random medium ($N = -d/2 = -1$, $a = +\infty$) of correlation length $a_G = 0.005H$, where H is the dimension of the square sample. The figure looks like a white noise if observed from a distance of 1.2 meter ($\approx 20H$) or more.

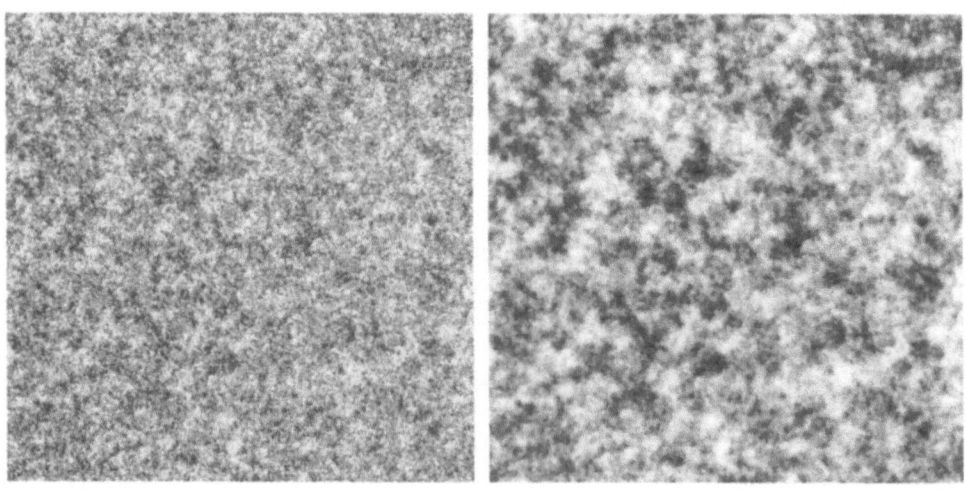

Figure 4

A 2-D representation of the von Kármán random medium ($a_G = 0$) described by Hurst exponent $N = -0.2$ and correlation length $a = 0.020H$, where H is the dimension of the square sample. Compare with Figure 7 imagined without large heterogeneities.

Figure 5

A 2-D representation of the exponential medium ($N = 1/2$, $a_G = 0$) described by correlation length $a = 0.020H$, where H is the dimension of the square sample.

Figure 6
A 2-D representation of the von Kármán random medium ($a_G = 0$) described by zero Hurst exponent $N = 0$ and correlation length $a = 0.020\,H$, where H is the dimension of the square sample.

Figure 7
A 2-D representation of the self-affine random medium ($a_G = 0$, $a = +\infty$) described by Hurst exponent $N = -0.2$.

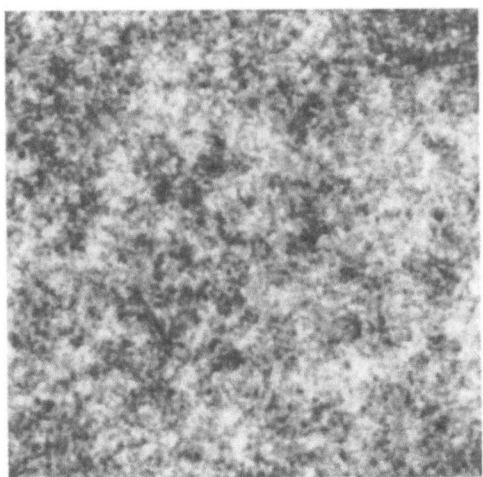

Figure 8
A 2-D representation of the Kummer random medium ($a = +\infty$) described by Hurst exponent $N = -0.2$ and correlation length $a_G = 0.005\,H$, where H is the dimension of the square sample. Compare with Figure 7 imagined without small heterogeneities.

For $0 < d$, the Hankel transform of $k^{\frac{d-1}{2}} \exp\left(-\frac{a_G^2 k^2}{4}\right)$ is

$$\int_0^{+\infty} \mathrm{d}k \sqrt{kx} \, \mathrm{J}_{\frac{d-2}{2}}(kx) k^{\frac{d-1}{2}} \exp\left(-\frac{a_G^2 k^2}{4}\right) = 2^{\frac{d}{2}} a_G^{-d} x^{\frac{d-1}{2}} \exp\left(-a_G^{-2} x^2\right) . \tag{41}$$

Inserting (40) and (41) into (29), we arrive at the Gaussian correlation function

$$c_G(\mathbf{x}) = \kappa^2 \pi^{-\frac{d}{2}} a_G^{-d} \exp\left(-a_G^{-2} x^2\right) , \tag{42}$$

or simply

$$c_G(\mathbf{x}) = \sigma_G^2 \exp\left(-a_G^{-2} x^2\right) \tag{43}$$

if we put

$$\kappa = \sigma_G \pi^{\frac{d}{4}} a_G^{\frac{d}{2}} . \tag{44}$$

The Gaussian correlation function describes, in principle, a low-pass filtered white noise, is too smooth to represent actual inhomogeneities in the earth (WU and AKI, 1985; SATO and FEHLER, 1998), and cannot explain observations of both seismic wave scattering and travel-time variations (FRANKEL and CLAYTON, 1986). White noise is approached for $a_G \to 0+$. The author thus does not see any relation of the Gaussian correlation function to the properties of geological structures.

For the illustration of the random medium with the Gaussian correlation function see Figure 3.

4.2. Von Kármán Correlation Functions ($a_G = 0$)

For $a_G = 0$, filter (34) simplifies to

$$\hat{f}_{K_N}(k) = \kappa \left[a^{-2} + k^2\right]^{-\frac{d}{4} - \frac{N}{2}} . \tag{45}$$

For $0 \le d$ and $-\frac{d+1}{2} < 2N$, the Hankel transform of $k^{\frac{d-1}{2}}[a^{-2} + k^2]^{-\frac{d}{4} - N}$ is

$$\int_0^{+\infty} \mathrm{d}k \sqrt{kx} \, \mathrm{J}_{\frac{d-2}{2}}(kx) k^{\frac{d-1}{2}} [a^{-2} + k^2]^{-\frac{d}{4} - N} = 2^{\frac{2-d}{2} - N} \left[\Gamma\left(\frac{d}{2} + N\right)\right]^{-1} a^N x^{\frac{d-1}{2} + N} \mathrm{K}_{-N}\left(\frac{x}{a}\right) .$$

$$\tag{46}$$

Inserting (45) and (46) into (29), and using relation

$$\mathrm{K}_{-N}(\xi) = \mathrm{K}_N(\xi) \tag{47}$$

for MacDonald functions K_N, we arrive at the von Kármán correlation function

$$c_{K_N}(\mathbf{x}) = \kappa^2 2^{1-d-N} \pi^{-\frac{d}{2}} \left[\Gamma\left(\frac{d}{2} + N\right)\right]^{-1} [ax]^N \mathrm{K}_N\left(a^{-1} x\right) . \tag{48}$$

Choosing, for Hurst exponent $N \ne 0$,

$$\kappa^2 = \sigma_K^2 2^d \pi^{\frac{d}{2}} [\Gamma(|N|)]^{-1} \Gamma\left(\frac{d}{2} + N\right) a^{-2N} , \tag{49}$$

we obtain the von Kármán correlation function in the form

$$c_{K_N}(\mathbf{x}) = \sigma_K^2 \frac{2}{\Gamma(|N|)} \left[\frac{x}{2a}\right]^N K_N\!\left(\frac{x}{a}\right) . \tag{50}$$

Right-hand side of equation (50) is σ_K^2 times the von Kármán function of argument $a^{-1}x$, see FRANKEL and CLAYTON (1986), WU (1982, 1989) or KORN (1993). The von Kármán correlation functions are frequently used in the geophysical literature, but unfortunately, ignoring the negative Hurst exponents and often using inappropriate correlation lengths. For the illustration of the von Kármán random medium see Figure 4.

The asymptotic behavior of the MacDonald functions K_N for very large arguments is

$$\underset{\xi \to +\infty}{K_N(\xi)} \approx \sqrt{\frac{\pi}{2}} \frac{\exp(-\xi)}{\sqrt{\xi}} \tag{51}$$

and, consequently, the asymptotic behavior of the von Kármán correlation functions (50) for very large arguments is

$$\underset{x \to +\infty}{c_{K_N}(\mathbf{x})} \approx \sigma_K^2 \sqrt{\pi} \, [\Gamma(|N|)]^{-1} \left[\frac{x}{2a}\right]^{N-\frac{1}{2}} \exp\!\left(-\frac{x}{a}\right) . \tag{52}$$

Note that relations (51) and (52) are exact for $N = \pm 1/2$. The asymptotic behavior of the MacDonald functions K_N for very small arguments is

$$\underset{\xi \to 0+}{K_N(\xi)} \approx \frac{1}{2} \Gamma(|N|) \left[\frac{\xi}{2}\right]^{-|N|} \tag{53}$$

and, consequently, the asymptotic behavior of the von Kármán correlation functions (50) for very small arguments is

$$N > 0 : \underset{x \to 0+}{c_{K_N}(\mathbf{x})} \approx \sigma_K^2 ,$$
$$N < 0 : \underset{x \to 0+}{c_{K_N}(\mathbf{x})} \approx \sigma_K^2 \left[\frac{x}{2a}\right]^{2N} . \tag{54}$$

The von Kármán medium is thus self-affine at distances considerably shorter than correlation length a.

In seismic practice, the von Kármán correlation functions with negative Hurst exponent $N < 0$ may be appropriate for the description of the self-affine random medium in which the large heterogeneities (larger than the correlation length) are fitted well enough by mean value $u_0(\mathbf{x})$. In such a case, the von Kármán correlation functions describe small heterogeneities, unresolved by mean value $u_0(\mathbf{x})$, or by the seismic model provided that $u_0(\mathbf{x})$ stands for the seismic model.

4.2.1. Exponential correlation function ($N = \frac{1}{2}$, $a_G = 0$)

For the special case of Hurst exponent $N = 1/2$, equation (45) takes the form

$$\hat{f}_E(k) = \hat{f}_{K_{\frac{1}{2}}}(k) = \kappa \, [a^{-2} + k^2]^{-\frac{d+1}{4}} \; . \tag{55}$$

For $0 < d$, the Hankel transform of $k^{\frac{d-1}{2}}[a^{-2} + k^2]^{-\frac{d+1}{2}}$ is

$$\int\limits_{0}^{+\infty} dk \sqrt{kx} \; J_{\frac{d-2}{2}}(kx) k^{\frac{d-1}{2}}[a^{-2} + k^2]^{-\frac{d+1}{2}} = 2^{-\frac{d}{2}} \pi^{\frac{1}{2}} \left[\Gamma\left(\frac{d+1}{2}\right)\right]^{-1} a \, x^{\frac{d-1}{2}} \exp(-a^{-1}x) \; . \tag{56}$$

Inserting (55) and (56) into (29), we arrive at

$$c_E(\mathbf{x}) = c_{K_{\frac{1}{2}}}(\mathbf{x}) = \kappa^2 2^{-d} \pi^{\frac{1-d}{2}} \left[\Gamma\left(\frac{d+1}{2}\right)\right]^{-1} a \exp(-a^{-1}x) \; . \tag{57}$$

Choosing

$$\kappa^2 = \sigma_K^2 2^d \pi^{\frac{d-1}{2}} \Gamma\left(\frac{d+1}{2}\right) a^{-1} \; , \tag{58}$$

we obtain the exponential correlation function in the form

$$c_E(\mathbf{x}) = c_{K_{\frac{1}{2}}}(\mathbf{x}) = \sigma_K^2 \exp(-a^{-1}x) \; . \tag{59}$$

Equations (57) and (59) are special cases of (48) and (50) respectively, with

$$K_{\frac{1}{2}}(\xi) = \sqrt{\frac{\pi}{2}} \, \xi^{-\frac{1}{2}} \exp(-\xi) \; . \tag{60}$$

The exponential random medium is illustrated by Figure 5. It cannot explain observations of both seismic wave scattering and travel-time variations (FRANKEL and CLAYTON, 1986), because both small-scale and large-scale heterogeneities are considerably suppressed. If the Hurst exponent is decreased from $N = 1/2$ to $N = 0$ and the von Kármán correlation length a is increased towards infinity, the fit of the arrival-time and amplitude fluctuations at the NORSAR is improving (FLATTÉ and WU, 1988).

4.2.2. Zero von Kármán correlation function ($N = 0$, $a_G = 0$)

The right-hand side of (49) goes to infinity for Hurst exponent $N \to 0$. This may be removed by rescaling (49) by factor $(1/2)\Gamma(|N|)$,

$$\kappa^2 = \sigma_{K_0}^2 2^{d-1} \pi^{\frac{d}{2}} \Gamma\left(\frac{d}{2} + N\right) a^{-2N} \; . \tag{61}$$

The rescaled correlation function (50) then takes the form

$$c_{K_N}(\mathbf{x}) = \sigma_{K_0}^2 \left[\frac{x}{2a}\right]^N K_N\left(\frac{x}{a}\right) . \tag{62}$$

For the special case of zero Hurst exponent, $N = 0$,

$$c_{K_0}(\mathbf{x}) = \sigma_{K_0}^2 K_0(a^{-1}x) , \tag{63}$$

see FRANKEL and CLAYTON (1986). The random medium with the zero von Kármán correlation function is illustrated in Figure 6.

4.3. Self-affine Random Medium ($a_G = 0$, $a = +\infty$)

The von Kármán medium is not self-affine over all scales, it is asymptotically self-affine at small distances $x \ll a$. The self-affine random medium may thus be obtained as the limiting case of the von Kármán medium for $a \to +\infty$,

$$\hat{f}_{S_N}(k) = \kappa k^{-\frac{d}{2}-N} , \tag{64}$$

see (45). For $-d/2 < 2N < 0$, the Hankel transform of $k^{\frac{d-1}{2}}k^{-d-2N}$ is

$$\int\limits_0^{+\infty} dk\sqrt{kx}\, J_{\frac{d-2}{2}}(kx)\, k^{-\frac{d+1}{2}-2N} = 2^{-\frac{d}{2}-2N}\frac{\Gamma(|N|)}{\Gamma(\frac{d}{2}+N)}x^{\frac{d-1}{2}+2N} . \tag{65}$$

Inserting (64) and (65) into (29), we arrive at the power-law correlation function

$$c_{S_N}(\mathbf{x}) = \kappa^2 2^{-d-2N}\pi^{-\frac{d}{2}}\frac{\Gamma(|N|)}{\Gamma(\frac{d}{2}+N)}x^{2N} \tag{66}$$

of the self-affine random medium. Choosing

$$\kappa^2 = \sigma_S^2 2^d \pi^{\frac{d}{2}}\left[\Gamma(|N|)\right]^{-1}\Gamma\left(\frac{d}{2}+N\right)\left[\frac{2}{L}\right]^{2N} , \tag{67}$$

we obtain the power-law correlation function in the form of

$$c_{S_N}(\mathbf{x}) = \sigma_S^2 \left[\frac{x}{L}\right]^{2N} . \tag{68}$$

Here L is some reference distance supplemented to preserve the correct physical units in the expressions. Multiplication factor $(2/L)^{2N}$ on the right-hand side of (67) substitutes here multiplication factor a^{-2N} which would approach zero on the right-hand side of (49) for $a \to +\infty$, see (54). Reference distance L is unnecessary, $L = 1$, if equations (33) are used to define x and k, because scaling matrix $\mathbf{L}$ can play the role of L^2.

The self-affine random medium seems to be the most appropriate description of the material properties of the geological structures, because the structures contain heterogeneities of all scales. Indeed it has been found that power spectra of a variety of geophysical variables show the power-law behavior over considerably large ranges of wavenumbers (PILKINGTON and TODOESCHUCK, 1990). Since correlation function (68) should decrease with distance x (CHERNOV, 1960), the Hurst exponent should be negative,

$$N < 0 , \tag{69}$$

in the self-affine random medium. Further limitations of the acceptable values of the Hurst exponent are imposed by the influence of heterogeneities of various scales on the properties of seismic waves.

In the inversion of travel times, correlation functions are required for the calculation of the geometrical travel-time covariances. The line integrals of correlation function (68) across $\mathbf{x} = \mathbf{0}$, representing travel-time variances along infinitely thin rays, are finite for

$$-1/2 < N \tag{70}$$

(KLIMEŠ 1996, 2002b). Although "thick" rays, corresponding to finite frequencies, allow also of $-d/2 < N \leq -1/2$, the travel-time variances would increase considerably with frequency for these values of Hurst exponent N. The author does not consider such a strong velocity dispersion to be realistic.

On smaller scales, correlation functions enable to approximately upscale a rough distribution of material parameters and estimate the relation to a smoother distribution of effective material parameters. MÜLLER and SHAPIRO (2001) derived the expressions for the mean travel-time delay and amplitude decay due to random perturbations of the squared slowness in the weak fluctuation regime. Assuming the self-affine random medium, the integrals in their expressions for the mean logarithmic acoustic wavefield are finite for $-1/2 < N < 1/2$ in 2-D and for $-1/2 < N < 0$ in 3-D. Restriction (70) on the Hurst exponent is thus useful also on small scales.

FLATTÉ and WU (1988) demonstrated that decreasing the Hurst exponent from $N = 1/2$ to $N = 0$ improves the fit of the arrival-time and amplitude fluctuations at the NORSAR. They did not continue with the negative Hurst exponents, but found the combination of two layers with the Hurst exponents of $N = -3/2$ and $N = 1/2$ yielding better fit than $N = 0$. Note that their combination of the two correlation functions can reasonably be fit within the relevant inner and outer cutoff scales by a single correlation function with the negative Hurst exponent $-1/2 < N < 0$. Let us also remark that the experimental values of the correlation function calculated from the velocity log in the granite at Fenton Hill at scales from 2.4 m to 24 m (WU, 1982) can be fit with the relative r.m.s. difference around 8% by the power-law correlation function with Hurst exponent $N = -0.3$.

Hurst exponent $N = -d/2$ yields a white noise. Negative Hurst exponents $-1/2 < N < 0$ are suitable to characterize the material parameters of the geological structures. The limiting case of $N = 0$ is called a *flicker noise* (SCHOTTKY, 1926).

Fractional Brown noises obtained for positive Hurst exponent $0 < N < 1$ (ADDISON, 1997) may be divided into *antipersistent fractional Brown noises* with $0 < N < 1/2$ (MANDELBROT, 1977), *Brown noise* with $N = 1/2$ (e.g., MANDELBROT, 1977; TURCOTTE, 1989; ADDISON, 1997), and *persistent fractional Brown noises* with $1/2 < N < 1$ (MANDELBROT, 1977), called also *black noises* (ADDISON, 1997). In spite of that, in 2-D space, $d = 2$, CROSSLEY and JENSEN (1989) refer to the case of $N = -1/2$ as flicker noise and to the case of $N = 0$ as brown noise.

The self-affine random medium, with Hurst exponent N acceptable for the Western Bohemia region (KLIMEŠ, 2002b), is illustrated in Figure 7.

4.4. Kummer Correlation Functions $(a = +\infty)$

On the other hand, the effective material parameters may be appropriately described in terms of the self-affine random medium, filtered by the low-pass Gaussian smoothing function,

$$\hat{f}_{F_N}(k) = \kappa \, k^{-\frac{d}{2}-N} \exp\left(-\frac{a_G^2 k^2}{8}\right) , \qquad (71)$$

see (37) and (64). The Hankel transform of $k^{\frac{d-1}{2}} k^{-d-2N} \exp(-\frac{1}{4} a_G^2 k^2)$ is

$$\int\limits_0^{+\infty} dk \sqrt{kx} \, J_{\frac{d-2}{2}}(kx) k^{-\frac{d+1}{2}-2N} \exp\left(-\frac{a_G^2 k^2}{4}\right) = \frac{\Gamma(|N|)}{\Gamma(\frac{d}{2})} \frac{a_G^{2N}}{2^{\frac{d}{2}+2N}} x^{\frac{d-1}{2}} {}_1F_1\left(-N; \frac{d}{2}; -x^2 a_G^{-2}\right) , \qquad (72)$$

where ${}_1F_1(\alpha; \beta; \xi)$ is the *confluent hypergeometric function*, called also the *Kummer function*. Inserting (71) and (72) into (29), we arrive at correlation function

$$c_{F_N}(\mathbf{x}) = \kappa^2 \, 2^{-d-2N} \pi^{-\frac{d}{2}} \frac{\Gamma(|N|)}{\Gamma(\frac{d}{2})} a_G^{2N} {}_1F_1\left(-N; \frac{d}{2}; -x^2 a_G^{-2}\right) , \qquad (73)$$

which we shall call the *Kummer correlation function*. Choosing

$$\kappa^2 = \sigma_F^2 2^{d+2N} \pi^{\frac{d}{2}} \left[\Gamma(|N|)\right]^{-1} \Gamma\left(\frac{d}{2}\right) a_G^{-2N} , \qquad (74)$$

we obtain the Kummer correlation function in the form of

$$c_{F_N}(\mathbf{x}) = \sigma_F^2 \, {}_1F_1\left(-N; \frac{d}{2}; -x^2 a_G^{-2}\right) . \qquad (75)$$

The Kummer medium is self-affine at distances considerably longer than correlation length a_G, with

$$\sigma_F^2 = \frac{\Gamma(\frac{d}{2}+N)}{\Gamma(\frac{d}{2})} \left[\frac{a_G}{L}\right]^{2N} \sigma_S^2 , \qquad (76)$$

where σ_S corresponds to the self-affine random medium, see (67). Conversely, the Kummer correlation function is smooth at short distances, with $c_{F_N}(0) = \sigma_F^2$. The Kummer correlation functions may thus be very useful for the discrete numerical approximation of the self-affine correlation function (68), in order to avoid the aliasing due to the singularity related to infinitely small heterogeneities when using the fast Fourier transform.

For the illustration of the random medium with the Kummer correlation function see Figure 8.

4.5. Low-pass Filtered Self-affine Random Medium

Box-window low-pass filtered, self-affine random medium spectral filter (64) takes the form

$$k < K: \quad \hat{f}_{S_N]}(k) = \kappa\, k^{-\frac{d}{2}-N},$$
$$k > K: \quad \hat{f}_{S_N]}(k) = 0 . \tag{77}$$

Correlation function (29) then reads

$$c_{S_N]}(\mathbf{x}) = \kappa^2 (2\pi)^{-\frac{d}{2}} x^{\frac{1-d}{2}} \int_0^K dk \sqrt{kx}\, J_{\frac{d-2}{2}}(kx) k^{\frac{-d-1}{2} - 2N} . \tag{78}$$

Changing the integration variable, this may be turned into

$$c_{S_N]}(\mathbf{x}) = \kappa^2 (2\pi)^{-\frac{d}{2}} x^{\frac{1-d}{2}} K^{\frac{-d+1}{2} - 2N} \int_0^1 dl \sqrt{l\,Kx}\, J_{\frac{d-2}{2}}(lKx) l^{\frac{-d-1}{2} - 2N} . \tag{79}$$

The low-pass filtered self-affine random medium has been considered, e.g., by FLATTÉ and WU (1988).

4.5.1. Low-pass filtered self-affine medium: $N = -1$
For the special case of Hurst exponent $N = -1$,

$$k < K: \quad \hat{f}_{S_{-1}]}(k) = \kappa\, k^{1-\frac{d}{2}},$$
$$k > K: \quad \hat{f}_{S_{-1}]}(k) = 0 , \tag{80}$$

the Hankel transform of $l^{\frac{-d+2}{2}+\frac{1}{2}}$ in equation (79) may be calculated analytically to yield

$$c_{S_{-1}]}(\mathbf{x}) = \kappa^2 (2\pi)^{-\frac{d}{2}} x^{\frac{1-d}{2}} K^{\frac{-d+1}{2}+2} \left[\frac{2^{\frac{-d+2}{2}+1}(Kx)^{\frac{d-2}{2}-\frac{3}{2}}}{\Gamma\left(\frac{d-2}{2}\right)} - \frac{1}{\sqrt{Kx}} J_{\frac{d}{2}}(Kx) \right] , \tag{81}$$

which finally reads

$$c_{S_{-1}}(\mathbf{x}) = \kappa^2 (2\pi)^{-\frac{d}{2}} \left[\frac{2^{2-\frac{d}{2}} x^{-2}}{\Gamma(\frac{d-2}{2})} - K^{2-\frac{d}{2}} x^{-\frac{d}{2}} J_{\frac{d}{2}}(Kx) \right] . \tag{82}$$

In three-dimensional space, $d = 3$, this equation reads

$$c_{S_{-1}}(\mathbf{x}) = \kappa^2 (2\pi)^{-\frac{3}{2}} \left[\frac{\sqrt{2}}{\sqrt{\pi} x^2} - \frac{\sqrt{2}}{\sqrt{\pi}} \frac{\cos(Kx)}{x^2} \right] = \frac{\kappa^2}{2\pi^2} \frac{1}{x^2} [1 - \cos(Kx)] . \tag{83}$$

For

$$\kappa = \sigma_{S_1} 2\pi K^{-1} , \tag{84}$$

see (67), it takes the form

$$c_{S_{-1}}(\mathbf{x}) = \sigma_{S_1}^2 \frac{2[1 - \cos(Kx)]}{(Kx)^2} . \tag{85}$$

In two-dimensional space, $d = 2$, the special case discussed here converges to the low-pass filtered white noise of the next section.

4.5.2. Low-pass filtered white noise

For the special case of Hurst exponent $N = -d/2$, equation (77) reads

$$\begin{aligned} k < K : \quad & \hat{f}_{W_1}(k) = \hat{f}_{S_{-\frac{d}{2}}}(k) = \kappa , \\ k > K : \quad & \hat{f}_{W_1}(k) = \hat{f}_{S_{-\frac{d}{2}}}(k) = 0 , \end{aligned} \tag{86}$$

and the Hankel transform of $I^{\frac{d-2}{2}+\frac{1}{2}}$ in equation (79) may be calculated analytically to arrive at

$$c_{W_1}(\mathbf{x}) = \kappa^2 (2\pi)^{-\frac{d}{2}} x^{\frac{1-d}{2}} K^{\frac{d+1}{2}} \frac{1}{\sqrt{Kx}} J_{\frac{d}{2}}(Kx) , \tag{87}$$

which finally reads

$$c_{W_1}(\mathbf{x}) = \kappa^2 (2\pi)^{-\frac{d}{2}} \left(\frac{K}{x} \right)^{\frac{d}{2}} J_{\frac{d}{2}}(Kx) . \tag{88}$$

5. Conclusions

Several simple types of medium correlation functions, their mutual relation and the physical meaning of their parameters has been briefly discussed. Among them, the self-affine random medium is the most appropriate for a description of the material properties of geological structures, in that the earth contains heterogeneities on all scales.

The Kummer correlation functions, with the Gaussian correlation length as small as the computing equipment allows, are recommended for numerical realizations of the self-affine random medium if we need to avoid the singularity related to infinitely small heterogeneities.

The von Kármán correlation functions with Hurst exponent $-1/2 < N < 0$ may be useful for the description of small-scale random heterogeneities if the large-scale heterogeneities are known, i.e., are not random.

It is difficult to justify the application of the frequently used Gaussian correlation function and von Kármán correlation functions with a positive Hurst exponent (e.g., exponential correlation function).

Acknowledgements

I am indebted to Paul Spudich who provided me with his perfect code I used to calculate the representations of the random media presented here.

This research has been supported by the Grant Agency of the Czech Republic under Contracts 205/95/1465, 205/01/0927 and 205/01/D097, by the Grant Agency of the Charles University under Contract 237/2001/B-GEO/MFF, by the Ministry of Education of the Czech Republic within Research Project J13/98 113200004, and by the members of the consortium "Seismic Waves in Complex 3-D Structures" (see "http://sw3d.mff.cuni.cz").

References

ADDISON, P.S., *Fractals and Chaos: An Illustrated Course* (IOP Publishing, London 1997).

BASTIAANS, M.J. (1981), *Signal description by means of a local frequency spectrum*. In *Transformations in Optical Signal Processing* (eds. Rhodes, W.T. *et al.*) (Proc. SPIE, vol. 373, Soc. Photo-Opt. Instrum. Eng., Bellingham) pp. 49–62.

CHERNOV, L.A., *Wave Propagation in a Random Medium* (McGraw-Hill, New York 1960).

CROSSLEY, D.J. and JENSEN, O.G. (1989), *Fractal Velocity Models in Refraction Seismology*, Pure appl. geophys. *131*, 61–76.

FLATTÉ, S.M. and WU, R-S. (1988), *Small-scale Structure of the Litosphere and Astenosphere Deduced from Arrival Time and Amplitude Fluctuations*, J. geophys. Res. *93B*, 6601–6614.

FRANKEL, A. and CLAYTON, R.W. (1986), *Finite Difference Simulations of Seismic Scattering: Implications for the Propagation of Short-period Seismic Waves in the Crust and Models of Crustal Heterogeneity*, J. geophys. Res. *91B*, 6465–6489.

GOFF, J.A. and JORDAN, T.H. (1988): *Stochastic Modeling of Seafloor Morphology: Inversion of Sea Beam Data for Second-order Statistics*, J. geophys. Res. *93B*, 13,589–13,608.

KLIMEŠ, L. (1996), *Correlation function of a self-affine random medium*. In *Seismic Waves in Complex 3-D Structures, Report 4* (Dep. Geophys. Charles Univ., Prague) pp. 25–38 (online at "http://sw3d.mff.cuni.cz").

KLIMEŠ, L. (2002a), *Application of the Medium Covariance Functions to Travel-time Tomography*, Pure appl. geophys. *159*, 1791–1810.

KLIMEŠ, L. (2002b), *Estimating the Correlation Function of a Self-affine Random Medium*, Pure appl. geophys. *159*, 1833–1853.

KORN, M. (1993), *Seismic Waves in Random Media*, J. appl. Geophys. *29*, 247–269.

MANDELBROT, B.B., *The Fractal Geometry of Nature* (W.H. Freeman and Co., New York 1977).

MAURER, H., HOLLIGER, K. and BOERNER, D.E. (1998), *Stochastic Regularization: Smoothness or Similarity?*, Geophys. Res. Lett. *25*, 2889–2892.

MÜLLER, T.M. and SHAPIRO, S.A. (2001), *Most Probable Seismic Pulses in Single Realizations of Two- and Three-dimensional Random Media*, Geophys. J. int. *144*, 83–95.

PILKINGTON, M. and TODOESCHUCK, J.P. (1990), *Stochastic Inversion for Scaling Geology*, Geophys. J. int. *102*, 205–217.

RYZHIK, L., PAPANICOLAOU, G. and KELLER, J.B. (1996), *Transport Equations for Elastic and Other Waves in Random Media*, Wave Motion *24*, 327–370.

SATO, H. and FEHLER, M., *Seismic Wave Propagation and Scattering in the Heterogeneous Earth* (Springer-Verlag, New York 1998).

SCHOTTKY, W. (1926), *Small-shot Effect and Flicker Effect*, Phys. Rev. *28*, 74.

SHAPIRO, S.A. and KNEIB, G. (1993), *Seismic Attenuation by Scattering: Theory and Numerical Results*, Geophys. J. int. *114*, 373–191.

SHAPIRO, S.A., SCHWARZ, R. and GOLD, N. (1996), *The Effect of Random Isotropic Inhomogeneities on the Phase Velocity of Seismic Waves*, Geophys. J. int. *127*, 783–794.

TARANTOLA, A., *Inverse Problem Theory* (Elsevier, Amsterdam 1987).

TURCOTTE, D.L. (1989), *Fractals in Geology and Geophysics*, Pure appl. geophys. *131*, 171–196.

WU, R-S. (1982), *Attenuation of Short Period Seismic Waves Due to Scattering*, Geophys. Res. Lett. *9*, 9–12.

WU, R-S. (1989), *The Perturbation Method in Elastic Wave Scattering*, Pure appl. geophys. *131*, 605–637.

WU, R-S. and AKI, K. (1985), *Elastic Wave Scattering by a Random Medium and the Small-scale Inhomogeneities in the Lithosphere*, J. geophys. Res. *90B*, 10,261–10,273.

(Received October 27, 2000, revised March 2, 2001, accepted May 17, 2001)

Pure appl. geophys. 159 (2002) 1833–1853
0033–4553/02/081833–21 $ 1.50 + 0.20/0

Estimating the Correlation Function
of a Self-affine Random Medium

Luděk Klimeš[1]

Summary — The medium covariance function is of principal importance in refraction travel-time tomographic inversion, especially when estimating the accuracy of the seismic model, its relation to the geological structure, or the covariance matrix describing the statistics of synthetic travel times. The medium correlation function for the travel-time tomography should be obtained from travel times.

Since a geological structure contains heterogeneities of all sizes, very similar on various scales, a self-affine random medium is a mathematical model very suitable for approximating the statistics of a geological structure. A particular class of self-affine random media, composed of a heterogeneous mean value and a stationary self-affine random function, is considered. The self-affine random function is assumed to be realized in terms of a white noise filtered by the power-law spectral filter of amplitude proportional to a reasonable power of the wavenumber. The corresponding power-law medium correlation function depends on two parameters: the Hurst exponent and the reference standard deviation.

The corresponding geometrical travel-time covariances are derived. The geometrical travel-time variances are proportional to the power of ray lengths. A method designed to estimate the parameters of the medium correlation function using field travel times is proposed, and applied to data from the Western Bohemia region.

The determination of the Hurst exponent from field travel times is very difficult and sensitive to numerical parameters selected for the inversion. The medium correlation functions with the values of the Hurst exponent like $N = -0.1$ or $N = -0.2$ are equally acceptable to statistically describe the travel times measured in the Western Bohemia region. On the other hand, for the fixed Hurst exponent, the determination of the reference standard deviation of slowness is easy and reliable.

Key words: Travel times, self-affine random medium, correlation function, scaling geology, fractal geology, inversion.

1. Introduction

In order to estimate the relation of a seismic model to the geological structure and, consequently, to estimate the relation of the synthetic quantities, calculated in the seismic model, to reality, it is important to obtain an estimate of the medium covariance function. The medium covariance function is of principal importance in refraction travel-time tomographic inversion (TARANTOLA, 1987; MAURER *et al.*,

[1] Department of Geophysics, Charles University, Ke Karlovu 3, 121 16 Praha 2, Czech Republic. E-mail: klimes@seis.karlov.mff.cuni.cz

1998; KLIMEŠ, 2002a), especially when estimating the accuracy of the seismic model, its relation to the geological structure, or the covariance matrix describing the statistics of synthetic travel times.

In a self-affine random medium, the material parameters may be scaled simultaneously with scaling the spatial dimensions in such a way that the statistical properties remain unchanged by the scaling. Since a geological structure contains heterogeneities of all sizes, very similar on various scales, a self-affine random medium is a mathematical model very suitable for approximating the statistics of a geological structure (PILKINGTON and TODOESCHUCK, 1990). We thus assume the slowness distribution in the geological structure to be a particular representation of the self-affine random medium in this paper. For an overview and brief discussion of other types of random media used in geophysics, refer to KLIMEŠ (2002b).

Assuming a stationary (statistically homogeneous) medium, the medium covariance function may be expressed in terms of the medium correlation function. In Section 2, a particular class of self-affine random media, composed of a heterogeneous mean value and a stationary random function, is considered, and the corresponding medium correlation function is derived. The medium correlation function depends on two parameters: the Hurst exponent and the corresponding reference standard deviation.

Section 3 is devoted to the dependence of the *a priori* geometrical covariance matrix of field travel times (TARANTOLA, 1987) on the medium covariance function. The *a priori* geometrical covariance matrix of field travel times describes the deviations of travel times from the mean travel-time curve. The deviations are caused by the heterogeneities, especially the lateral ones.

The scales of applicability of the empirical medium correlation function are limited by the inner and outer scales of the correlation function (MANDELBROT, 1977), which depend on the nature of experimental data used to estimate the correlation function. For example, the correlation function determined from well logs (e.g., WU, 1982; WU et al., 1994; HOLLIGER, 1997; GOFF and HOLLIGER, 1999), attenuation or scattering (e.g., WU, 1982, 1989a, b; WU and AKI, 1985b; 1988; SHAPIRO, 1992; SHAPIRO and KNEIB, 1993; KNEIB and SHAPIRO, 1995; SHAPIRO et al., 1996; SATO and FEHLER, 1998; MÜLLER and SHAPIRO, 2001) generally has too small an outer scale to be applied in the travel-time tomography. The medium correlation function for the travel-time tomography should be obtained from travel times. Application of the correlation functions estimated from quite different data, e.g., from the surface geological maps (HOLLIGER and LEVANDER, 1994; LEVANDER et al., 1994) can hardly be statistically justified.

The medium correlation function derived in Section 2 depends on two parameters: the Hurst exponent and the corresponding reference standard deviation. Section 4 is devoted to the method for estimation of these two parameters, essential for travel-time tomography, from field travel times. The determination of the reference standard deviation for the fixed Hurst exponent is easy, and there are numerous geophysical papers doing it, e.g., MINTZER (1953), AKI (1973), BERTEUSSEN et al. (1974), CAPON (1974), GUDMUNDSSON et al. (1990), WU and XIE (1991), or ROTH (1997). The problem is to estimate the Hurst exponent on the regional scales. FLATTÉ and WU (1988) and WU and FLATTÉ (1990) demonstrated that decreasing the Hurst exponent from $N = 1/2$ to $N = 0$ improves the fit of the arrival-time and amplitude fluctuations at the NORSAR. They did not continue with the negative Hurst exponents, but found the combination of two layers with the Hurst exponents of $N = -3/2$ and $N = 1/2$ yielding better fit than $N = 0$. Note that their combination of the two correlation functions can reasonably be fit within the relevant inner and outer cutoff scales by a single correlation function with the negative Hurst exponent $-1/2 < N < 0$. On the other hand, WU and AKI (1985a) used the seismic coda waves to determine the Hurst exponents from $N = 0.23$ for shallow events to $N = 0.5$ for deep events.

The method proposed in Section 4 is applied to field data from the Western Bohemia region in Section 5 to demonstrate the possibilities of estimating the medium correlation function.

The reader should be aware that the Einstein summation does not apply to the equations anywhere in this paper.

2. Correlation Function of a Self-affine Random Medium

Random medium $u(\mathbf{x})$ is *stationary* if mean value $\langle u(\mathbf{x}) \rangle$ is constant and the *medium covariance function*

$$C(\mathbf{x}_1, \mathbf{x}_2) = \langle u(\mathbf{x}_1)\, u(\mathbf{x}_2) \rangle - \langle u(\mathbf{x}_1) \rangle \langle u(\mathbf{x}_2) \rangle \tag{1}$$

depends only on distance $\mathbf{x}_1 - \mathbf{x}_2$,

$$C(\mathbf{x}_1, \mathbf{x}_2) = c(\mathbf{x}_1 - \mathbf{x}_2) \ , \tag{2}$$

where $c(\mathbf{x})$ is the *medium correlation function* (TARANTOLA, 1987).

Realizations of a statistically isotropic stationary *self-affine random medium*, uniformly scalable over all lengths, may be obtained by multiplying the Fourier transform of the realizations of a stationary *white noise* by spectral filter

$$\widehat{F}(\mathbf{k}) = \kappa \, k^{-\frac{d}{2} - N} \tag{3}$$

with

$$k = (\mathbf{k}^{\mathrm{T}} \mathbf{k})^{\frac{1}{2}} \ , \tag{4}$$

and inversely Fourier transforming the products back into the space domain. Here d is the Euclidean dimension of the space, $d = 3$ in 3-D, constant N is called the *Hurst exponent*, and κ is a constant proportional to the reference standard deviation of resulting self-affine random functions.

Assuming that the white noise has a unit standard deviation, the medium correlation function is then

$$c(\mathbf{x}) = (2\pi)^{-d} \int\limits_{-\infty}^{+\infty} dk_1 \cos(k_1 x_1) \cdots \int\limits_{-\infty}^{+\infty} dk_d \cos(k_d x_d) \, [\widehat{F}(\mathbf{k})]^2 \; . \tag{5}$$

Since spectral filter (3) is rotationally symmetric, we may rotate, before integrating, the k_1 axis into the direction of vector $\mathbf{x}$ to arrive at

$$c(\mathbf{x}) = \kappa^2 (2\pi)^{-d} \int\limits_{-\infty}^{+\infty} dk_1 \cos(k_1 x) \int\limits_{-\infty}^{+\infty} dk_2 \cdots \int\limits_{-\infty}^{+\infty} dk_d \, k^{-d-2N} \; , \tag{6}$$

where

$$x = (\mathbf{x}^{\mathrm{T}} \mathbf{x})^{\frac{1}{2}} \; . \tag{7}$$

For $d > 1$, we introduce distance r from the k_1 axis,

$$r = [(k_2)^2 + \cdots + (k_d)^2]^{\frac{1}{2}} \; , \tag{8}$$

and recall the equation

$$V_d(r) = \frac{\pi^{\frac{d}{2}}}{\Gamma(\frac{d}{2} + 1)} \, r^d \tag{9}$$

for the volume of the d-dimensional sphere of radius r. Differentiating the volume with respect to the radius, we derive the surface of the sphere,

$$S_d(r) = \frac{d\pi^{\frac{d}{2}}}{\Gamma(\frac{d}{2} + 1)} \, r^{d-1} = \frac{2\pi^{\frac{d}{2}}}{\Gamma(\frac{d}{2})} \, r^{d-1} \; . \tag{10}$$

Integrating (6) over the surface of the $(d-1)$-dimensional sphere of radius r in subspace $k_1 = constant$, and taking into account that the integrands are constant along the surface, we arrive at

$$c(\mathbf{x}) = \kappa^2 (2\pi)^{-d} \frac{2 \, \pi^{\frac{d-1}{2}}}{\Gamma(\frac{d-1}{2})} \int\limits_{-\infty}^{+\infty} dk_1 \cos(k_1 x) \int\limits_{0}^{+\infty} dr \, r^{d-2} [(r)^2 + (k_1)^2]^{-\frac{d}{2} - N} \; . \tag{11}$$

For $0 < d - 1 < d + 2N$, the integral with respect to r may be calculated,

$$c(\mathbf{x}) = \kappa^2 (2\pi)^{-d} \pi^{\frac{d-1}{2}} \frac{\Gamma(\frac{1}{2} + N)}{\Gamma(\frac{d}{2} + N)} \int\limits_{-\infty}^{+\infty} dk_1 \cos(k_1 x) |k_1|^{-1-2N} \; . \tag{12}$$

For $d = 1$, equation (6) reads

$$c(\mathbf{x}) = \kappa^2 (2\pi)^{-1} \int\limits_{-\infty}^{+\infty} dk_1 \cos(k_1 x) |k_1|^{-d-2N} \ . \tag{13}$$

For $0 \leq d - 1 < d + 2N$, equations (12) and (13) take the common form of

$$c(\mathbf{x}) = \kappa^2 2^{-d} \pi^{-\frac{d+1}{2}} \frac{\Gamma(\frac{1}{2} + N)}{\Gamma(\frac{d}{2} + N)} \int\limits_{-\infty}^{+\infty} dk_1 \cos(k_1 x) |k_1|^{-1-2N} \ . \tag{14}$$

For $-1/2 < N < 0$, the integral with respect to k_1 comes out as

$$c(\mathbf{x}) = \kappa^2 2^{-d} \pi^{-\frac{d+1}{2}} \frac{\Gamma(N + \frac{1}{2})}{\Gamma(N + \frac{d}{2})} 2\Gamma(-2N) \sin[\pi(N + \frac{1}{2})] \ x^{2N} \ . \tag{15}$$

We define *reference standard deviation* σ, related to arbitrarily selected *reference distance L*, by equation

$$\sigma^2 = \kappa^2 L^{2N} 2^{-d+1} \pi^{-\frac{d+1}{2}} \frac{\Gamma(N + \frac{1}{2})}{\Gamma(N + \frac{d}{2})} \Gamma(-2N) \cos(-\pi N) \ . \tag{16}$$

Reference distance L is introduced here to define σ consistently with respect to physical units. The medium covariance function (2) may then be expressed in the form of

$$\boxed{C(\mathbf{x}_1, \mathbf{x}_2) = \sigma^2 \left(\frac{|\mathbf{x}_1 - \mathbf{x}_2|}{L} \right)^{2N} \ , \tag{17}}$$

where the Hurst exponent satisfies

$$-1/2 < N < 0 \ . \tag{18}$$

In 3-D space, $d = 3$, which is of particular interest in travel-time tomography, equation (16) reads

$$\sigma^2 = \kappa^2 L^{2N} 2^{-1} \pi^{-2} \frac{\Gamma(-2N)}{1 + 2N} \cos(-\pi N) \ . \tag{19}$$

Note that the limiting case of Hurst exponent $N = -1/2$ and the cases of $-d/2 < N < -1/2$ are somewhat unstable (see the above derivation) and their statistical properties resemble a *white noise* produced by $N = -d/2$ in (3), with the medium covariance function

$$C(\mathbf{x}_1, \mathbf{x}_2) = \kappa^2 \delta(\mathbf{x}_1 - \mathbf{x}_2) \ . \tag{20}$$

Negative Hurst exponents $-1/2 < N < 0$ are suitable to characterize the material parameters of the geological structures.

For Hurst exponents $N \geq 0$, the medium cannot be self-affine at all scales, but only at scales sufficiently smaller than a finite *correlation length*, which may be selected arbitrarily large. The limiting case of $N = 0$ is called a *flicker noise* (SCHOTTKY, 1926). *Fractional Brown noises* obtained for positive Hurst exponent $0 < N < 1$ (ADDISON, 1997) may be divided into *antipersistent fractional Brown noises* with $0 < N < 1/2$ (MANDELBROT, 1977), *Brown noise* with $N = 1/2$ (e.g., MANDELBROT, 1977; TURCOTTE, 1989; ADDISON, 1997), and *persistent fractional Brown noises* with $1/2 < N < 1$ (MANDELBROT, 1977), called also *black noises* (ADDISON, 1997). In spite of that, in 2-D space, $d = 2$, CROSSLEY and JENSEN (1989) refer to the case of $N = -1/2$ as flicker noise and to the case of $N = 0$ as brown noise. If the correlation length approaches infinity, all self-affine medium covariance functions with Hurst exponents $N \geq 0$ converge to covariance function

$$C(\mathbf{x}_1, \mathbf{x}_2) = \sigma^2 \tag{21}$$

of a *random homogeneous medium*, obtained as the limiting case of (17) for $N = 0$.

Hereinafter, we shall assume a statistically homogeneous medium covariance function, see (2), as in a stationary random medium, but allow for a *spatially variable mean value*. We shall assume that the medium covariance function is of isotropic power-law form (17).

3. A priori Geometrical Covariance Matrix of Travel Times

Assuming the ray-theory linearization approach, field travel times may be expressed in the form of

$$T_I = \tau_I + \delta T_I \tag{22}$$

where

$$\tau_I = \int_0^{s_I} \mathrm{d}s \; u(\mathbf{x}(s)) \tag{23}$$

is the integral of the slowness $u(\mathbf{x})$ along the corresponding ray of length s_I, and δT_I is the error in determining field travel time T_I.

The *a priori geometrical covariance* (TARANTOLA, 1987)

$$\Theta_{KL} = \left\langle (\tau_K - \tau_K^0)(\tau_L - \tau_L^0) \right\rangle , \tag{24}$$

of the Kth and Lth travel times is then given by

$$\Theta_{KL} = \int_0^{s_K} \mathrm{d}s_K' \int_0^{s_L} \mathrm{d}s_L' \, C(\mathbf{x}(s_K'), \mathbf{x}(s_L')) \tag{25}$$

(CHERNOV, 1960), where the integration is performed along the corresponding rays of lengths s_K and s_L. Here τ_K^0 are the reference travel times corresponding to mean value $u^0(\mathbf{x}) = \langle u(\mathbf{x}) \rangle$ of the random slowness. Notice that, for medium covariance function (17), $\sigma^{-2}\Theta_{KL}$ is independent of σ and is determined by a single medium parameter, N.

The derivative of geometrical covariance Θ_{KL} with respect to N is then

$$\frac{\partial \Theta_{KL}}{\partial N} = \int_0^{s_K} \mathrm{d}s_K' \int_0^{s_L} \mathrm{d}s_L' C\big(\mathbf{x}(s_K'), \mathbf{x}(s_L')\big) \, 2\ln\left(\frac{|\mathbf{x}(s_K') - \mathbf{x}(s_L')|}{L}\right) \tag{26}$$

and $\sigma^{-2}\frac{\partial \Theta_{KL}}{\partial N}$ is again independent of σ.

3.1. A priori Geometrical Variances of Travel Times

If we approximate the distance between ray points $\mathbf{x}(s_1)$ and $\mathbf{x}(s_2)$ of the same ray by

$$|\mathbf{x}(s_1) - \mathbf{x}(s_2)| \approx |s_1 - s_2| \tag{27}$$

(BERGMANN, 1946), equation (17) may be inserted into (25) to arrive at

$$\Theta_{KK} \approx \sigma^2 L^{-2N} \int_0^{s_K} \mathrm{d}s_1 \int_0^{s_K} \mathrm{d}s_2 \, |s_1 - s_2|^{2N} \; . \tag{28}$$

For $-1/2 < N$, the integrals may be calculated to read

$$\Theta_{KK} \approx \sigma^2 L^{-2N} \int_0^{s_K} \mathrm{d}s_1 \left[\int_0^{s_1} \mathrm{d}s_2 (s_2)^{2N} + \int_0^{s_K - s_1} \mathrm{d}s_2 (s_2)^{2N} \right]$$

$$= \sigma^2 L^{-2N} \int_0^{s_K} \mathrm{d}s_1 \left[\frac{(s_1)^{2N+1}}{2N+1} + \frac{(s_K - s_1)^{2N+1}}{2N+1} \right] , \tag{29}$$

and finally

$$\Theta_{KK} \approx \frac{2\sigma^2 L^2}{(2N+1)(2N+2)} \left(\frac{s_K}{L}\right)^{2N+2} . \tag{30}$$

The derivative of variance Θ_{KK} with respect to N is then

$$\frac{\partial \Theta_{KK}}{\partial N} \approx \frac{2\sigma^2 L^2}{(2N+1)(2N+2)} \left(\frac{s_K}{L}\right)^{2N+2} 2\ln\left(\frac{s_K}{L}\right) . \tag{31}$$

Unfortunately, off-diagonal elements $\Theta_{K \neq L}$ of the geometrical travel-time covariance matrix must be calculated numerically.

4. Determination of the Medium Correlation Function
from the Field Travel Times

4.1. Differences of the Relative Field Travel Times and their Statistical Moments

Let us study the mutual differences of the field travel times. Since the travel times are strongly dependent on hypocentral distances, it is possible to compare only the travel times T_K and T_L along the rays of similar lengths s_K and s_L. This restriction may, to some extent, be reduced if we relate the travel times to some reference travel-time curve

$$\tau^0 = \tau^0(s) \ . \tag{32}$$

We may then compare the relative travel times T_K / τ_K^0 and T_L / τ_L^0, where

$$\tau_I^0 = \tau^0(s_I) \ . \tag{33}$$

The differences of the relative travel times then depend on the local value of the reference travel-time curve in terms of a multiplication factor which has practically no influence on the statistics. The differences of the relative travel times are distorted especially by the error in the derivative of the reference travel-time curve multiplied by $|s_K - s_L|$. We assume that the local error in the derivative of the reference travel-time curve compared to the exact mean travel-time curve is locally negligible in intervals defined by

$$qT_L < T_K < T_L \tag{34}$$

with given parameter q, $0 \leq q < 1$.

We now define the variances of the relative travel-time differences

$$D_{KL,KL} = \left\langle \left[\frac{T_K}{\tau_K^0} - \frac{T_L}{\tau_L^0} \right]^2 \right\rangle \ , \tag{35}$$

and the fourth-order variances

$$D_{KL,KL,KL,KL} = \left\langle \left[\left(\frac{T_K}{\tau_K^0} - \frac{T_L}{\tau_L^0} \right)^2 - D_{KL,KL} \right]^2 \right\rangle = \left\langle \left(\frac{T_K}{\tau_K^0} - \frac{T_L}{\tau_L^0} \right)^4 \right\rangle - \left[D_{KL,KL} \right]^2 \ , \tag{36}$$

describing the standard deviations of the squared differences of the relative field travel times from variances (35).

We assume picking errors δT_I statistically independent of travel times τ_I. The variances

$$D_{KL,KL} = \left\langle \left[\frac{\tau_K + \delta T_K}{\tau_K^0} - \frac{\tau_L + \delta T_L}{\tau_L^0} \right]^2 \right\rangle \ , \tag{37}$$

of the relative travel-time differences may then be expressed in terms of the first two statistical moments of the relative travel times τ_I/τ_I^0,

$$\theta_K = \left\langle \frac{\tau_K}{\tau_K^0} \right\rangle \ , \quad \theta_{KL} = \left\langle \frac{\tau_K}{\tau_K^0} \frac{\tau_L}{\tau_L^0} \right\rangle \ , \tag{38}$$

and the first two statistical moments of the relative picking errors $\delta T_I/\tau_I^0$,

$$t_K = \left\langle \frac{\delta T_K}{\tau_K^0} \right\rangle \ , \quad t_{KL} = \left\langle \frac{\delta T_K}{\tau_K^0} \frac{\delta T_L}{\tau_L^0} \right\rangle \ , \tag{39}$$

as

$$D_{KL,KL} = \theta_{KK} - 2\theta_{KL} + \theta_{LL} + t_{KK} - 2t_{KL} + t_{LL} + 2[\theta_K t_K - \theta_K t_L - \theta_L t_K + \theta_L t_L] \ . \tag{40}$$

We assume zero mean value of picking errors δT_K,

$$\langle \delta T_K \rangle = 0 \ . \tag{41}$$

Then

$$t_K = 0 \tag{42}$$

and variances (40) become independent of the mean values θ_K of the reduced travel times,

$$D_{KL,KL} = \theta_{KK} - 2\theta_{KL} + \theta_{LL} + t_{KK} - 2t_{KL} + t_{LL} \ . \tag{43}$$

We assume the *data covariance matrix*

$$\langle \delta T_K \ \delta T_L \rangle = T_{KL} \tag{44}$$

to be known, at least approximately. Inserting (38) with (24) for θ_{MN} and (39) with (44) for t_{MN} into (43), we see that variances (43) depend on parameters σ and N of the medium covariance function (17) through

$$D_{KL,KL} = D_{KL,KL}^0 + \sigma^2 D_{KL,KL}^1(N) \tag{45}$$

with

$$D_{KL,KL}^0 = \frac{T_{KK}}{\tau_K^0 \tau_K^0} - 2\frac{T_{KL}}{\tau_K^0 \tau_L^0} + \frac{T_{LL}}{\tau_L^0 \tau_L^0} \tag{46}$$

and

$$D_{KL,KL}^1(N) = \frac{\sigma^{-2}\Theta_{KK}}{\tau_K^0 \tau_K^0} - 2\frac{\sigma^{-2}\Theta_{KL}}{\tau_K^0 \tau_L^0} + \frac{\sigma^{-2}\Theta_{LL}}{\tau_L^0 \tau_L^0} \ . \tag{47}$$

Values (46) are constants with respect to N and σ. Functions (47) of N are independent of σ.

To be able to approximate the fourth-order moments in (36) using the second-order moments, we assume Gaussian probability distributions for both the self-affine random medium and the picking errors. If all probability distributions in the problem are Gaussian, the marginal probability distribution describing the relative travel-time differences is also Gaussian. If the probability distribution is Gaussian,

$$\left\langle \left(\frac{T_K}{\tau_K^0} - \frac{T_L}{\tau_L^0} \right)^4 \right\rangle = 3 \left\langle \left(\frac{T_K}{\tau_K^0} - \frac{T_L}{\tau_L^0} \right)^2 \right\rangle^2 \tag{48}$$

(BERAN, 1968; GOFF and JORDAN, 1988), and equation (36) reads

$$D_{KL,KL,KL,KL} = 2[D_{KL,KL}]^2 . \tag{49}$$

4.2. Objective Function

We select the objective function in the form of

$$y = \left[\sum_{K,L} 1 \right]^{-1} \sum_{K,L} \left[\left(\frac{T_K}{\tau_K^0} - \frac{T_L}{\tau_L^0} \right) - D_{KL,KL} \right]^2 \left[D_{KL,KL,KL,KL} \right]^{-1} , \tag{50}$$

and minimize it with respect to parameters σ and N of the medium covariance function (17). We emphasize that the minimum has to be sought for constant fourth-order variances $D_{KL,KL,KL,KL}$.

Inserting (45), and (49) with constant $\sigma = \sigma_0$ and $N = N_0$, objective function (50) reads

$$y(\sigma, N) = \frac{1}{2} \left[\sum_{K,L} 1 \right]^{-1} \sum_{K,L} \left[\left(\frac{T_K}{\tau_K^0} - \frac{T_L}{\tau_L^0} \right)^2 - D_{KL,KL}^0 - \sigma^2 D_{KL,KL}^1(N) \right]^2$$

$$\times \left[D_{KL,KL}^0 + (\sigma_0)^2 D_{KL,KL}^1(N_0) \right]^{-2} . \tag{51}$$

Here parameters σ_0 and N_0 are fixed during the minimization, however they should be selected close to the final solution,

$$\sigma_0 \approx \sigma, \quad N_0 \approx N . \tag{52}$$

Objective function (51) has its minimum with respect to σ for

$$\sigma^2(N) = \frac{F_1(N)}{F_2(N)} \tag{53}$$

with

$$F_0(N) = \sum_{K,L} \left[\left(\frac{T_K}{\tau_K^0} - \frac{T_L}{\tau_L^0} \right)^2 - D_{KL,KL}^0 \right]^2 \left[D_{KL,KL}^0 + (\sigma_0)^2 D_{KL,KL}^1 (N_0) \right]^{-2} , \qquad (54)$$

$$F_1(N) = \sum_{K,L} \left[\left(\frac{T_K}{\tau_K^0} - \frac{T_L}{\tau_L^0} \right)^2 - D_{KL,KL}^0 \right] D_{KL,KL}^1 (N) [D_{KL,KL}^0 + (\sigma_0)^2 D_{KL,KL}^1 (N_0)]^{-2} , \quad (55)$$

and

$$F_2(N) = \sum_{K,L} [D_{KL,KL}^1 (N)]^2 [D_{KL,KL}^0 + (\sigma_0)^2 D_{KL,KL}^1 (N_0)]^{-2} . \qquad (56)$$

The minimum value of the objective function with respect to σ is

$$y(N) = \frac{1}{2} \left[\sum_{K,L} 1 \right]^{-1} \left[F_0(N) - \frac{[F_1(N)]^2}{F_2(N)} \right] . \qquad (57)$$

Since inaccurate field travel times may severely distort the estimated statistics of the geological structure, it is reasonable to restrict the summation only to travel times satisfying inequality

$$T_{KK} \leq (\sigma_{\mathrm{err}})^2 \sigma^{-2} \Theta_{KK} , \qquad (58)$$

where σ_{err} is a given constant. The right-hand side of (58) must be calculated at fixed $N = N_0$ in order not to alter the data set during the minimization of the objective function.

4.3. Minimization of the Objective Function

First we select reasonable values of constants q and σ_{err}. Then we select the value of N_0. The corresponding value of σ_0 may be found iteratively: for an initial estimate of σ_0 we calculate new $\sigma_0 = \sigma(N_0)$ using (53) to gain a better estimate, rapidly approaching the value of σ_0 consistent with the chosen value of N_0. For fixed N_0 and σ_0 we calculate the values of the objective function (57) at different values of N in order to find the minimum. If the values of N depart from N_0, we should select new N_0 and determine new σ_0.

The minimum of the objective function with respect to N is not very pronounced and is very sensitive to serious mistakes in the data. It is also influenced by artificial numerical parameters such as q or σ_{err}. This behavior is due to the sensitivity of N to the fourth statistical moment of the field travel times. That is why N cannot be determined very accurately. An accuracy of the order of ± 0.05 in N may be thought to be an excellent result, difficult to achieve in practice. However, the author hopes that some minor uncertainty in N will not considerably influence the travel-time inversion, see PILKINGTON and TODOESCHUCK (1990).

On the other hand, for given N, reference standard deviation σ depends on the second statistical moment of the field travel times and may be determined very accurately.

For the Gaussian probability distribution, the resulting minimum objective function should be close to 1.

5. Example: Western Bohemia

An attempt is made to estimate the medium correlation function for the region of Western Bohemia and the surrounding part of Germany, using the travel times from the refraction measurements performed from 1989 to 1991 (BUCHA et al., 1992), see Figure 1.

5.1. Reference Travel-time Curve

The mean dependence of the travel times on the hypocentral distance has been roughly approximated by a rational function of the form

$$\tau^0(s) = \frac{as + bs^2}{c + s} \ . \tag{59}$$

At large distances s, the reference travel time (59) approaches the asymptotic line given by slowness b and the travel-time delay of $a - bc$. Constants a, b, and c have been fitted using least squares,

$$a = 0.50 \text{ s}, \quad b = 0.17 \text{ s km}^{-1}, \quad c = 1.25 \text{ km} \ . \tag{60}$$

The deviations $T_K - \tau(s_K)$ of field travel times T_K with respect to reference travel-time curve (59) are shown in Figure 2. The area of the dots in Figure 2 is proportional to the weights w'_K used in the least squares. Weights

$$w_K = \frac{1}{1 + \frac{T_{KK}}{(\delta_{\text{err}} + \rho_{\text{err}} T_K)^2}} \tag{61}$$

with

$$\delta_{\text{err}} = 0.010 \text{ s}, \quad \rho_{\text{err}} = 0.005 \tag{62}$$

have been normalized separately in each interval of length

$$\Delta s = 1 \text{ km} \tag{63}$$

using formula

$$w'_K = w_K \left[1 + \sum_L w_L \right]^{-1} \tag{64}$$

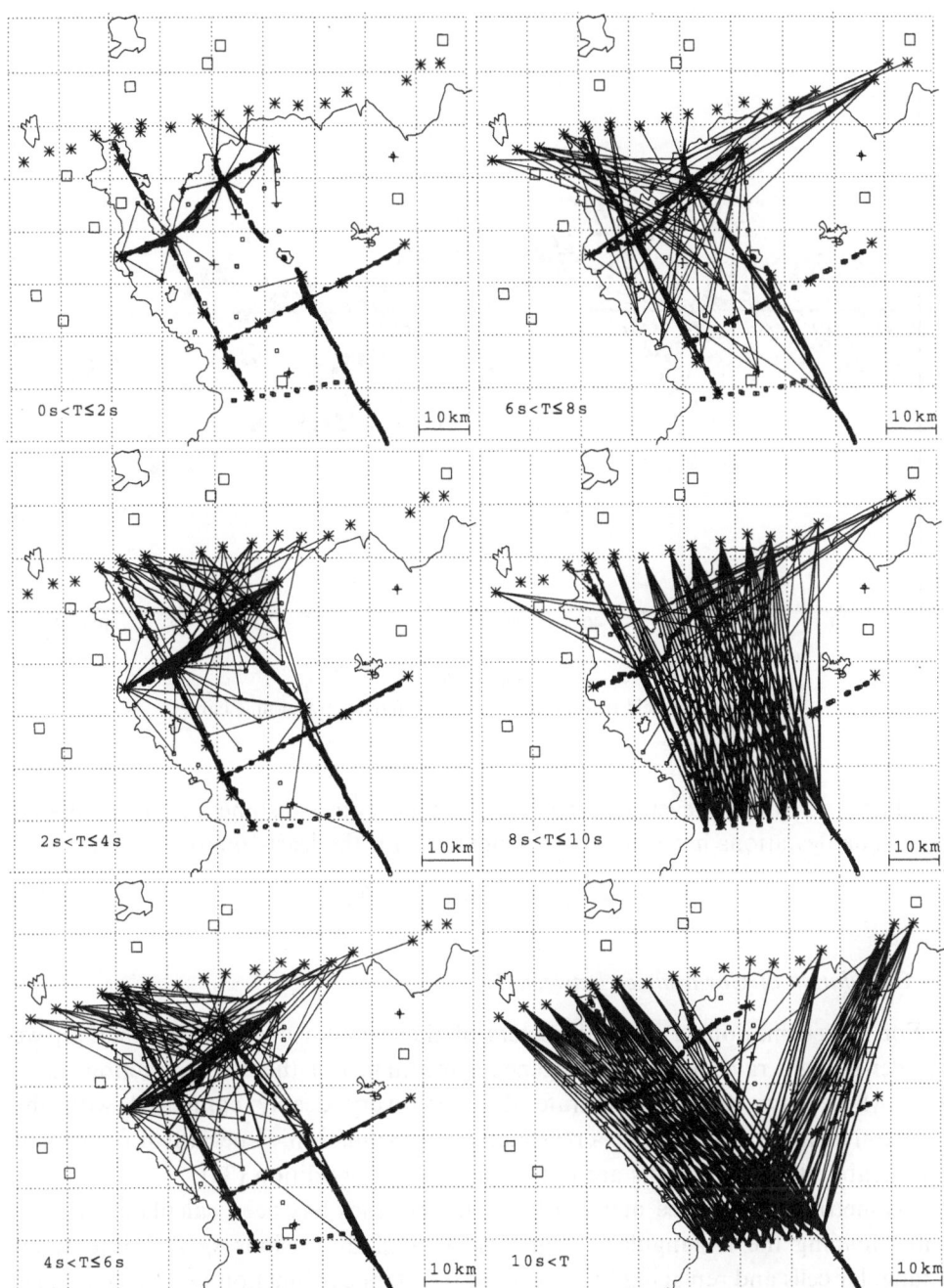

Figure 1

Two-point travel times T measured in the region of Western Bohemia and the surrounding part of Germany during the years 1989 to 1991, graphically represented by segments connecting sources (asterisks) and receivers (small squares). The travel times are sorted according to their length. The state border and greater towns have actual shapes, small towns are represented by the greater squares.

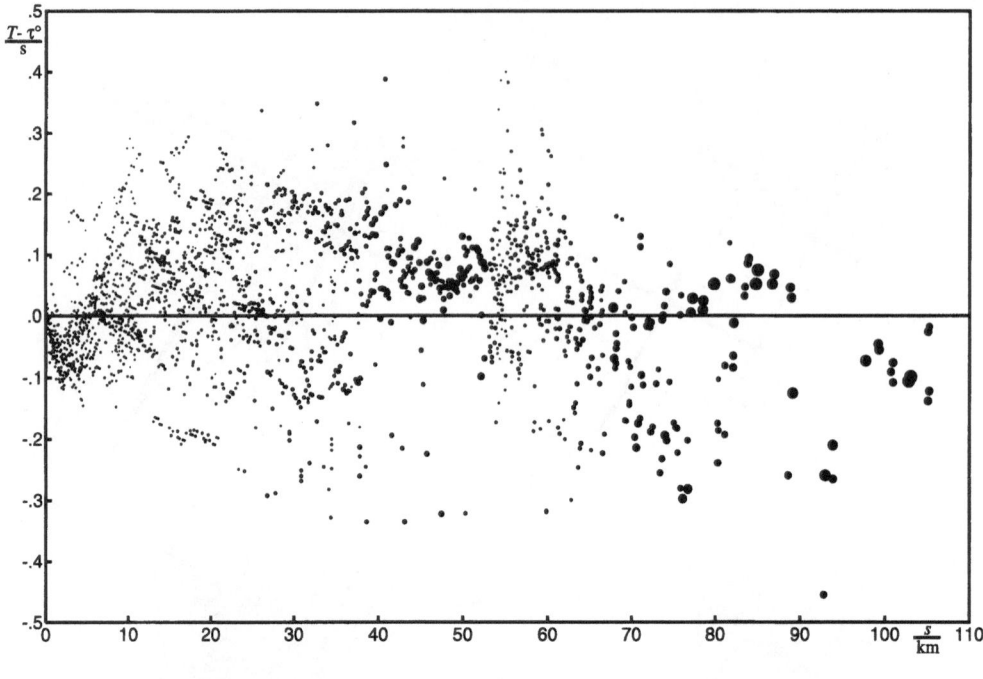

Figure 2
Deviations of field travel times from the reference travel-time curve.

to achieve a relatively even coverage of all hypocentral distances s. The squared travel-time deviations have then been weighted with the least-squares weights

$$w_K'' = w_K' s^{-p} \quad \text{where } p = 0.5 \ . \tag{65}$$

5.2. Calculation of Covariances between Travel Times

For the estimation of the parameters of the medium correlation function, we consider straight rays, as in a homogeneous medium. For the used refraction travel times whose rays do not penetrate the earth very deeply compared with the epicentral distance, it should be a reasonable approximation. Especially if the rays of considerably different lengths are not compared, see condition (34).

Geometrical covariance matrix (25) of travel times has been calculated numerically, dividing the rectangular integration area of dimensions $s_K \times s_L$ into small rectangular cells and replacing the integrand by a bilinear function in each cell. Since the integrand may reach infinity at some points, the integrand has been limited from above at each grid point in such a way as to secure exact values of the integral in all square cells touched by the diagonal for the special case of variances Θ_{KK}.

Unfortunately, the first version of the code used for these tests is not sufficiently debugged, is slow and not sufficiently accurate. This may influence the reliability of the presented numerical results. However, the numerical tests will be improved further.

5.3. Medium Correlation Function

The reference distance of $L = 1$ km is used.

First we attempted to find a good value of numerical parameter σ_{err} which selects the set of field travel times measured with sufficient accuracy, see (58). Table 1 shows the dependence of objective function $y(N)$ and number $\sum_K 1$ of the field travel times used on the selection of σ_{err}, for $q = 0.90$, $N_0 = -0.10$, and $N = -0.11$.

Here the value of $\sigma_0 = 0.011330$ s km^{-1} has been determined for $\sigma_{err} = 0.001558$ s km^{-1} and has been kept fixed in calculating $y(N)$ for different σ_{err}. The value of the objective function is relatively stable for 0.0010 s km$^{-1} \le \sigma_{err} \le 0.0017$ s km^{-1} and increases considerably for larger σ_{err}. Such an increase probably indicates the influence of bad travel-time data. The author has chosen $\sigma_{err} = 0.001558$ s km^{-1} for the subsequent calculations.

The next task is to choose a reasonable value of the other numerical parameter q, which selects the pairs of field travel times according to (34). Unfortunately, the position $N = N_{min}$ of the minimum of objective function $y(N)$ is influenced considerably by the choice of numerical parameter q. Table 2 shows the dependence of the minimum of objective function $y(N)$ on the selection of q, for $\sigma_{err} = 0.001558$ s km^{-1} and σ_0 corresponding to our choice of N_0.

Table 1

The dependence of number $\sum_K 1$ of the field travel times used and of the value of objective function $y(N)$ on the selection of σ_{err}, for $q = 0.90$, $N_0 = -0.10$, and $N = -0.11$

σ_{err}	$\sum_K 1$	$y(N)$
0.0008	1091	0.750
0.0009	1266	0.849
0.0010	1406	0.999
0.0011	1489	1.037
0.0012	1549	1.054
0.0013	1602	1.061
0.0014	1657	1.078
0.0015	1698	1.086
0.0016 ←	1728	1.084
0.0017	1759	1.095
0.0018	1792	1.138
0.0019	1854	1.222
0.0020	1881	1.274
0.0025	1980	1.719
0.0030	2054	2.007

Table 2

The dependence of the minimum of objective function $y(N)$ on the selection of q, for $\sigma_{err} = 0.001558$ s km^{-1} and σ_0 corresponding to our choice of N_0. N_{min} and $\sigma(N_{min})$ are the respective parameters of the medium correlation function. The arrows denote the range of acceptable values in the author's estimation

q	N_0	N_{min}	$\sigma(N_{min})$	$y(N_{min})$
0.50	−0.24	−0.24	0.0092	0.805
0.60	−0.22	−0.23	0.0092	0.850
0.70	−0.18	−0.21	0.0093	0.905
0.75 ←	−0.19	−0.20	0.0094	0.921
0.80	−0.14	−0.17	0.0097	1.006
0.85	−0.12	−0.15	0.0100	1.053
0.90 ←	−0.10	−0.12	0.0106	1.099
0.95	−0.10	−0.09	0.0117	1.121
0.98	−0.10	−0.08	0.0122	1.196

There are at least three different drawbacks of small values of q:

(a) For decreasing q, inaccurate reference travel-time curve $\tau^0(s)$ may begin to influence the results considerably.

(b) The number of differences $T_K/\tau_K^0 - T_L/\tau_L^0$ of the relative travel times is much greater than the number of field travel times T_K, whereas we treat the differences as independent data in objective function (50). This processing need not be correct from the statistical point of view and may worsen for smaller values of q.

(c) Here we substituted curved rays with straight ones. It is probably a reasonable approximation for rays of similar lengths, however for rays of different lengths the straight approximations of rays may be much closer together than the correct rays, separated in depth, are. For small q, some geometrical covariances (25) may thus be calculated greater than correct, which may result in compensation by smaller (more negative) values of N.

On the other hand, the drawback of q approaching 1 consists in exclusion of field travel times not surrounded by other travel times, and consequently in considerable limitation of the amount of available information. This may be the case of the results obtained for values of $q = 0.95$ and $q = 0.98$.

The estimated statistical properties of the medium are applicable at distances between the *inner and outer cutoff scales* (MANDELBROT, 1977), determined here by the epicentral distances. The reference travel-time curve (59) with (60) has been estimated using travel times at epicentral distances from 0.1 km to 105 km which serve as the inner and outer cutoff scales of the reference travel-time curve. The same travel times are used to estimate the medium covariance function. However, a small number of statistically independent travel times at the shortest and longest epicentral distances may require the *inner cutoff scale* of the medium correlation function to be increased towards 0.3 km and the *outer cutoff scale* to be decreased towards 60 km.

A priori geometrical standard deviations of travel times from the mean travel-time curve are

$$\sqrt{\Theta_{KK}} = \sigma L \sqrt{\frac{2}{(1+2N)(2+2N)}} \left(\frac{|\mathbf{x}_1 - \mathbf{x}_2|}{L}\right)^{1+N} , \qquad (66)$$

see (30). The dependence of geometrical standard deviations (66) on the selection of q is displayed in Table 3 for several epicentral distances $s = |\mathbf{x}_1 - \mathbf{x}_2|$.

Taking into account the above considerations, the author considers the medium correlation functions obtained for the values of q from $q = 0.90$ (KLIMEŠ, 1996) to $q = 0.75$ as acceptable.

For $q = 0.90$, we have got Hurst exponent $N = -0.12$ and $\sigma = 0.0106$ s km^{-1}. Medium covariance function (17) then reads

$$C(\mathbf{x}_1, \mathbf{x}_2) \approx (0.0106 \, \text{s km}^{-1})^2 \left(\frac{|\mathbf{x}_1 - \mathbf{x}_2|}{\text{km}}\right)^{-0.24} , \qquad (67)$$

and geometrical standard deviations (66) are

$$\sqrt{\Theta_{KK}} \approx 0.0130 \, \text{s} \left(\frac{s_K}{\text{km}}\right)^{0.88} . \qquad (68)$$

For $q = 0.75$, we have got Hurst exponent $N = -0.20$ and $\sigma = 0.0094$ s km^{-1}. Medium covariance function (17) then reads

$$C(\mathbf{x}_1, \mathbf{x}_2) \approx (0.0094 \, \text{s km}^{-1})^2 \left(\frac{|\mathbf{x}_1 - \mathbf{x}_2|}{\text{km}}\right)^{-0.40} , \qquad (69)$$

and geometrical standard deviations (66) are

Table 3

The dependence of the geometrical standard deviations $\sqrt{\Theta_{KK}}$ of travel times on the selection of q. The geometrical standard deviations $\sqrt{\Theta_{KK}}$ in seconds are displayed for several epicentral distances. N_{min} is the respective coefficient of the medium correlation function. The arrows denote the range of acceptable values in the author's estimation

q	$N_{\min}$	0.1 km	0.3 km	1 km	3 km	10 km	60 km	100 km
0.50	−0.24	0.0025	0.0058	0.0146	0.034	0.084	0.33	0.48
0.60	−0.23	0.0024	0.0057	0.0143	0.033	0.084	0.33	0.50
0.70	−0.21	0.0022	0.0053	0.0137	0.033	0.084	0.35	0.52
0.75 ←	−0.20	0.0021	0.0052	0.0136	0.033	0.086	0.36	0.54
0.80	−0.17	0.0019	0.0048	0.0131	0.033	0.089	0.39	0.60
0.85	−0.15	0.0018	0.0047	0.0130	0.033	0.092	0.42	0.65
0.90 ←	−0.12	0.0017	0.0045	0.0130	0.034	0.099	0.48	0.75
0.95	−0.09	0.0017	0.0045	0.0135	0.037	0.110	0.56	0.89
0.98	−0.08	0.0017	0.0046	0.0139	0.038	0.116	0.60	0.96

$$\sqrt{\Theta_{KK}} \approx 0.0136\,\text{s} \left(\frac{s_K}{\text{km}}\right)^{0.80}. \tag{70}$$

For the dependence of geometrical standard deviations (66) on $N = N_{\min}$ refer to Table 3. However, the tomographic inversion of the travel times should be performed with several different values of N, and the dependence of the resulting models on the uncertainty in N should be studied. Note that at least the inversion for 1-D electrical resistivities yields very similar results for the exact Hurst exponent and that decreased by 0.25 (PILKINGTON and TODOESCHUCK, 1990).

Because the medium covariance function has been determined using the straight-ray approximation, it is applicable to horizontal directions, but not vertical. The vertical behavior of the medium correlation function has been supplemented under the assumption of a statistically isotropic medium, which is obviously not the case of geological structures.

Note that standard deviations (66) also describe the accuracy of the synthetic travel times in the hypothetical best 1-D model of the 3-D geological structure under Western Bohemia, and may, e.g., be used to estimate the accuracy of the kinematic hypocenter determination in such a 1-D model.

WU et al. (1994) studied the medium correlation function using the well-log data from the German continental deep-drilling project (KTB), situated very close to the Western Bohemia region (in the left-hand bottom corner of the region displayed in Figure 1). The relative standard velocity deviation on the sampling interval of 0.1524 m in the well-log data is smaller than the relative standard travel-time deviation on the epicentral distance of 1 km in the Western Bohemia region, see Table 3. This may occur if the KTB locality is considerably less heterogeneous than the Western Bohemia region. WU et al. (1994) determined the Hurst exponent for the vertical direction close to $N = 0$, the Hurst exponent for the horizontal direction close to $N = 1/2$ and the aspect ratio of the horizontal to vertical reference distances around 1.8. GOFF and HOLLIGER (1999) studied the same well-log data and determined the Hurst exponent for the P-wave velocity in the vertical direction also close to $N = 0$.

6. Conclusions

Since a geological structure contains heterogeneities of all sizes, very similar on various scales, we have considered a particular class of self-affine random media, composed of a heterogeneous mean value and a stationary self-affine random function. The power-law medium correlation function then depends on two parameters, the Hurst exponent and the corresponding reference standard deviation.

A method designed to estimate the parameters of the medium correlation function of slowness distribution using field travel times has been proposed in Section 4, and applied to data from the Western Bohemia region in Section 5.

The determination of the Hurst exponent from field travel times is very difficult and sensitive to numerical parameters selected for the inversion. The medium correlation functions with the values of the Hurst exponent such as $N = -0.1$ or $N = -0.2$ are equally acceptable to statistically describe the travel times measured in the Western Bohemia region. On the other hand, for the fixed Hurst exponent, the determination of the reference standard deviation of slowness is easy and reliable. The resulting minimum values of the objective function, chosen for the determination of the Hurst exponent and the corresponding reference standard deviation, indicate that the probability distribution of slowness in the Western Bohemia region may be close to Gaussian.

Acknowledgements

This research has been supported by the Grant Agency of the Czech Republic under Contracts 205/95/1465, 205/01/0927 and 205/01/D097, by the Grant Agency of the Charles University under Contract 237/2001/B-GEO/MFF, by the Ministry of Education of the Czech Republic within Research Project J13/98 113200004, and by the members of the consortium "Seismic Waves in Complex 3-D Structures" (see "http://sw3d.mff.cuni.cz").

REFERENCES

ADDISON, P.S., *Fractals and Chaos: An Illustrated Course* (IOP Publishing, London 1997).

AKI, K. (1973), *Scattering of P Waves under the Montana LASA*, J. geophys. Res. *78*, 1334–1346.

BERAN, M.J., *Statistical Continuum Theories* (Wiley-Interscience, New York 1968).

BERGMANN, P.G. (1946), *Propagation of Radiation in a Medium with Random Inhomogeneities*, Phys. Rev. *70*, 486–492.

BERTEUSSEN, K.A., CHRISTOFFERSON, A., HUSEBYE, E.S., and DAHLE, A. (1974), *Wave Scattering Theory in Analysis of P-wave Anomalies at NORSAR and LASA*, Geophys. J. R. astr. Soc. *42*, 403–417.

BUCHA, V., KLIMEŠ, L., DVOŘÁK, V., and SÝKOROVÁ, Z., *Refraction 3-D seismic measurements in Western Bohemia*. In *Proc. XXIII General Assembly Eur. Seismol. Comm.* (Geoph. Inst. Czechosl. Acad. Sci., Praha 1992) pp. 155–158.

CAPON, J. (1974), *Characterization of Crust and Upper Mantle Structure under LASA as a Random Medium*, Bull. Seismol. Soc. Am. *64*, 235–266.

CHERNOV, L.A., *Wave Propagation in a Random Medium* (McGraw-Hill, New York 1960).

CROSSLEY, D.J., and JENSEN, O.G. (1989), *Fractal Velocity Models in Refraction Seismology*, Pure appl. geophys. *131*, 61–76.

FLATTÉ, S.M., and WU, R-S. (1988), *Small-scale Structure of the Lithosphere and Astenosphere Deduced from Arrival Time and Amplitude Fluctuations*, J. geophys. Res. *93B*, 6601–6614.

GOFF, J.A., and HOLLIGER, K. (1999), *Nature and Origin of Upper Crustal Seismic Velocity Fluctuations and Associated Scaling Properties: Combined Stochastic Analyses of KTB Velocity and Lithology Logs*, J. geophys. Res. *104B*, 13,169–13,182.

GOFF, J.A., and JORDAN, T.H. (1988), *Stochastic Modeling of Seafloor Morphology: Inversion of Sea Beam Data for Second-order Statistics*, J. geophys. Res. *93B*, 13,589–13,608.

GUDMUNDSSON, O., DAVIES, J.H., and CLAYTON, R.W. (1990), *Stochastic Analysis of Global Traveltime Data: Mantle Heterogeneity and Random Errors in the ISC Data*, Geophys. J. int. *102*, 25–43.

HOLLIGER, K. (1997), *Seismic Scattering in the Upper Crystalline Crust Based on Evidence from Sonic Logs*, Geophys. J. int. *128*, 65–72.

HOLLIGER, K., and LEVANDER, A. (1994), *Seismic Structure of Gneissic/Granitic Upper Crust: Geological and Petrophysical Evidence from the Strona-Ceneri Zone (Northern Italy) and Implications for Crustal Seismic Exploration*, Geophys. J. int. *119*, 497–510.

KLIMEŠ, L., *Correlation function of a self-affine random medium*, In *Seismic Waves in Complex 3-D Structures. Report 4* (Dep. Geophys. Charles Univ., Prague 1996) pp. 25–38 (online at "http://sw3d.mff.cuni.cz").

KLIMEŠ, L. (2002a), *Application of the Medium Covariance Functions to Travel-time Tomography*, Pure appl. geophys. *159*, 1791–1810.

KLIMEŠ, L. (2002b), *Correlation Functions of Random Media*, Pure appl. geophys. *159*, 1811–1831.

KNEIB, G., and SHAPIRO, S.A. (1995), *Viscoacoustic Wave Propagation in 2-D Random Media and Separation of Absorption and Scattering Attenuation*, Geophysics *60*, 459–467.

LEVANDER, A., ENGLAND, R.W., SMITH, S.K., HOBBS, R.W., GOFF, J.A., and HOLLIGER, K. (1994), *Stochastic Characterization and Seismic Response of Upper and Middle Crustal Rocks Based on the Lewisian Gneiss Complex, Scotland*, Geophys. J. int. *119*, 243–259.

MANDELBROT, B.B., *The Fractal Geometry of Nature* (W.H.Freeman and Co., New York 1977).

MAURER, H., HOLLIGER, K., and BOERNER, D.E. (1998), *Stochastic Regularization: Smoothness or Similarity?*, Geophys. Res. Lett. *25*, 2889–2892.

MINTZER, D. (1953), *Wave Propagation in a Randomly Inhomogeneous Medium*, J. Acoust. Soc. Am. *25*, 922–927.

MÜLLER, T.M., and SHAPIRO, S.A. (2001): *Most Probable Seismic Pulses in Single Realizations of Two- and Three-dimensional Random Media*, Geophys. J. int. *144*, 83–95.

PILKINGTON, M., and TODOESCHUCK, J.P. (1990), *Stochastic Inversion for Scaling Geology*, Geophys. J. int. *102*, 205–217.

ROTH, M. (1997), *Statistical Interpretation of Traveltime Fluctuations*, Phys. Earth planet. Interiors *104*, 213–228.

SATO, H., and FEHLER, M., *Seismic Wave Propagation and Scattering in the Heterogeneous Earth* (Springer-Verlag, New York 1998).

SCHOTTKY, W. (1926), *Small-shot Effect and Flicker Effect*, Phys. Rev. *28*, 74.

SHAPIRO, S.A. (1992), *Elastic Wave Scattering and Radiation by Fractal Inhomogeneity of a Medium*, Geophys. J. int. *110*, 591–600.

SHAPIRO, S.A., and KNEIB, G. (1993), *Seismic Attenuation by Scattering: Theory and Numerical Results*, Geophys. J. int. *114*, 373–391.

SHAPIRO, S.A., SCHWARZ, R., and GOLD, N. (1996), *The Effect of Random Isotropic Inhomogeneities on the Phase Velocity of Seismic Waves*, Geophys. J. int. *127*, 783–794.

TARANTOLA, A., *Inverse Problem Theory* (Elsevier, Amsterdam 1987).

TURCOTTE, D.L. (1989), *Fractals in Geology and Geophysics*, Pure appl. geophys. *131*, 171–196.

WU, R-S. (1982), *Attenuation of Short Period Seismic Waves Due to Scattering*, Geophys. Res. Lett. *9*, 9–12.

WU, R-S., *Seismic wave scattering*. In *Encyclopedia of Geophysics* (ed. James, D.E.) (Van Nostrand Reinhold 1989a) pp. 1166–1187.

WU, R-S. (1989b), *The Perturbation Method in Elastic Wave Scattering*, Pure appl. geophys. *131*, 605–637.

WU, R-S., and AKI, K. (1985a), *The Fractal Nature of the Inhomogeneities in the Lithosphere Evidenced from the Seismic Wave Scattering*, Pure appl. geophys. *123*, 805–818.

WU, R-S., and AKI, K. (1985b), *Elastic Wave Scattering by a Random Medium and the Small-scale Inhomogeneities in the Lithosphere*, J. geophys. Res. *90B*, 10,261–10,273.

WU, R-S., and AKI, K. (1988), *Multiple Scattering and Energy Transfer of Seismic Waves — Separation of Scattering Effect from Intrinsic Attenuation. II. Application of the Theory to Hindu Kush Region*, Pure appl. geophys. *128*, 49–80.

Wu, R-S., and Flatté, S.M. (1990), *Transmission Fluctuations across an Array and Heterogeneities in the Crust and Upper Mantle*, Pure appl. geophys. *132*, 175–196.

Wu, R-S., and Xie, X-B. (1991), *Numerical Tests of Stochastic Tomography*, Phys. Earth planet. Interiors *67*, 180–193.

Wu, R-S., Xu, Z., and Li, X-P. (1994), *Heterogeneity Spectrum and Scale-anisotropy in the Upper Crust Revealed by the German Continental Deep-Drilling (KTB) Holes*, Geophys. Res. Lett. *21*, 911–914.

(Received October 17, 2000, revised April 2, 2001, accepted May 17, 2001)

To access this journal online:
http://www.birkhauser.ch

Pure appl. geophys. 159 (2002) 1855–1879
0033–4553/02/081855–25 $ 1.50 + 0.20/0

▌Pure and Applied Geophysics

P-wave Tomography in Inhomogeneous Orthorhombic Media

THOMAS MENSCH[1] and VÉRONIQUE FARRA[1]

Abstract—A *P*-wave tomographic method for 3-D complex media (3-D distribution of elastic parameters and curved interfaces) with orthorhombic symmetry is presented in this paper. The technique uses an iterative linear approach to the nonlinear travel-time inversion problem. The hypothesis of orthorhombic anisotropy and 3-D inhomogeneity increases the set of parameters describing the model dramatically compared to the isotropic case. Assuming a Factorized Anisotropic Inhomogeneous (FAI) medium and weak anisotropy, we solve the forward problem by a perturbation approach. We use a finite element approach in which the FAI medium is divided into a set of elements with polynomial elastic parameter distributions. Inside each element, analytical expressions for rays and travel times, valid to first-order, are given for *P* waves in orthorhombic inhomogeneous media. More complex media can be modeled by introducing interfaces separating FAI media with different elastic properties. Simple formulae are given for the Fréchet derivatives of the travel time with respect to the elastic parameters and the interface parameters. In the weak anisotropy hypothesis the *P*-wave travel times are sensitive only to a subset of the orthorhombic parameters: the six *P*-wave elastic parameters and the three Euler angles defining the orientation of the mirror planes of symmetry. The *P*-wave travel times are inverted by minimizing in terms of least-squares the misfit between the observed and calculated travel times. The solution is approached using a Singular Value Decomposition (SVD). The stability of the inversion is ensured by making use of suitable *a priori* information and/or by applying regularization. The technique is applied to two synthetic data sets, simulating simple Vertical Seismic Profile (VSP) experiments. The examples demonstrate the necessity of good 3-D ray coverage when considering complex anisotropic symmetry.

Key words: Seismic anisotropy, orthorhombic symmetry, *qP*-wave, reflection tomography, transmission tomography.

1. Introduction

There is a large body of evidence supporting the existence of widespread seismic anisotropy at all scales. The determination of seismic anisotropy can provide a valuable clue to the structure, lithology and possible deformation processes in subsurface rocks. Moreover, an estimate of the anisotropy parameters can help in providing a propagation model that can be used, for example, in depth imaging where neglecting the presence of anisotropy may result in inaccurate estimations of reflector depths.

[1] Département de Sismologie, Institut de Physique du Globe de Paris, 4 Place Jussieu, 75252 Paris cedex 05, France. E-mail: farra@ipgp.jussieu.fr

Orthorhombic symmetry may be common in certain geophysical structures, *e.g.*, fractured reservoirs. The increased use of multi-azimuthal seismic surveys can allow the generalization of inversion techniques to 3-D orthorhombic media. However, the hypothesis of anisotropy and 3-D inhomogeneity considerably increases the set of parameters describing the model compared to the isotropic case. Assumptions on the model such as known symmetry system, Factorized Anisotropic Inhomogeneous (FAI) media (ČERVENÝ, 1989) and weak anisotropy (THOMSEN, 1986; MENSCH and RASOLOFOSAON, 1997; PŠENČÍK and GAJEWSKI, 1998) are often made to reduce the model parameter set and to simplify the ray-tracing system (see among others JECH and PŠENČÍK, 1992; CHAPMAN and PRATT, 1992; LE BÉGAT and FARRA, 1997).

Fortunately, geological media are often weakly anisotropic (THOMSEN, 1986), which makes the perturbation technique relevant. First-order perturbation techniques can be used to trace rays which propagate in the vicinity of a given reference ray, but in a different medium. A perturbation approach is very attractive for the determination of rays in anisotropic models. FARRA (1989) proposed a method to determine rays and seismograms in a medium with transversely isotropic anisotropy. NOWACK and PŠENČÍK (1991) extended the perturbation method proposed by FARRA and MADARIAGA (1987) in order to obtain rays in general weak anisotropic media. Recently MENSCH and FARRA (1999) presented an approach in which a medium with ellipsoidal anisotropy is used as a reference medium to obtain P-wave travel times in 3-D orthorhombic media.

We present a tomographic method in orthorhombic media that makes use of the Fréchet derivatives of the travel times. The method is applied to two synthetic examples, simulating VSP experiments; as only qP waves are considered, only the P-wave elastic parameters are estimated from travel-time information.

2. qP-wave Ray Tracing in Orthorhombic Media

We compute the qP-wave travel time in orthorhombic media by using a perturbation ray method. Let us recall the Hamiltonian formulation used for ray tracing in anisotropic media (*e.g.* FARRA, 1993).

The ray equations can be written very simply by using the Hamiltonian formulation. In the high frequency approximation, the elastodynamic equation yields a nonlinear first-order partial differential equation for the travel time (the eikonal equation) which has the general form:

$$H(\mathbf{x}, \mathbf{p}) = 0 . \tag{1}$$

The function H is called the Hamiltonian, $\mathbf{x}$ is the position vector, $\mathbf{p} = \nabla T$ is the slowness vector, and T is the travel time. The most common way of solving equation (1) is to use the ray-tracing method. A ray is defined by its canonical vector

$\mathbf{y}(\tau) = \begin{bmatrix} \mathbf{x}(\tau) \\ \mathbf{p}(\tau) \end{bmatrix}$, where $\mathbf{x}(\tau)$ is the position along the ray, $\mathbf{p}(\tau)$ is the slowness vector of the wavefront at position $\mathbf{x}(\tau)$ and τ is a sampling parameter which depends on the chosen form of the Hamiltonian (see ČERVENÝ, 1989, for a discussion). The canonical vector of the rays satisfies Hamilton's equations:

$$\dot{\mathbf{x}} = \nabla_\mathbf{p} H,$$
$$\dot{\mathbf{p}} = -\nabla_\mathbf{x} H \; , \tag{2}$$

where $\nabla_\mathbf{x}$ and $\nabla_\mathbf{p}$ denote the gradients with respect to the vectors $\mathbf{x}$ (position vector) and $\mathbf{p}$ (slowness vector), respectively; dots indicate derivatives with respect to the sampling parameter τ. The travel time is obtained by simple integration along the ray path:

$$T(\mathbf{x}) = \int_{\text{ray}} \mathbf{p} \cdot \dot{\mathbf{x}} \, d\tau \; . \tag{3}$$

The ray equations (2) are independent of the chosen form of the Hamiltonian and are valid in isotropic as well as anisotropic media.

Let us introduce the Christoffel matrix $\mathbf{\Gamma}(\mathbf{x}, \mathbf{p})$, whose elements are given by: $\Gamma_{jk} = a_{ijkl} p_i p_l$. The parameters $a_{ijkl}(\mathbf{x}) = c_{ijkl}(\mathbf{x})/\rho(\mathbf{x})$ are the density normalized elastic parameters and $p_i = \partial T/\partial x_i$ are the components of the slowness vector $\mathbf{p}$. The matrix $\mathbf{\Gamma}(\mathbf{x}, \mathbf{p})$ has three eigenvalues $G_m(\mathbf{x}, \mathbf{p})$ $(m = 1, 2, 3)$.

Three waves, the qP wave $(m = 1)$ and the two qS waves $(m = 2, 3)$, can propagate in the anisotropic solid defined by the elastic parameters a_{ijkl}. For each wave, the eikonal equation may be written in the following form (ČERVENÝ, 1972):

$$G_m(\mathbf{x}, \mathbf{p}) = 1 \; , \tag{4}$$

where $\mathbf{p} = \nabla T$ is the slowness vector of the considered wave. The eikonal equation describes implicitly the phase slowness surface at position $\mathbf{x}$.

For general anisotropic media, the Hamiltonian may be written as follows (see FARRA, 1993):

$$H(\mathbf{x}, \mathbf{p}) = \tfrac{1}{2} h(\mathbf{x})[G_m(\mathbf{x}, \mathbf{p}) - 1] \; , \tag{5}$$

where $h(\mathbf{x})$ is any positive function of position. $h(\mathbf{x})$ is a scaling parameter which is related to the sampling parameter τ along the ray. We can show from Euler's theorem (see ČERVENÝ, 1989) that $dT/d\tau = h(\mathbf{x})$. Several sampling parameters τ have been used in the literature: τ may represent the arclength, the travel time, etc. We take $h(\mathbf{x}) = u^2(\mathbf{x})$, where u is the phase slowness along a given direction and the sampling parameter τ is related to the travel time T by $dT/d\tau = u^2$.

In general anisotropic media the problem is to obtain an explicit form of the Hamiltonian and its partial derivatives. ČERVENÝ (1989) gave analytical expressions for the derivatives of the eigenvalue $G_m(\mathbf{x}, \mathbf{p})$ that require the computation of many terms at each step of integration. For the qP wave, following MENSCH and FARRA

(1999), we use an explicit form of the eigenvalue $G_m(\mathbf{x}, \mathbf{p})$ which is valid to first-order in the perturbation of some anisotropy parameters.

We restrict our study to qP-wave propagation in orthorhombic media. An orthorhombic medium is defined by nine independent density-normalized elastic parameters a_{ijkl} and three mutually orthogonal mirror planes of symmetry (i.e., three additional parameters, the Euler angles ψ_a, θ_a and ϕ_a, are needed to define the orientation of these planes). Note that we adopt the classical Voigt notation of contracted indices for the components of the fourth-rank tensor a_{ijkl}.

Let us assume that the coordinate planes of the system (x, y, z) coincide with the symmetry planes of anisotropy. We introduce the dimensionless parameters:

$$\epsilon_x = \frac{A_{11} - A_{33}}{A_{33}}, \quad \epsilon_y = \frac{A_{22} - A_{33}}{A_{33}}, \tag{6a}$$

$$\Xi_x = 2\frac{\widehat{A}_{23}}{A_{33}}, \quad \Xi_y = 2\frac{\widehat{A}_{13}}{A_{33}}, \quad \Xi_z = 2\frac{\widehat{A}_{12}}{A_{33}}, \tag{6b}$$

where

$$\begin{aligned}
\widehat{A}_{12} &= A_{12} - \frac{A_{11} + A_{22}}{2} + 2A_{66}, \\
\widehat{A}_{13} &= A_{13} - \frac{A_{11} + A_{33}}{2} + 2A_{55}, \\
\widehat{A}_{23} &= A_{23} - \frac{A_{22} + A_{33}}{2} + 2A_{44}.
\end{aligned} \tag{7}$$

Assuming small values of these parameters (i.e., a weakly orthorhombic medium), we can obtain an explicit form of the Hamiltonian (5) of the qP wave, which is valid to first order (MENSCH and FARRA, 1999)

$$H(\mathbf{x}, \mathbf{p}) = \frac{1}{2}\left[(1 + \epsilon_x)p_x^2 + (1 + \epsilon_y)p_y^2 + p_z^2 + \Xi_x\frac{p_y^2 p_z^2}{\mathbf{p}^2} + \Xi_y\frac{p_x^2 p_z^2}{\mathbf{p}^2} + \Xi_z\frac{p_x^2 p_y^2}{\mathbf{p}^2} - u^2(\mathbf{x})\right], \tag{8}$$

where $u^2(\mathbf{x}) = A_{33}^{-1}$ is the vertical squared slowness. If we assume that the medium is a weakly orthorhombic medium, the qP-wave kinematics is described by six parameters: the squared vertical phase-slowness $u^2 = A_{33}^{-1}$ and five dimensionless anisotropy parameters (namely ϵ_x, ϵ_y, Ξ_x, Ξ_y and Ξ_z), instead of nine parameters in an arbitrary orthorhombic medium. These anisotropy parameters present the advantage of having a simple physical interpretation: ϵ_x and ϵ_y are the so-called P anisotropy, i.e., the difference between the horizontal squared qP-wave velocity in the x- and y- directions and the vertical squared qP-wave velocity, normalized by the latter quantity. The parameters Ξ_x, Ξ_y and Ξ_z describe the anellipticity of the qP-wave slowness surface in the three planes of symmetry yOz, xOz and xOy, respectively.

In the general case, where the Cartesian coordinate system (x, y, z) does not coincide with the crystal coordinate system denoted by (x_a, y_a, z_a), the rotation matrix between the two coordinate systems has to be introduced (MENSCH and FARRA, 1999). The first order form of the Hamiltonian of the qP wave is given by:

$$H(\mathbf{x}, \mathbf{p}) = \frac{1}{2} \left[\mathbf{p}^2 + \epsilon_x \mathbf{p}^T \mathbf{M}_x \mathbf{p} + \epsilon_y \mathbf{p}^T \mathbf{M}_y \mathbf{p} + \Xi_x \frac{\mathbf{p}^T \mathbf{M}_y \mathbf{p} \, \mathbf{p}^T \mathbf{M}_z \mathbf{p}}{\mathbf{p}^2} \right.$$
$$\left. + \Xi_y \frac{\mathbf{p}^T \mathbf{M}_x \mathbf{p} \, \mathbf{p}^T \mathbf{M}_z \mathbf{p}}{\mathbf{p}^2} + \Xi_z \frac{\mathbf{p}^T \mathbf{M}_x \mathbf{p} \, \mathbf{p}^T \mathbf{M}_y \mathbf{p}}{\mathbf{p}^2} - u^2(\mathbf{x}) \right] , \tag{9}$$

where the superscript ^{T} in $\mathbf{p}^T$ denotes the transposed vector. $u^2(\mathbf{x})$ is the squared slowness along the crystal coordinate axis z_a, and the anisotropy parameters ϵ_x, ϵ_y, Ξ_x, Ξ_y and Ξ_z are still defined in the crystal coordinate system.

When the crystal coordinate system and the Cartesian coordinate system coincide, the matrices $\mathbf{M}_x$, $\mathbf{M}_y$ and $\mathbf{M}_z$ are given by:

$$\mathbf{M}_x = \begin{pmatrix} 1 & 0 & 0 \\ 0 & 0 & 0 \\ 0 & 0 & 0 \end{pmatrix}, \quad \mathbf{M}_y = \begin{pmatrix} 0 & 0 & 0 \\ 0 & 1 & 0 \\ 0 & 0 & 0 \end{pmatrix}, \quad \mathbf{M}_z = \begin{pmatrix} 0 & 0 & 0 \\ 0 & 0 & 0 \\ 0 & 0 & 1 \end{pmatrix} , \tag{10}$$

and the expression (9) reduces to (8). If the two coordinate systems do not coincide, the matrices $\mathbf{M}_x$, $\mathbf{M}_y$ and $\mathbf{M}_z$ can be obtained from (10) by using the rotation matrix $\mathbf{R}$ (see Appendix).

The first-order expression (9) of the Hamiltonian may be used to compute rays by integration of expressions (2). However, as we want to achieve fast ray tracing, we prefer to use a ray perturbation approach.

The orthorhombic anisotropy and 3-D considerations considerably increase the set of parameters describing the qP-wave propagation, compared to the isotropic case. In order to decrease the number of parameters and to simplify the ray computation, it is extremely convenient to use a factorized anisotropic inhomogeneous (FAI) medium (ČERVENÝ, 1989): in such a medium the density-normalized elastic constants a_{ijkl} share the same spatial variations. As a consequence in a FAI orthorhombic medium, the symmetry planes are invariant and the five anisotropy dimensionless parameters (ϵ_x, ϵ_y, Ξ_x, Ξ_y, Ξ_z) are constant. The concept of the FAI medium not only reduces the number of parameters describing the model, but also simplifies considerably the ray computations.

In FAI orthorhombic media, rays and traveltimes may be computed using a perturbation approach similar to that proposed by FARRA (1989, 1990) for transversely isotropic media. In a medium with ellipsoidal anisotropy ($\Xi_x = \Xi_y = \Xi_z = 0$; but ϵ_x and ϵ_y can be non-zero) and a constant gradient of the squared slowness u^2, exact analytical expressions for rays and traveltimes can be obtained (MENSCH and FARRA, 1999). For a more complex distribution of squared slowness and non-zero anisotropy parameters Ξ_x, Ξ_y and Ξ_z, we use a perturbation approach: a reference medium with ellipsoidal anisotropy and constant gradient of squared slowness u^2 is constructed. The difference between this reference medium and the exact medium is used as a perturbation in order to obtain analytical expressions of rays and traveltimes valid to first-order in weakly orthorhombic media (MENSCH and FARRA, 1999).

FAI media with smooth variations of the squared slowness can be modeled by dividing the medium into a series of cells (the finite element approach) with polynomial expressions of the squared slowness. Ray tracing in models interpolated by local functions like B-splines reduces to connecting analytical solutions at the vertices of parallelepipedic cells. More complex media can be divided into a set of FAI layers separated by interfaces, across which the elastic parameters (slowness squared u^2 and anisotropy parameters) may be discontinuous. The introduction of interfaces may be achieved by using the method explained in FARRA (1989). Some of these interfaces may be artificially introduced to allow a variation in anisotropy parameters and symmetry directions between different parts of the same geological layer.

MENSCH and FARRA (1999) illustrated the accuracy of the perturbation ray tracing for the computation of qP-wave travel times and slowness vectors in orthorhombic media. The results obtained by this technique have been compared with those given by the program package ANRAY written by GAJEWSKI and PŠENČÍK (1990). The ANRAY code computes the exact ray trajectories in anisotropic media by a Runge-Kutta integration. For a model with a qP anisotropy of about 14%, comparisons of the results show a relative travel-time error of less than 0.5 percent and errors in the slowness vector direction of less than 0.5 degrees.

3. Fréchet Derivatives of Travel Time

In order to perform the inversion, we need to estimate the Fréchet derivatives of the travel time with respect to the different model parameters. The model can be divided into a set of layers separated by interfaces, across which the elastic parameters (slowness squared u^2 and anisotropy parameters) may be discontinuous. Within each layer, the slowness squared varies three-dimensionally. Moreover, the finite element ray-tracing approach requires that the slowness squared has continuous first derivatives (FARRA, 1990). For these reasons, the slowness squared is defined by a quadratic 3-D B-spline series (DE BOOR, 1978):

$$u^2(x,y,z) = \sum_{j,k,l} u^2_{jkl}\alpha_j(x)\alpha_k(y)\alpha_l(z) \ , \tag{11}$$

where u^2_{jkl} ($j = 1,\ldots,J$, $k = 1,\ldots,K$, $l = 1,\ldots,L$) are the coefficients of B-spline interpolation and the functions α are the quadratic basis functions. Within each layer the orthorhombic anisotropy is defined by five anisotropy parameters and the three Euler angles ψ_a, θ_a and ϕ_a.

The interfaces separating each layer are 2-D curved surfaces and are represented by an explicit function $f(x, y, z) = z - Z(x, y) = 0$, where $Z(x, y)$ is the interface depth given as a function of the x and y coordinates. In order to assure continuity of the first and second order partial derivatives, the function Z is defined by a cubic 2-D B-spline expansion:

$$Z(x, y) = \sum_{p,q} z_{pq} \beta_p(x) \beta_q(y) \ , \tag{12}$$

where z_{pq} $(p = 1, \ldots, P, \ q = 1, \ldots, Q)$ are the coefficients of B-spline interpolation and the functions β are the cubic basis functions.

Then the model in the ith layer is described by the elastic parameter set:

$$\mathbf{m}^{E_i} = \{u^2_{111}, \ldots, u^2_{JKL}, \epsilon_x, \epsilon_y, \Xi_x, \Xi_y, \Xi_z, \psi_a, \theta_a, \phi_a\}$$

and the i-th interface by a set of interface parameters:

$$\mathbf{m}^{I_i} = \{z^i_{11}, \ldots, z^i_{PQ}\} \ .$$

The full model composed of n layers is defined by the vector:

$$\mathbf{m} = \{\mathbf{m}^{I_1}, \mathbf{m}^{E_1}, \ldots, \mathbf{m}^{I_i}, \mathbf{m}^{E_i}, \ldots, \mathbf{m}^{I_{n+1}}\} \tag{13}$$

containing the total set of parameters.

Let us consider a ray between one source and one receiver in a reference medium. This ray is described by its position $\mathbf{x}_0(\tau)$ and its slowness vector $\mathbf{p}_0(\tau)$. If we consider elastic and interface perturbations of the medium, the travel-time perturbation between the source and the receiver can be written to first order (FARRA and LE BÉGAT, 1995):

$$\Delta T(\mathbf{x}_s, \mathbf{x}_r) = -\int_{\tau_s}^{\tau_r} \Delta H \, d\tau + \sum_{i=1}^{N} \frac{[\mathbf{p}_0(\tau_i^+) - \mathbf{p}_0(\tau_i^-)] \cdot \nabla f_0^i}{\nabla f_0^i \cdot \nabla f_0^i} \Delta f_i \ , \tag{14}$$

where the integration is made along the reference ray path between the source and the receiver and τ_s and τ_r are the sampling parameters of the reference ray at positions $\mathbf{x}_s$ and $\mathbf{x}_r$, respectively. The perturbation of Hamiltonian ΔH contains the elastic parameter perturbations. The last term in equation (14) is due to the perturbation Δf_i of the i-th interface, which is described in the reference medium by the equation $f_0^i(\mathbf{x}) = 0$, ∇f_0^i being the normal vector at the incident point of the reference ray. $\mathbf{p}_0(\tau_i^-)$ and $\mathbf{p}_0(\tau_i^+)$ are the slowness vectors of the incident and reflected/transmitted reference ray at the i-th interface, respectively. N is the number of perturbed interfaces. At the interface, the slowness vector $\mathbf{p}_0(\tau_i^+)$ of the reflected/transmitted ray can be obtained from the relations:

$$\mathbf{x}_0(\tau_i^+) = \mathbf{x}_0(\tau_i^-) \ , \tag{15a}$$

$$\mathbf{p}_0(\tau_i^+) \times \mathbf{\nabla}f_0 = \mathbf{p}_0(\tau_i^-) \times \mathbf{\nabla}f_0 \ , \tag{15b}$$

$$H(\mathbf{x}_0(\tau_i^+), \mathbf{p}_0(\tau_i^+)) = H(\mathbf{x}_0(\tau_i^-), \mathbf{p}_0(\tau_i^-)) = 0 \ , \tag{15c}$$

where the cross-product is denoted by $\times$. The first equation (15a) imposes the ray continuity on the interface. The second equation (15b) corresponds to Snell's law: the projections of the slowness vector of the incident and generated waves along the interface are equal. Lastly the third equation (15c) describes the conservation of the Hamiltonian along the ray and is used to deduce the normal components of slowness vectors of generated waves.

From expression (14) we can write the partial derivative of the travel time with respect to the elastic parameter m_i as:

$$\frac{\partial T}{\partial m_i} = -\int_{\tau_s}^{\tau_r} \frac{\partial H}{\partial m_i} \, d\tau \ , \tag{16}$$

where the Hamiltonian H is given to first order by equation (9).

Therefore the partial derivatives of the travel time with respect to the elastic (slowness and anisotropy) parameters are explicitly given by the expressions following from (9) and (11):

$$\frac{\partial T}{\partial u_{jkl}^2} = \frac{1}{2}\int_{\tau_0^i}^{\tau_1^i} \alpha_j(x)\alpha_k(y)\alpha_l(z) \, d\tau \tag{17}$$

and

$$\frac{\partial T}{\partial \epsilon_x} = -\frac{1}{2}\int_{\tau_0^i}^{\tau_1^i} \mathbf{p}_0^T \mathbf{M}_x \mathbf{p}_0 \, d\tau, \quad \frac{\partial T}{\partial \epsilon_y} = -\frac{1}{2}\int_{\tau_0^i}^{\tau_1^i} \mathbf{p}_0^T \mathbf{M}_y \mathbf{p}_0 \, d\tau,$$

$$\frac{\partial T}{\partial \Xi_x} = -\frac{1}{2}\int_{\tau_0^i}^{\tau_1^i} \frac{\mathbf{p}_0^T \mathbf{M}_y \mathbf{p}_0 \, \mathbf{p}_0^T \mathbf{M}_z \mathbf{p}_0}{\mathbf{p}_0^2} \, d\tau, \quad \frac{\partial T}{\partial \Xi_y} = -\frac{1}{2}\int_{\tau_0^i}^{\tau_1^i} \frac{\mathbf{p}_0^T \mathbf{M}_x \mathbf{p}_0 \, \mathbf{p}_0^T \mathbf{M}_z \mathbf{p}_0}{\mathbf{p}_0^2} \, d\tau, \tag{18}$$

$$\frac{\partial T}{\partial \Xi_z} = -\frac{1}{2}\int_{\tau_0^i}^{\tau_1^i} \frac{\mathbf{p}_0^T \mathbf{M}_x \mathbf{p}_0 \, \mathbf{p}_0^T \mathbf{M}_y \mathbf{p}_0}{\mathbf{p}_0^2} \, d\tau \ .$$

The integration is only made along the ray path contained in the i-th layer. The Fréchet derivatives of the travel time with respect to the Euler angles ψ_a, θ_a and ϕ_a can be expressed in terms of the partial derivatives of the rotation matrix $\mathbf{R}$ and are linear functions of the reference anisotropy parameters. Therefore, they have zero values in an isotropic reference model. In order to obtain perturbations for these angles, the reference model should be anisotropic.

Note that expressions (18) show that the Fréchet derivatives of the travel time with respect to the anisotropy parameters are sensitive to the propagation direction in different ways, consequently the estimations of these parameters will be strongly dependent on the angular coverage of the experiment. For reasons of convenience, let us use the coordinate system (x, y, z) whose coordinate planes coincide with the symmetry planes of anisotropy. In this coordinate system, the matrices $\mathbf{M}_x$, $\mathbf{M}_y$ and $\mathbf{M}_z$ are given by (10). From equations (18) the following conclusions can be obtained; the travel time is most sensitive to the perturbation of ϵ_x and ϵ_y for propagations along x-axis and y-axis, respectively, and to the perturbation of Ξ_x, Ξ_y and Ξ_z for propagations along the diagonals of the yOz, xOz and xOy planes, respectively. In order to study the behavior of these derivatives as a function of the slowness vector direction, let us consider the propagation plane xOz in a homogeneous medium. In this plane, certain partial derivatives are zero, namely $\partial H/\partial \epsilon_y$, $\partial H/\partial \Xi_x$ and $\partial H/\partial \Xi_z$. The remaining derivatives $\partial H/\partial \epsilon_x$ and $\partial H/\partial \Xi_y$ can be expressed (in the plane xOz) as a function of the angle θ between the slowness vector direction and the z-axis:

$$\frac{\partial H}{\partial \epsilon_x} = \frac{1}{2} \mathbf{p}_0^2 \sin^2 \theta \quad \text{and} \quad \frac{\partial H}{\partial \Xi_y} = \frac{1}{2} \mathbf{p}_0^2 \sin^2 \theta \cos^2 \theta \ ,$$

where $\mathbf{p}_0^2$ is the reference phase slowness squared in the direction given by θ. Because the derivative $\partial H/\partial \Xi_y$ is always smaller than the derivative $\partial H/\partial \epsilon_x$ for any value of θ, then the travel time is more sensitive to ϵ_x than to Ξ_y. Consequently, the parameter ϵ_x will always be recovered more accurately than Ξ_y from travel-time inversion. Moreover, in order to obtain independent information about ϵ_x, Ξ_y and the slowness squared, all incidence directions between 0 and 90 degrees should be scanned. Therefore, if the mirror planes of symmetry are known, the ideal experiment should contain propagation in three planes, two of them close to two symmetry planes.

Considering Snell's law (15b) and our choice of interface parameterization, the travel-time perturbation due to the i-th interface changes [the last term in equation (14)] can be rewritten in the following form:

$$\frac{[\mathbf{p}_0(\tau_i^+) - \mathbf{p}_0(\tau_i^-)] \cdot \nabla f_0^i}{\nabla f_0^i \cdot \nabla f_0^i} \Delta f_i = [p_{0z}(\tau_i^-) - p_{0z}(\tau_i^+)] \Delta Z^i \ , \tag{19}$$

where ΔZ^i is the perturbation of the i-th interface depth and $p_{0z}(\tau_i^-)$ and $p_{0z}(\tau_i^+)$ are the incident and reflected/transmitted vertical slowness components, respectively, at the i-th interface. From equation (19) the partial derivatives of the travel time with respect to the interface parameter $m_j^{l_i}$, which is the j-th component of the vector $\mathbf{m}^{l_i}$, have the following form

$$\frac{\partial T}{\partial z_{pq}^i} = \beta_p(x_i)\beta_q(y_i)[p_{0z}(\tau_i^-) - p_{0z}(\tau_i^+)] \ , \tag{20}$$

where x_i and y_i are the x and y coordinates of the intersection point. From expression (20) we can see that the travel time will be particularly sensitive to the perturbation of the interface depth in the case of reflected waves (i.e., the vertical components of the slowness vector $p_{0z}(\tau_i^-)$ and $p_{0z}(\tau_i^+)$ are opposite signs) or in the case of strong velocity contrasts at the interface for transmitted waves. Then reflection experiments would be generally more favorable than transmission experiments to determine the interface parameters.

4. Inversion

Following LE BÉGAT and FARRA (1997), the inverse problem consists of estimating the set of parameters $\mathbf{m}$ that explains the observed travel times in a least-squares sense, i.e., that minimizes the least-squares misfit function S defined as:

$$S(\mathbf{m}) = [\mathbf{T}_{\mathrm{obs}} - \mathbf{T}(\mathbf{m})]^T \mathbf{C}_T^{-1} [\mathbf{T}_{\mathrm{obs}} - \mathbf{T}(\mathbf{m})] + [\mathbf{F}_m - \mathbf{F}(\mathbf{m})]^T \mathbf{C}_F^{-1} [\mathbf{F}_m - \mathbf{F}(\mathbf{m})] , \quad (21)$$

where $\mathbf{T}_{\mathrm{obs}}$ and $\mathbf{T}(\mathbf{m})$ are vectors containing the observed and calculated qP travel times, respectively; $\mathbf{C}_T$ represents the covariance matrix of observed travel-time data. To constrain the solution, we introduce *a priori* values $\mathbf{F}_m$ of some parameter combinations $\mathbf{F}(\mathbf{m})$. The model covariance matrix $\mathbf{C}_F$ describes the uncertainties in the *a priori* values $\mathbf{F}_m$ (see FARRA and MADARIAGA, 1988). The *a priori* information can have several forms and can be used for the different model parameter types. For instance, we may have access to knowledge of the velocity distribution from borehole information (e.g., velocity logs) or previous surface seismic experiments. Geological information, e.g. layering, cracks, fractures, can be used to introduce *a priori* information on the anisotropy parameters. Another type of information is the stability conditions of the elastic matrix.

Regularization (or constraint) is usually needed to make the inverse problem nonsingular. Tomography is intrinsically unstable due to the aperture limitation of the experiments and has a poor resolution near the borders of the model (see PRATT and CHAPMAN, 1992; PRATT *et al.*, 1993). An example of constraint is that the velocity field can be smoothed in some layers by minimizing the first and/or second partial derivatives of the squared slowness function u^2 (LE BÉGAT and FARRA, 1997). A function $\mathbf{F}(\mathbf{m})$ can be introduced in expression (21) to set the first or second derivatives equal to zero with a given uncertainty.

The nonlinear least-squares problem (21) can be solved iteratively by the Gauss-Newton method, which linearizes the functions $\mathbf{T}(\mathbf{m})$ and $\mathbf{F}(\mathbf{m})$ in expression (21) around a current model $\mathbf{m}_c$ to obtain the quadratic approximation $E(\Delta\mathbf{m})$ of the least-squares misfit function S (e.g., TARANTOLA, 1987):

$$E(\Delta\mathbf{m}) = [\Delta\mathbf{T} - \mathbf{A}_T\Delta\mathbf{m}]^T \mathbf{C}_T^{-1} [\Delta\mathbf{T} - \mathbf{A}_T\Delta\mathbf{m}] + [\Delta\mathbf{F}_m - \mathbf{A}_F\Delta\mathbf{m}]^T \mathbf{C}_F^{-1} [\Delta\mathbf{F}_m - \mathbf{A}_F\Delta\mathbf{m}] .$$
$$(22)$$

$\Delta\mathbf{m}$ is a m-vector containing the parameter perturbations, $\Delta\mathbf{T} = \mathbf{T}_{\mathrm{obs}} - \mathbf{T}(\mathbf{m}_c)$ is a n-vector containing the travel-time residuals, i.e., the difference between the observed and calculated travel times, and $\mathbf{A}_T = \partial\mathbf{T}/\partial\mathbf{m}$ a $n \times m$ matrix containing the partial derivatives of travel time with respect to model parameters. The vector

$\Delta\mathbf{F}_m = \mathbf{F}_m - \mathbf{F}(\mathbf{m}_c)$ and the matrix $\mathbf{A}_F = \partial\mathbf{F}/\partial\mathbf{m}$ describe additional constraints such as *a priori* information and regularization.

As the data are considered to be uncorrelated, only the diagonal terms of matrices $\mathbf{C}_T$ and $\mathbf{C}_F$ are nonzero. These terms are computed from uncertainties of travel-time measurements and *a priori* values of the combinations of the parameters $\mathbf{F}(\mathbf{m})$.

Let us introduce the diagonal matrices $\boldsymbol{\sigma}_T$ and $\boldsymbol{\sigma}_F$ defined as $\boldsymbol{\sigma}_T^2 = \mathbf{C}_T$ and $\boldsymbol{\sigma}_F^2 = \mathbf{C}_F$ and the matrix $\mathbf{D}$ as $\mathbf{D} = \begin{pmatrix} \boldsymbol{\sigma}_T^{-1}\mathbf{A}_T \\ \boldsymbol{\sigma}_F^{-1}\mathbf{A}_F \end{pmatrix}$. The least-square solution of the misfit function (22)

$$\Delta\mathbf{m} = (\mathbf{D}^T\mathbf{D})^{-1}\mathbf{D}^T \begin{pmatrix} \boldsymbol{\sigma}_T^{-1}\Delta\mathbf{T} \\ \boldsymbol{\sigma}_F^{-1}\Delta\mathbf{F}_m \end{pmatrix} \tag{23}$$

is calculated using a Singular Value Decomposition (SVD) or conjugate gradient algorithm, this choice being dependent on the number of model parameters. According to NOLET (1993) it is generally admitted that the SVD method efficiently solves problems containing less than about one thousand parameters, otherwise other numerical methods should be used. The process is repeated until the data residuals drop to the level of measurement error. The iterations include ray tracing in the current model.

The *a posteriori* error analysis is performed in order to evaluate a model confidence interval. This estimate is obtained by deriving the *a posteriori* covariance matrix $\mathbf{C}'_M$ (TARANTOLA, 1987):

$$\mathbf{C}'_M = (\mathbf{A}_T^T\mathbf{C}_T^{-1}\mathbf{A}_T + \mathbf{A}_F^T\mathbf{C}_F^{-1}\mathbf{A}_F)^{-1} \ . \tag{24}$$

The stability of the inversion process requires the availability of *a priori* information on the solution and sufficient confidence in this *a priori* information. However, these conditions may not be sufficient and adequate regularization constraints may have to be introduced. The manner in which this regularization is introduced in the misfit function is of prime importance. DELPRAT-JANNAUD and LAILLY (1993) demonstrated that adequate regularizations are those which make use of derivatives of the function describing the model (see also ORY and PRATT, 1995). The choice of the constraint weight lies in a trade-off between the data fit and the solution stability. Stability is related to the condition number (ratio of the largest to the smallest singular value) of matrix $\mathbf{D}$. The condition number measures the sensitivity of the solution to perturbations in the data and matrix $\mathbf{D}$. A stable solution will be obtained if the condition number is less than a certain value, which is related to the computational errors in the elements of matrix $\mathbf{D}$ and to noise in the data. Numerical tests reveals that the eigenvectors associated with singular values, whose ratio with respect to the largest singular value is less than 5×10^{-4}, are poorly determined; therefore, in a practical application, the condition number must be less than 2×10^3. If *a priori* information is not sufficient, the condition number may be too large and errors in the data could cause strong fluctuations in the solution: regularization techniques would then have to be used.

5. Synthetic Examples

5.1. Transmission Tomography

The technique was applied to a simple synthetic multi-azimuthal multiple source offset experiment simulating a VSP experiment. The acquisition geometry is depicted in Figure 1. Travel times were generated from 20 sources distributed evenly along four profiles (azimuth 0°, 45°, 90° and 135°) intersecting at the mouth of a vertical borehole. The origin of the coordinate system was chosen at the borehole mouth. The minimum and maximum offsets for the profiles along the coordinate axes (azimuth 0° and 90°) were −0.95 km and 0.95 km respectively and −1.343 km and 1.343 km, respectively for the two other profiles (azimuth 45° and 135°). Nineteen receivers were distributed every 50 m inside the borehole between the depths 0.05 km and 0.95 km. In order to obtain information about P-wave elastic parameters, only direct qP-wave arrivals were used giving travel-time measurements.

We considered a 3-D model of a FAI orthorhombic medium with a constant gradient of vertical qP-wave phase velocity (Fig. 2):

$$v = (3.75 + 0.25x + 0.5z) \, \text{km/s} \ , \qquad (25)$$

and orthorhombic parameters:

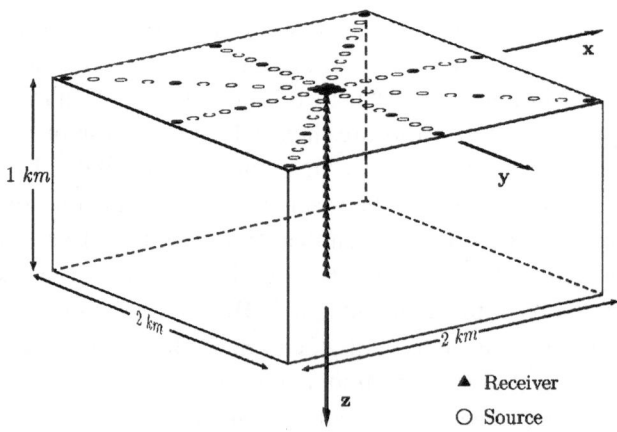

Figure 1
Acquisition geometry: 80 sources (black and white circles) are located on the surface along four lines and 19 receivers (black triangles) along a vertical borehole.

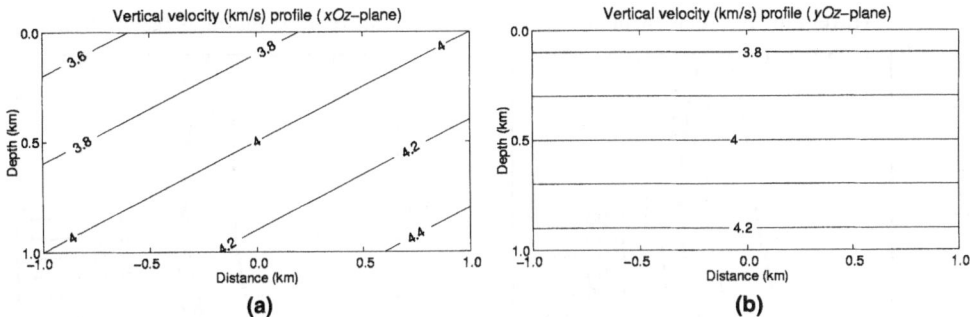

Figure 2
Vertical phase velocity profiles of the model in the xOz (a) and yOz (b) planes.

$$\epsilon_x = 0.10, \quad \epsilon_y = 0.20,$$

$$\Xi_x = -0.20, \quad \Xi_y = -0.10, \quad \Xi_z = -0.10 \ .$$

The coordinate planes of the system (x, y, z) coincided with the symmetry planes.

The synthetic data set from the configuration described above was generated using the ANRAY program package (GAJEWSKI and PŠENČÍK, 1990) and used as observed travel times in the inversion process.

The starting model was homogeneous and isotropic with velocity 4.0 km/s; the anisotropy parameters were set to zero. Rays were traced through this initial model and travel times computed. In Figure 3 we show the initial travel-time residuals ΔT as a function of the depth for six selected source positions along each profile (azimuth 0°, 45°, 90° and 135°). These six sources (denoted by black circles in Fig. 1) are located symmetrically on both sides of the borehole with offsets −0.95 km, −0.55 km, −0.05 km, 0.05 km, 0.55 km and 0.95 km respectively, for the azimuthal planes 0° and 90° (−1.34 km, −0.78 km, −0.07 km, 0.07 km, 0.78 km and 1.34 km, respectively, for the azimuthal planes 45° and 135°). The root mean square of the whole set of travel-time residuals is about $\|\Delta T\| = 5.6$ ms, however, the maximum error reaches 22 ms at the bottom of the well. These plots exhibit the symmetry of the medium (the xOz-plane is a mirror plane of symmetry): we distinguish only three curves in the yOz-plane (azimuth 90°) and we observe identical travel-time residuals in the vertical planes of azimuth 45° and 135°.

The inverse problem consisted of estimating the following set of parameters: the velocity field described by the squared vertical slowness and the five orthorhombic parameters. We assumed that the orientation of the mirror planes of symmetry was known and the coordinate planes of the system (x, y, z) coincided with the symmetry planes. In our case, the square of the slowness was represented by B-splines with a regular mesh of 252 grid points $(6 \times 6 \times 7)$. Then the model was fully described by 257 parameters. The B-spline coefficients of the slowness squared have units in s^2/km^2. Because the parameter perturbations were expected to be of the same order

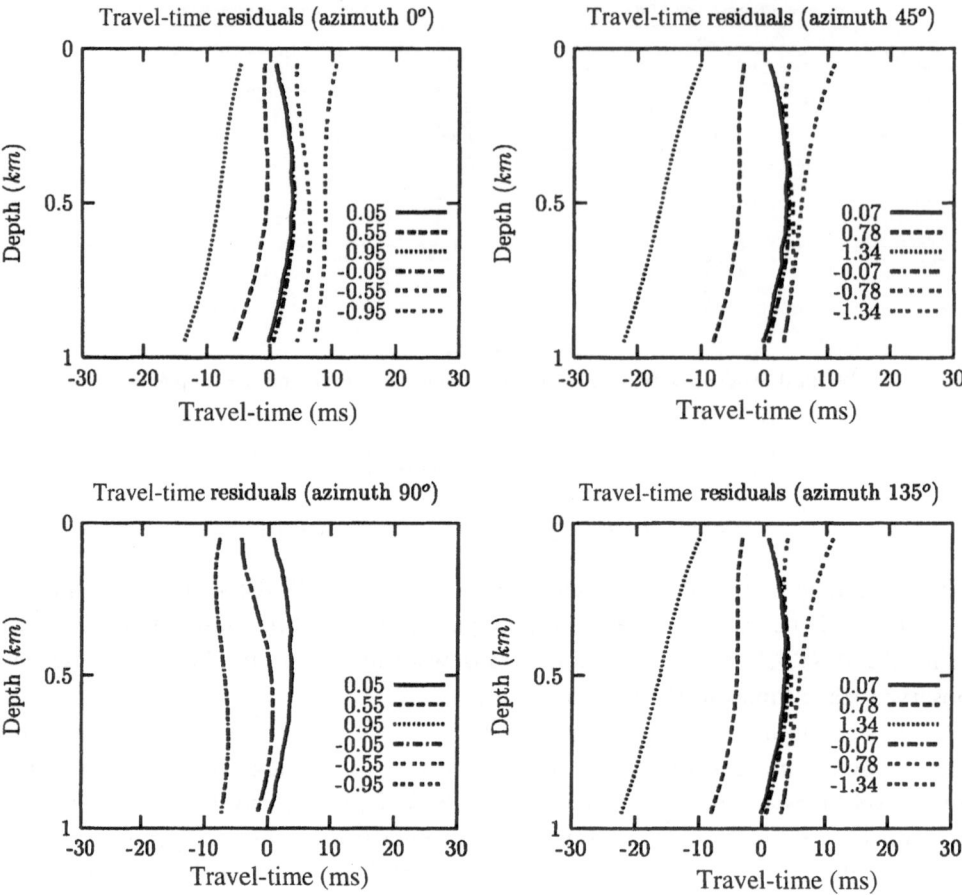

Figure 3

Initial travel-time residuals of the direct waves as a function of depth. Each panel corresponds to a different azimuthal plane (0°, 45°, 90° and 135°). In each plane the curves correspond to six selected source positions (denoted by black circles in Fig. 1) on both sides of the mouth of the borehole with offsets −0.95 km, −0.55 km, −0.05 km, 0.05 km, 0.55 km and 0.95 km for the azimuthal planes 0° and 90° (−1.34 km, −0.78 km, −0.07 km, 0.07 km, 0.78 km and 1.34 km for the azimuthal planes 45° and 135°).

of magnitude, it was not useful to apply a normalization according to the type of parameter (slowness squared or anisotropy parameters); normalization would have been necessary if we used other units for the slowness squared, such as s^2/m^2. Due to the small number of parameters, we used the Singular Value Decomposition method which presents two remarkable aspects: numerical stability and surprising efficiency (PARKER, 1994). We assumed an uncertainty of 1 ms in the travel-time data.

In the inversion process, constraints were added. As only direct P-waves were used and considering the acquisition geometry, we expected that only the center of the model would be correctly sampled by the ray paths. Therefore, to constrain the velocity nodes on the edges of the model we controlled the variations of the velocity distribution. These constraints were introduced by minimizing the first and second finite difference derivatives in the horizontal and vertical directions. No *a priori* information on the anisotropy was used, other than the assumption of FAI orthorhombic anisotropy with known symmetry planes.

The procedure was divided into several steps:

- At the first iteration we looked for one homogeneous isotropic medium; strong constraints on the first finite differences of the B-spline parameters u_i^2 were used.
- For the next few iterations the constraints on the velocity field were progressively relaxed; we kept some constraints on the first and second derivatives in order to assure the stability of the inversion. Due to the acquisition geometry, i.e., the receivers in the well, the constraints applied in the vertical direction (along the z-axis) were smaller than those applied in the horizontal directions (along the x- and y-axes).
- At the second iteration we introduced the anisotropy parameters in the inversion process.

The process was stopped when the travel-time residuals no longer decreased. After four iterations the final model was obtained. The anisotropy was well resolved by the inversion and we obtained the following anisotropy parameters: $\epsilon_x = 0.10$, $\epsilon_y = 0.20$, $\Xi_x = -0.20$, $\Xi_y = -0.10$ and $\Xi_z = -0.10$. The root-mean-square of travel-time residuals (Fig. 4a) was reduced to 0.2 ms. During the inversion process the condition number (Fig. 4b) was always smaller than 2.0×10^2, i.e. the SVD expression (23) was used without any truncation (damping) to compute the solution. The final travel-time residuals are less than 0.5 ms. The final velocity model and the corresponding deviation with respect to the true model are plotted in Figure 5, the maximum error is about 3% at the bottom of the model.

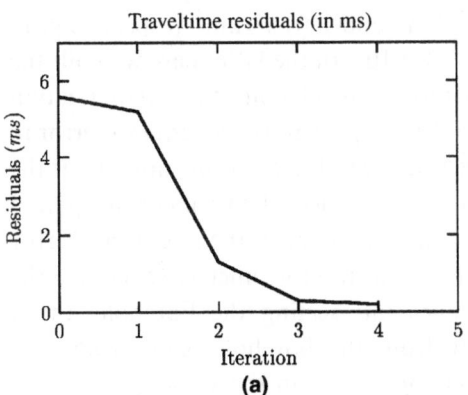
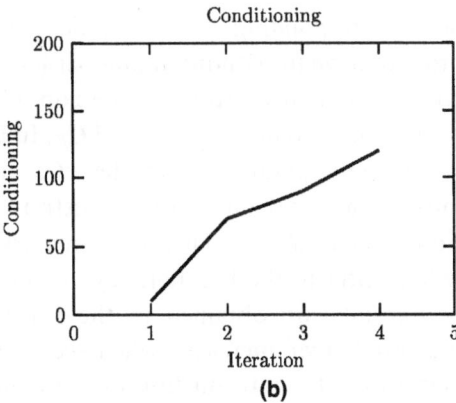

(a)   (b)

Figure 4

Transmission tomography: (a) travel-time residuals and (b) condition number as a function of the number of iteration.

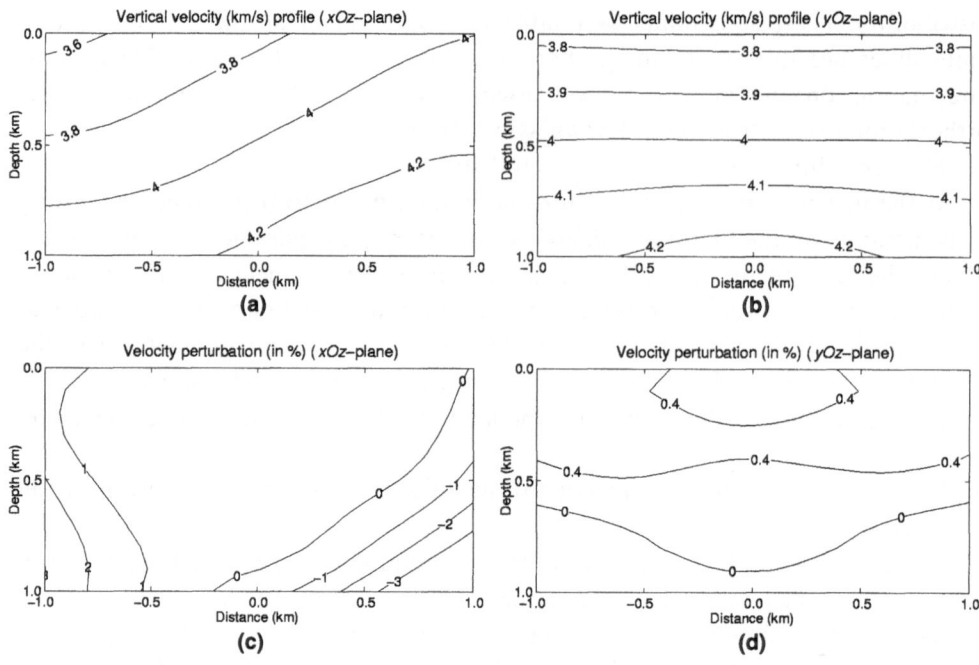

Figure 5

Transmission tomography: (a) vertical velocity profiles of the final model in the xOz (a) and yOz (b) planes and [(c) and (d)] their corresponding normalized velocity deviations (in %) with respect to the true model (Fig. 2).

An estimate of the model confidence was obtained by calculating the *a posteriori* covariance matrix (24), whose diagonal terms are variances of estimated parameters. For an estimated error of 1 ms in the travel-time data, we determined the final model confidence with and without the use of the additional constraints. In the first case we applied a regularization of the first and second derivatives of the squared slowness field, corresponding to a σ_F value of $0.02 \, \text{s}^2/\text{km}^3$ and $0.01 \, \text{s}^2/\text{km}^4$ respectively, in order to have the condition number less than 2.0×10^2. In the latter case we built the covariance matrix from expression (24) without introducing the regularization constraints into matrices $\mathbf{A}_F$ and $\mathbf{C}_F$. In Figure 6 we represent the *a posteriori* error in the velocity parameters in the xOz and yOz planes. First let us note that the introduction of regularization constraints drastically reduces the variance of squared slowness parameters. Then let us remark upon the symmetry of the *a posteriori* error with respect to the borehole; this is due to the acquisition geometry. Of course the best results are obtained in the vertical planes intersecting the borehole; the *a posteriori* error increases when we withdraw from the borehole. Concerning the anisotropy, the introduction of the additional constraints in the velocity field does not change significantly the *a posteriori* error of the anisotropy parameters: the error bars for these parameters, i.e. ϵ_x, ϵ_y, Ξ_x, Ξ_y and Ξ_z, are about ± 0.01 in the case of the use of additional constraints and between ± 0.01 and ± 0.02 otherwise. The weak

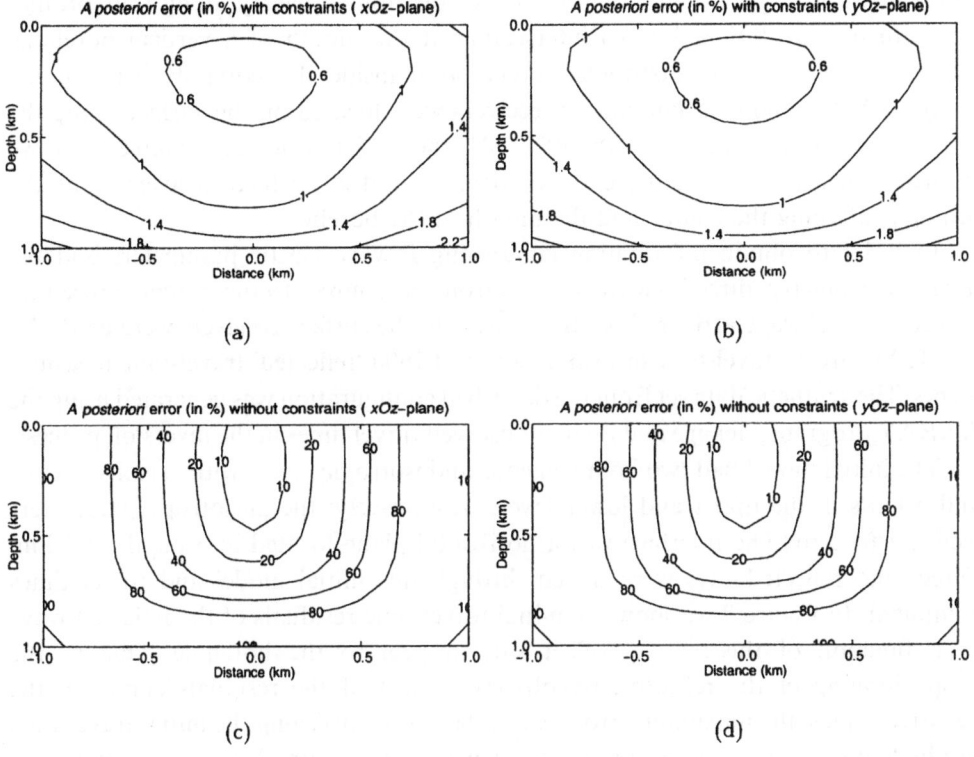

Figure 6

Transmission tomography: *a posteriori* error in the velocity field, normalized by the true velocity, in the *xOz* (a) and *yOz* (b) planes with the use of additional constraints. The same without the use of additional constraints [(c) and (d)].

dependency of error in anisotropy parameters on the additional constraints is due to FAI assumption, which assumes a constant anisotropy in the layer, and the choice of the acquisition geometry: the ray paths sample the space directions well and in a symmetrical way.

As regards the use of regularization (Figs. 6a and 6b), the condition number characterizing the stability of this inverse system is about equal to 1.2×10^2. Without these additional constraints (Figs. 6c and 6d) the estimated errors are large and the condition number is larger than 1.0×10^5.

5.2. Reflection-transmission Tomography

We considered a model with two layers separated by a horizontal interface at a depth of 0.91 km. The upper layer was the FAI orthorhombic medium described in the previous example. The lower layer was isotropic and homogeneous with velocity 5 km/s. In this model an experiment was performed considering both direct and reflected travel times. The acquisition geometry is depicted in Figure 1: as previously, travel times were generated from 20 sources distributed evenly along four profiles (azimuth 0°, 45°, 90° and 135°) intersecting at the mouth of a vertical borehole. Nineteen receivers were distributed every 50 m inside the borehole between the depths 0.05 km and 0.95 km and 20 receivers were located on the surface along the four source profiles (azimuth 0°, 45°, 90° and 135°). For each source shot we recorded the arrivals at the borehole receivers and at 20 surface receivers along the profile containing the source and the mouth of the borehole.

In order to obtain information concerning P-wave elastic parameters and the interface geometry, direct P-wave arrivals from the sources to the borehole receivers and reflected P-wave arrivals from the sources to the surface receivers were used. We had 1520 'direct' travel-time measurements and 1600 'reflected' travel-time measurements. The synthetic data set from the described configuration was generated using the ANRAY program package and used as observed travel times in the inversion process.

The initial model had two homogeneous and isotropic layers with velocity 4 km/s and 5 km/s in the upper and lower layers, respectively; the anisotropy parameters were set to zero. The interface was a horizontal plane located at a depth of 1 km. Direct and reflected rays were traced through this initial model and travel times computed. In Figure 7 we show the initial travel-time residuals of the reflected wave as a function of the receiver offset with respect to the borehole. Due to the mispositioning of the reflector, we observe a shift of the residuals curves in the negative values; the maximum error reaches 60 ms. Concerning the initial travel-time residuals of the direct wave, the reader can refer to Figure 3, as we use the same velocity model as in the previous section. The root-mean-square of the whole set of travel-time residuals is about $\|\Delta T\| = 30\,\mathrm{ms}$.

In this case the inverse problem consisted of estimating the following set of parameters: the velocity field, the five orthorhombic parameters in the upper layer and the interface geometry separating the two layers. We did not invert the elastic parameters (velocity and anisotropy) describing the lower layer because the rays did not go through it. In the upper layer, the velocity field was represented by B-splines with a regular mesh of 252 grid points ($6 \times 6 \times 7$) and the anisotropy by five parameters. The interface geometry was represented by B-splines with a regular mesh of 49 grid points (7×7). The model was therefore fully described by 306 parameters. Because of the units used for the B-spline coefficients, i.e. s^2/km^2 for the slowness squared and km for the interface depth, the parameter (slowness squared, anisotropy and interface) perturbations were expected to be of the same order of magnitude, and normalization was not needed. We used the SVD method and assumed an uncertainty of 1 ms in the travel-time data.

At the first iteration strong constraints were applied on the first and second derivatives in the velocity field and in the interface geometry. For the next few iterations the constraints on the first derivatives were progressively relaxed, however we imposed large constraints on the second derivatives. The anisotropy parameters were introduced at the second iteration.

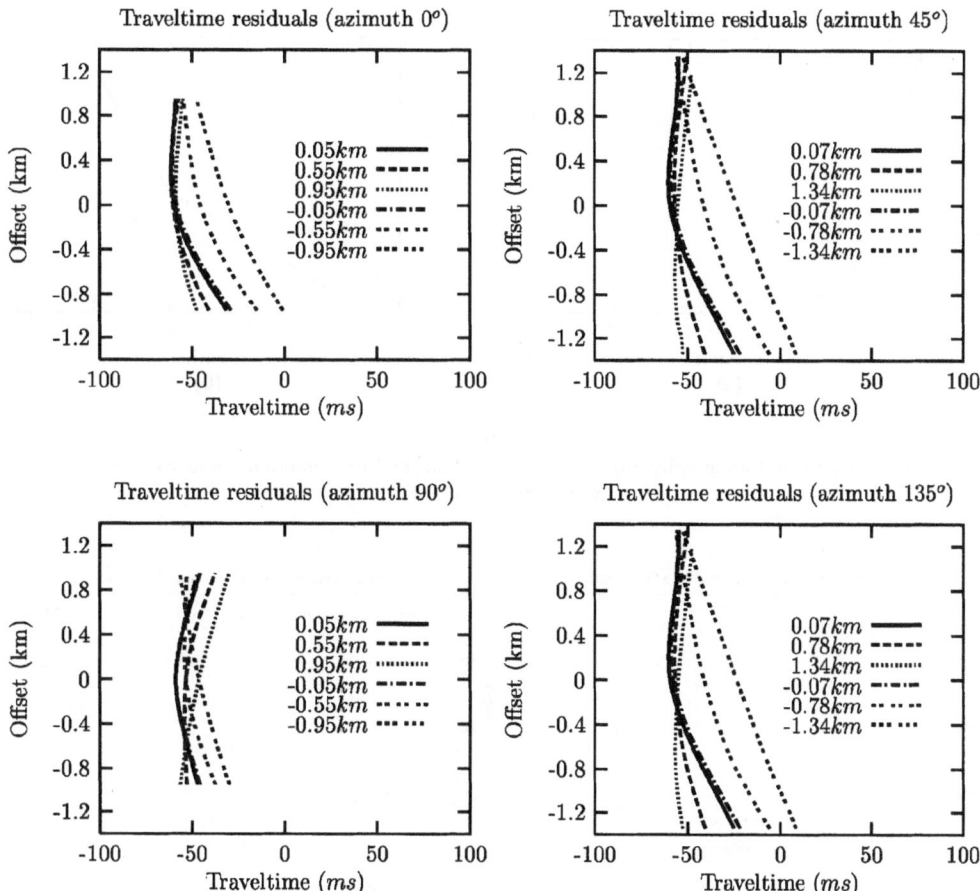

Figure 7

Initial travel-time residuals of the reflected wave as a function of the receiver offset with respect to the borehole. Each panel corresponds to a different azimuthal plane (0°, 45°, 90° and 135°). In each plane the curves correspond to six selected source positions (denoted by black circles in Fig. 1) on both sides of the mouth of the borehole with offsets −0.95 km, −0.55 km, −0.05 km, 0.05 km, 0.55 km and 0.95 km for the azimuthal planes 0° and 90° (−1.34 km, −0.78 km, −0.07 km, 0.07 km, 0.78 km and 1.34 km for the azimuthal planes 45° and 135°).

The process was stopped when the travel-time residuals no longer decreased. After four iterations the final model was obtained. The anisotropy was well resolved by the inversion and we obtained the following anisotropy parameters: $\epsilon_x = 0.10$, $\epsilon_y = 0.20$, $\Xi_x = -0.20$, $\Xi_y = -0.10$ and $\Xi_z = -0.10$. The root-mean-square of final travel-time residuals (Fig. 8a) was less than 0.2 ms. During the inversion process the condition number (Fig. 8b) was always smaller than 4.0×10^2. The final travel-time residuals are less than 0.5 ms. The final velocity model and the corresponding deviation with respect to the true model are plotted in Figure 9, the maximum error is less than 0.5%. The interface geometry is well resolved by the inversion: the corresponding deviation with respect to the true interface position is less than 0.2%.

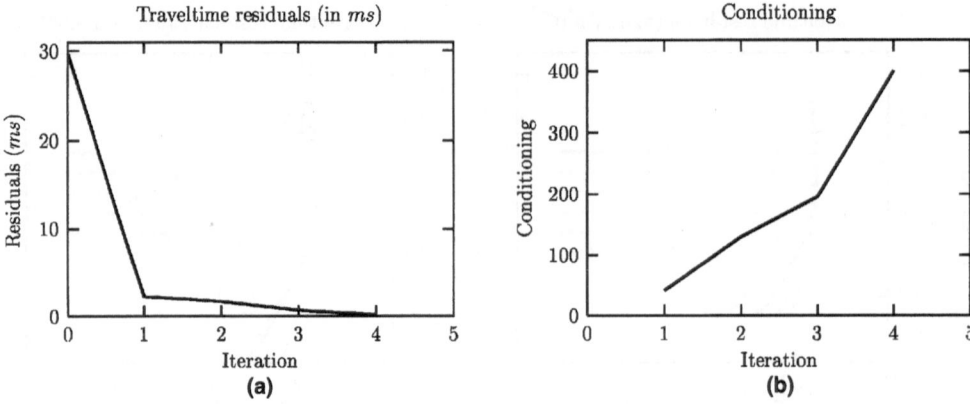

Figure 8
Reflection-transmission tomography: (a) travel-time residuals and (b) condition number as a function of the number of iteration.

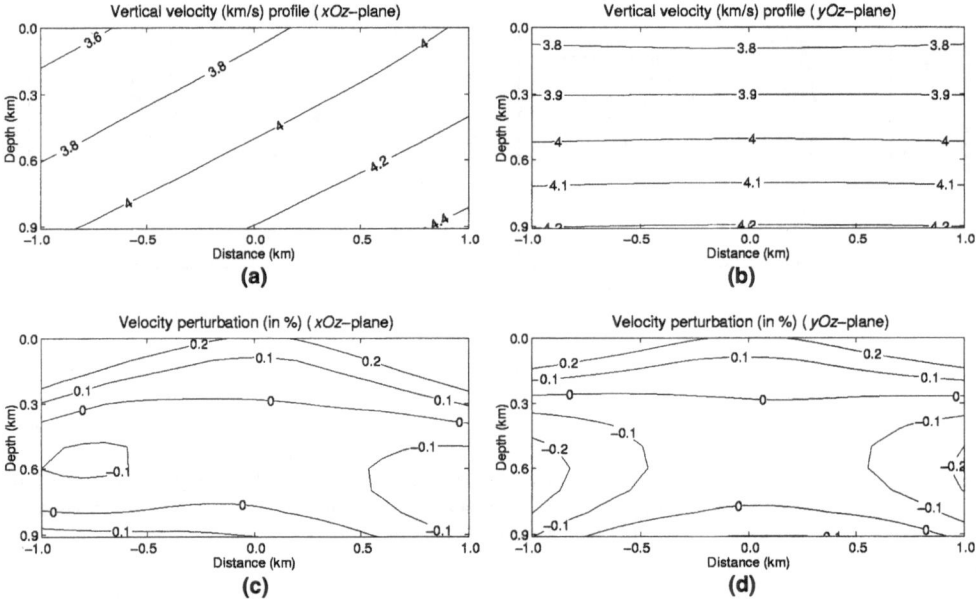

Figure 9
Reflection-transmission tomography: (a) vertical velocity profiles of the final model in the xOz (a) and yOz (b) planes and [(c) and (d)] their corresponding normalized velocity deviations (in %) with respect to the true model (Fig. 2).

An estimate of the model confidence was obtained by calculating the *a posteriori* covariance matrix (24). For an estimated error of 1 ms in the travel-time data we determined the final model confidence for the velocity field (Fig. 10) and for the interface position (Fig. 11) with and without the use of the additional constraints. In

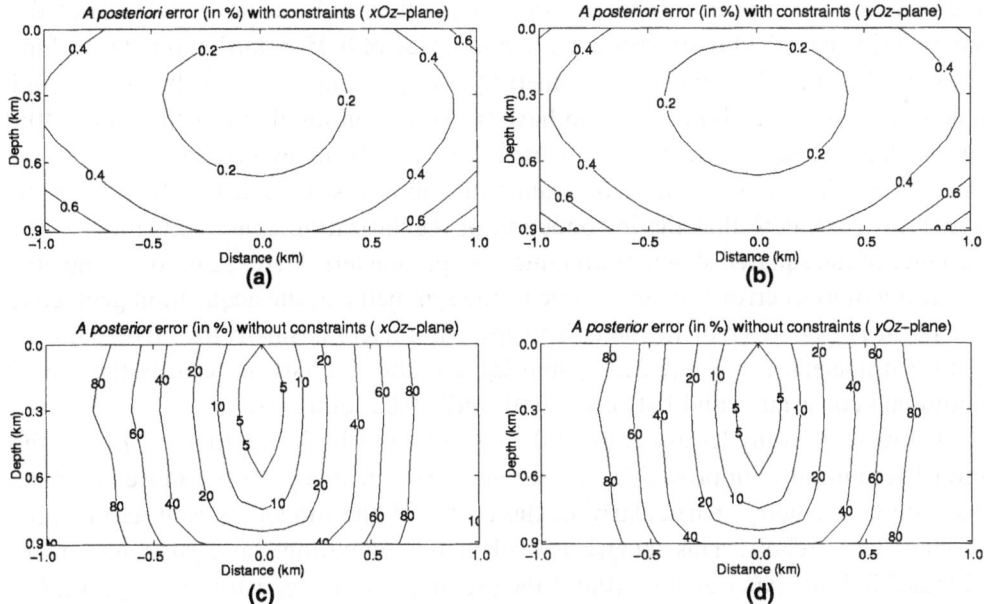

Figure 10

Reflection-transmission tomography: *a posteriori* error in the velocity field, normalized by the true velocity, in the *xOz* (a) and *yOz* (b) planes with the use of additional constraints. The same without the use of additional constraints [(c) and (d)].

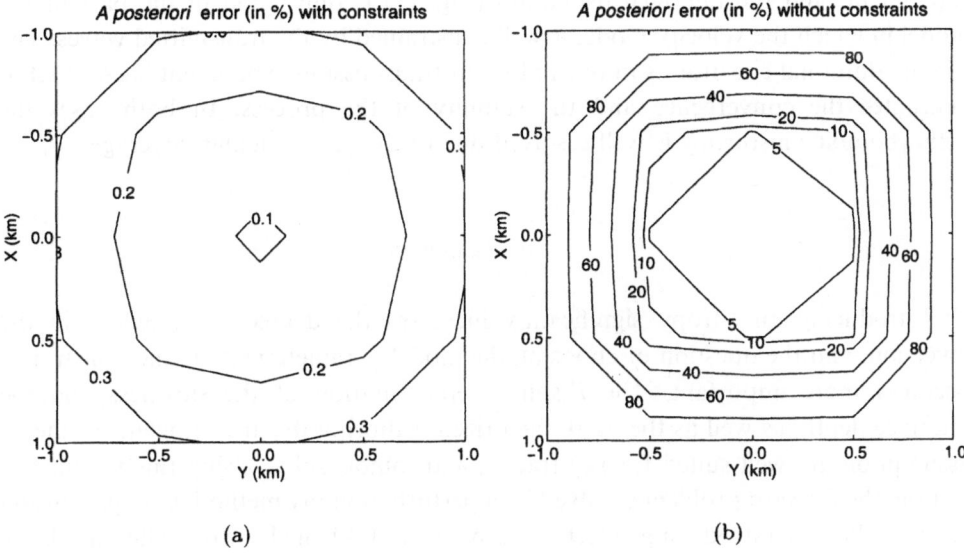

Figure 11

Reflection-transmission tomography: *a posteriori* error in the interface position, normalized by the true interface depth with (a) and without (b) the use of additional constraints.

the first case we applied a regularization of the first and second derivatives of the squared slowness field (corresponding to a σ_F value of $0.05\,\mathrm{s}^2/\mathrm{km}^3$ and $0.003\,\mathrm{s}^2/\mathrm{km}^4$ respectively) and of the interface geometry (corresponding to a σ_F value of 0.05 and $0.003\,\mathrm{km}^{-1}$, respectively) in order to have the condition number less than 4.0×10^2. In the latter case we built the covariance matrix from expression (24) without introducing the regularization constraints into matrices $\mathbf{A}_F$ and $\mathbf{C}_F$. In Figures 10 and 11, we can note that the introduction of regularization constraints reduces the variance of the squared slowness and interface parameters. Here, again the symmetry of the *a posteriori* error function is due to the symmetry of the acquisition geometry.

The *a posteriori* error in the anisotropy parameters is small: the error bars for these parameters, i.e. ϵ_x, ϵ_y, Ξ_x, Ξ_y and Ξ_z, are about ±0.01 as regards the use of additional constraints and between ±0.01 and ±0.02, otherwise.

Before concluding let us compare the resolution of both experiments. Firstly the model resolution is improved in the second experiment due to the more complete acquisition geometry, particularly in the center of the model, as well as in depth around the borehole. This remark is evident by comparing the *a posteriori* errors obtained in both experiments without the use of additional constraints (Figs. 6c, 6d, 10c and 10d). The *a posteriori* errors obtained with the use of additional constraints (Figs. 6a, 6b, 10a and 10b) are more difficult to compare because we do not use the same regularization in the squared slowness parameters. However the reflection experiments do not allow the velocity vertical variations to be obtained (see FARRA and MADARIAGA, 1988): the resolution of the velocity model is not actually better outside the central part of the model in the reflection-transmission tomography experiment, and the interface resolution (Fig. 11) is only good in the part of the model in which the velocity model is well constrained by the transmitted waves. The use of additional constraints in our reflection-transmission experiment is essential to guarantee the convergence and the stability of the process. In both cases the orthorhombic anisotropy is well resolved due to the good angular coverage.

Conclusions

Introducing anisotropy significantly increases the degrees of freedom in the inversion, and the question of choosing the model parameterization and constraints becomes more important. The *B*-spline representation of the slowness and the interface depth, as well as the weak factorized orthorhombic inhomogeneous media assumption are well suited for ray tracing and tomography. Using this parameterization, the forward problem is solved by a perturbation ray method. Complex media are modelled by introducing interfaces separating FAI media with different elastic properties. This approach allows efficient computation of the travel time and its Fréchet derivatives in orthorhombic media. In the weak anisotropy hypothesis, the *P*-wave travel times are sensitive only to a subset of the orthorhombic parameters:

the six *P*-wave elastic parameters and the three Euler angles defining the orientation of the mirror planes of symmetry.

The synthetic examples show the necessity of having a good 3-D ray coverage for considering 3-D orthorhombic media and the need to constrain the inverse problem by introducing independent information. Fortunately, tomography experiments are rarely performed in 'virgin' territory and information from previous seismic experiments, wells and geology are generally available, and can be used as *a priori* constraints. The inversion matrix conditioning, which determines the inversion stability, depends strongly on the experiment geometry. It is therefore necessary to design the survey geometry in order to optimize the resolution of the wanted parameters. The evaluation of the orthorhombic anisotropy parameters is difficult and requires information from various azimuth and incidence angles. Moreover in order to determine the interface positions and lateral velocity variations away from the borehole, direct and reflected waves are needed. Recently LE BÉGAT and FARRA (1997) showed that due to the limited angular coverage of most of the VSP experiments, the travel times alone are not sufficient to study the vertical variations of anisotropic properties of the structure. The addition of independent information, such as polarizations, can provide useful information relative to geological structures and rock properties.

Acknowledgements

We would like to thank the consortium SW3-D for allowing us use of the program package ANRAY for non-commercial purposes. We thank Ivan Pšenčík, R. Gerhard Pratt and Alberto Michelini for their critical reviews of the manuscript which facilitated improvement of the text. Comments by Steve Horne and James Hobro are also appreciated. This work was partially supported by the European Commission in the framework of the Joule project 'Reservoir-oriented delineation technology' and the Institut Français du Pétrole.

This is IPG contribution No 1727.

Appendix

The transformation from a rectangular coordinate system (x, y, z) to a new rectangular coordinate system (x', y', z') may be defined by the three Euler's angles ψ, θ and ϕ. The rotation matrix from the Cartesian coordinate system (x, y, z) to the new coordinate system (x', y', z') is given by:

$$\mathbf{R} = \begin{pmatrix} -s_\psi s_\phi + c_\psi c_\phi c_\theta & -s_\psi c_\phi - c_\psi s_\phi c_\theta & s_\theta c_\psi \\ c_\psi s_\phi + s_\psi c_\phi c_\theta & c_\psi c_\phi - s_\psi s_\phi c_\theta & s_\theta s_\psi \\ -c_\phi s_\theta & s_\phi s_\theta & c_\theta \end{pmatrix} , \tag{51}$$

here $s_\psi = \sin\psi$, $c_\psi = \cos\psi$, $s_\theta = \sin\theta$, $c_\theta = \cos\theta$, $s_\phi = \sin\phi$ and $c_\phi = \cos\phi$. This rotation corresponds to the three following elementary rotations: A rotation by an angle ψ about the z-axis, a subsequent rotation by an angle θ about the new y-axis, and then a final rotation of the rotated system once more about the z'-axis by an angle ϕ.

The relation between the old coordinates (x, y, z) and the new ones (x', y', z') is given by:

$$\begin{pmatrix} x \\ y \\ z \end{pmatrix} = \mathbf{R} \begin{pmatrix} x' \\ y' \\ z' \end{pmatrix} . \tag{52}$$

Let us consider a quadratic form defined by the matrix $\mathbf{M}$ in the Cartesian coordinate system (x, y, z) and the matrix $\mathbf{M}'$ in the Cartesian coordinate system (x', y', z'). The matrix $\mathbf{M}$ is related to the matrix $\mathbf{M}'$ by the relation:

$$\mathbf{M} = \mathbf{R}\mathbf{M}'\mathbf{R}^T . \tag{53}$$

REFERENCES

ČERVENÝ, V. (1972), Seismic Rays and Rays Intensities in Inhomogeneous Anisotropic Media, Geophys. J. R. Astr. Soc. 29, 1–13.

ČERVENÝ, V. (1989), Ray Tracing in Factorized Anisotropic Inhomogeneous Media, Geophys. J. Int. 99, 91–100.

CHAPMAN, C. H. and PRATT, R. G. (1992), Travel-time Tomography in Anisotropic Media–I. Theory, Geophys. J. Int. 109, 1–9.

DE BOOR, C., A Practical Guide to Splines (Springer-Verlag, New York 1978).

DELPRAT-JANNAUD, F. and LAILLY, P. (1993), Ill-posed and Well-posed Formulations of the Reflection Travel-Time Tomography Problem, J. Geophys. Res. 98, 6589–6605.

FARRA, V. (1989), Ray Perturbation Theory for Heterogeneous Hexagonal Anisotropic Medium, Geophys. J. Int. 99, 723–738.

FARRA, V. (1990), Amplitude Computation in Heterogeneous Media by Ray Perturbation Theory: A Finite Element Method Approach, Geophys. J. Int. 103, 341–354.

FARRA, V. (1993), Ray Tracing in Complex Media, J. Appl. Geophys. 30, 55–73.

FARRA, V. and LE BÉGAT, S. (1995), Sensitivity of qP-wave Travel Times and Polarization Vectors to Heterogeneity, Anisotropy and Interfaces, Geophys. J. Int. 121, 371–384.

FARRA, V. and MADARIAGA, R. (1987), Seismic Waveform Modeling in Heterogeneous Media by Ray Perturbation Theory, J. Geophys. Res. 92, 2697–2712.

FARRA, V. and MADARIAGA, R. (1988), Non-linear Reflection Tomography, Geophys. J. 95, 135–147.

GAJEWSKI, D. and PŠENČÍK, I. (1990), Vertical Seismic Profile Synthetics by Dynamic Ray Tracing in Laterally Varying Layered Anisotropic Structures, J. Geophys. Res. 95, 11,301–11,315.

JECH, J. and PŠENČÍK, I. (1992), Kinematic Inversion for qP- and qS-waves in Inhomogeneous Hexagonally Symmetric Structures, Geophys. J. Int. 108, 604–612.

LE BÉGAT, S. and FARRA, V. (1997), P-wave Travel-time and polarization tomography of VSP, Geophys. J. Int. 131, 100–114.

MENSCH, T. and FARRA, V. (1999), Computation of P-wave Rays, Travel Times and Slowness in Orthorhombic Media, Geophys. J. Int. 138, 244–256.

MENSCH, T. and RASOLOFOSAON, P. (1997), Elastic-wave Velocities in Anisotropic Media of Arbitrary Symmetry – Generalization of Thomsen Parameters ε, δ and γ, Geophys. J. Int. 128, 43–64.

NOLET, G., *Solving large linearized tomographic problems.* In *Seismic Tomography, Theory and Practice* (eds. Iyer, H. M. and Hirahara, K.) (Chapman and Hall, London 1993) pp. 227–247.

NOWACK, R. L. and PŠENČÍK, I. (1991), *Perturbation from Isotropic to Anisotropic Heterogeneous Media in Ray Approximation,* Geophys. J. Int. *106*, 1–10.

ORY, J. and PRATT, R. G. (1995), *Are our Parameter Estimators Biased? The Significance of Finite-Difference Regularization Operators,* Inverse Problems *11*, 397–424.

PARKER, R. L., *Geophysical Inverse Theory* (Princeton University Press, Princeton 1994).

PRATT, R. G. and CHAPMAN, C. H. (1992), *Travel-time Tomography in Anisotropic Media–II. Application,* Geophys. J. Int. *109*, 20–37.

PRATT, R. G., McGAUGHLEY, W. J., and CHAPMAN, C. H. (1993), *Anisotropic Velocity Tomography: A Case Study in a Near-Surface Rock Mass,* Geophysics *58*, 1748–1763.

PŠENČÍK, I. and GAJEWSKI, D. (1998), *Polarization, Phase Velocity and NMO Velocity of qP Waves in Arbitrary Weakly Anisotropic Media,* Geophysics *63*, 1754–1766.

TARANTOLA, A. (1987), *Inverse Problem Theory: Methods for Data Fitting and Parameter Estimation* (Elsevier, Amsterdam 1987).

THOMSEN, L. (1986), *Weak Elastic Anisotropy,* Geophysics *51*, 1954–1966.

(Received November 2, 2000, revised January 1, 2001, accepted February 22, 2001)

To access this journal online:
http://www.birkhauser.ch

Pure appl. geophys. 159 (2002) 1881–1905
0033–4553/02/081881–25 $ 1.50 + 0.20/0

© Birkhäuser Verlag, Basel, 2002

▌Pure and Applied Geophysics

Local Determination of Weak Anisotropy Parameters from qP-wave Slowness and Particle Motion Measurements

XUYAO ZHENG[1] and IVAN PŠENČÍK[1]

Abstract — We propose an algorithm for local evaluation of weak anisotropy (WA) parameters from measurements of slowness vector components and/or of particle motions of qP waves at individual receivers in a borehole in a multi-azimuthal multiple-source offset VSP experiment. As a byproduct the algorithm yields approximate angular variation of qP-wave phase velocity. The formulae are derived under assumption of weak but arbitrary anisotropy and lateral inhomogeneity of the medium. The algorithm is thus independent of structural complexities between the source and the receiver. If complete slowness vector is determinable from observed data, then the information about polarization can be used as an independent additional constraint. If only the component of the slowness along the borehole can be determined from observations (which is mostly the case), the inversion without information about polarization is impossible. We present several systems of equations which can be used when different numbers of components of the slowness vector are available. The SVD algorithm is used to solve an overdetermined system of linear equations for WA parameters for two test examples of synthetic multi-azimuthal multiple-source offset VSP data. The system of equations results from approximate first-order perturbation equations for the slowness and polarization vectors of the qP wave. Analysis of singular values and of variances of WA parameters is used for the estimation of chances to recover the sought parameters. Effects of varying number of profiles with sources and of noise added to "observed" data are illustrated. An important observation is that although, due to insufficient data, we often cannot recover all individual WA parameters with sufficient accuracy, angular phase velocity variation can be recovered rather well.

Key words: Weak anisotropy, weak anisotropy parameters, qP waves, slowness vector, polarization vector, local inversion, multi-azimuthal VSP.

Introduction

Let us consider a multi-azimuthal multiple-source offset VSP experiment with three-component downhole recordings not affected by a free surface. Attempts to determine anisotropy parameters from data measured in such an experiment are not new. In the first attempts, VTI anisotropy (transverse isotropy with vertical axis of symmetry) was considered and measured components of the slowness vector were used. The vertical component of the slowness vector was determined from the travel

[1] Geophysical Institute, Acad. Sci. of Czech Republic, Boční II, Praha 4, Czech Republic. E-mail: zheng@ig.cas.cz, ip@ig.cas.cz

times recorded in a vertical borehole. For the determination of the horizontal component, WHITE *et al.* (1983) used qP- and qS-wave travel times from two neighboring boreholes. For the data recorded in a single borehole, GAISER (1990) proposed an approach, in which he assumed lateral homogeneity of the overburden. This assumption allowed him to make use of the travel-time reciprocity and to determine the horizontal components of the qP-wave slowness vector from travel times between sources and a given receiver. DE PARSCAU (1991), HSU *et al.* (1991) and recently WILLIAMSON *et al.* (2000) also assumed a VTI medium and generalized Gaiser's approach in several respects. They used polarization data as an additional source of information on the anisotropy. In addition to data related to qP waves, DE PARSCAU (1991) and HSU *et al.* (1991) also used the qSV-wave data. De Parscau removed Gaiser's assumption of lateral homogeneity of the overburden. PŠENČÍK and ZHENG (1998) generalized Gaiser's approach for qP waves in anisotropic media of arbitrary symmetry, assuming that the anisotropy is weak and the linearized formulae can be used. As shown by MENSCH and RASOLOFOSAON (1997) or PŠENČÍK and GAJEWSKI (1998), the linearized formulae yield sufficiently accurate results for up to 20% strong anisotropy. As DE PARSCAU (1991), PŠENČÍK and ZHENG (1998) also used polarization as an independent source of information relating to the local WA parameters distribution. As GAISER (1990), they assumed that the medium was laterally homogeneous. HORNE *et al.* (1998) used qP-wave slowness vector components and polarization information to reveal parameters of arbitrarily strong TI medium. Their method had the potential to be applied even to media of lower symmetry than TI. To solve a nonlinear problem resulting from the use of exact formulae, they used search strategy based on simulated annealing. HORNE and LEANEY (2000) extended a procedure similar to the above one to include qS waves and reflected waves. For qP waves, their procedure can be used for arbitrary anisotropy. In both cases, the authors successfully applied their approach to real data. WILLIAMSON *et al.* (2000) used an approach similar to HORNE *et al.* (1998) and applied it to real data from a 3-D VSP experiment performed over an assumingly VTI medium.

The formulae given in this paper represent a generalization of the above approaches for anisotropic media of arbitrary symmetry, assuming that their anisotropy is weak. For higher symmetries, they represent their alternatives. Thanks to the assumption of weak anisotropy (WA), which is common for many real materials (THOMSEN, 1986), the formulae have a simple, explicit and transparent form and broad applicability. They can be used for the local determination of up to 15 qP-wave WA parameters (generalization of THOMSEN's, 1986, parameters) in close proximity of receivers situated in a borehole. The medium between source and receiver can be arbitrarily complex. As the input data we use qP-wave slownesses and polarizations. Our approach is as follows. For each receiver we find a reference isotropic medium, from which the sought weakly anisotropic medium differs only slightly. Then we can express the qP-wave slowness and the polarization vector in the weakly anisotropic medium using the formulae relating them linearly to the WA parameters specifying anisotropy of the medium (PŠENČÍK and GAJEWSKI, 1998). For the estimation of components of the slowness vector and polarization, the parametric wavefield decomposition proposed by ESMERSOY (1990) and extended by LEANEY (1990) can be used. In this and related techniques, often only a component of the slowness vector along the borehole can be determined from observations. In such a case, the use of the polarization information becomes inevitable. With the information regarding slowness and polarization available, an overdetermined system of linear equations at each receiver can be constructed, from which the WA parameters can be determined.

In the next two sections, we derive basic equations for qP waves. In further sections, we illustrate the use of these equations to invert data from two synthetic experiments, see Figure 1. The SVD (singular value decomposition) algorithm is used for solving the system of linear equations in both cases. We call the first model VTI and the second TI. In the first, the studied medium is assumed to be laterally homogeneous, of VTI symmetry. In the second experiment, a synthetic multi-azimuthal multiple-source offset VSP experiment is carried out over a laterally inhomogeneous TI medium with tilted axis of symmetry. Since the axis of symmetry is nonvertical and unknown, all 15 qP-wave WA parameters are sought in the process of inversion in the TI model. The "observed" seismograms were generated by the ANRAY package (GAJEWSKI and PŠENČÍK, 1990). Analysis of singular values and variances of WA parameters is used to estimate chances to recover individually sought parameters. Effects of the ray coverage and of noise in data are briefly described.

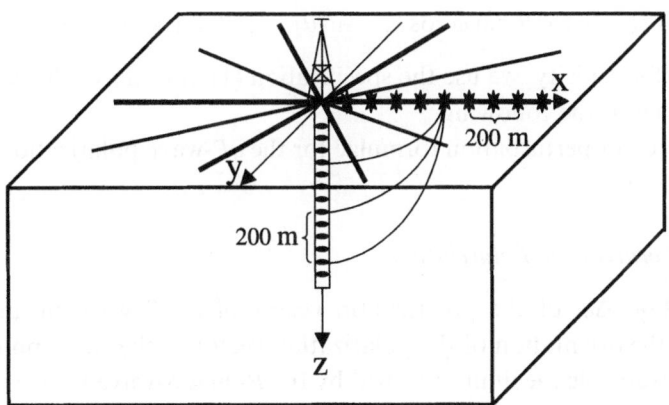

Figure 1
Multi-azimuthal multiple-source offset VSP experiment configuration: 5 radial surface profiles (the 3 profiles from the 3-profile experiment by solid lines) with 9 shots to each side of the borehole, with 0.1 km separation; 13 three-component receivers in the borehole with 0.05 km separation.

Perturbation Formulae

Let us consider a Cartesian coordinate system (x, y, z), with positive z-axis pointing down along the borehole, see Figure 1. Furthermore, in the vicinity of a receiver let us define a reference isotropic medium, from which the sought local WA medium differs only slightly. The reference medium is specified by the P- and S-wave velocities α and β, which can differ from receiver to receiver. In the reference isotropic medium, we introduce three mutually perpendicular unit vectors (in the component notation) $e_i^{(1)}$, $e_i^{(2)}$ and $e_i^{(3)}$ so that the vector $e_i^{(3)}$ is parallel to the polarization vector of the P wave in the reference isotropic medium. The vector $e_i^{(3)}$ is also parallel to the P-wave wave normal n_i since in an isotropic medium the wave normal and polarization vector of a P wave are identical, i.e., $e_i^{(3)} = n_i$. The vectors $e_i^{(1)}$ and $e_i^{(2)}$ can be chosen arbitrarily in the plane perpendicular to n_i. A practical choice of vectors $e_i^{(1)}$, $e_i^{(2)}$ expressed in terms of components of the vector $e_i^{(3)}$ seems to be the following one, see PŠENČÍK and GAJEWSKI (1998):

$$\vec{e}^{(1)} = D^{-1}(n_1 n_3, n_2 n_3, n_3^2 - 1), \quad \vec{e}^{(2)} = D^{-1}(-n_2, n_1, 0), \quad \vec{e}^{(3)} = \vec{n} = (n_1, n_2, n_3) , \quad (1)$$

where

$$D = (n_1^2 + n_2^2)^{1/2}, \quad n_1^2 + n_2^2 + n_3^2 = 1 . \quad (2)$$

Let us specify the phase normal as

$$\vec{n} = (\cos \varphi \sin \theta, \sin \varphi \sin \theta, \cos \theta) . \quad (3)$$

Here φ denotes an azimuthal angle, θ a polar angle, $(0 \leq \varphi \leq 2\pi, 0 \leq \theta \leq \pi)$. In this case we have $D = \sin \theta$ and vectors $\vec{e}^{(1)}$ and $\vec{e}^{(2)}$ from Eq. (1) can be rewritten in the form

$$\vec{e}^{(1)} = (\cos \varphi \cos \theta, \sin \varphi \cos \theta, -\sin \theta), \quad \vec{e}^{(2)} = (-\sin \varphi, \cos \varphi, 0) . \quad (4)$$

For the sake of simplicity, we use the specification (1) in terms of the components of the wave normal in the following.

First, we present perturbation formulae for the qP-wave polarization vector, then for the slowness vector.

Perturbation Formula for Polarization

The deviation Δg_i of the polarization vector of a qP wave in an anisotropic medium from the orientation of the polarization vector of the corresponding P wave in a reference isotropic medium, specified by the P- and S-wave velocities α and β, is given by the formula, see PŠENČÍK (1998)

$$\Delta g_i = \alpha \frac{\partial \Delta \tau}{\partial q_I} e_i^{(I)} + \frac{B_{13} e_i^{(1)}}{\alpha^2 - \beta^2} + \frac{B_{23} e_i^{(2)}}{\alpha^2 - \beta^2} . \quad (5)$$

The symbols B_{mn} denote elements of the *weak anisotropy* matrix,

$$B_{mn} = \Delta a_{ijkl} e_i^{(m)} e_j^{(3)} e_l^{(3)} e_k^{(n)} \ . \tag{6}$$

The symbol Δa_{ijkl} in (6) denotes the deviation of elastic parameters of the anisotropic medium from elastic parameters defining the isotropic reference medium. Explicit expressions for B_{13} and B_{23} for a general case and for the case of a VTI medium are given in the Appendix. The symbols q_I $(I = 1, 2)$ in (5) denote local Cartesian coordinates (the ray-centered coordinates, see Pšenčík, 1998) in the plane perpendicular to the wave normal n_i. The coordinate q_I is specified by the vector $\vec{e}^{(I)}$. The symbol $\Delta\tau$ in (5) denotes a difference between the travel times in the anisotropic and reference isotropic media.

The first term on the RHS of Eq. (5) represents a correction of the slowness vector at a receiver from its orientation in the reference isotropic medium. This correction is due to the perturbation of the phase front, which, in turn, is due to the perturbation Δa_{ijkl}. This term can be determined by integration along the ray in the reference isotropic medium, see, e.g., Farra and Le Bégat (1995). Here, we evaluate this term from the observed data.

The second and third terms on the RHS of Eq. (5) represent deviations of the polarization vector of the qP wave from its slowness vector. They can also be obtained by the use of the perturbation method, see Jech and Pšenčík (1989), Pšenčík and Gajewski (1998).

All terms in (5) lead to the deviation of the polarization vector from the direction $e_i^{(3)} = n_i$ of the polarization vector in the reference isotropic medium. The terms in Eq. (5) remain non-zero even in homogeneous media.

Using (5), we can write the first-order expression for the polarization vector in an arbitrary WA medium in the following form

$$g_i \sim n_i + \Delta g_i = n_i + \alpha \frac{\partial \Delta\tau}{\partial q_I} e_i^{(I)} + \frac{B_{13} e_i^{(1)}}{\alpha^2 - \beta^2} + \frac{B_{23} e_i^{(2)}}{\alpha^2 - \beta^2} \ . \tag{7}$$

Multiplying Eq. (7) by $e_i^{(K)}$, $K = 1, 2$, we get the following two approximate equations

$$\alpha \frac{\partial \Delta\tau}{\partial q_K} + \frac{B_{K3}}{\alpha^2 - \beta^2} = g_i e_i^{(K)} \ . \tag{8}$$

The RHS of Eq. (8) can be determined from observations. The second term on the LHS of Eq. (8) is related to the sought WA parameters through the term B_{K3}, see Eqs. (A1b) and (A1c). The partial derivative in the first term on the LHS of Eq. (8) can be rewritten as follows:

$$\frac{\partial \Delta\tau}{\partial q_K} = p_i e_i^{(K)} - p_i^0 e_i^{(K)} = p_i e_i^{(K)} \ . \tag{9}$$

Here p_i is the slowness vector of the qP wave in the anisotropic medium, p_i^0 is the slowness vector of the corresponding P wave in the reference isotropic medium. The vector p_i^0 is parallel to n_i and thus perpendicular to $e_i^{(K)}$. Equation (8) can be now rewritten into the following form

$$\alpha p_i e_i^{(K)} + \frac{B_{K3}}{\alpha^2 - \beta^2} = g_i e_i^{(K)} \ . \tag{10}$$

The slowness vector at a receiver can be expressed in the following way:

$$p_i = p_i^0 + \Delta\xi i_i + \Delta\zeta j_i + \Delta\eta k_i = (\xi + \Delta\xi)i_i + (\zeta + \Delta\zeta)j_i + (\eta + \Delta\eta)k_i \ . \tag{11}$$

The symbols i_i, j_i and k_i represent i-th components of unit vectors along the x-, y- and z-axis, respectively. The symbols ξ, ζ and η denote sizes of projections of the slowness vector p_i^0 onto vectors i_i, j_i and k_i. The slowness vector p_i^0 reads

$$p_i^0 = \alpha^{-1} n_i \ . \tag{12}$$

This means that ξ, ζ and η in Eq. (11) are

$$\xi = \frac{n_1}{\alpha}, \quad \zeta = \frac{n_2}{\alpha}, \quad \eta = \frac{n_3}{\alpha} \ . \tag{13}$$

Inserting Eq. (11) into Eq. (10), we get the following set of two equations

$$\frac{B_{K3}}{\alpha^2 - \beta^2} + \alpha\Delta\xi e_1^{(K)} + \alpha\Delta\zeta e_2^{(K)} + \alpha\Delta\eta e_3^{(K)} = g_i e_i^{(K)} \ . \tag{14}$$

Equations (14) are equations for 15 WA parameters and three quantities specifying deviation of p_i from $p_i^{(0)}$.

Perturbation Formula for Slowness

We can express the square of the qP-wave slowness c^{-2} in a WA medium using the following approximate formula (see, e.g., PŠENČÍK and GAJEWSKI, 1998)

$$c^{-2} = \alpha^{-2}\left(1 - \frac{B_{33}}{\alpha^2}\right) \ . \tag{15}$$

Neglecting second-order terms of $\Delta\xi$, $\Delta\zeta$ and $\Delta\eta$, Eq. (11) yields

$$c^{-2} = p_i p_i = \alpha^{-2}(1 + 2\alpha^2\xi\Delta\xi + 2\alpha^2\zeta\Delta\zeta + 2\alpha^2\eta\Delta\eta) \ . \tag{16}$$

Comparison of the RHS of Eqs. (15) and (16) yields the following formula

$$B_{33} + 2\alpha^4\xi\Delta\xi + 2\alpha^4\zeta\Delta\zeta + 2\alpha^4\eta\Delta\eta = 0 \ . \tag{17}$$

Expressions for B_{33} for general anisotropy and for a VTI medium can be found in the Appendix, Eqs. (A1a) and (A3a).

Equations For Inversion – General Weak Anisotropy

Equations (14) and (17) represent a set of equations for the WA parameters and for perturbations $\Delta\xi$, $\Delta\zeta$ and $\Delta\eta$ of the slowness vector, for each source-receiver pair. In most applications we can assume the quantity $\Delta\eta$ to be known since $(\eta + \Delta\eta)k_i$ is, approximately, the z-component of the slowness vector p_i, which can be determined from travel-time measurements in a borehole.

Let us first consider the case in which all the components of the slowness vector are determinable from observations, i.e., in addition to $\Delta\eta$, also the deviations $\Delta\xi$ and $\Delta\zeta$ can be considered to be known. The slowness vector components could be determined, for example, from measurements in nearby wells, see WHITE *et al.* (1983), under the assumption of lateral homogeneity of the medium (GAISER, 1990) or under another assumption controlling behavior of the slowness vector. Knowledge of all the three components implies knowledge of the square of the slowness c^{-2}, see Eq. (16). In such a case, Eqs. (14) and (17) (or (15)) can be rewritten into the form

$$B_{K3} = (\alpha^2 - \beta^2)\left(g_i e_i^{(K)} - \alpha\Delta\xi e_1^{(K)} - \alpha\Delta\zeta e_2^{(K)} - \alpha\Delta\eta e_3^{(K)}\right) , \qquad (18)$$

$$B_{33} = \alpha^2(1 - \alpha^2 c^{-2}) . \qquad (19)$$

For N_S sources generating considered waves recorded at a given receiver, Eqs. (18) and (19) represent a set of $3\,N_S$ linear equations for 15 WA parameters (A2). The WA parameters enter Eqs. (18) and (19) through the elements of the WA matrix B_{mn} given in Eqs. (A1). In this special formulation we can consider equations for the slowness separately from equations for the polarization. Use of all equations collectively allows reduction of the number of sources N_S.

We now consider the case, in which, in addition to $\Delta\eta$, only the deviation $\Delta\zeta$ can be determined from observations; the perturbation $\Delta\xi$ being unknown. Then, instead of solving a large set of $3\,N_S$ equations for 15 WA parameters and N_S values of $\Delta\xi$, it is more efficient to eliminate $\Delta\xi$ from Eqs. (14) and (17). In such a way, we arrive at the following two equations

$$(\alpha^2 - \beta^2)^{-1}\xi B_{K3} - \tfrac{1}{2}\alpha^{-3}B_{33}e_1^{(K)} = \xi g_i e_i^{(K)} - \alpha X^{(K)}\Delta\zeta - \alpha\Delta\eta\left(\xi e_3^{(K)} - \eta e_1^{(K)}\right) , \qquad (20)$$

where

$$X^{(K)} = \xi e_2^{(K)} - \zeta e_1^{(K)} . \qquad (21)$$

For N_S sources, we thus have $2\,N_S$ equations for 15 WA parameters (A2) at each receiver. In this formulation it is not possible to deal with information about the slowness vector and polarization separately. Without the information about polarization the inversion is impossible.

If none of the deviations $\Delta\xi$ and $\Delta\zeta$ is known, which is the case in laterally inhomogeneous media, we can eliminate $\Delta\zeta$ from Eqs. (20) and get a single equation

$$(\alpha^2 - \beta^2)^{-1}(B_{13}X^{(2)} - B_{23}X^{(1)}) - \tfrac{1}{2}\alpha^{-2}B_{33}\eta = \left(X^{(2)}e_i^{(1)} - X^{(1)}e_i^{(2)}\right)g_i$$
$$+ \alpha\Delta\eta\left(X^{(1)}e_3^{(2)} - X^{(2)}e_3^{(1)} + \alpha\eta^2\right) . \ (22)$$

For N_S sources we have a system of N_S equations for 15 WA parameters. If a higher anisotropy symmetry is not assumed, N_S must be larger than 15. Information on the slowness vector and polarization must be used together.

Synthetic Examples

Model and Configuration of the Experiment

For the following tests we use the VSP configuration shown schematically in Figure 1. The whole model is confined in a model box whose horizontal dimensions are 10×10 km. The borehole is situated in the center of the model. We consider two types of models. First, we test the above described approach on a simple model of a vertically inhomogeneous VTI medium. This type of anisotropic symmetry has been studied by, for example, WHITE et al. (1983), GAISER (1990), DE PARSCAU (1991), WILLIAMSON et al. (2000). Then we test the approach on a model of a laterally inhomogeneous TI medium with an inclined axis of symmetry. We call the above models VTI and TI, respectively. For simplicity, the same density-normalized elastic parameters are used at the top and the bottom surfaces of the VTI and TI model. Anisotropy of the models is about 8%. At the top surface of each model, the density-normalized elastic parameters A_{ij}, in $(km/s)^2$, are: $A_{11} = A_{22} = 15.71$, $A_{33} = 13.39$, $A_{12} = 5.05$, $A_{13} = A_{23} = 4.46$, $A_{44} = A_{55} = 4.98$, $A_{66} = \tfrac{1}{2}(A_{11} - A_{12}) = 5.33$. At each bottom surface: $A_{11} = 35.348$, $A_{33} = 30.128$, $A_{12} = 11.363$, $A_{13} = 10.035$, $A_{44} = 11.205$. In the VTI model, the top and the bottom surfaces are the horizontal surfaces at depths of 0 and 5 km. In the TI model, the axis of symmetry at the horizontal top surface at the depth of 0 km is rotated by 80° from the z-axis in the (x,z)-plane, and then by 25° around the z-axis. The axis of symmetry at the bottom surface is rotated by 90° from the z-axis in the (x,z)-plane (thus the symmetry of the medium at the bottom surface is HTI). The bottom surface is an inclined plane intersecting the vertical border of the model $x = -5$ km at the depth of 9 km and the border $x = 5$ at the depth of 1 km. Between the top and the bottom surfaces the elastic parameters are determined by linear interpolation. Since both surfaces are horizontal in the case of the VTI model, the variation of elastic parameters with depth is the same everywhere and the model is laterally homogeneous. The inclined bottom surface of the TI model introduces lateral inhomogeneity into the model. Also note that while anisotropy in the VTI model is the same throughout the model (only its strength varies with depth), anisotropy of the TI model varies with depth.

Due to the axial symmetry of the VTI model, a single, arbitrarily oriented, surface profile with sources to one side of the borehole is sufficient. In the TI model, we consider 3 or 5 surface profiles with their center at the mouth of the borehole, see Figure 1. The three profiles are distributed uniformly, with the step in the angle between them equal 60°. This means that the three profiles can be associated successively with 0°, 60° and 120°, measured from the positive x-axis. The five profiles are associated with angles 0°, 30°, 60°, 120° and 150°.

Each profile contains nine sources at each side of the borehole, starting at 0.1 km from the mouth of the borehole and separated by 0.1 km. The vertical borehole contains 13 three-component receivers starting at a depth of 0.1 km and separated by 0.05 km, see Figure 1. The free surface effects are neglected throughout the experiment. From the above configuration we can see that observations along the single profile in the VTI model yield 18 independent equations (see Eqs. (A3) and Eqs. (18) and (19)) for 3 WA parameters at each receiver. In the TI model, the system of three profiles yields 54 equations (see Eq. (22)) for 15 WA parameters at each receiver. The system of five profiles yields 90 equations for 15 WA parameters. Let us note that independently of the number of sources used along the three profiles, we cannot, in principle, recover all the 15 WA parameters. Some of the WA parameters can be recovered only in a linear combination with others; see the discussion after Eq. (17) of Pšenčík and Gajewski (1998). The case of the five profiles represents, according to the mentioned authors, the minimum of profiles allowing, theoretically, a complete recovery of all 15 WA parameters.

Data

For both models, synthetic seismograms were generated using the program package ANRAY (Gajewski and Pšenčík, 1990). From the seismograms, arrival times were picked and particle motion diagrams were constructed. Picked travel times from five neighboring receivers were used to determine the best fitting hyperbola, from which the partial derivative of the travel time with respect to the z-coordinate (coordinate along the borehole) was found. In this way $\Delta\eta$ used in (18), (19), (20) and (22) could be determined. The above described method of determining the components of the slowness vector along the borehole causes inaccuracies in its determination at the shallowest and deepest receivers, see further. To find the polarization vector g_i, a straight line best fitting (in the least-squares sense) the observed particle motion was sought. Alternatively, the parametric inversion method of Esmersoy (1990), extended by Leaney (1990), could be used to recover $\Delta\eta$ and the vertical component of the polarization vector g_i.

The observed travel times were used to determine an approximate 1-D distribution of the P-wave velocity α. The S-wave velocity β was taken as $\beta = \alpha/\sqrt{3}$. In this way, the reference isotropic model was built. Next the rays from the source to individual receivers were traced in the reference medium. Normalized slowness vectors at receivers yielded the wave normal n_i, see Eq. (3), with respect to which, the vectors $e_i^{(1)}$ and $e_i^{(2)}$ were defined.

To solve the resulting system of equations, the SVD method was used. The singular values and variances were used to estimate chances to recover individual WA parameters. The variance $\sigma(\epsilon_i)$ of a WA parameter ϵ_i was calculated from the singular values λ_i ($i = 1, \ldots, 15$) and the 15×15 orthogonal matrix V_{ij} that spans the model space of the WA parameters ($V_{ji}V_{jk} = I_{ik}$, I_{ik} being the 15×15 identity matrix):

$$\sigma^2(\epsilon_i) = \sigma_0^2 \sum_{j=1}^{15} (V_{ji}/\lambda_j)^2 \ . \tag{23}$$

The symbol σ_0^2 denotes a variance caused by variances of a polarization vector and of a vertical component of a slowness vector. The variance is assumed to be unit in the following. The variances depend on the number of used profiles, on the illumination of receivers and also on the chosen reference velocity. For analysis of the data we used the relative singular values. At each receiver they were calculated as ratios of the given singular value and the maximum one.

In the following, we present results of inversion for the above-mentioned two experiments.

VTI Model

Equations (18) and (19) with the elements of the WA matrix B_{mn} specified for a VTI medium, see (A3), are used for this model. This means that the number of sought WA parameters reduces from 15 to only 3: ϵ_x, ϵ_z and δ_x. Because of the model symmetry, the slowness vector is confined to a vertical plane and has only two non-zero components. The vertical component is determined by numerical differentiation of travel times along the borehole. The horizontal component is determined with the use of reciprocity of travel times by a similar numerical differentiation along a surface profile of sources. Since the polarization formula (18) is sensitive to the choice of the reference medium, we use the reference velocity determined from the slowness vector. In other words, we use the velocity c determined from Eq. (19) as the reference velocity for inverting Eq. (18). Because a different reference isotropic medium is used at each receiver, we show inverted values of elastic parameters A_{11}, A_{33} and $A_{13} + 2A_{55}$ in Figure 2 instead of the above-mentioned WA parameters.

Except for the "polarization" method, the relative singular values are relatively close to each other at each receiver, indicating a small condition number. This means that chances to retrieve all three WA parameters independently are high. The smallest relative singular values are of the order 10^{-1}. In the "polarization" method, one singular value is very small, indicating interdependency of the 3 WA parameters. Nonetheless, even in this case, with the small singular value ignored, the inversion yields acceptable results; see below.

The three frames in Figure 2 display, from left to right, the results obtained with the "slowness" equation (19) (left), the results obtained with the "polarization" equation (18) (middle), and the results obtained from the use of both equations

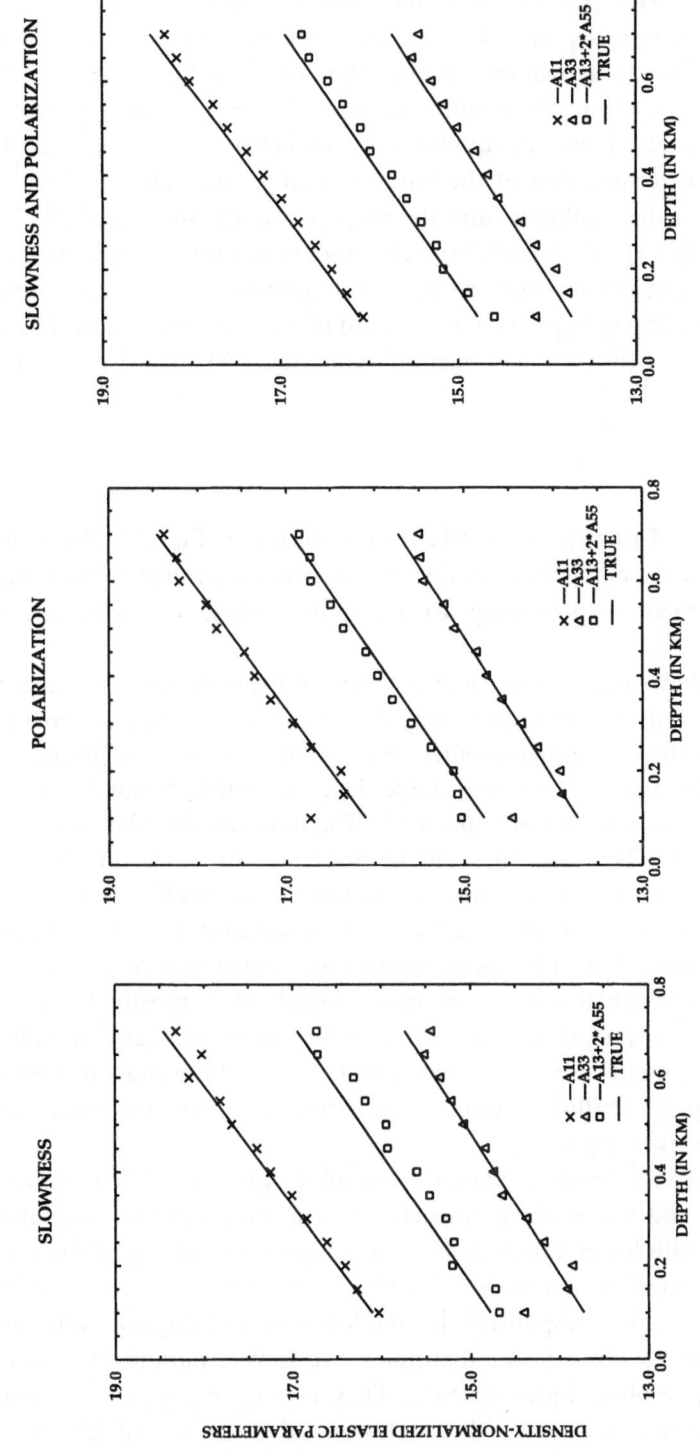

Figure 2

VTI model: Comparison of exact (solid lines) and inverted (symbols) density-normalized elastic parameters using slowness equation (19) (left), polarization equations (18) (middle) and both equations (right).

together (right). The solid lines represent the exact model, the symbols correspond to the specified inverted parameters. In the "slowness" frame we can see relatively well inverted parameters A_{11} and A_{33}, the determination of $A_{13} + 2A_{55}$ not being so satisfactory. The "polarization" frame gives considerably better fit although, as mentioned above, one of the singular values had to be ignored. The best results are, however, obtained when both equations (18) and (19) are used together. The relative errors in the determination of the parameters A_{11}, A_{33} and $A_{13} + 2A_{55}$, for all the receivers except the shallowest and the deepest one, do not exceed 2%.

Remarkably large deviations from the exact values for the shallowest and deepest receivers are caused by inaccuracies in the determination of the slowness vector at the ends of the best-fitting hyperbola, mentioned in the previous section. This also affects results obtained with the "polarization" equation (18) since Eq. (18) contains the slowness vector.

TI Model

The results of this section are based on the use of Eq. (22). Now the observed data yield only the component of the slowness vector parallel to the borehole. Thus without information pertaining to the polarization, the inversion cannot be performed.

Figure 3 shows the relative singular values of the problem as a function of depth. The upper plot shows the relative singular values for the case of three profiles, the lower plot for the case of five profiles. We can see that for three profiles, the span of the relative singular values is very large, i.e., the condition number is large. Three singular values are smaller or equal 10^{-6}. The next two singular values are between 10^{-3} and 10^{-2}. This indicates interdependency of some of the sought WA parameters. This is not a surprise because three profiles are insufficient for independent recovery of all 15 qP-wave WA parameters. The situation changes dramatically when two additional profiles are added (lower plot in Fig. 3). The lowest relative singular values are now around 10^{-3}, mostly for the shallowest receivers. This means that chances for independent recovery of individual WA parameters are much larger using five profiles. From both plots in Figure 3 we can see that, except for shallow receivers, the distribution of singular values depends only slightly on receiver depth.

Figures 4a and 4b show variances of all 15 qP-wave WA parameters for the case of three and five profiles, respectively. The variances are calculated from the formula (23) with lower values, indicating a higher probability of the determination of the corresponding parameter. In Figure 4a, we can clearly see which WA parameters are mainly responsible for the low values of singular values in the upper plot of Figure 3. We can see that there are 6 WA parameters, which have no chance to be resolved independently. They are: ϵ_y, δ_z, χ_x, ϵ_{16}, ϵ_{24} and ϵ_{26}. This observation is in agreement with conclusions of PŠENČÍK and GAJEWSKI (1998) if

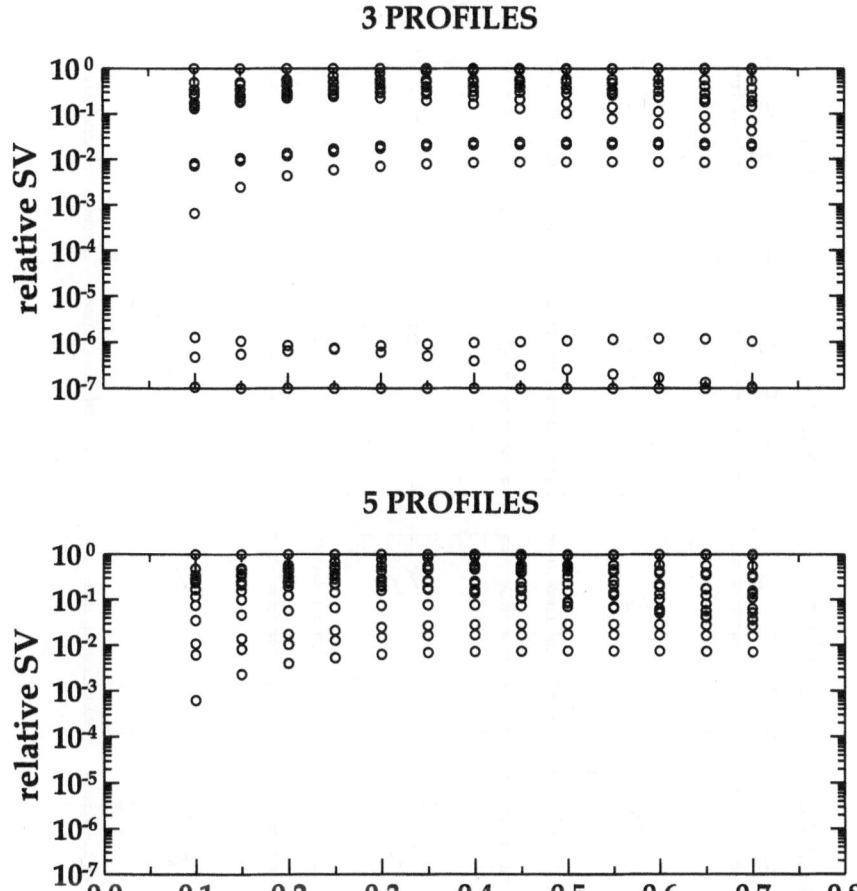

Figure 3
TI model: Singular values corresponding to 15 qP-wave WA parameters as a function of depth obtained for 3 profiles and 5 profiles.

we take into account that none of the three considered profiles is situated in the plane (y, z). All the above listed parameters are related to prevailingly horizontal propagation, and in a horizontal plane to the propagation along the y-axis, see Eqs. (A1). This is a consequence of insufficient horizontal illumination, especially of deeper receivers. Another consequence is the increase of variances with depth for most of the WA parameters. Extension of profiles or, better, increase of their number can reduce this problem, see Figure 4b. The variances dramatically change when two additional profiles are considered. The variances are now generally small and differences between them are substantially smaller. We can observe slightly larger variances for the parameters ϵ_x, ϵ_y, δ_x, δ_y, δ_z and χ_z. The addition of the two

3 PROFILES

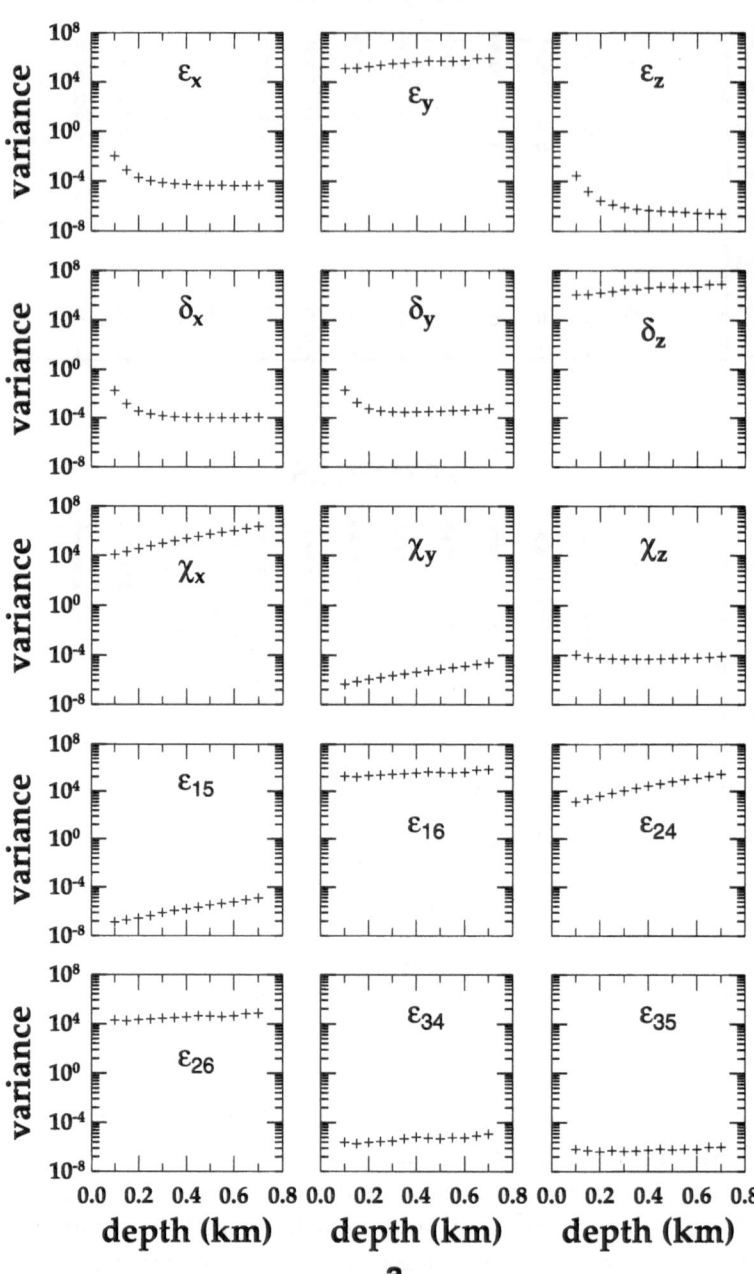

a

Figure 4

TI model: Variances of 15 qP-wave WA parameters as a function of depth obtained from system of equations (22) for 3 profiles and 5 profiles. The lower the value of variance, the better the chance to determine the corresponding parameter.

5 PROFILES

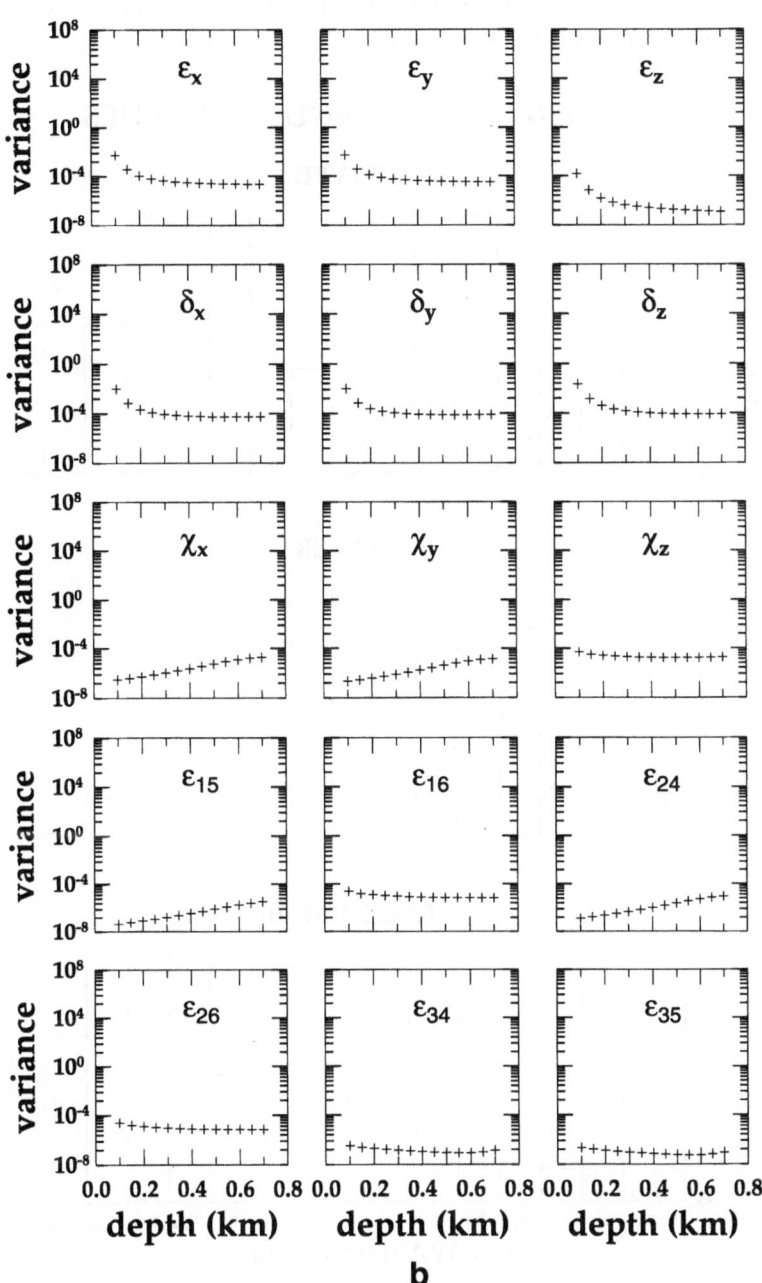

Figure 4
Continuation.

profiles improved illumination of deeper receivers so that only a few variances increase with depth. Let us note that Figure 4b would only slightly change if an additional profile were added.

3 PROFILES, ALL SINGULAR VALUES

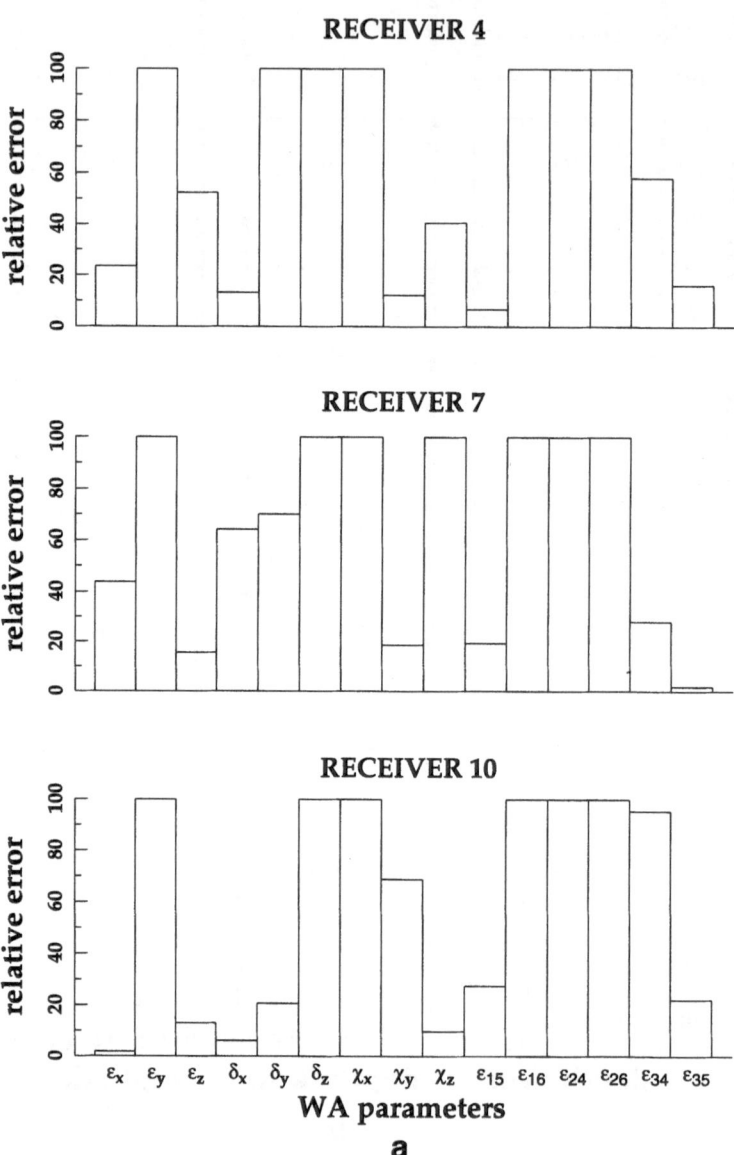

Figure 5
TI model: Relative errors (%) of 15 qP-wave WA parameters for noise-free data from 3 profiles with all singular values, with the 3 smallest singular values ignored and from 5 profiles with all singular values, for receivers 4, 7 and 10. Equation (22) used.

Figure 5 indicates the relative errors of the inverted parameters. Each figure shows 3 bar charts corresponding to the receiver 4 (top), 7 (middle) and 10 (bottom). The values of the exact WA parameters for these receivers are given in Table 1. Figure 5a shows the results of the inversion for the case of three profiles with all

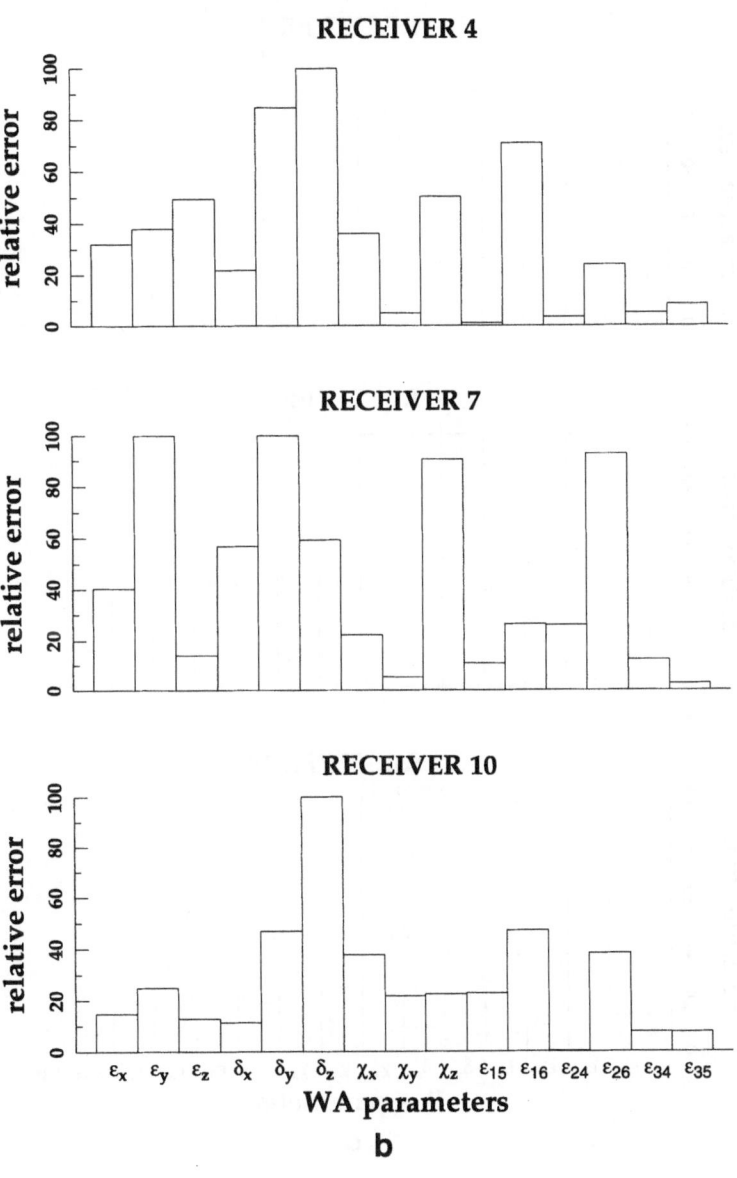

Figure 5
Continuation.

singular values, including the small ones in Figure 3a, considered. As we could expect from the analysis of singular values and variances, some relative errors reach values of 100% and more (all values larger than 100% are displayed as 100% in Figure 5). This concerns the 6 WA parameters with extremely high variances in Figure 4a. In

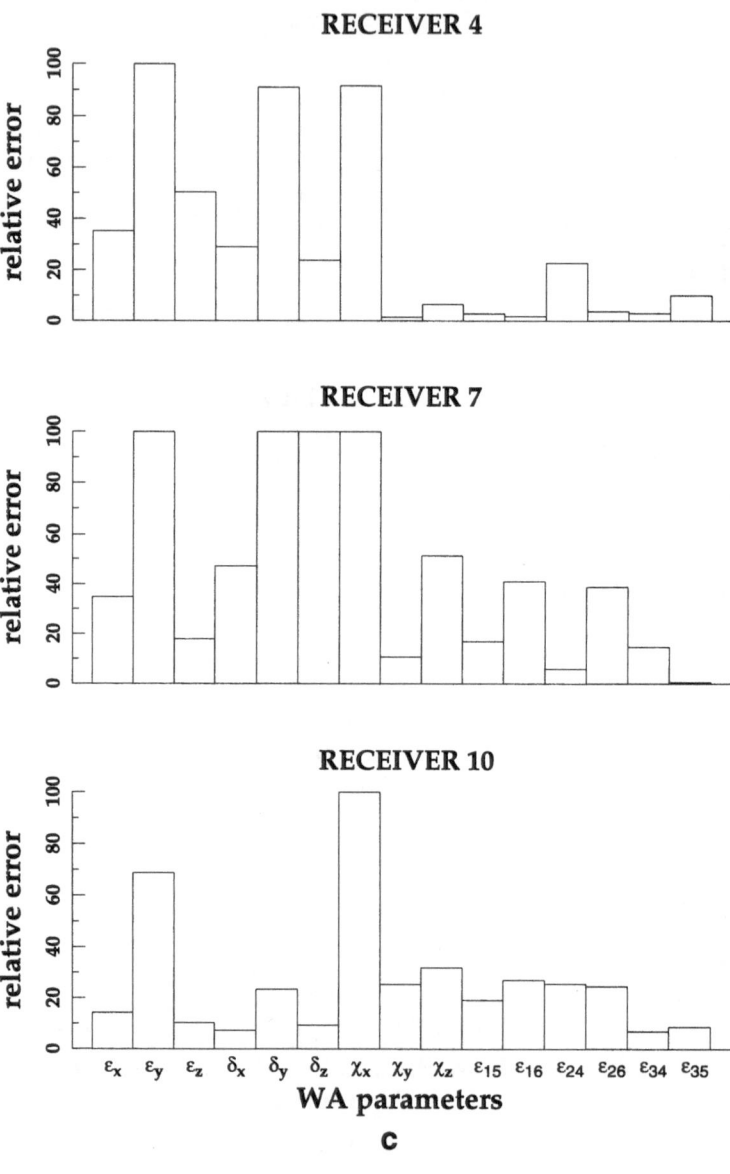

Figure 5
Continuation.

Table 1

Exact values of the WA parameters at receivers 4, 7 and 10 in the TI model

WA parameters	Receiver 4	Receiver 7	Receiver 10
ϵ_x	−0.04370	−0.04455	−0.04534
ϵ_y	0.00742	0.00823	0.00899
ϵ_z	0.01813	0.01825	0.01836
δ_x	−0.03027	−0.03125	−0.03216
δ_y	0.02555	0.02648	0.02735
δ_z	−0.03594	−0.03655	−0.03712
χ_x	−0.00286	−0.00268	−0.00250
χ_y	−0.01060	−0.00992	−0.00928
χ_z	−0.02787	−0.02607	−0.02439
ϵ_{15}	−0.00994	−0.00930	−0.00870
ϵ_{16}	−0.02383	−0.02229	−0.02085
ϵ_{24}	−0.00533	−0.00499	−0.00466
ϵ_{26}	−0.02740	−0.02563	−0.02397
ϵ_{34}	−0.00549	−0.00513	−0.00480
ϵ_{35}	−0.01177	−0.01101	−0.01030

addition to them, high relative errors can also be observed for the parameters δ_y, ϵ_{34}, χ_y and χ_z. This is again in agreement with observations of PŠENČÍK and GAJEWSKI (1998) for the case of the number of profiles less than 5, of which none is situated in the plane (y, z). The results of the inversion can be considerably improved if we ignore the lowest three singular values in the upper frame of Figure 3. This is shown in Figure 5b. Further substantial improvement can be reached if two additional profiles are considered, see Figure 5c. Even in this case, however, the above listed six WA parameters with high variances still have large relative errors. This could be fixed by extension of the profiles or by their different orientation, in other words by a better illumination of receivers. This is a task for future studies.

In Figure 6, equal area plots of isolines of the qP-wave phase velocity obtained from inversion (thin lines) and from the exact model (thick lines) are compared. The solid circles indicate positions of the axes of symmetry, which are identical with the longitudinal directions (directions, in which slowness and polarization vectors are parallel, HELBIG, 1994), in the exact model. The open circles indicate longitudinal directions obtained from inversion. The left column shows results for three profiles, the right column for five profiles. In each column, results for receivers 4, 7 and 10 situated at depths $z = 0.25$, 0.4 and 0.55 km are shown. The results of inversion for three profiles would make no sense unless at least three of the smallest singular values shown in the upper frame of Figure 3 were ignored. In Figure 5b we illustrated how removal of small singular values reduces relative errors. For this reason we ignored all relative singular values substantially smaller than 10^{-2} in the following results. Actually it means that we ignored four singular values in the case of three profiles and one singular value in the case of five profiles.

We can see that the thin line curves in Figure 6 have a more complicated form than the solid line curves, which correspond to the TI model. This means that the inverted phase velocity no longer corresponds to the TI symmetry. For this reason we compare the longitudinal directions instead of the axes of symmetry. As expected, the fit between exact and inverted phase velocities is better for five profiles. The best overall fit can be observed for receiver 4. Isolines of the exact and inverted model fit each other well not only in the centre of the circle, i.e., for vertical direction, but also at the border of the circle, i.e., for horizontal direction. This picture changes when moving to deeper receivers. For receiver 10, a good fit can be observed at the center

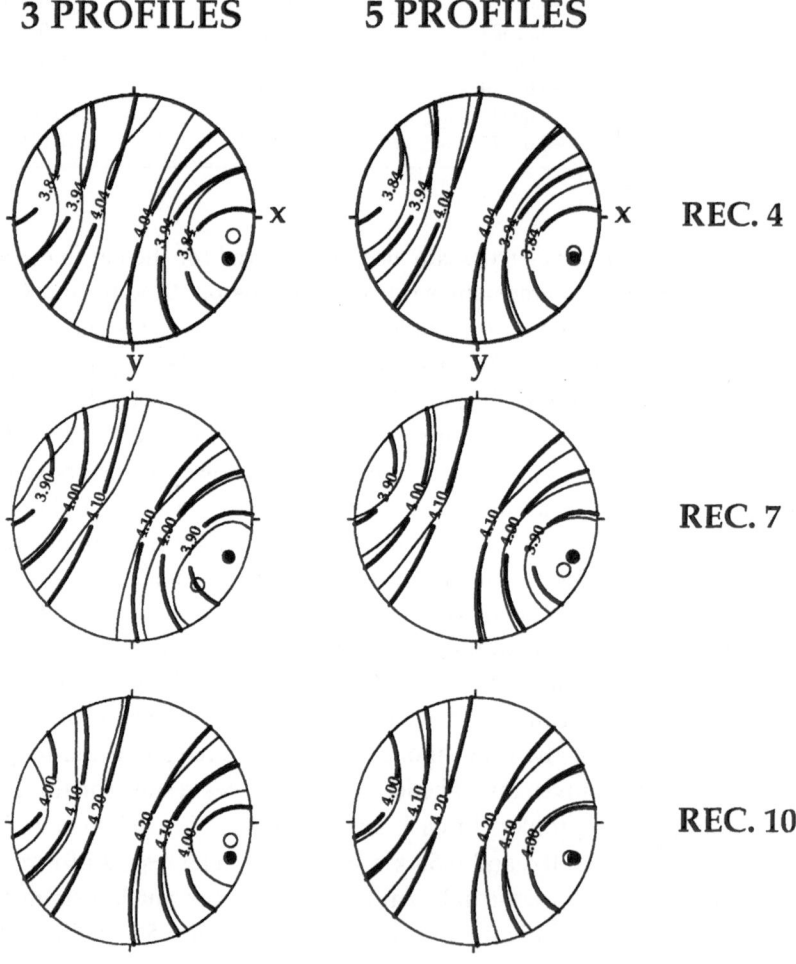

Figure 6

TI model: Equal area plots of phase velocity obtained from inversion (thin lines) and for the exact model (thick lines). The open and solid circles: longitudinal axes. Four and one of the smallest singular values ignored for 3 and 5 profiles, respectively.

of the circle. Towards the border of the circle, misfit increases. This indicates insufficient illumination of receivers with increasing depth. While the angles of rays illuminating receiver 4 vary from 23° to 79° (0° specifies vertical direction), the angles of rays illuminating receiver 10 vary from 11° to 63°. Thus receiver 10 is insufficiently illuminated by nearly horizontally propagating rays. Longitudinal directions are obviously better determined for five profiles.

Figure 7 shows, in the same display and for the same specification as Figure 6, effects of the 10% random noise added separately to each component of the observed seismograms. The noise affects significantly angular variation of the phase velocity

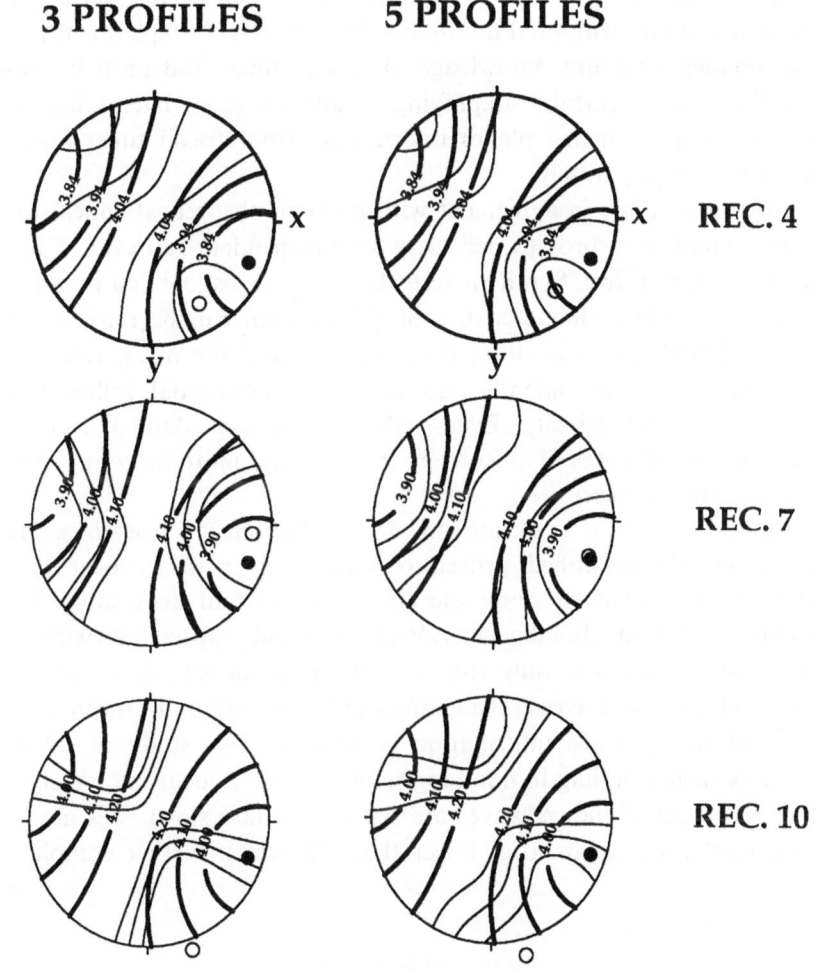

Figure 7

TI model: 10% random noise added to the "observed" data. Equal area plots of phase velocity obtained from inversion (thin lines) and for the exact model (thick lines). The open and solid circles: longitudinal axes. Four and one of the smallest singular values ignored for 3 and 5 profiles, respectively.

and also the position of the minimum velocity. The results for receiver 10 are unacceptable for both three and five profiles. This is because of the insufficient illumination of the receiver mentioned in the discussion of Figure 6. At receivers 4 and 7, slightly better results, especially longitudinal directions, are found for five profiles.

Conclusions

We present an algorithm for the local determination of the 15 WA parameters specifying qP-wave propagation in laterally varying, weakly anisotropic media of arbitrary symmetry. We show that for the determination of the WA parameters from observations in a multi-azimuthal multiple-source offset VSP experiment performed in such a complex structure, knowledge of arrival times and particle motions is necessary. Only under certain simplifying conditions (e.g., lateral homogeneity, measurements in a symmetry plane) information from travel times and particle motions can be used separately.

Performed tests show, in agreement with previous theoretical conclusions, that the minimum number of profiles necessary for independent retrieval of all 15 qP-wave WA parameters is five. Synthetic tests, however, also show that even with three profiles some interesting characteristics of the medium can be retrieved, e.g., an estimate of the longitudinal direction. This occurs because the WA parameters which cannot be obtained independently, do not have substantial influence on the determination of phase velocity. Even with 10% noise in data, it is possible to make approximate estimates of anisotropy. Further and more thorough tests of the effects of noise are necessary.

In the presented tests, a moderate anisotropy of about 8% was considered. It is desirable to test the described procedure on models with stronger or weaker anisotropy. We also plan to study the effects of different acquisition geometry, deviated wells and lateral inhomogeneity. We have already started tests with real data.

In the present study, our only source of information was the direct qP wave. Extension to reflected qP waves is straightforward. Derivation of formulae necessary for the use of the qS-wave information is planned. The study of qS waves is complicated by their coupling in inhomogeneous weakly anisotropic media, see e.g. PŠENČÍK (1998). Even if the qS waves are separated, their slownesses are given by nonlinear formulae for symmetries lower than TI, which further complicates the procedure.

Acknowledgements

The authors appreciate very much the stimulating reviews of Steve Horne and Soazig le Bégat. Critical comments of M. Baan, P. Bakker, V. Farra and V. Vavryčuk are also appreciated. This work was supported by the consortium project "Seismic waves in complex 3-D structures (SW3D)" and by the Grant Agency of the Czech Republic under the contract No. 205/00/1350. XZ is grateful for the support of the National Natural Science Foundation of China, No. 49774230. The final part of this work was accomplished during the stay of IP at CPGG, Universidade Federal da Bahia, Salvador, Brazil.

Appendix

The elements B_{33}, B_{13} and B_{23} of the weak anisotropy matrix for an anisotropic medium of arbitrary symmetry read, see PŠENČÍK and GAJEWSKI (1998):

$$
\begin{aligned}
B_{33} = 2\alpha^2 \big[& \epsilon_z n_3^4 + 2n_3^3(\epsilon_{34}n_2 + \epsilon_{35}n_1) + n_3^2(\delta_x n_1^2 + \delta_y n_2^2 + 2\chi_z n_1 n_2) \\
& + 2n_3(\chi_x n_1^2 n_2 + \chi_y n_1 n_2^2 + \epsilon_{15}n_1^3 + \epsilon_{24}n_2^3) + \epsilon_x n_1^4 + \delta_z n_1^2 n_2^2 \\
& + \epsilon_y n_2^4 + 2\epsilon_{16}n_1^3 n_2 + 2\epsilon_{26}n_1 n_2^3 \big] \ ,
\end{aligned} \tag{A1a}
$$

$$
\begin{aligned}
B_{13} = \alpha^2 D^{-1} \Big[& 2\epsilon_z n_3^5 + n_3^4(\epsilon_{34}n_2 + \epsilon_{35}n_1) + n_3^3(\delta_x n_1^2 + \delta_y n_2^2 + 2\chi_z n_1 n_2 - 2\epsilon_z) \\
& + n_3^2[(4\chi_x - 3\epsilon_{34})n_1^2 n_2 + (4\chi_y - 3\epsilon_{35})n_1 n_2^2 + (4\epsilon_{15} - 3\epsilon_{35})n_1^3 + (4\epsilon_{24} - 3\epsilon_{34})n_2^3] \\
& + n_3[(2\delta_z - \delta_x - \delta_y)n_1^2 n_2^2 + 2(2\epsilon_{16} - \chi_z)n_1^3 n_2 + 2(2\epsilon_{26} - \chi_z)n_1 n_2^3 \\
& + (2\epsilon_x - \delta_x)n_1^4 + (2\epsilon_y - \delta_y)n_2^4] - \chi_x n_1^2 n_2 - \chi_y n_1 n_2^2 - \epsilon_{15}n_1^3 - \epsilon_{24}n_2^3 \Big]
\end{aligned} \tag{A1b}
$$

and

$$
\begin{aligned}
B_{23} = \alpha^2 D^{-1} \Big[& n_3^3(\epsilon_{34}n_1 - \epsilon_{35}n_2) + n_3^2[(\delta_y - \delta_x)n_1 n_2 + \chi_z n_1^2 - \chi_z n_2^2] \\
& + n_3[(2\chi_y - 3\epsilon_{15})n_1^2 n_2 - (2\chi_x - 3\epsilon_{24})n_1 n_2^2 + \chi_x n_1^3 - \chi_y n_2^3] \\
& + (\delta_z - 2\epsilon_x)n_1^3 n_2 + (2\epsilon_y - \delta_z)n_1 n_2^3 + 3(\epsilon_{26} - \epsilon_{16})n_1^2 n_2^2 + \epsilon_{16}n_1^4 - \epsilon_{26}n_2^4 \Big] \ . \tag{A1c}
\end{aligned}
$$

The quantity D is defined in Equation (2) and the WA parameters appearing in (7) are defined as follows

$$
\begin{array}{lll}
\epsilon_x = \dfrac{A_{11} - \alpha^2}{2\alpha^2}, & \epsilon_y = \dfrac{A_{22} - \alpha^2}{2\alpha^2}, & \epsilon_z = \dfrac{A_{33} - \alpha^2}{2\alpha^2}, \\[2mm]
\delta_x = \dfrac{A_{13} + 2A_{55} - \alpha^2}{\alpha^2}, & \delta_y = \dfrac{A_{23} + 2A_{44} - \alpha^2}{\alpha^2}, & \delta_z = \dfrac{A_{12} + 2A_{66} - \alpha^2}{\alpha^2}, \\[2mm]
\chi_x = \dfrac{A_{14} + 2A_{56}}{\alpha^2}, & \chi_y = \dfrac{A_{25} + 2A_{46}}{\alpha^2}, & \chi_{z'} = \dfrac{A_{36} + 2A_{45}}{\alpha^2}, \\[2mm]
\epsilon_{15} = \dfrac{A_{15}}{\alpha^2}, \quad \epsilon_{16} = \dfrac{A_{16}}{\alpha^2}, & \epsilon_{24} = \dfrac{A_{24}}{\alpha^2}, \quad \epsilon_{26} = \dfrac{A_{26}}{\alpha^2}, & \epsilon_{34} = \dfrac{A_{34}}{\alpha^2}, \quad \epsilon_{35} = \dfrac{A_{35}}{\alpha^2} \ .
\end{array} \tag{A2}
$$

The formulae (A1) simplify if a higher symmetry anisotropic media are considered. For the VTI medium considered in our experiments, Equations (A1) reduce to

$$B_{33} = 2\alpha^2[\epsilon_z n_3^4 + \delta_x n_3^2(1 - n_3^2) + \epsilon_x(1 - n_3^2)^2] \ , \tag{A3a}$$

$$B_{13} = \alpha^2 n_3(1 - n_3^2)^{1/2}[(\delta_x - 2\epsilon_z)n_3^2 - (\delta_x - 2\epsilon_x)(1 - n_3^2)] \tag{A3b}$$

and

$$B_{23} = 0 \ . \tag{A3c}$$

REFERENCES

DE PARSCAU, J. (1991), *P- and SV-wave Transversely Isotropic Phase Velocities Analysis from VSP Data*, Geophys. J. Int. *107*, 629–638.

ESMERSOY, C. (1990), *Inversion of P and SV Waves from Multicomponent Offset Vertical Seismic Profiles*, Geophysics *55*, 39–50.

FARRA, V., and LE BÉGAT, S. (1995), *Sensitivity of qP-wave Travel Times and Polarization Vectors to Heterogeneity, Anisotropy and Interfaces*, Geophys. J. Int. *121*, 371–384.

GAISER, J. E. (1990), *Transversely Isotropic Phase Velocity Analysis from Slowness Estimates*, J. Geophys. Res. *95*, 11,241–11,254.

GAJEWSKI, D., and PŠENČÍK, I. (1990), *Vertical Seismic Profile Synthetics by Dynamic Ray Tracing in Laterally Varying Layered Anisotropic Structures*, J. Geophys. Res. *95*, 11,301–11,315.

HELBIG, K., *Foundations of Anisotropy for Exploration Seismics* (Pergamon, Oxford 1994).

HORNE, S. A., and LEANEY, S. (2000), *Polarization and Slowness Component Inversion for TI Anisotropy*, Geophys. Prospect. *48*, 779–788.

HORNE, S. A., MCGARRITY, J. P., SAYERS, C. M., SMITH, R. L., and WIJNANDS, F., *Fractured reservoir characterisation using multi-azimuthal walkaway VSPs*. In *Expanded Abstracts, 68th Ann. Internat. Mtg.* (Soc. Expl. Geophys., Tulsa 1998) pp. 1640–1643.

HSU, K., SCHOENBERG, M., and WALSH, J., *Anisotropy from polarization and moveout*. In *Expanded Abstracts, 61th Ann. Internat. Mtg.* (Soc. Expl. Geophys., Tulsa 1991) pp. 1526–1529.

JECH, J., and PŠENČÍK, I. (1989), *First-order Perturbation Method for Anisotropic Media*, Geophys. J. Int. *99*, 369–376.

LEANEY, W. S., *Parametric wavefield decomposition and applications*. In *Expanded Abstracts, 60th Ann. Internat. Mtg.* (Soc. Expl. Geophys., Tulsa 1990) pp. 1097–1100.

MENSCH, T., and RASOLOFOSAON, P. (1997), *Elastic Wave Velocities in Anisotropic Media of Arbitrary Symmetry — Generalization of Thomsen's Parameters ε, δ and γ*, Geophys. J. Int. *128*, 43–64.

PŠENČÍK, I. (1998), *Green's Functions for Inhomogeneous Weakly Anisotropic Media*, Geophys. J. Int. *135*, 279–288.

PŠENČÍK, I., and GAJEWSKI, D. (1998), *Polarization, Phase Velocity and NMO Velocity of qP Waves in Arbitrary Weakly Anisotropic Media*, Geophysics *63*, 1754–1766.

PŠENČÍK, I., and ZHENG, X., *Determination of weak anisotropy parameters from the qP-wave particle motion measurements*, In *Seismic Waves in Complex 3-D Structures*, Report 7 (Dept. of Geophysics, Charles University, Prague 1998) pp. 299–306 (available online at "http://seis.karlov.mff.cuni.cz/papers/r7ip3.htm).

THOMSEN, L. (1986), *Weak Elastic Anisotropy*, Geophysics *51*, 1954–1966.

WHITE, J. E., MARTINEAU-NICOLETIS, L., and MONASH, C. (1983), *Measured Anisotropy in Pierre Shale*. Geophys. Prospect. *31*, 709–725.

WILLIAMSON, P., MAOCEC. E., and BOELLE, J-L., *Local estimation of anisotropy parameters from well seismic polarisation data.*, In *Expanded Abstracts, 70th Ann. Internat. Mtg.* (Soc. Expl. Geophys., Tulsa 2000) pp. 2241–2244.

(Received October 18, 2000, revised February 25, 2001, accepted April 6, 2001)

 To access this journal online:
http://www.birkhauser.ch